Fault Diagnosis for Linear Discrete Time-Varying Systems and Its Applications

Maiying Zhong · Ting Xue · Steven X. Ding ·
Donghua Zhou

Fault Diagnosis for Linear Discrete Time-Varying Systems and Its Applications

Springer

Maiying Zhong
The College of Electrical Engineering and Automation
Shandong University of Science and Technology
Qingdao, Shandong, China

Ting Xue
The College of Electrical Engineering and Automation
Shandong University of Science and Technology
Qingdao, Shandong, China

Steven X. Ding
The Institute for Automatic Control and Complex Systems
University of Duisburg-Essen
Duisburg, Germany

Donghua Zhou
The College of Electrical Engineering and Automation
Shandong University of Science and Technology
Qingdao, Shandong, China

ISBN 978-981-19-5440-5 ISBN 978-981-19-5438-2 (eBook)
https://doi.org/10.1007/978-981-19-5438-2

This Springer imprint is published by the registered company Springer Nature Singapore Pte Ltd.
The registered company address is: 152 Beach Road, #21-01/04 Gateway East, Singapore 189721, Singapore

Preface

Over the past 40 years, a great deal of attention in research has been devoted to model-based fault diagnosis and a large number of achievements have been obtained. One of the key issues of a model-based fault diagnosis system is concerned with its robustness. Different from such a concept in robust control, the robustness of a fault diagnosis system involves two aspects, i.e., robustness to unknown inputs, model uncertainties and sensitivity to the concerned faults. It is often the nature of industrial systems that the effects of the possible faults and unknown inputs are coupled. The core of a fault diagnosis system should therefore be robust to unknown inputs and, at the same time, sensitive to faults.

Regarding fault detection (FD), the major stages of model-based FD consist of residual generation and evaluation. According to the techniques of residual generation, the main results of model-based FD can be classified into the unknown input observer (UIO) approach, the H_∞-optimization approach, the H_∞-filtering formulation and the parity space approach. Unknown input decoupling design is a standard technique, which can be achieved by using eigenstructure assignment, i.e., the so-called UIO scheme. If a complete unknown input decoupling is impossible, one can consider to design a residual generator by minimizing a performance index containing a measure of the effects of both unknown inputs and faults. H_∞-optimization and H_∞-filtering are two widely used approaches to robust FD for linear time-invariant (LTI) systems subject to l_2-norm bounded unknown inputs. It has been demonstrated that the robustness to unknown input can be indicated by an H_∞ norm criterion, while the sensitivity to faults be denoted by an induced H_- index, H_2 or H_∞ norm. Then the robust FD problem under different performance indices H_-/H_∞, H_2/H_∞ and H_∞/H_∞ can be solved by a unified optimal solution, i.e., the so-called H_i/H_∞-optimization-based FD. The main objective of H_∞-filtering-based FD is to design a fault detection filter (FDF) such that the error between the residual and the fault is made as small as possible, which can be solved by applying the existing H_∞ filtering techniques such as linear inequality matrix (LMI). However, an H_∞-FDF is difficult to balance the robustness to unknown input and the sensitivity to faults. As for the parity space approach to FD, its major advantage is that the generated residual can be completely decoupled from control input, reference input as well as unknown initial state.

It is worth pointing out that most of the existing achievements in model-based fault diagnosis are dedicated to LTI systems. In many practical applications, however, most of the monitored systems are inherently time-varying and/or nonlinear, such as chemical processes, automobile plants and networked control systems with event-triggering data transmission mechanisms. Moreover, a widely used way of dealing with nonlinear systems is approximate linearization along with an actual or nominal trajectory of the systems, which generally leads to linear time-varying systems. It is thus clear that the linear discrete time-varying (LDTV) system can be used as an appropriate model for many practical processes. Especially, due to the time-varying characteristics, the existing methods of robust FD for LTI systems have some limitations. To mention a few of them, for instance, the above-mentioned unified solution to the H_i/H_∞-optimization-based FD was obtained by using the co-inner-outer factorization technique in the frequency domain, which is not applicable for solving the FD problem for LDTV systems. On the other hand, to perform H_∞-filtering-based FD for time-varying systems, it is required to solve a set of LMIs at every time step. Hence, the computational tasks of both the H_∞-filtering-based FD and parity space method are considerably heavy. Moreover, only sufficient conditions have been obtained so that the LMI-based FD methods are more conservative. Besides, aiming to estimate the fault for more information such as type, size and location, fault estimation is another important task of fault diagnosis, which remains an active and open topic for LDTV systems. In summary, issues of fault diagnosis for LDTV systems subject to unknown inputs remain significant and challenging.

The authors' study on fault diagnosis for LDTV systems is strongly motivated by the recent development in the fault diagnosis theory research and from the practical application requirements. Following the funding support from the Chinese National Science Foundation (Grant Nos. 61333005, 61873149, 60774071 and 61733009) and the Research Fund for the Taishan Scholar Project of Shandong Province of China, much efforts of the authors have been dedicated to the study of fault diagnosis for LDTV systems and its applications. A lot of technical papers have been published in the open literature.

The book begins with an introduction of the state-of-the art developments in fault diagnosis for LDTV systems and then focuses on the detailed works carried out by the authors. The main contents cover the H_2- and H_∞-optimization-based fault diagnosis for LDTV systems, the H_∞-filtering-based fault diagnosis for LDTV systems, the parity space-based fault diagnosis for LDTV systems and some applications in discrete-time nonlinear systems.

The first author of the book would like to express her sincere appreciation to Profs. Qing-Long Han, Eve L. Ding and Hao Ye for their consistent support. She would also like to thank her students, namely Jia Guo, Yang Song, Dingfei Guo, Quan Cao,

Hai Liu and Shuai Liu. Many thanks to the collaborations with Dr. Ligang Zhang, Dr. Chengrui Liu and Dr. Wenbo Li in application studies.

Qingdao, China
May 2022

Maiying Zhong

Contents

Acronyms

Abbreviations

AT	Aerial triangulation
ATEKF	Adaptive two-stage extended Kalman filter
CCA	Canonical correlation analysis
DG	Direct georeferencing
DTG	Dynamically tuned gyroscope
EKF	Extended Kalman filter
ENU	East-north-up
EO	Exterior orientation
FAR	False alarm rate
FD	Fault detection
FDF	Fault detection filter
FDI	Fault detection and isolation
FDR	Fault detection rate
FE	Fault estimation
GPS	Global position system
HJ	Hamilton-Jacobi
HJI	Hamilton-Jacobi inequality
i.i.d	Independent identically distributed
IFA	In-flight alignment
IMU	Inertial measurement unit
INS	Inertial navigation system
ISP	Inertial stable platform
KF	Kalman filter
LDTV	Linear discrete time-varying
LMI	Linear matrix inequality
LMS	Least mean square
LPV	Linear parameter-varying
LQG	Linear quadratic Gaussian
LS	Least square

LTI	Linear time-invariant
MDR	Missed detection rate
MRA	Multiscale resolution analysis
MVA	Multivariate analysis
PDF	Probability density function
PMI	Proportional and multi-integral
POS	Position and orientation measurement system
PSD	Power spectral density
RLG	Ring laser gyroscope
RMS	Root mean square
STF	Strong tracking filter
SVD	Singular value decomposition
SWT	Stationary wavelet transform
UAV	Unmanned aerial vehicles
UIO	Unknown input observer
UKF	Unscented Kalman filter

Mathematical Notations

$\forall$	For all
$\exists$	Exists
$\in$	Belong to
$\subset$	Subset
$\cup$	Union
$\cap$	Intersection
$\perp$	Orthogonal
$\lceil \cdot \rceil$	Ceiling function
$\approx$	Approximately equal
$:=$	Defined as
$\Longrightarrow$	Implies
$\Longleftrightarrow$	Equivalent to
$\lvert\lvert \cdot \rvert\rvert$	Euclidean norm of a vector
$\lvert\lvert \cdot \rvert\rvert_2$	H_2 norm of a signal
$\lvert\lvert \cdot \rvert\rvert_F$	Frobenius norm of a matrix
$\lvert\lvert X \rvert\rvert_\infty$	H_∞ norm of X
$\mathrm{diag}(\cdots)$	A block-diagonal matrix
$\dim(x)$	The dimension of x
$\Pr\{\cdot\}$	The probability of $\{\cdot\}$
$\mathrm{Tr}(X)$	The trace of X
$\mathrm{rank}(X)$	The rank of X
max	Maximum
min	Minimum

sup	Supremum
inf	Infimum
$\max\{\cdot\}$	The maximal element in $\{\cdot\}$
$\text{cov}(\cdot)$	The covariance of $(\cdot)$
$\mathcal{RH}_\infty$	The set of all stable transfer matrices
$\mathbb{C}$	Space of complex numbers
$\mathbb{R}^n$	Space of n-dimensional vectors
$\mathbb{R}^{n\times m}$	Space of n by m matrices
$\mathbb{E}[\cdot]$	Expectation of $[\cdot]$
$\|(\cdot)\|_E$	The mean value of $\|(\cdot)\|$
$x*y$	Convolution of x and y
X^T	Transpose of X
X^{-1}	Inverse of X
X^-	Pseudo-inverse of X
$X^\perp$	Orthogonal complement of X
X^*	Conjugate transpose of (complex) matrix X
$X > 0$	X is a real positive definite matrix
$X < 0$	X is a real negative definite matrix
$\lambda(X)$	Eigenvalues of X
$\lambda_{\max}(X)$	The maximum eigenvalue of X
$\lambda_-(X)$	The minimum eigenvalue of X
$\bar{\sigma}(X)$	The maximal singular value of X
$\underline{\sigma}(X)$	The minimal singular value of X
$l_{2,[0,N]}$	$\mathcal{L}_2$ norm of a signal within the time interval $[0,\ N]$
$\mathcal{N}(a,\ \Sigma)$	Gaussian distribution with mean a and covariance Σ
$\mathcal{U}[a,b]$	Uniform distribution over $[a,b]$
$\mathcal{L}\left\{\{\theta(i)\}_{i=m}^{j}\right\}$	The linear space spanned by $\theta(m),\ \ \theta(m+1),\ \ \ldots,\ \ \theta(j)$
$\text{Proj}\{x\|y\}$	Orthogonal projection of x on linear space spanned by y
$[R(i,\ j)]_{i,j=0:k}$	$(k+1)\times(k+1)$ Block matrix with element $R(i,\ j)$
$\{\theta(k)\}_{k=0}^N$	A set with elements $\theta(k)$, $k=1,\ 2,\ \ldots,\ N$

Part I
Introduction and Preliminaries

Chapter 1
Introduction

In this introduction we give an overview of model-based fault diagnosis by paying special attention to linear discrete time-varying (LDTV) systems. The fundamentals of model-based fault diagnosis and the state-of-the-art achievements for LDTV systems are informally introduced. The organization of the book is also outlined at the end.

1.1 Motivations

Ever-increasing complexity of modern control systems has stimulated rising demands for high levels of safety and reliability, especially for the safety critical systems such as chemical plants, control systems in aircraft and satellites, etc. In response to this requirement, fault diagnosis technique has attracted worldwide attention over the past decades and great efforts have been made in this research area both in academia communities and industrial applications.

As one of the most important mainstreams of fault diagnosis, model-based technique emerged in the early 1970s and, associated with the rapid development of modern control theory, mathematics and computer science, has been extensively studied since then. Although in recent years tremendous attention has been paid to data-driven process monitoring with the advent of data and information explosion age, model-based technique remains much to be preferred in many practical occasions thanks to its efficiency in dealing with fault diagnosis issues for dynamic systems.

Model-based fault diagnosis aims at detecting the occurrence of faults and diagnosing their location and size by making use of a mathematical model of the supervised system. The research topics cover fault detection, fault isolation and fault estimation. After four decades development, a variety of methods and techniques have been developed for model-based fault diagnosis, that can be roughly categorized into

M. Zhong et al., *Fault Diagnosis for Linear Discrete Time-Varying Systems and Its Applications*, https://doi.org/10.1007/978-981-19-5438-2_1

the observer-based method, the parity space-based scheme and the parameter estimation approach. Especially, concerning the mismatch of a real system and its analytical model due to modeling uncertainties, robust fault diagnosis has attracted considerable attention over the last two decades. To pursue a trade-off between the robustness of fault diagnosis system against uncertainties and the sensitivity to faults, robust fault diagnosis techniques such as unknown input decoupling, H_2-, H_∞-, and H_i/H_∞-optimization were developed successively and have been successfully applied to various engineering systems.

On the other hand, many well-established results on model-based fault diagnosis are dedicated for linear time invariant (LTI) systems. In real applications, however, most of the engineering systems are inherently time-varying and/or nonlinear. A common remedy is to approximate practical processes with linear discrete time-varying (LDTV) systems by means of discretization and trajectory linearization. Due to the time-varying characteristics of the dynamics of LDTV systems, existing fault diagnosis results for LTI systems could hardly be directly applied to LDTV systems in most cases. For these observations, fault diagnosis for LDTV systems is a significant and nontrivial topic both in theoretical study and practical applications. Even though many efforts have been made on this research topic and a rich body literature have been reported up to now, a comprehensive introduction on fault diagnosis for LDTV systems and its applications in engineering systems is still unavailable. This book is organized to close such a gap.

1.2 Model-Based Fault Diagnosis

We start in this section with introducing the basic concepts involved in model-based fault diagnosis. Then an overview of fault diagnosis for LDTV systems is given.

1.2.1 Basic Concepts

In fault diagnosis, a "fault" is considered to be an undesired behavior that eventually hampers normal operation and causes intolerable performance deterioration of the supervised system. Fault diagnosis mainly consists of the following three tasks:

- *Fault detection* (FD): detection of the occurrence of faults in the functional units of the process;
- *Fault isolation*: localization or classification of different faults;
- *Fault estimation* (FE): estimation of the fault so as to determine its type, size and location.

FD is the most simple but basic task of fault diagnosis, that is generally regarded as a binary decision-making problem, i.e., the supervised system is either faulty or fault-free. By means of FD, we can decide if and when a fault has happened. Fault isolation and FE are generally performed to show where and which fault of what size occurs.

Hardware redundancy based fault diagnosis is a traditional technique that addresses fault detection and isolation (FDI) issues based on redundant hardware components. The basic idea behind it is to compare the outputs of process components with the redundant ones, a fault can then be diagnosed by a voting scheme. Compelling advantages of this technique lie in its high reliability and the direct fault isolation, while the conflict of high reliability and the cost of redundant hardware hinder its applications in practice.

Model-based fault diagnosis deals with fault diagnosis issues by making use of an analytical model of the monitored system without equipping additional components. In this strategy, the process dynamics are described by a mathematical model, that is then applied to reconstruct process behaviors online, i.e., the so-called analytical redundancies. During process operation, the analytical model will run in parallel to the process and be driven by the same system inputs. The so-called residual signal can then be generated to check the consistency of measured process variables and their estimates obtained based on the analytical model. The procedure of creating the estimates of process outputs and building the difference of them with the measured outputs is called *residual generation*, which plays a central role in fault diagnosis for dynamic systems and is also the core issue that will be discussed in this book.

Because of the inevitable modeling uncertainties (e.g., linearization and discretization errors, process and measurement noises, and external unknown inputs, etc.), an accurate and complete mathematical model of a practical system is inaccessible, which results in a residual signal containing not only fault information but also uncertainties. Such uncertainties constitute the main source of false and missed alarms in FD and estimation errors in FE. To alleviate this deficiency, a residual generator is usually designed to be robust against uncertainties while keeping its sensitiveness to fault. This basic principle builds the core of robust fault diagnosis.

Apart from residual generation, *residual evaluation* is another essential part of fault diagnosis, that attempts to extract fault information of interest from residual and serves as a post-processing and decision-making unit. Especially, for FD purpose, the norm-based method and the hypothesis testing scheme are two most popular residual evaluation techniques. In these two strategies, the so-called *threshold* (associated with uncertainties) should be determined such that the occurrence of a fault can be detected if the evaluation function exceeds it. By noting that once a fault is estimated, FDI issues can be solved as well, a recent trend is to address FDI problems by directly performing fault estimation, despite sometimes optimal FDI is unachievable.

In model-based fault diagnosis, faults are usually categorized into the *additive faults* and *multiplicative faults*, according to the way of how they influence the system dynamics. Roughly speaking, additive faults such as offsets in actuators or sensors would not affect the system stability, while multiplicative faults like malfunctions of components would lead to the change of model parameters and directly affect the system stability. In certain scenarios, multiplicative faults can be modeled as additive

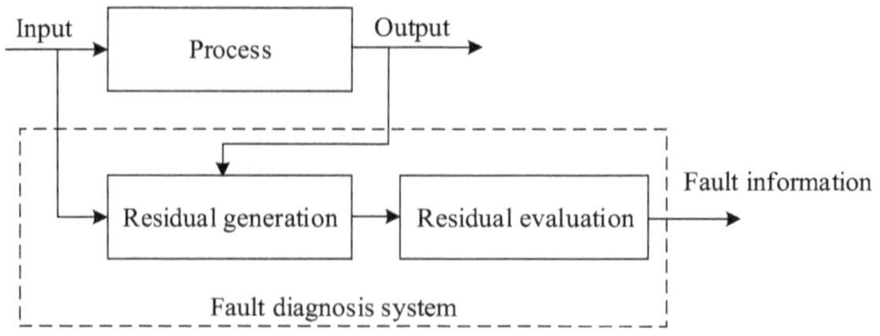

Fig. 1.1 Schematic diagram of model-based fault diagnosis

faults. In this book, we mainly focus on additive fault diagnosis in LDTV systems subject to stochastic and/or norm-bounded unknown inputs.

1.2.2 Classical Methods

As depicted in Fig. 1.1, a model-based fault diagnosis system consists of residual generation and residual evaluation units. According to the way of residual generation, existing model-based fault diagnosis techniques fall into three classes, namely the observer-based method, parity space-based scheme and the parameter estimation approach. Below we give a brief introduction of these schemes.

Observer-Based Method

The basic idea behind observer-based fault diagnosis is to construct a state observer driven by system input to estimate the system output, so as to apply the output estimation error as a residual signal. Since the fault detection filter (FDF) based FDI proposed by Beard and Jones in 1970s', observer-based fault diagnosis has made great progress over the past decades, closely coupled with the development of advanced control theory. Among the fruitful results, much attention has been paid to the observer-based robust fault diagnosis under consideration of inevitable modeling uncertainties, for example, the unknown input observer (UIO) based disturbances decoupling method, the observer-based H_∞-filtering method, and the H_i/H_∞-optimal FDF approaches, etc.

The UIO technique goes back to 1980s and was initially developed for control purpose. A UIO is a typical Luenberger observer that delivers a state estimation error independent of unknown input. Since the first work on UIO based residual generation by Wünnenberg and Frank in 1987, various full-order and reduced-order UIOs have been put forward to cope with FDI issues in LTI systems and the existence conditions of them have been already well studied [27]. In recent years, much more attention on UIO-based fault diagnosis has been paid to extend it to more complex systems such as linear parameter varying (LPV) systems [1, 5, 24, 73], linear time-varying systems [48], and the systems with uncertainty time delay [69], etc.

With the development of H_2- and H_∞-optimization in control theory in the 1980s and early 1990s, Ding and Frank first proposed to apply H_2- and H_∞-optimization techniques to the design of observer-based FDI systems in 1989 and 1991 [47, 68], which start another mainstream of robust fault diagnosis. In the frameworks of H_2- and H_∞-optimization, the robustness of residual to unknown input can be indicated by H_2 or H_∞ index, while the fault sensitivity is measured by H_2, H_- and/or H_∞ indices. Then the residual generation issue can be formulated as a multi-objective optimization problem towards a trade-off between the robustness and fault sensitivity criteria. Along this research line, methodologies of observer-based robust fault diagnosis in the sense of H_2/H_2-, H_∞/H_∞-, H_2/H_∞-, H_-/H_∞-, and mixed H_i/H_∞-optimization have been extensively studied over the past few decades. Particularly, a unified solution for LTI systems has been proposed by Ding et al. in [59]. The result was further extended to linear continuous time-varying systems by Li and Zhou in [77], and then to LDTV systems by Zhong and Ding in [39].

Parity Space-Based Method

Parity space-based residual generation is characterized by its straightforward realization form and the complete decoupling from system initial state and input. Since the initial work by Chow and Willsky in the 1980s, parity space-based fault diagnosis have been deeply investigated. The topics cover

- establishing connections between the parity relation based residual generator and the observer-based and parameter estimation based residuals, which has been well done in the 1990s [28, 49, 54, 56, 57, 61];
- improving the performance of parity space-based fault diagnosis systems (e.g., perfect decoupling of residual from unknown inputs, enhancing the robustness of residual to unknown input and sensitivity to fault) with acceptable computational load [37, 45, 60, 61], and
- extending the well-developed results of parity space-based fault diagnosis for LTI systems to more complex systems like Markovian jump systems, time-varying systems, nonlinear systems, and event-triggered systems, [12, 36, 38, 44], etc.

Very recently, the idea of parity space-based residual generation has also been applied in subspace identification aided data-driven fault diagnosis studies, see e.g., [29, 62, 66, 67, 71, 81, 83].

We would like to call reader's attention that, different from the observer-based residual generator of recursive implementation form, a parity relation based residual is generated in a non-recursive form by using system input and output data within a finite time interval (the length of which is the so-called parity space order). Meanwhile, parity relations between system input and output are derived in static algebraic forms, that makes the design and performance analysis of fault diagnosis system much easier in comparison with the observer-based method. Indeed, it has been shown in [61] that an s-order parity space based residual generator equals to an s-step dead-beat observer and all design approaches proposed for parity space-based residual generator can also be applied to an observer-based residual generation, and vice verse. Regarding of this, it is an alternative strategy to design a residual generator based on parity space method while realize it online recursively via an observer.

Parameter Estimation Approach

Parameter estimation approach to FDI, first demonstrated by Bakiotis et al. in 1979 [10] and Geiger in 1982 [18], is developed based on parameter or system identification techniques. It is often used to diagnose faults in system parameters that are unmeasurable directly. The core is to construct a parameter estimator to online identify the real parameters of the monitored system, and then the estimation error (i.e., the difference between estimates and the reference parameters in fault-free situation) can be obtained as a residual signal for fault diagnosis purpose.

By means of joint estimation of system state and parameters, the applications of Kalman filter (KF) and its derivatives such as extend Kalman filter (EKF) and strong tracking filter (STF) have been widely reported in parameter estimation based fault diagnosis [23, 32, 33, 51]. The adaptive observer based algorithms have also been reported recently, e.g., the proportional integral observer [9], interval observer [55], sliding mode observer [80] and adaptive observer [14, 50, 85], etc. With the rapid development of machine learning techniques and their applications in system identification nowadays, rising attention has been paid to cope with parameter estimation based fault diagnosis issues via machine learning algorithms, see e.g., [4, 19, 34, 58].

1.2.3 Overview of Fault Diagnosis for LDTV Systems

After four decades development, research on model-based fault diagnosis for LTI systems has been well studied. In practical applications, the fact is that most of engineering systems are inherently time-varying and nonlinear. Note that a nonlinear system can be mathematically approximated by a linear time-varying process with trajectory linearization technique and the continuous processes are commonly discretized for computerized processing. LDTV model is thus a common choice to describe practical processes. Yet, because of the time-varying characteristics of LDTV systems, existing fault diagnosis methods for LTI processes could hardly be directly applied to LDTV situations. Study on fault diagnosis for LDTV systems is of both practical and theoretical significance.

Substantial numbers of achievements on fault diagnosis for LDTV systems sprang up in the early 21s' and great progress has been made over the past two decades. The related results scatter in observer-based methods [11, 15, 20, 24, 43, 65, 73, 79], parity space-based approaches [25, 36, 70], parameter estimation methods [52, 72, 76], etc. Among them, many efforts have been dedicated into robust fault diagnosis for LDTV systems subject to stochastic and/or deterministic disturbances, e.g., UIO-based unknown input decoupling, H_∞-filtering, and H_i/H_∞-optimization, etc.

However, due to the time-varying characteristic of system parameters, the difficulties of robust fault diagnosis for LDTV systems, compared with LTI processes, mainly lie in the following aspects.

- More rigid condition for complete unknown input decoupling is required. A typical example is the UIO-based method. A UIO that exhibits satisfactory performance in LTI systems may not even exist for the LDTV counterparts. Some recent efforts have been made to modify standard UIO to suit LDTV systems, see e.g., [5, 15, 24, 73].
- Many techniques in frequency domain for LTI systems cannot be applicable in LDTV scenarios. For instance, coprime factorization defined originally in frequency domain plays a central role in the design and analysis of H_2- and H_∞-optimization based fault diagnosis in LTI systems, that, unfortunately, is no longer applicable or should be modified (see [78]) in LDTV situations.
- Online realization of fault diagnosis for LDTV systems is more computation expensive because of the time-varying system matrices. This deficiency is of particularly critical in parity space-based scheme with high parity space order and the H_∞-optimization based observer approaches with respect to solving Riccati equations.

Up to now, many efforts have been made to handle the above difficulties, with particular attention being paid to the last two issues.

In the context of H_2- and H_∞-optimization, finite horizon criteria have been introduced to indicate the robustness of residual to unknown input and the fault sensitivity. On this basis, H_2-, H_∞-, and multi-objective (e.g., H_2/H_2, H_-/H_∞, H_∞/H_∞, and H_i/H_∞) optimization approaches to fault diagnosis in LDTV systems have been widely investigated, see e.g., [13, 39, 44, 46, 79]. Particularly, in the work [63] by Ding, a unified framework of H_i/H_∞-optimal fault detection and estimation in the least mean square estimation sense for LDTV systems has been established.

On the other hand, by noting the interesting connections between linear estimation under H_2 and H_∞ criteria with searching for a minimum of certain indefinite quadratic form, a unified argument of Kalman filtering and H_∞ estimation in Krein space has been proposed by Hassibi et al. in the later 1990s [6–8]. In comparison with Hilbert space, the merits of estimation in Krein space lie in its computation attractiveness and less conservativeness. Motivated by these advantages, Zhong et al. [43] first proposed to apply Krein space theory into fault estimation for LDTV systems in 2008, in which the H_∞ fault estimation issue was equated to a minimization problem of a scalar quadratic form in Krein space and then addressed by recursive Riccati equation computations. Since then, Krein space aided design of H_∞-filtering, H_i/H_∞-FDFs and parity space-based fault detection and estimation in LDTV systems have been successively investigated [35, 42, 46]. Extensive applications of the results in nonlinear systems [41], event-triggered systems [36], 2-dimension LDTV systems [13], etc., have also been reported. In this book, we will introduce the preliminaries of Krein space theory in Chap. 4 and, on this basis, Krein space aided approaches to fault diagnosis for LDTV systems will be later discussed detailedly, that build the main bonus of Parts II–IV.

Very recently, concerning the limited applicability of traditional coprime factorization technique in LDTV situations, an operator based co-inner-outer factorization technique has been proposed by Ding and applied to the design and analysis of fault diagnosis systems in LDTV situations (referring to Chapter 7 in the monograph [64]). This technique provides us with a powerful tool to handle optimal fault diagnosis issues for LDTV systems and a new perspective to interpret the corresponding solutions. In Chap. 5 of this book, this point has also been demonstrated to a better understanding of H_2-optimal FD and FE.

In parity space-based fault diagnosis for LDTV systems, much research attention has been paid to searching for a compromise between computation cost and the fault diagnosis performance [38, 74, 84]. Following the fact that a parity space-based residual generator structurally equals to an observer, Wan et al. proposed a parity space-based design scheme towards satisfactory FD performance and an observer-based recursive algorithm for online implementation has been developed [82]. Zhong et al. formulated the design of a parity relation based fault estimator as to find a minimum for a matrix quadratic form and an analytical solution has been derived [40]. A Krein space projection aided scheme was further applied to reduce the computation load in [44]. Besides, the combination of signal processing techniques with parity space-based residual generation suggests another way to balance the FD performance and computation load, see e.g., [22, 74].

Aside from the above mentioned observer-based schemes and parity space-based approaches, some other types of diagnosis observers have also been studied for the fault diagnosis in LDTV systems, e.g., the proportional integral observer [9], interval observers [55], sliding mode observers [21, 53, 80] and adaptive observers [50, 85].

Applications of fault diagnosis techniques for LDTV systems can be found across all technical fields, including the chemical processes [9, 13, 26], aerospace vehicles [3, 30], power systems [75], wireless sensor networks [2], and cyber-physical systems [16, 17, 31], etc., even most of them are actually nonlinear systems. In this book, the utilization of fault diagnosis methods in engineering systems, e.g., unmanned aerial vehicle (UAV) control systems, satellite attitude control systems and integrated inertial navigation system and global position systems (INS/GPS), is demonstrated in Part V.

1.3 Outline of the Contents

In this book, we attempt to give a comprehensive introduction of the fundamental theory, methodologies, algorithms and typical engineering applications of model-based fault diagnosis for LDTV systems. We divide the contents into five parts, entitled respectively introduction and preliminaries (Part I: Chaps. 1–4), H_2- and H_∞-optimization based fault diagnosis for LDTV systems (Part II: Chaps. 5–10), H_∞-filtering based fault diagnosis for LDTV systems (Part III: Chaps. 11–13), parity space-based fault diagnosis for LDTV systems (Part IV: Chaps. 14–17), and applications in discrete-time nonlinear systems (Part V: Chaps. 18–23).

Part I: Introduction and Preliminaries

In Part I, fundamental knowledge that is essential for organizing and formulating the topics in later parts is introduced. That covers the basic concepts in model-based fault diagnosis, the preliminaries of LDTV systems and the fault diagnosis for LDTV systems, as well as the basis of Krein spaces. We would like to mention that most of the results in this part can be found in other standard textbooks on fault diagnosis for dynamic systems such as [27, 61, 64]. So the reader can also refer to these monographs for possible more detailed explanation and deeper interpretation.

Chapter 2 confines to conventional paradigm of model-based fault diagnosis for dynamic systems. With special attention being paid to LDTV systems, system description and typical fault diagnosis schemes for LDTV systems are introduced in Chap. 3. Chapter 4 gives an overview of Krein spaces and Krein space based optimization techniques that serve as an essential mathematical tools in the subsequent studies on Krein space based approaches to optimal fault diagnosis for LDTV systems.

Part II: H_2- and H_∞-Optimization Based Fault Diagnosis for LDTV Systems

Part II focuses on fault detection and estimation issues for LDTV systems in the frameworks of H_2- and H_∞-optimization.

In Chap. 5, H_2-optimization based FD for LDTV systems is demonstrated. Basic concepts and problems of FD and least mean square (LMS) estimation are first reviewed. Considering the system is corrupted by stochastic disturbances, a Kalman filter based solution to optimal FD is demonstrated. For the case of the unknown inputs being norm-bounded deterministic signals, a Krein space based solution to H_2-optimal FD is derived. Moreover, an alternative interpretation of the solution is given by virtue of co-inner-outer factorization.

Chapter 6 is dedicated to optimal FD for LDTV systems in the framework of H_i/H_∞-optimization. The core idea is to construct an FDF-based residual generator, in which the gain matrix and the post-filter are determined towards maximizing the finite horizon H_-/H_∞ and/or H_∞/H_∞ indices. The solution is proved to be not only applicable for the LDTV systems with norm-bounded unknown inputs but also the ones corrupted by stochastic disturbances.

Chapter 7 presents a projection based approach to optimal FD for LDTV systems with stochastic disturbances. By means of innovation analysis, a residual generator is constructed by an arbitrary linear combination of output estimation error sequence, in which a post-filter is designed in the sense of H_∞/H_∞- and H_-/H_∞-optimization. A recursive implementation algorithm in state space is developed by equating the projection based residual generator with H_∞/H_∞- and H_-/H_∞-FDFs.

Concerning LDTV systems with $l_{2,[0,N]}$-norm bounded unknown inputs, Chap. 8 confines to an H_i/H_∞-optimization based FD scheme by virtue of Krein space projection. Different from Chaps. 6 and 7, the FD issue in this chapter is formulated as an estimation problem of a quadratic form evaluation function. To reduce the online computation burden, a recursive algorithm is derived via Krein space projection. Equality of the formulated estimation problem with observer-based H_i/H_∞-optimal FD problem is also proved.

By modeling the event-triggered system as a special type of LDTV process, the objective of Chap. 9 is to deal with event-triggered FD issues in the sense of H_i/H_∞-optimization, the key of which lies in decoupling the event-triggered transmission errors from the residual signal while achieving an optimal trade-off between the robustness of residual to unknown input and the sensitivity to fault.

Chapter 10, following the idea of Chap. 8, is dedicated to the H_i/H_∞-optimization based FD via Krein space projection for LDTV systems with delayed state.

Part III: H_∞-Filtering Based Fault Diagnosis for LDTV Systems

In Part III, we demonstrate H_∞-filtering approaches to robust fault detection and estimation for LDTV systems, with main attention being paid to handling H_∞-filtering problems via Krein space techniques.

Chapter 11 focuses on addressing H_∞ fault estimation issues for LDTV systems with $l_{2,[0,N]}$-norm bounded unknown input. The targeting H_∞-optimal fault estimation problem is equated into a minimization issue of a quadratic function in Krein space and a recursive solution is derived in terms of Riccati equations. In Chap. 12, a design scheme of H_∞-optimal FDF is presented for FD purpose by means of orthogonal projection and innovation analysis in Krein space. Chapter 13 extends the results in Chap. 12 to the LDTV systems with delayed state.

Part IV: Parity Space-Based Fault Diagnosis for LDTV Systems

Part IV confines to parity space-based approaches to fault detection and estimation for LDTV systems with unknown inputs, focusing on alleviating the conflict of improving fault diagnosis performance and decreasing the online realization cost.

In Chap. 14, we formulate the parity space-based optimal FD problem under finite horizon H_2/H_2 criterion as finding a minimum of quadratic form of unknown input. A Krein space projection based solution is derived that can deliver an optimal residual generator in the sense of minimizing the robustness/sensitivity criteria ratio and, simultaneously, a more computation efficient recursive implementation algorithm is available.

In Chap. 15, parity space-based fault estimation issues for LDTV systems are addressed with respect to solving a minimization problem of a quadratic form over parity space matrix.

In Chap. 16, we apply the parity space-based approach to cope with FD issues in event-triggered systems, the core of which, similar to Chap. 9, lies in the complete decoupling of transmission errors from the residual signal.

Chapter 17 demonstrates a stationary wavelet transform aided design and analysis scheme of parity space-based FD for LDTV systems, which suggests an alternative way to balance FD performance with computation cost.

Part V: Applications in Discrete-Time Nonlinear Systems

In Part V, we illustrate typical engineering applications of the schemes in former parts in discrete-time nonlinear systems, including the UAV control systems (Chap. 18), satellite attitude control systems (Chap. 19) and INS/GPS integrated systems (Chaps. 20–23). Our aims are to show the reader that the derived theoretical results for standard LDTV systems can be extended to nonlinear systems and, how to apply these methods to address fault diagnosis problems in practical engineering systems.

References

1. Aguilera-Gonzalez, A., Theilliol, D., Adam-Medina, M., et al. (2012). Sensor fault and unknown input estimation based on proportional integral observer applied to LPV descriptor systems. *IFAC Proceedings Volumes, 45*(20), 1059–1064.
2. Mahapatro, A., & Khilar, P. M. (2013). Fault diagnosis in wireless sensor networks: A survey. *IEEE Communications Surveys & Tutorials, 15*(4), 2000–2026.
3. Zolghadri, A., Henry, D., Cieslak, J., et al. (2014). *Fault diagnosis and fault-tolerant control and guidance for aerospace vehicles*. Berlin: Springer.

4. Wang, A., & Wang, H. (2005). Fault diagnosis for nonlinear systems via neural networks and parameter estimation. In *Proceedings of the 5th International Conference on Control and Automation (ICCA'05)*, June 26–29, Budapest, Hungary (pp. 559–563).
5. Hassanabadi, A. H., Shafiee, M., & Puig, V. (2016). UIO design for singular delayed LPV systems with application to actuator fault detection and isolation. *International Journal of Systems Science, 47*(1), 107–121.
6. Hassibi, B., Sayed, A. H., & Kailath, T. (1999). *Indefinite-quadratic estimation and control: A unified approach to H_2 and H_∞ theories*. Philadelphia: SIAM.
7. Hassibi, B., Sayed, A. H., & Kailath, T. (1996). Linear estimation in Krein spaces-Part I: Theory. *IEEE Transactions on Automatic Control, 41*(1), 18–33.
8. Hassibi, B., Sayed, A. H., & Kailath, T. (1996). Linear estimation in Krein spaces-Part II: Applications. *IEEE Transactions on Automatic Control, 41*(1), 34–49.
9. Rabaoui, B., Hamdi, H., Rodrigues, M., et al. (2017). A polytopic proportional integral observer design for fault diagnosis. In *Proceeding of the International Conference on Advanced Systems and Electrical Technologies*, January 14–17, Hammamet, Tunisia (pp. 14–17).
10. Baskiotis, C., Raymond, J., & Rault, A. (1979). Parameter identification and discriminant analysis for jet engine machanical state diagnosis. In *Proceedings of the 18th IEEE Conference on Decision and Control including the Symposium on Adaptive Processes*, December 12–14, Fort Lauderdale, FL, USA (Vol. 2, pp. 648–650).
11. Zhang, C., Zhao, H., & Li, T. (2016). Krein space-based H_∞ adaptive smoother design for a class of Lipschitz nonlinear discrete-time systems. *Applied Mathematics and Computation, 287*, 134–148.
12. Guernez, C., Cassar, J., & Staroswiecki, M. (1997). Extension of parity space to non linear polynomial dynamic systems. *IFAC Proceedings Volumes, 30*(18), 857–862.
13. Zhao, D., Wang, Y., Li, Y., et al. (2017). H_∞ fault estimation for 2-D linear discrete time-varying systems based on Krein space method. *IEEE Transactions on Systems, Man, and Cybernetics: Systems*, accepted for publication, *48*(12), 2070–2079.
14. Efimov, D., Ralssi, T., & Zolghadri, A. (2010). Adaptive set observers design for nonlinear continuous-time systems: Application to fault detection and diagnosis. arXiv:1012.0684.
15. Ichalal, D., & Mammar, S. (2015). On unknown input observers for LPV systems. *IEEE Transactions on Industrial Electronics, 62*(9), 5870–5880.
16. Ding, D., Wang, Z., Ho, D. W. C., et al. (2017). Observer-based event-triggering consensus control for multiagent systems with lossy sensors and cyber-attacks. *IEEE Transactions on Cybernetics, 47*(8), 1936–1947.
17. Ding, D., Wang, Z., Wei, G., et al. (2016). Event-based security control for discrete-time stochastic systems. *IET Control Theory & Applications, 10*(15), 1808–1815.
18. Geiger, G. (1982). Monitoring of an electrical driven pump using continuous-time parameter estimation methods. *IFAC Proceedings Volumes, 15*(4), 603–608.
19. Xu, G., & Zhou, D. (2010). Fault prediction for state-dependent fault based on online learning neural network. *Journal of Zhejiang University, 44*(7), 1251–1320.
20. Dong, H., Wang, Z., Bu, X., et al. (2016). Distributed fault estimation with randomly occurring uncertainties over sensor networks. *International Journal of General Systems, 45*(5), 662–674.
21. Ríos, H., Punta, E., & Fridman, L. (2017). Fault detection and isolation for nonlinear non-affine uncertain systems via sliding-mode techniques. *International Journal of Control, 90*(2), 218–230.
22. Ye, H., Wang, G., Ding, S. X., et al. (2002). An IIR filter based parity space approach for fault detection. *IFAC Proceedings Volumes, 35*(1), 155–160.
23. Wang, H., & Huang, H. (2009). Fault estimation based on PI strong tracking filter. *Aerospace Control and Application, 35*(3), 34–48.
24. Behzad, H., Casavola, A., Tedesco, F., et al. (2016). A fault-tolerant sensor reconciliation scheme based on LPV unknown input observers. In *Proceeding of the 55th Conference on Decision and Control (CDC)*, December 12–14, Las Vegas, NV, USA (pp. 2158–2163).
25. Odendaal, H. M., & Jones, T. (2014). Actuator fault detection and isolation: An optimised parity space approach. *Control Engineering Practice, 26*, 222–232.

26. Ali, J. M., Hoang, N. H., Hussain, M. A., et al. (2015). Review and classification of recent observers applied in chemical process systems. *Computers & Chemical Engineering, 76*, 27–41.
27. Chen, J., & Patton, R. J. (1999). *Robust model-based fault diagnosis for dynamic systems*. Boston, MA: Kluwer Academic Publishers.
28. Gertler, J. (1995). Diagnosing parametric faults: From parameter estimation to parity relation. In *Proceedings of 1995 American Control Conference (ACC'95)*, June 21–23, Seattle, WA, USA (pp. 1615–1620).
29. Wang, J., & Qin, S. J. (2006). Closed-loop subspace identification using the parity space. *Automatica, 42*(2), 315–320.
30. Marzat, J., Piet-Lahanier, H., Damongeot, F., et al. (2012). Model-based fault diagnosis for aerospace systems: A survey. *Proceedings of the Institution of Mechanical Engineers, Part G: Journal of aerospace engineering, 226*(10), 1329–1360.
31. Manandhar, K., Cao, X., Hu, F., et al. (2014). Detection of faults and attacks including false data injection attack in smart grid using kalman filter. *IEEE Transactions on Control of Network Systems, 1*(4), 370–379.
32. Gou, L., Sun, R., & Han, X. (2021). FDIA system for sensors of the aero-engine control system based on the immune fusion kalman filter. *Mathematical Problems in Engineering, 2021*(1), 1–17.
33. Almasri, M., Tricot, N., & Lenain, R. (2020). Parameter estimation based-FDI method enhancement with mixed particle filter. *Neurocomputing, 403*, 441–451.
34. Unal, M., Onat, M., Demetgul, M., et al. (2014). Fault diagnosis of rolling bearings using a genetic algorithm optimized neural network. *Measurement, 58*, 187–196.
35. Zhong, M., Ding, S. X., Han, Q.-L., et al. (2022). A Krein space-based approach to event-triggered H_∞ filtering for linear discrete time-varying systems, *135*, 110001.
36. Zhong, M., Du, X., Song, Y., et al. (2021). Event-triggered parity space approach to fault detection for linear discrete-time systems. *IEEE Transactions on Systems, Man, and Cybernetics: Systems*. https://doi.org/10.1109/TSMC.2021.3103816.
37. Zhong, M., Xue, T., Song, Y., et al. (2021). Parity space vector machine approach to robust fault detection for linear discrete-time systems. *IEEE Transactions on Systems, Man, and Cybernetics: Systems, 51*(7), 4251–4261.
38. Zhong, M., Ding, Q., & Shi, P. (2009). Parity space-based fault detection for Markovian jump systems. *International Journal of Systems Science, 40*(4), 421–428.
39. Zhong, M., Ding, S. X., & Ding, E. L. (2010). Optimal fault detection for linear discrete time-varying systems. *Automatica, 46*(8), 1395–1400.
40. Zhong, M., Ding, S. X., Han, Q.-L., et al. (2010). Parity space-based fault estimation for linear discrete time-varying systems. *IEEE Transactions on Automatic Control, 55*(7), 1726–1731.
41. Zhong, M., Ding, S. X., & Zhou, D. (2016). A new scheme of fault detection for linear discrete time-varying systems. *IEEE Transactions on Automatic Control, 61*(9), 2597–2602.
42. Zhong, M., Guo, D., & Zhou, D. (2014). A Krein space approach to H_∞ filtering of discrete-time nonlinear systems. *IEEE Transactions on Circuits and Systems I: Regular Papers, 61*(9), 2644–2652.
43. Zhong, M., Liu, S., & Zhao, H. (2008). Krein space-based H_∞ fault estimation for linear discrete time-varying systems. *Acta Automatica Sinica, 34*(12), 1529–1533.
44. Zhong, M., Song, Y., & Ding, S. X. (2015). Parity space-based fault detection for linear discrete time-varying systems with unknown input. *Automatica, 59*, 120–126.
45. Zhong, M., Xue, T., Ding, S. X., et al. (2017). A wavelet-based parity space approach to fault detection of linear discrete time-varying systems. *IFAC-PapersOnLine, 50*(1), 2836–2841.
46. Zhong, M., Zhou, D., & Ding, S. X. (2010). On designing H_∞ fault detection filter for linear discrete time-varying systems. *IEEE Transactions on Automatic Control, 55*(7), 1689–1695.
47. Frank, P. M. (1991). Enhancement of robustness in observer-based fault detection. In *Proceedings of the IFAC SAFEPROCESS'91*, September 10–13, Baden-Baden, Germany (pp. 99–111).
48. Lu, P., van Kampen, E.-J., de Visser, C. C., et al. (2016). Framework for state and unknown input estimation of linear time-varying systems. *Automatica, 73*, 145–154.

49. Zhang, P., Ye, H., Ding, S. X., et al. (2006). On the relationship between parity space and H_2 approaches to fault detection. *Systems & Control Letters, 55*(2), 94–100.
50. Zhang, Q. (2002). Adaptive observer for multiple-input-multiple-output (MIMO) linear time-varying systems. *IEEE Transactions on Automatic Control, 47*(3), 525–529.
51. Zhang, Q. (2017). Adaptive Kalman filter for actuator fault diagnosis. *IFAC-PapersOnLine, 50*(1), 14272–14277.
52. Zhang, Q., & Basseville, M. (2014). Statistical detection and isolation of additive faults in linear time-varying systems. *Automatica, 50*(10), 2527–2538.
53. Galván-Guerra, R., Fridman, L., & Dávila, J. (2017). High-order sliding-mode observer for linear time-varying systems with unknown inputs. *International Journal of Robust and Nonlinear Control, 27*(14), 2338–2356.
54. Patton, R. J., & Chen, J. (1991). A re-examination of the relationship between parity space and observer-based approaches in fault diagnosis. *European Journal of Diagnosis and Safety in Automation (Revue européenne Diagnostic et sûreté de fonctionnement), 1*(2), 183–200.
55. Lamouchi, R., Amairi, M., Raissi, T, et al. (2016). Interval observer design for linear parameter-varying systems subject to component faults. In *Proceeding of the 24th Conference on Control and Automation (MED)*, June 21–24, Athens, Greece (pp. 707–712).
56. Patton, R. J., & Chen, J. (1991). A review of parity space approaches to fault diagnosis. *IFAC Proceedings Volumes, 24*(6), 65–81.
57. Patton, R. J., & Chen, J. (1991). A robust parity space approach to fault diagnosis based on optimal eigenstructure assignment. In *Proceedings of the International Conference on Control*, Edinburgh, UK, March 25–28 (pp. 1056–1061).
58. Rahimilarki, R., Gao, Z., Zhang, A., et al. (2019). Robust neural network fault estimation approach for nonlinear dynamic systems with applications to wind turbine systems. *IEEE Transactions on Industrial Informatics, 15*(12), 6302–6312.
59. Ding, S. X., Ding, E. L., Jeinsch, T., et al. (2000). An approach to a unified design of FDI systems. In *Proceedings of the 3rd Asian Control Conference*, Shanghai China (pp. 2812–2817).
60. Ding, S. X., Guo, L., & Jeinsch, T. (1999). A characterization of parity space and its application to robust fault detection. *IEEE Transactions on Automatic Control, 44*(2), 337–343.
61. Ding, S. X. (2013). *Model-based fault diagnosis techniques: Design schemes, algorithms, and tools* (2nd ed.). London, U.K.: Springer.
62. Ding, S. X. (2014). *Data-driven design of fault diagnosis and fault-tolerant control systems*. London, U.K.: Springer.
63. Ding, S. X. (2018). Optimal fault detection and estimation: A unified scheme and least squares solutions. *IFAC PapersOnLine, 51*(24), 465–472.
64. Ding, S. X. (2020). *Advanced methods for fault diagnosis and fault-tolerant control*. Berlin, Germany: Springe.
65. Ding, S. X., Shen, B., Wang, Z., et al. (2014). A fault detection scheme for linear discrete-time systems with an integrated online performance evaluation. *International Journal of Control, 87*(12), 2511–2521.
66. Ding, S. X., Yang, Y., Zhang, Y., et al. (2014). Data-driven realizations of kernel and image representations and their application to fault detection and control system design. *Automatica, 50*(10), 2615–2623.
67. Ding, S. X., Zhang, P., Naik, A., et al. (2009). Subspace method aided data-driven design of fault detection and isolation systems. *Journal of Process Control, 19*, 1496–1510.
68. Ding, S. X., & Frank, P. (1989). Fault detection via optimally robust detection filters. In *Proceedings of the 28th IEEE Conference on Decision and Control (CDC)*, December 13–15, Tampa, FL, USA, INS. 3695202.
69. Ahmadizadeh, S., Zarei, J., & Karimi, H. R. (2014). Robust unknown input observer design for linear uncertain time delay systems with application to fault detection. *Asian Journal of Control, 16*(4), 1006–1019.
70. Varrier, S., Koenig, D., & Martinez, J. J. (2012). A parity space-based fault detection on LPV systems: Approach for vehicle lateral dynamics control system. *IFAC Proceedings Volumes, 45*(20), 1191–1196.

71. Yin, S., Naik, A. S., & Ding, S. X. (2009). Adaptive process monitoring based on parity space methods. *IFAC Proceedings Volumes, 42*(8), 947–952.
72. Yin, S., & Zhu, X. (2015). Intelligent particle filter and its application to fault detection of nonlinear system. *IEEE Transactions on Industrial Electronics, 62*(6), 3852–3861.
73. Li, S., Wang, H., Aitouche, A., et al. (2016). Robust unknown input observer design for state estimation and fault detection using linear parameter varying model. In *Proceeding of the 13th European Workshop on Advanced Control and Diagnosis (ACD)*, November 17–18, Lille, France (Vol. 783, pp. 012001).
74. Xue, T., Zhong, M., Ding, S. X., et al. (2018). Stationary wavelet transform aided design of parity space vectors for fault detection in LDTV systems. *IET Control Theory & Applications, 12*(7), 857–864.
75. Qiao, W., & Lu, D. (2015). A survey on wind turbine condition monitoring and fault diagnosis–Part I: Components and subsystems. *IEEE Transactions on Industrial Electronics, 62*(10), 6536–6545.
76. Chen, X., Sun, R., Jiang, W., et al. (2016). A novel two-stage extended Kalman filter algorithm for reaction flywheels fault estimation. *Chinese Journal of Aeronautics, 29*(2), 462–469.
77. Li, X., & Zhou, K. (2009). A time domain approach to robust fault detection of linear time-varying systems. *Automatica, 45*(1), 94–102.
78. Li, X. (2009). *Fault detection filter design for linear systems*. Ph.D. dissertation (pp. 66–82). Louisiana State University, USA, August.
79. Li, Y., & Zhong, M. (2015). Fault estimation for discrete-time singular systems with random sensor failures. *IFAC-PapersOnLine, 48*(21), 701–706.
80. Zhou, Y., Soh, Y. C., & Shen, J. (2014). High-gain observer with higher order sliding mode for state and unknown disturbance estimations. *International Journal of Robust and Nonlinear Control, 24*(15), 2136–2151.
81. Wang, Y., Gao, B., & Chen, H. (2015). Data-driven design of parity space-based FDI system for AMT vehicles. *IEEE/ASME Transactions on Mechatronics, 20*(1), 405–415.
82. Wan, Y., Dong, W., & Ye, H. (2013). Integrated trade-off design of fault detection system for linear discrete time-varying systems. *IET Control Theory & Applications, 7*(3), 455–463.
83. Jiang, Y., Yin, S., & Kaynak, O. (2021). Optimized design of parity relation-based residual generator for fault detection: Data-driven approaches. *IEEE Transactions on Industrial Informatics, 17*(2), 1449–1458.
84. Li, Y., & Zhong, M. (2009). Parity space-based fault detection for linear time-varying systems with multiple packet dropouts. In *Proceeding of the 4th International Conference on Innovative Computing, Information and Control (ICICIC)*, December 07–09, Kaohsiung, Taiwan (pp. 236–239).
85. Wang, Z., Rodrigues, M., Theilliol, D., et al. (2015). Actuator fault estimation observer design for discrete-time linear parameter-varying descriptor systems. *International Journal of Adaptive Control and Signal Processing, 29*(2), 242–258.

Chapter 2
Paradigm of Model-Based Fault Diagnosis

In this chapter the paradigm of model-based fault diagnosis for linear dynamic systems is introduced. Basic concepts, tasks, design setup problems, as well as the involved mathematical tools are introduced.

2.1 Basic Tasks and Problems of Fault Diagnosis

For the ease of subsequent description, in this chapter the following linear dynamic system with unknown input and fault are considered

$$y = \mathcal{T}_u(u) + \mathcal{T}_d(d) + \mathcal{T}_f(f), \tag{2.1}$$

where $u \in \mathbb{R}^l$ and $y \in \mathbb{R}^q$ are the system input and output vectors, respectively, $d \in \mathbb{R}^{k_d}$ denotes the unknown input that can be stochastic and/or deterministic signals, $f \in \mathbb{R}^{k_f}$ is the fault vector to be diagnosed, linear operators

$$\mathcal{T}_u : \mathbb{R}^l \to \mathbb{R}^q,\ \mathcal{T}_d : \mathbb{R}^{k_d} \to \mathbb{R}^q,\ \mathcal{T}_f : \mathbb{R}^{k_f} \to \mathbb{R}^q$$

represent linear mappings from spaces $\mathbb{R}^l$, $\mathbb{R}^{k_d}$ and $\mathbb{R}^{k_f}$ to $\mathbb{R}^q$, respectively.

Given system model (2.1), mathematical formulations of model-based fault detection and estimation problems are introduced in what follows.

M. Zhong et al., *Fault Diagnosis for Linear Discrete Time-Varying Systems and Its Applications*, https://doi.org/10.1007/978-981-19-5438-2_2

2.1.1 Basic Problems of Fault Detection and Estimation

The objective of FD is to detect the occurrence of a fault in the process to be monitored. Residual generation and residual evaluation are two essential parts of a model-based FD system.

In residual generation stage, a residual generator driven by system input and output should be first constructed, serving as a fault information carrier, i.e.,

$$r = \mathcal{K}(u, y), \tag{2.2}$$

where $\mathcal{K}(\cdot)$ is an operator that maps the space spanned by measurable vectors u and y onto the residual space with residual vector r. For FD purpose, the residual signal in ideal case should satisfy

$$\forall u, \begin{cases} r = 0, \text{ if } f = 0, \\ r \neq 0, \text{ if } f \neq 0. \end{cases} \tag{2.3}$$

Unfortunately, the existence of modeling uncertainties makes it difficult to achieve a zero residual in fault-free case. Regarding of this, denote by $\mathcal{M}_d$ and $\mathcal{M}_f$ the mappings from unknown input d and fault f to residual r, respectively. The dynamics of residual generator (2.2) is governed by

$$r = \mathcal{K}(u, y) = \mathcal{M}_f(f) + \mathcal{M}_d(d). \tag{2.4}$$

In fault-free case, $r = \mathcal{M}_d(d)$ holds. Indeed, operators $\mathcal{M}_d$ and $\mathcal{M}_f$ constitute the image spaces of unknown input d and fault f, respectively [12]. A residual generator is expected to be designed achieving a complete decoupling from unknown input or, alternatively, guaranteeing good robustness against unknown input and, at the same time, keeping its sensitivity to fault.

Following residual generation, another important task of FD is residual evaluation. In this stage, an evaluation function $J(r)$ and an appropriate threshold J_{th} should be determined such that the occurrence of a fault can be detected by performing the following decision logic

$$\begin{cases} J(r) \leq J_{th}, & \text{no alarm}, \\ J(r) > J_{th}, & \text{fault alarm}. \end{cases} \tag{2.5}$$

Fault estimation aims to estimate the fault for more information such as type, size and location. To this end, a fault estimator $\mathcal{G}$ can be constructed by

$$\hat{f} = \mathcal{G}(y) \text{ or } \hat{f} = \mathcal{G}(u, y). \tag{2.6}$$

In short, the main tasks of FD are the construction of residual generator (2.2) and the determination of evaluation function $J(r)$ and threshold J_{th}. The core of fault estimation lies in the design of fault estimator (2.6).

2.1.2 Optimal Fault Detection and Estimation Problems

Consider the unknown inputs in the monitored system are stochastic signals. The indices of false alarm rate (FAR), fault detection rate (FDR) and/or missed detection rate (MDR) are generally used to evaluate the performance of FD. Their definitions in the probabilistic context are given below.

Definition 2.1 Given a test statistic $J(r)$, a threshold J_{th} and detection logic (2.5), FAR is defined as the probability

$$\text{FAR} = \Pr\{J(r) > J_{th} | f = 0\}. \tag{2.7}$$

The MDR is defined as the probability

$$\text{MDR} = \Pr\{J(r) \le J_{th} | f \ne 0\}. \tag{2.8}$$

The FDR is defined as the probability

$$\text{FDR} = \Pr\{J(r) > J_{th} | f \ne 0\}. \tag{2.9}$$

which satisfies FDR $= 1 -$ MDR.

It is notable that the evaluation function $J(r)$ is related to unknown input and fault. We could hardly decrease the FAR and MDR simultaneously. A remedy is to design an FD system towards an optimal trade-off between FAR and MDR (or FDR). In this regard, optimal FD problems can be formulated as follows [12].

Definition 2.2 A solution to this FD problem is said to be optimal, when, for a given acceptable FAR α

$$\{J(r), J_{th}\} = \arg \min_{J(r),\, J_{th},\, \text{FAR} \le \alpha} \text{MDR}. \tag{2.10}$$

Definition 2.3 A solution to this FD problem is said to be optimal, when, for a given acceptable MDR β,

$$\{J(r), J_{th}\} = \arg \min_{J(r),\, J_{th},\, \text{MDR} \le \beta} \text{FAR}. \tag{2.11}$$

The optimal FD problem in Definition 2.2 seeks a minimum FRA with a given MDR level. It is a dual formulation of the problem in Definition 2.3 where a minimum MDR is pursued for a preset upper bound of FAR.

For systems with norm-bounded disturbances, the FD performance cannot be appropriately assessed with the indices FAR, MDR and/or FDR. We directly refer to [12] for the definitions of optimal FD that are derived by measuring the fault detectability and false fault alarms based on the concepts of "set", as given below.

Definition 2.4 A solution $\{J(r), J_{th}\}$ to the FD problem with the presence of disturbance d is called optimal in the sense of maximizing the fault detectability when

$$\forall r \in \mathcal{S}_d,\ f = 0,\ J(r) \leq J_{th}, \tag{2.12}$$

$$\forall f \notin \mathcal{S}_{f,undect},\ d = 0,\ J(r) > J_{th}, \tag{2.13}$$

where $\mathcal{S}_d = \left\{ r_d \left| r_d = \mathcal{M}_d(d),\ \forall d \in \mathbb{R}^{k_d},\ \|d\|_p \leq \delta_d \right. \right\}$ is the image of the vector d, $\|d\|_p$ denotes the p-type norm of d, $\mathcal{S}_{f,undect} = \left\{ f \,|\, f \in \mathbb{R}^{k_f},\ r_f = \mathcal{M}_f(f) \in \mathcal{S}_d \right\}$ is the set of undetectable faults.

It is seen that the solution to optimal FD problem in Definition 2.4 guarantees a zero FAR and the maximal fault detectability can be achieved when all faults outside the set $\mathcal{S}_{f,undect}$ can be detected. As a dual formulation of Definition 2.4, the optimal FD problem in the sense of minimizing the number of false alarms while ensuring the detectability of certain faults of interest is given as follows.

Definition 2.5 A solution $\{J(r), J_{th}\}$ to the FD problem with the presence of disturbance d is called optimal in the sense of maximizing the fault detectability when

$$\forall f \notin \mathcal{S}_{f,\delta_f},\ d = 0,\ J(r) > J_{th}, \tag{2.14}$$

$$\forall r \in \mathcal{S}_{d,\delta_f},\ f = 0,\ J(r) \leq J_{th}, \tag{2.15}$$

where $\mathcal{S}_{f,\delta_f} = \left\{ f \,|\, f \in \mathbb{R}^{k_f},\ \|f\|_p \leq \delta_f \right\}$ is the set of faults of not interest, $\mathcal{S}_{d,\delta_f} = \left\{ r_d | r_d = \mathcal{M}_d(d) \in \bar{\mathcal{S}}_{f,\delta_f},\ d \in \mathbb{R}^{k_d} \right\}$, $\bar{\mathcal{S}}_{f,\delta_f} = \left\{ r_f | r_f = \mathcal{M}_f(f),\ f \in \mathcal{S}_{f,\delta_f} \right\}$.

Regarding fault estimation, optimal fault estimation problems in the sense of least mean square (LMS) and least square (LS) are defined as follows.

Definition 2.6 An $\hat{f}$ is called an optimal estimate of f if it is unbiased, i.e., $\mathbb{E}[\hat{f}] = f$, and in the sense of LMS it solves

$$\min_{f, \mathbb{E}[f] = \bar{f}} \mathbb{E}\left[\left(f - \bar{f} \right) \left(f - \bar{f} \right)^T \right]. \tag{2.16}$$

Definition 2.7 For the system with deterministic disturbances, in the sense of LS the estimate $\hat{f}$ is optimal if it solves

$$\min_{\hat{f}} \left\| f - \hat{f} \right\|_2^2. \tag{2.17}$$

Remark 2.1 It is worth mentioning that a traditional way for fault detection and estimation is to carry out FD at first, then the fault estimator is activated after a fault is detected. In certain cases, the fault estimator can be embedded in the residual generator. A recent trend for fault estimation argues to direct construct a fault estimator, then FDI issues can be handled intuitively. Despite the simple idea, this strategy, unfortunately, may not deliver the optimal FD in the context of optimizing a trade-off between the fault detectability and false fault alarm performance, as defined in Definitions 2.2–2.5.

Remember the central role of residual generation in FD for dynamic systems. In the forthcoming sections, we first review the principles of residual generation towards perfect unknown input decoupling, and then optimal residual generation in the sense of maximizing the sensitivity/robustness criteria is briefly introduced. Finally, our attention turns to residual evaluation issues.

2.2 Perfect Unknown Input Decoupling

Generally, the error between real system output and its estimate driven by the same system input is used as a residual signal, i.e.,

$$r = \mathcal{W}(y - \hat{y}), \tag{2.18}$$

where $\hat{y} \in \mathbb{R}^q$ is the estimate of output driven by system input u, $\mathcal{W}$ is the so-called post-filter that provides us with an additional design freedom to improve the FD performance. The design of a residual generator thus lies in the design of output estimator and the post-filter $\mathcal{W}$. Recalling the dynamics of residual generator (2.4) for system (2.1), we can further rewrite the dynamics of (2.18) as follows

$$r = \mathcal{W} \circ \left(\mathcal{M}_d(d) + \mathcal{M}_f(f)\right). \tag{2.19}$$

On account of (2.3), an ideal residual generator is desired that the residual signal is completely decoupled from the unknown input and, at the same time, keeping sensitivity to fault, i.e.,

$$\mathcal{W} \circ \mathcal{M}_d = 0, \ \mathcal{W} \circ \mathcal{M}_f \neq 0. \tag{2.20}$$

Such an issue can be formulated as the following perfect unknown input decoupling (PUID) problem [11].

Definition 2.8 Residual generator (2.19) is called perfectly decoupled from the unknown input d if condition (2.20) is satisfied. The design of such a residual generator is called perfect unknown input decoupling problem.

To better understand the idea behind the PUID strategy, we take the following linear time invariant system as an example

$$y(z) = G_{yu}(z)u(z) + G_{yd}(z)d(z) + G_{yf}(z)f(z),$$

where $G_{yu}, G_{yd}, G_{yu} \in \mathcal{RH}_\infty$ denote the transfer functions from input u, unknown input d and fault f to output y. Let $\hat{y}(z) = G_{yu}(z)u(z)$. A residual generator is then constructed as

$$r(z) = W(z)\left(y(z) - \hat{y}(z)\right) = W(z)\left(G_{yd}(z)d(z) + G_{yf}(z)f(z)\right), \tag{2.21}$$

where $W(z) \in \mathcal{RH}_\infty$ is the post-filter. Towards a PUID, $W(z)$ should satisfy

$$W(z)G_{yd}(z) = 0, \;\; W(z)G_{yf}(z) \neq 0,$$

which holds if and only if

$$\text{rank}\left(G_{yd}(z)\right) < \text{rank}\left(\begin{bmatrix} G_{yd}(z) & G_{yf}(z) \end{bmatrix}\right) = q. \tag{2.22}$$

A physical interpretation of inequality (2.22) is that the unknown inputs observable from the system output should be smaller than the number of sensors. If it is the case, the measurements coupled with faults of interest is decoupled with the unknown input. However, despite a PUID design can completely eliminate the influence of unknown input to residual,

- the existence condition of a solution to PUID problem is hard to be satisfied in many practical physical systems. For example, when residual generator (2.21) is applied, enough sensors are required to meet $\text{rank}(G_{yd}(z)) < q$. In particular, when noises are fully spanned in the measurement space, this condition cannot be satisfied no matter how many sensors are used.
- Additionally, faults in the subspace spanned by the decoupled unknown input (a kernel space of d) cannot be detected any more, which results in the reduction of fault detectability.

To overcome these deficiencies, a more popular solution, as stated before, is to construct a residual generator towards an optimal trade-off between the robustness of residual to unknown inputs and the sensitivity to faults. In this context, optimal design issues of residual generator in the frameworks of H_2- and H_∞-optimization have been extensively studied.

Remark 2.2 The PUID problem for LTI systems have been systematically discussed in [11]. Solutions by using the diagnosis observer scheme, the matrix pencil method, parity space method, and the eigenstructure assignment approach, etc., as well as their existence conditions, have been well summarized. When it comes to the linear time-varying systems, UIO-based methods have been widely accepted to achieve PUID, with main efforts being made to the existence conditions of UIOs, improvement of fault diagnosis performance [6, 7], and extensive applications to nonlinear and LPV processes [1, 3, 5, 15]. In the next chapter, we will give a UIO-based solution to PUID problem for LDTV system.

2.3 H_2- and H_∞-Optimal Design of Residual Generator

Pioneer works on H_2- and H_∞-optimization for robust control purpose can date back to the 1980s s and early 1990s. Following these works, applications of H_2- and H_∞-optimization techniques to the design of robust fault diagnosis systems have been extensively reported. In this section, we briefly introduce the principles of optimal residual generation in the frameworks of H_2- and H_∞-optimization. More detailed discussion can be found in the later chapters.

Different from the idea of PUID, H_2- and H_∞-optimization based residual generation aims at a trade-off between the robustness of residual to unknown input and the sensitivity to fault. In this regard, the residual generator design can be formulated as a multi-objective optimization problem, i.e.,

$$\max P_f, \ \min P_d. \tag{2.23}$$

where P_d and P_f denote the robustness index and fault sensitivity index that can be defined in terms of H_2-norm, H_∞-norm and/or H_- index, respectively.

Depending on the type of robustness and fault sensitivity indices, various optimal fault detection schemes have been proposed towards (2.23), including

- H_2- and H_∞-filtering methods;
- H_2/H_2- and H_∞/H_∞-optimization methods;
- mixed H_2/H_∞-, H_-/H_∞- and H_i/H_∞-optimization approaches, etc.

So far, in the frameworks of H_2- and H_∞-optimization, widespread study on robust fault diagnosis for LTI systems have been carried out over the past few decades. Elaborated solutions based on diagnosis observers, FDFs, and parity space relations have been extensively reported. Focusing on LDTV systems, we will detailedly discuss optimal fault diagnosis issues in the sense of H_2- and H_∞-optimization in Parts II–IV in this book.

Remark 2.3 Different from the H_2- or H_∞ robust control where only the disturbance attenuation is required, the design objective of H_2- and H_∞-filtering based fault diagnosis systems is to minimize the effect of unknown input on residual whilst keep sensitive to faults [8].

2.4 Residual Evaluation

Apart from residual generation, another important task for FD is residual evaluation that involves the evaluation function and threshold settings. Existing residual evaluation strategies can be roughly classified into two categories, namely the norm-based method and the statistical hypothesis testing scheme, depending on the type of unknown input in the monitored system is either deterministic or stochastic signals.

2.4.1 Norm-Based Method

Norm-based residual evaluation is usually adopted for FD in systems with l_2-norm bounded unknown inputs. For ease of notation, we define the l_2-norm over finite time horizon $[k_1, k_2]$, $(k_2 \geq k_1)$ for a vector $\xi \in \mathbb{R}^{k_\xi}$ as

$$\|\xi(k)\|^2_{2,[k_1,k_2]} = \sum_{i=k_1}^{k_2} \|\xi(i)\|_2^2 = \sum_{i=k_1}^{k_2} \xi^T(i)\xi(i).$$

On this basis, the following norm-based evaluation functions are mostly used

$$J_{2,[0,N]}(r) = \|r(k)\|^2_{2,[0,N]} = \sum_{j=0}^{N} r^T(j)r(j), \tag{2.24}$$

$$J_{2,[k-N,k]}(r) = \|r(k)\|^2_{2,[k-N,k]} = \sum_{j=k-N}^{k} r^T(j)r(j), \; N < k, \tag{2.25}$$

$$J_{ave,N}(r) = \|r(k)\|^2_{ave,N} = \frac{1}{N+1}\sum_{j=k-N}^{k} r^T(j)r(j), \tag{2.26}$$

where $J_{2,[0,N]}(r)$ gives the cumulative energy of the residual in time interval $[0, N]$, $J_{2,[k-N,k]}(r)$ measure the energy of residual in sliding window $[k-N, k]$ of length $N+1$, and $J_{ave,N}(r)$ is the average energy of residual over $[k-N, k]$.

A threshold is generally set, together with the evaluation function, towards an optimal FD in the sense of maximizing the fault detectability with a zero false fault alarm (see Definition 2.4) or minimizing the number of false alarms while guaranteeing certain fault detectability (see Definition 2.5). Regarding of this, we recall residual generator (2.19) with $||d(k)||_{2,[0,N]} \leq \delta_d$. Using residual evaluation function $J_{2,[0,N]}(r)$ yields

$$J_{2,[0,N]}(r) = \left\|\mathcal{W} \circ \left(\mathcal{M}_d(d) + \mathcal{M}_f(f)\right)\right\|^2_{2,[0,N]}, \tag{2.27}$$

which delivers in fault-free case that

$$J_{2,[0,N]}(r) = \|\mathcal{W} \circ \mathcal{M}_d(d)\|^2_{2,[0,N]} \leq \|\mathcal{W} \circ \mathcal{M}_d\|_2^2 \,\delta_d^2. \tag{2.28}$$

A zero false fault alarm can then be obtained by setting the threshold as

$$J_{th} = \|\mathcal{W} \circ \mathcal{M}_d\|_2^2 \,\delta_d^2. \tag{2.29}$$

Furthermore, under assumption of $\mathcal{M}_d$ is left invertible, i.e., $\mathcal{M}_d^- \circ \mathcal{M}_d = I$, we can set the post-filter $\mathcal{W}$ as

$$\mathcal{W} = \mathcal{M}_d^-, \tag{2.30}$$

which delivers

$$J_{th} = \delta_d^2. \tag{2.31}$$

It is remarkable that (2.31) solves the optimal FD problem in Definition 2.4. This result has been proven in [12]. Also, an optimal solution to the dual problem in Definition 2.5 can be derived. Here we directly refer to [12] for the following corollary.

Corollary 2.1 *Given residual generator (2.19) and the margin of detectable faults δ_f, let $\mathcal{W} = \mathcal{M}_f^-$ satisfy*

$$\forall f \in \mathbb{R}^{k_f},\ r_f = \mathcal{M}_f^- \circ \mathcal{M}_f(f),\ \|r_f\|_{2,[0,N]} = \|f\|_{2,[0,N]}. \tag{2.32}$$

Then $\{J(r),\ J_{th}\}$ given by

$$J(r) = J_{2,[0,N]}(r),\ \ J_{th} = \delta_f^2. \tag{2.33}$$

solves the optimal FD problem defined in Definition 2.5.

2.4.2 Statistical Hypothesis Testing Method

Statistical hypothesis testing methods are commonly applied for residual evaluation in stochastic systems. In this research line, an evaluation function is set as a test statistic of residual sequence and a threshold is then determined towards satisfactory detection performance. Criteria like Bayes' criterion, Neyman–Pearson criterion and maximal posteriori probability criterion, to name just a few, can be applied to this end. Among them, the Neyman–Pearson criterion is one of the most popular choices, thanks to its non-essentiality of prior probability for hypothesis.

Given residual sample $r(k),\ k = 1, 2, \ldots$ following probability density (or mass) function $p_\theta(r)$ with parameter θ carrying fault information, define a likelihood function of r as $\mathcal{L}(r|\theta)$. Denote $\theta = \theta_0$ in fault-free case and $\theta = \theta_f$ in faulty case. A residual evaluation function $J(r)$ can be defined as

$$J(r) = \frac{\mathcal{L}(r|\theta_0)}{\mathcal{L}(r|\theta_f)}, \tag{2.34}$$

which is a so-called likelihood ratio function. By introducing a null hypothesis $\mathcal{H}_0$ and an alternative hypothesis $\mathcal{H}_1$ as follows

$$\mathcal{H}_0 : \theta = \theta_0,\ \ \mathcal{H}_1 : \theta = \theta_f, \tag{2.35}$$

a threshold J_{th} is then determined with respect to decision logic

$$\begin{cases} J(r) \leq J_{th,\alpha} & \Rightarrow \mathcal{H}_0 : \text{ no alarm} \\ J(r) > J_{th,\alpha} & \Rightarrow \mathcal{H}_1 : \text{ fault alarm} \end{cases} \tag{2.36}$$

such that

$$\Pr\left\{J(r) > J_{th,\alpha}|\mathcal{H}_0\right\} = \alpha, \tag{2.37}$$

where α is an acceptable significance level of the probability rejecting $\mathcal{H}_0$ in favor of $\mathcal{H}_1$. In FD fashion, α is termed as the FAR. According to the well-established Neyman–Pearson lemma, such a generalized likelihood ratio scheme regarding evaluation function $J(r)$ in (2.34) and threshold $J_{th,\alpha}$ delivers the maximal fault detectability with an FAR level α.

A typical example is the χ^2-test based residual evaluation scheme under Gaussian distributed noises. Consider the residual signal follows Gaussian distribution, i.e., $r(k) \sim \mathcal{N}(0, \Sigma_r)$. Construct an evaluation function

$$J(r) = r^T(k)\Sigma_r^{-1}r(k) \sim \chi^2(q),$$

where q is the degree of freedom. By setting the threshold as

$$J_{th,\alpha} = \chi_\alpha^2(q),$$

the occurrence of a fault can be detected with FAR $= \alpha$ by performing (2.36).

Remark 2.4 Hypothesis testing schemes generally requires that the distribution for stochastic disturbances is known exactly in prior, which, unfortunately, might be unavailable in practical applications. As remedies, increasing attention has been paid to apply randomized algorithms and the so-called distributionally robust optimization techniques for residual evaluation purpose, see for instance [2, 4, 10, 14, 16]. In this book, applications of randomized algorithms to threshold setting can be found in Chaps. 18 and 19.

2.5 Conclusion

The paradigm of model-based fault diagnosis has been introduced in this chapter, focusing on the design setup formulations of fault detection and estimation problems for linear dynamic systems with stochastic and/or deterministic disturbances. For the systems subject to stochastic disturbances, indices FAR, MDR and FDR are applied to evaluate the FD performance, optimal FD problems towards a trade-off between FAR and MDR/FDR criteria have been formulated in the probabilistic context. For the systems with norm-bounded disturbances, optimal FD problems in the context of maximizing the fault detectability with a zero FAR have also been defined by introducing appropriate sets. Fault estimation issues in the context of LMS and LS were also formulated. By noting the central role of residual generator in fault detection and estimation, issues of PUID and H_2- and H_∞-optimal design of residual generator have also been briefly reviewed. As another important part of FD, typical residual evaluation schemes have been introduced in the end. It is remarkable that, the content of this chapter is very basic for model-based fault diagnosis that can also be found in the monographs [9, 11, 12].

References

1. Hassanabadi, A. H., Shafiee, M., & Puig, V. (2016). UIO design for singular delayed LPV systems with application to actuator fault detection and isolation. *International Journal of Systems Science, 47*(1), 107–121.
2. Shang, C., Ding, S. X., & Ye, H. (2021). Distributionally robust fault detection design and assessment for dynamical systems. *Automatica, 125*, 109434.
3. Ichalal, D., & Mammar, S. (2015). On unknown input observers for LPV systems. *IEEE Transactions on Industrial Electronics, 62*(9), 5870–5880.
4. Xu, G., & Burer, S. (2018). A data-driven distributionally robust bound on the expected optimal value of uncertain mixed 0–1 linear programming. *Computational Management Science, 15*(1), 111–134.
5. Behzad, H., Casavola, A., Tedesco, F., et al. (2016). A fault-tolerant sensor reconciliation scheme based on LPV unknown input observers. In *Proceeding of the 55th Conference on Decision and Control (CDC)* (pp. 2158–2163). December 12–14, Las Vegas, NV, USA.
6. Hur, H., & Ahn, H. S. (2014). Unknown input H_∞ observer-based localization of a mobile robot with sensor failure. *IEEE/ASME Transactions on Mechatronics, 19*(6), 1830–1838.
7. Hur, H., & Ahn, H. S. (2014). Unknown input observer-based filtering for mobile pedestrian localization using wireless sensor networks. *IEEE Sensors Journal, 14*(8), 2590–2600.
8. Chen, J., & Patton, R. J. (2000). Standard H_∞ filtering formulation of robust fault detection. *IFAC Proceedings Volumes, 33*(11), 261–266.
9. Chen, J., & Patton, R. J. (1999). *Robust model-based fault diagnosis for dynamic systems*. Boston, MA: Kluwer Academic Publishers.
10. Krueger, M., Koenings, T., Liu, Y., et al. (2017). Randomized algorithm based fault detection system design for uncertain LTI systems *50*(1), 3711–3716
11. Ding, S. X. (2013). *Model-based fault diagnosis techniques: Design schemes, algorithms, and tools* (2nd ed.). London, U.K.: Springer.
12. Ding, S. X. (2020). *Advanced methods for fault diagnosis and fault-tolerant control*. Berlin Germany: Springer.
13. Ding, S. X., Jeinsch, T., Frank, P. M., et al. (2000). A unified approach to the optimization of fault detection systems. *International Journal of Adaptive Control and Signal Processing, 14*(7), 725–745.
14. Ding, S. X., Li, L., & Krüger, M. (2019). Application of randomized algorithms to assessment and design of observer-based fault detection systems. *Automatica, 107*, 175–182.
15. Li, S., Wang, H., Aitouche, A., et al. (2017). Robust unknown input observer design for state estimation and fault detection using linear parameter varying model. *Journal of Physics: Conference Series, 783*(1), 012001.
16. Xue, T., Zhong, M., Li, L., et al. (2020). An optimal data-driven approach to distribution independent fault detection. *IEEE Transactions on Industrial Informatics, 16*(11), 6826–6836.

Chapter 3
LDTV Systems and Fault Detection and Estimation for LDTV Systems

This chapter confines to essential preliminaries of fault detection and estimation for LDTV systems. State space models and an alternative input-output model of LDTV processes in nominal and faulty cases are first established. On this basis, classical FD techniques using Kalman filter, UIO, FDF, and parity space-based approach are briefly introduced. Finally, H_∞ fault estimation scheme for LDTV systems is reviewed.

3.1 Introduction

In the real world most of physical systems are naturally time-varying, for example, the sampled-data systems, multi-agent systems, networked control systems, and event-triggered systems, etc. LDTV systems have been widely accepted as models describing such kind of processes and, furthermore, approximating nonlinear systems. Unfortunately, due to the time-varying characteristics, fault diagnosis methods initially proposed for LTI systems could hardly be directly applied in most time-varying scenarios. Observing these, a surge of research on fault diagnosis for LDTV systems has emerged both in theoretical community and practical applications in the early 2000s and a rich body literature has been reported over the past two decades [1, 3, 4, 8, 13–15, 21, 22, 33, 34].

Concerning the central role of residual generator in fault diagnosis for dynamic systems, a majority of the related achievements for LDTV systems focus on residual generation issues, that can be roughly classified into the parameter estimation method, the observer-based scheme, and the parity space-based approach. Especially, to handle robust fault detection and estimation issues, methodologies of finite horizon H_∞-filtering and H_2/H_2-, H_∞/H_∞-, H_-/H_∞- and H_i/H_∞-optimization have been extensively studied, see e.g., [16, 17, 25]. Their applications in multi-agent

M. Zhong et al., *Fault Diagnosis for Linear Discrete Time-Varying Systems and Its Applications*, https://doi.org/10.1007/978-981-19-5438-2_3

distributed systems [26, 27], event-triggered systems [14, 22], processes with delayed state [4, 12, 18, 19], 2-dimensional systems [7–9], LPV systems [10, 31, 32], and nonlinear systems [6, 24], etc., have also been reported.

The objective of this chapter is to introduce the fundamentals of fault detection and estimation for LDTV systems. We begin with system modeling, and then introduce classical FD techniques including the Kalman filter scheme, UIO method, FDF method, and the parity space-based approach. Finally, fault estimation for LDTV systems is discussed in the sense of H_∞-optimization.

3.2 Mathematical Descriptions of LDTV Systems

Building an analytical model of the monitored system is the first task in model-based fault diagnosis. A dynamic system can be mathematically described in different ways. For LDTV processes, the system is commonly described by a state space model. Below we first give state space models of LDTV processes both in nominal (disturbance- and fault-free) case and the case with uncertainties and faults. And then an alternative input-output model is established.

3.2.1 Description of Nominal Systems

A standard state space model of an LDTV system in nominal case is given by

$$\begin{cases} x(k+1) = A(k)x(k) + B(k)u(k), x(0) = x_0, \\ y(k) = C(k)x(k) + D(k)u(k), \end{cases} \tag{3.1}$$

where $x(k) \in \mathbb{R}^n$, $u(k) \in \mathbb{R}^p$, and $y(k) \in \mathbb{R}^q$, are the state, system input and measurement output, respectively, x_0 is the initial state, $A(k),\ B(k),\ C(k)$ and $D(k)$ are known time-varying matrices with appropriate dimensions.

As mentioned before, LDTV systems can be applied to model many practical processes like nonlinear systems, time-varying periodic systems, delayed-state systems, LPV systems, and event-triggered systems, etc. The following two examples are demonstrated to show this point.

Example 3.1 Consider the system matrix $A(k)$ is modeled in terms of constant matrices A_i and functions $h_i(\cdot)$ of a time-varying parameter vector $\rho(k) \in \mathbb{R}^{k_\rho}$ with $i = 1, 2, \ldots t$, i.e.,

$$A(k) = \sum_{i=1}^{t} A_i h_i(\rho(k)). \tag{3.2}$$

The model (3.1) immediately becomes an LPV system.

Example 3.2 Consider the following nominal nonlinear system

$$x(k+1) = \varphi\left(x(k), u(k)\right), \quad y(k) = C(k)x(k),$$

where $\varphi(\cdot)$ is a smooth function of state $x(k)$ and input $u(k)$. By doing Taylor expansions of $\varphi\left(x(k), u(k)\right)$ on time-varying working point $\left(x_p(k), u_p(k)\right)$, it yields

$$x(k+1) \cong A_{x_p}(k)x(k) + B_{u_p}(k)u(k) + d(k),$$

where $A_{x_p}(k)$ and $B_{u_p}(k)$ are the Jacobian matrices of $\varphi\left(x(k), u(k)\right)$ on $x(k)$ and $u(k)$ with respect to the working point $\left(x_p(k), u_p(k)\right)$, respectively, i.e.,

$$\begin{aligned} A_{x_p}(k) &= \frac{\partial \varphi\left(x(k), u(k)\right)}{\partial x(k)} \Big|_{x(k)=x_p(k), u(k)=u_p(k)}, \\ B_{u_p}(k) &= \frac{\partial \varphi\left(x(k), u(k)\right)}{\partial u(k)} \Big|_{x(k)=x_p(k), u(k)=u_p(k)}, \end{aligned}$$

$d(k) = \varphi\left(x_p(k), u_p(k)\right) - A_{x_p}(k)x_p(k) - B_{u_p}(k)u_p(k)$ denotes the approximation error. In this way, the original nonlinear system is approximated by an LDTV process.

3.2.2 *Modeling of Dynamic Systems with Uncertainties and Faults*

Remarkably, an accurate and complete mathematical model of an engineering system is unachievable in practice, because of the inevitable modeling uncertainties such as the unmodeled system dynamics, process and measurement noises, environment disturbances, and the perturbation of system parameters, etc. Such uncertainties constitute the main source of false and missed fault alarms in FD and the estimation errors in fault estimation. Generally, the uncertainties in system parameters are called model uncertainties that are usually described in multiplicative forms. The so-called unknown inputs are external disturbances modeled in additive forms.

Faults in a plant, depending on their location, can be categorized into the actuator fault, sensor fault and component fault. According to weather or not they will influence the system stability, faults can also be classified into additive fault or multiplicative fault.

We first consider the system (3.1) corrupted by unknown input and additive fault. The process dynamics can be described by the following state space model

$$\begin{cases} x(k+1) = A(k)x(k) + B(k)u(k) + B_d(k)d(k) + B_f(k)f(k), x(0) = x_0, \\ y(k) = C(k)x(k) + D(k)u(k) + D_d(k)d(k) + D_f(k)f(k), \end{cases} \tag{3.3}$$

where $d(k) \in \mathbb{R}^{k_d}$ is the unknown input vector that could be stochastic or deterministic signal, $f(k) \in \mathbb{R}^{k_f}$ is the fault vector to be diagnosed, $B_d(k)$, $D_d(k)$, $B_f(k)$ and $D_f(k)$ are time-varying matrices of appropriate dimensions.

Remark 3.1 Additive faults will not affect the system stability. The faults in actuators and sensors are usually modeled as additive faults. For example, an actuator fault would cause changes in the actuator response to the input $u(k)$ and such a faulty system can be modeled as (3.3) by setting $B_f(k) = B(k)$ and $D_f(k) = D(k)$. When a sensor fault occurs, process measurement output $y(k)$ would change undesirably and the system with sensor faults can be modeled with $B_f(k) = 0$ and $D_f(k) = I$.

Model uncertainties and multiplicative faults are generally modeled in terms of the changes of nominal system parameters. In particular, multiplicative faults may cause instability of the monitored system. The LDTV system (3.3) with model uncertainties and multiplicative faults can be described by

$$\begin{cases} x(k+1) = \bar{A}(k)x(k) + \bar{B}(k)u(k) + B_d(k)d(k) + B_f(k)f(k), x(0) = x_0, \\ y(k) = \bar{C}(k)x(k) + \bar{D}(k)u(k) + D_d(k)d(k) + D_f(k)f(k), \end{cases} \tag{3.4}$$

where

$$\begin{aligned} \bar{A}(k) &= A(k) + \Delta A(k) + \Delta A_f(k), \;\; \bar{B}(k) = B(k) + \Delta B(k) + \Delta B_f(k), \\ \bar{C}(k) &= C(k) + \Delta C(k) + \Delta C_f(k), \;\; \bar{D}(k) = D(k) + \Delta D(k) + \Delta D_f(k), \end{aligned}$$

$\Delta A(k)$, $\Delta B(k)$, $\Delta C(k)$ and $\Delta D(k)$ represent model uncertainties, $\Delta A_f(k)$, $\Delta B_f(k)$, $\Delta C_f(k)$, and $\Delta D_f(k)$ are multiplicative faults in the plant.

Remark 3.2 Model uncertainties and multiplicative faults can be described in norm-bounded, polytopic or stochastic types. Especially, when polytopic-type model is applied, model uncertainties and multiplicative faults can be further expressed into additive forms under certain conditions.

In this book we focus on additive faults diagnosis for LDTV systems with (additive) unknown inputs (e.g., noises and norm-bounded deterministic disturbances), as modeled in (3.3). Additionally, the following common assumptions are always made unless otherwise stated.

Assumption 3.1 $(C(k), A(k))$ is uniformly detectable and $(A(k), B_d(k))$ is uniformly stabilizable.

Assumption 3.2 The fault $f(k)$ to be detected is $l_{2,[0,N]}$-norm bounded, i.e., $f(k) \in l_{2,[0,N]}$ and the unknown input $d(k)$, in case of known to be deterministic, is $l_{2,[0,N]}$-norm bounded, i.e., $d(k) \in l_{2,[0,N]}$.

3.2.3 Alternative Input-Output Model

Alternatively, an input-output model of the LDTV system (3.3) can be established.

Let

$$\begin{cases} \Phi(k,i) = A(k-1)A(k-2)\cdots A(i), \\ \Phi(i,i) = I, \ k = 1,2,\ldots \ \ i = 1,2,\ldots,k-1. \end{cases} \tag{3.5}$$

Define

$$y_k = \begin{bmatrix} y(0) \\ y(1) \\ \vdots \\ y(k) \end{bmatrix}, \ u_k = \begin{bmatrix} u(0) \\ u(1) \\ \vdots \\ u(k) \end{bmatrix}, \ d_k = \begin{bmatrix} d(0) \\ d(1) \\ \vdots \\ d(k) \end{bmatrix}, \ f_k = \begin{bmatrix} f(0) \\ f(1) \\ \vdots \\ f(k) \end{bmatrix} \tag{3.6}$$

for $k = 0, 1, 2, \ldots$. We can rewrite the state space model (3.3) as

$$y_k = H_{y0}(k)x_0 + H_{yu}(k)u_k + H_{yd}(k)d_k + H_{yf}(k)f_k, \tag{3.7}$$

where

$$H_{y0}(k) = \begin{bmatrix} h_0(0) \\ h_0(1) \\ \vdots \\ h_0(k) \end{bmatrix}, \ h_0(i) = C(i)\Phi(i,0), \tag{3.8}$$

$$H_{yd}(k) = \begin{bmatrix} h_d(0,0) & 0 & \cdots & 0 \\ h_d(1,0) & h_d(1,1) & \ddots & \vdots \\ \vdots & \ddots & \ddots & 0 \\ h_d(k,0) & \cdots & \cdots & h_d(k,k) \end{bmatrix}, \ \begin{matrix} h_d(i,j) = C(i)\Phi(i,j+1)B_d(j), \\ h_d(i,i) = D_d(i), \end{matrix} \tag{3.9}$$

$$H_{yu}(k) = \begin{bmatrix} h_u(0,0) & 0 & \cdots & 0 \\ h_u(1,0) & h_u(1,1) & \ddots & \vdots \\ \vdots & \ddots & \ddots & 0 \\ h_u(k,0) & \cdots & \cdots & h_u(k,k) \end{bmatrix}, \ \begin{matrix} h_u(i,j) = C(i)\Phi(i,j+1)B(j), \\ h_u(i,i) = D(i), \end{matrix} \tag{3.10}$$

$$H_{yf}(k)=\begin{bmatrix} h_f(0,0) & 0 & \cdots & 0 \\ h_f(1,0) & h_f(1,1) & \ddots & \vdots \\ \vdots & \ddots & \ddots & 0 \\ h_f(k,0) & \cdots & \cdots & h_f(k,k) \end{bmatrix}, \quad \begin{array}{l} h_f(i,j)=C(i)\Phi(i,j+1)B_f(j), \\ h_f(i,i)=D_f(i). \end{array} \tag{3.11}$$

It is notable that the matrices $H_{yu}(k)$ and $H_{yf}(k)$ are given by replacing $\{B_d(i),\ D_d(i)\}$ in $H_{yd}(i)$ with $\{B(i),\ D(i)\}$ and $\{B_f(i),\ D_f(i)\}$, respectively.

When $k=N$, the model (3.7) can be further specified as

$$y_N=H_{y0}(N)x_0+H_{yu}(N)u_N+H_{yd}(N)d_N+H_{yf}(N)f_N \tag{3.12}$$

with y_N, u_N, d_N and f_N being given in (3.6) by replacing k with N.

3.3 Fault Detection for LDTV Systems

In this section, classical schemes of FD for LDTV systems are introduced, including the Kalman filter method, UIO-based method, FDF method, as well as the parity space-based approach.

3.3.1 Kalman Filter Method

Consider the following LDTV system

$$\begin{cases} x(k+1)=A(k)x(k)+B(k)u(k)+B_f(k)f(k)+w(k),\ x(0)=x_0, \\ y(k)=C(k)x(k)+D(k)u(k)+D_f(k)f(k)+v(k), \end{cases} \tag{3.13}$$

where $w(k)\in\mathbb{R}^n$ and $v(k)\in\mathbb{R}^q$ are respectively process and measurement noise vectors that satisfy

$$w(k)\sim\mathcal{N}\left(0,\Sigma_w(k)\right),\ v(k)\sim\mathcal{N}\left(0,\Sigma_v(k)\right),\ x\left(0\right)\sim\mathcal{N}\left(0,\Pi_0\right), \tag{3.14}$$

$$\mathbb{E}\left[\begin{bmatrix} w(i) \\ v(i) \\ x(0) \end{bmatrix}\begin{bmatrix} w(j) \\ v(j) \\ x(0) \end{bmatrix}^T\right]=\begin{bmatrix} \begin{bmatrix} \Sigma_w(i) & S(i) \\ S^T(i) & \Sigma_v(i) \end{bmatrix}\delta_{ij} & 0 \\ 0 & \Pi_0 \end{bmatrix}. \tag{3.15}$$

$S(i)$, $\Sigma_w(i)$, $\Sigma_v(i)$, Π_0 are known of appropriate dimensions.

A well-known Kalman filter based residual generator can be constructed as follows

$$\begin{cases} \hat{x}(k+1|k) = A(k)\hat{x}(k|k-1) + B(k)u(k) + L(k)r(k) \\ \hat{y}(k) = C(k)\hat{x}(k|k-1) - D(k)u(k) \\ r(k) = y(k) - \hat{y}(k) \end{cases}, \tag{3.16}$$

where $\hat{x} \in \mathbb{R}^n$ is the estimate of state x, $L(k)$ is the gain matrix that minimize the covariance of state estimation error. Denote $e(k|k-1) = x(k) - \hat{x}(k|k-1)$. The gain matrix $L(k)$ is obtained by solving

$$\min_{L(k)} P(k|k-1) := \min_{L(k)} \mathbb{E}\left[e(k|k-1)e(k|k-1)^T\right]$$

and thus can be computed recursively by

$$L(k) = \left(A(k)P(k|k-1)C^T(k) + S(k)\right)\Sigma_r^{-1}(k), \tag{3.17}$$

$$\Sigma_r(k) = \Sigma_v(k) + C(k)P(k|k-1)C^T(k), \tag{3.18}$$

$$P(k+1|k) = A(k)P(k|k-1)A^T(k) - L(k)\Sigma_r(k)L^T(k) + \Sigma_w(k),$$
$$P(0) = \Pi_0, \tag{3.19}$$

where $\Sigma_r(k) = \mathbb{E}\left[r(k)r^T(k)\right]$ is the covariance matrix of residual $r(k)$ in fault-free case. Remarkably, the Kalman filter based residual generator (3.16) is optimal in the sense of LMS estimation, i.e., the mean square of the state estimation error is minimized with the gain matrix in (3.17).

Note that the residual signal (i.e., the innovation) is Gaussian distributed with $r(k) \sim \mathcal{N}(0, \Sigma_r(k))$. Thanks to the whiteness of the residual signal, the χ^2-test scheme at FAR level α can then be applied for residual evaluation purpose by setting

$$J(r) = r^T(k)\Sigma_r^{-1}(k)r(k) \sim \chi^2(q), \; J_{th} = \chi_\alpha^2(q),$$

which, according to Neyman–Pearson lemma, delivers the maximal fault detectability. Meanwhile, such $\{J(r), J_{th}\}$ solve the optimal FD problem in Definition 2.2.

Due to the popularity and maturity of Kalman filter algorithms both in theoretical study and practical applications, detailed derivation of the recursive algorithm (3.17)–(3.19) are not given here. Instead, we will demonstrate a new interpretation and derivation of Kalman filter in Krein space in Chap. 5.

3.3.2 Unknown Input Observer Method

Recall that the dynamics of a residual generator can be expressed as

$$r = \mathcal{K}(u, y) = \mathcal{W} \circ \left(\mathcal{M}_d(d) + \mathcal{M}_f(f)\right). \tag{3.20}$$

To handle the unknown input d, an intuitive way is to find a post-filter $\mathcal{W}$ satisfying $\mathcal{W} \circ \mathcal{M}_d = 0,\ \mathcal{W} \circ \mathcal{M}_f \neq 0$, such that the residual signal r can be decoupled completely from unknown input d while keeping the sensitivity to fault f, i.e., addressing the PUID problem defined in Definition 2.8.

A UIO is a Luenberger type observer that is primarily used for the reconstruction and estimation of state variables. One of its particular advantages lies in the capability of decoupling the estimated states from unknown inputs completely under certain conditions, and thus can be applied to cope with PUID problem. The underlying idea of UIO-based FD lies in expressing the state estimation error as a system that is free from any unknown inputs.

Consider the following LDTV system

$$\begin{cases} x(k+1) = A(k)x(k) + B(k)u(k) + B_d(k)d(k), \\ y(k) = C(k)x(k). \end{cases} \tag{3.21}$$

Reconstruct the state variable by the following UIO

$$\begin{cases} z(k+1) = N(k)z(k) + T(k)B(k)u(k) + L(k)y(k), \\ \hat{x}(k) = z(k) + H(k)y(k), \end{cases} \tag{3.22}$$

where $N(k)$, $T(k)$, $L(k)$ and $H(k)$ are matrices to be designed. Then a UIO-based residual generator is obtained as

$$r(k) = y(k) - \hat{y}(k),\ \ \hat{y}(k) = C(k)\hat{x}(k). \tag{3.23}$$

Denote by $e(k) = x(k) - \hat{x}(k)$ the state estimation error. It yields from (3.21) and (3.22) that

$$e(k) = x(k) - z(k) - H(k)y(k) = (I - H(k)C(k))x(k) - z(k).$$

Let $\Gamma(k) = I - H(k)C(k)$. We further have

$$\begin{aligned} e(k+1) &= \Gamma(k+1)x(k+1) - z(k+1) \\ &= N(k)e(k) + (\Gamma(k+1)A(k) - L(k)C(k) - N(k)\Gamma(k+1))x(k) + \\ &\quad (\Gamma(k+1) - T(k))B(k)u(k) + \Gamma(k+1)B_d(k)d(k). \end{aligned}$$

To ensure the stability of UIO (3.22) and decouple the estimation error $e(k)$ from $x(k)$, $u(k)$ and $d(k)$, matrices $N(k)$, $T(k)$, $L(k)$ and $H(k)$ should be designed to satisfy the following conditions:

- *Condition 1*: $e(k+1) = N(k)e(k)$ is asymptotically stable;
- *Condition 2*: $N(k)$, $T(k)$, $L(k)$ and $H(k)$ satisfy

$$\begin{cases} \Gamma(k+1) = T(k), \ \Gamma(k+1)B_d(k) = 0, \\ \Gamma(k+1)A(k) = L(k)C(k) + N(k)\Gamma(k+1). \end{cases} \tag{3.24}$$

In the stage of residual evaluation, the norm-based evaluation functions and the corresponding thresholds can be applied to an optimal FD.

As mentioned earlier, UIO-based residual generation suggests a solution to PUID problem, only if the above Conditions 1 and 2 are satisfied. However, in practical engineering systems it might be difficult to find $N(k)$, $T(k)$, $L(k)$ and $H(k)$ to guarantee equation (3.24) at each time step k. Meanwhile, faults lying the subspace spanned by $B_d(k)$ cannot be detected any more due to $\Gamma(k+1)B_d(k) = 0$, which means the reduction of fault detectability to some extent. To overcome these deficiencies, an alternative solution is to handle robust FD issues towards a trade-off between the robustness of residual to unknown input and the sensitivity to fault. In this research line, observer-based FDF is mostly applied as a residual generator.

3.3.3 Fault Detection Filter Method

FDF, as a full-order state observer, is one of the most popular choices for residual generation purpose. Given LDTV system (3.3), an FDF-based residual generator is constructed as

$$\begin{cases} \hat{x}(k+1) = A(k)\hat{x}(k) + B(k)u(k) + L(k)(y(k) - \hat{y}(k)), \\ \hat{y}(k) = C(k)\hat{x}(k) + D(k)u(k), \\ r(k) = W(k)\left(y(k) - \hat{y}(k)\right), \end{cases} \tag{3.25}$$

where $\hat{x}(k)$ is the estimate of state $x(k)$, $\hat{x}_0 = \hat{x}(0)$ is a guess of initial state, $r(k)$ is the residual, $L(k)$ is the observer gain matrix, $W(k)$ is the so-called post-filter.

Denote the state estimation error by $e(k) = x(k) - \hat{x}(k)$. Then $e(0) = x_0 - \hat{x}_0$. It follows from (3.3) and (3.25) that

$$\begin{cases} e(k+1) = (A(k) - L(k)C(k))e(k) + (B_d(k) - L(k)D_d(k))d(k) \\ \qquad\qquad + (B_f(k) - L(k)D_f(k))f(k), \\ r(k) = W(k)\left(C(k)e(k) + D_d(k)d(k) + D_f(k)f(k)\right). \end{cases} \tag{3.26}$$

The state transition matrix of system (3.26) is given by

$$\Phi_L(k,i) = \prod_{j=i}^{k-1}(A(j) - L(j)C(j)), \ \Phi_L(k,k) = I, \ i \le k-1. \tag{3.27}$$

Let

$$\begin{aligned}
d_k &= \left[e^T(0),\, d^T(0),\, d^T(1),\, \ldots,\, d^T(k)\right]^T, \\
f_k &= \left[f^T(0),\, f^T(1),\, \ldots,\, f^T(k)\right]^T, \\
g_e(k,0) &= C(k)\Phi_L(k,0), \\
g_d(k,i) &= C(k)\Phi_L(k,i+1)(B_d(i) - L(i)D_d(i)), \\
g_f(k,i) &= C(k)\Phi_L(k,i+1)(B_f(i) - L(i)D_f(i)), \\
g_{\varepsilon d}(k) &= \left[\, g_e(k,0)\ g_d(k,0)\ \cdots\ g_d(k,k-1)\ D_d(k)\,\right], \\
g_{\varepsilon f}(k) &= \left[\, g_f(k,0)\ \cdots\ g_f(k,k-1)\ D_f(k)\,\right].
\end{aligned}$$

The residual generator (3.26) takes the form

$$r(k) = W(k)(g_{\varepsilon d}(k)d_k + g_{\varepsilon f}(k)f_k) = r_d(k) + r_f(k) \tag{3.28}$$

with

$$r_d(k) = W(k)g_{\varepsilon d}(k)d_k,\ r_f(k) = W(k)g_{\varepsilon f}(k)f_k. \tag{3.29}$$

Remark 3.3 Under Assumptions 3.1 and 3.2, the FDF (3.25) with state transfer matrix (3.27) can be guaranteed exponentially stable.

It is noted that the FDF-based residual generation lies in the design of the gain matrix $L(k)$ and the post-filter $W(k)$. When the disturbance $d(k)$ is a stochastic sequence, the FDF (3.25) can be equated to a Kalman filter with $L(k)$ given in (3.17) and $W(k) = \Sigma_r^{-\frac{1}{2}}(k)$, then the χ^2-test scheme can be applied for residual evaluation to an optimal FD. In case that the disturbance $d(k)$ and fault $f(k)$ are $l_{2,[0,N]}$-norm bounded, optimal FDF design schemes in the frameworks of H_2-, H_∞- and H_i/H_∞-optimization are commonly used in order to enhance the fault sensitivity and the robustness to disturbances (as formulated in (2.23)). Bearing this idea in mind, we will derive a unified solution to FDF-based optimal FD in Chap. 6. A projection interpretation and recursive implementation algorithm will be given in Chap. 7. Moreover, by means of Krein space techniques, a Krein space approach to FDF-based optimal FD will be illustrated in Chap. 8. Applications of the results to event-triggered systems and delayed state LDTV systems are illustrated in Chaps. 9 and 10, respectively.

3.3.4 Parity Space-Based Method

Parity space approach is first proposed by Chow and Willsky in the 1980s. Different from observer-based methods, parity space-based residual generator is characterized by its straightforward construction form and complete independence from the system initial state that maybe unknown in practical applications. In this subsection, the idea behind the parity space based residual generation for LDTV systems is introduced.

For ease of notation, we denote $\xi_s(k)$ for a vector $\xi \in \mathbb{R}^{k_\xi}$ and integer $s > 0$ as

$$\xi_s(k) = \begin{bmatrix} \xi(k-s) \\ \xi(k-s+1) \\ \vdots \\ \xi(k) \end{bmatrix}, \tag{3.30}$$

Given LDTV system (3.3) and $s > 0$, it easily yields the following input-output model over the time interval $[k-s, k]$, i.e.,

$$y_s(k) = H_{os}(k)x(k-s) + H_{us}(k)u_s(k) + H_{ds}(k)d_s(k) + H_{fs}(k)f_s(k), \tag{3.31}$$

where

$$H_{os}(k) = \begin{bmatrix} C(k-s) \\ C(k-s+1)A(k-s) \\ \vdots \\ C(k)A(k-1)\cdots A(k-s+1)A(k-s) \end{bmatrix}, \tag{3.32}$$

$$\begin{aligned} H_{ds}(k) &= [h_{ds}(i,j)]_{(s+1)\times(s+1)}, \qquad (3.33)\\ h_{ds}(i,i) &= D_d(k-s+i-1), \ \text{for } 1 \le i \le s+1, \\ h_{ds}(i,i-1) &= C(k-s+i-1)B_d(k+i-s-2), \\ h_{ds}(i,j) &= 0, \ \text{for } j > i, \\ h_{ds}(i,j) &= C(k-s+i-1)A(k-s+i-2)\cdots \\ &\quad \cdots A(k-s+j)B_d(k-s+j-1), \ \text{for } j < i \quad 1. \end{aligned}$$

The matrices $H_{us}(k)$ and $H_{fs}(k)$ are constructed by replacing $\{B_d(k),\ D_d(k)\}$ in $H_{ds}(k)$ with $\{B(k),\ 0\}$ and $\{B_f(k),\ D_f(k)\}$, respectively.

The core of parity space-based FD is to generate a residual via linear combinations of measurement outputs and inputs taken over a finite time window, such that the residual is independent of initial state and system input. Based on (3.31), a parity relation based residual generator is constructed as

$$\begin{aligned} r(k) &= W_s(k)(y(k) - H_{us}(k)u_s(k)) && (3.34)\\ &= W_s(k)(H_{os}(k)x(k-s) + H_{ds}(k)d_s(k) + H_{fs}(k)f_s(k)). && (3.35) \end{aligned}$$

where $W_s(k) \in \mathbb{R}^{\gamma\times(s+1)q}$ stands for the parity space matrix of order s that belongs to the so-called parity space $\mathcal{P}_s(k)$ defined by

$$\mathcal{P}_s(k) = \{W_s(k) | W_s(k)H_{os}(k) = 0,\ \forall W_s(k) \neq 0\}. \tag{3.36}$$

The item $x(k-s)$ is thus decoupled completely from residual $r(k)$.

Remark 3.4 The parity space matrix $W_s(k) \in \mathcal{P}_s(k)$ exists on condition that the null space of $H_{os}(k) \in \mathbb{R}^{(s+1)q \times n}$ exists and $s > n$ guarantees

$$\text{rank}(H_{os}(k)) \leq n < (s+1)q. \tag{3.37}$$

Hence, there exists at least a vector $W_s(k) \in \mathbb{R}^{(s+1)q}$ delivering $W_s(k)H_{os}(k) = 0$ on condition that the parity space order satisfies $s > n$.

Denote the basis matrix of parity space $\mathcal{P}_s(k)$ by $N_{bs}(k) \in \mathbb{R}^{\gamma \times (1+s)q}$. Rewrite

$$W_s(k) = P_s(k)N_{bs}(k). \tag{3.38}$$

Let

$$\begin{aligned}
\bar{H}_{ds}(k) &= N_{bs}(k)H_{ds}(k), \\
\bar{H}_{fs}(k) &= N_{bs}(k)H_{fs}(k), \\
\bar{H}_{us}(k) &= N_{bs}(k)H_{us}(k).
\end{aligned}$$

The residual generator (3.34) becomes

$$\begin{aligned}
r(k) &= P_s(k)(N_{bs}(k)y_s(k) - \bar{H}_{us}(k)u_s(k)) && (3.39)\\
&= P_s(k)(\bar{H}_{ds}(k)d_s(k) + \bar{H}_{fs}(k)f_s(k)) && (3.40)\\
&= r_d(k) + r_f(k), \\
r_d(k) &= P_s(k)\bar{H}_{ds}(k)d_s(k),\ r_f(k) = P_s(k)\bar{H}_{fs}(k)f_s(k).
\end{aligned}$$

In general, Eq. (3.39) is applied for online residual generation, while (3.40) is used for offline design purpose.

Note that parity space-based residual generation lies in the design of the parity space matrix $P_s(k)$, which can be addressed in the sense of finite horizon H_2/H_2- or H_i/H_∞-optimization towards enhancing the fault sensitivity and the robustness to disturbances. For ease of organization, parity space-based approaches to optimal fault detection and estimation will be detailedly discussed in Part III of this book. What we would like to mention here are:

- In comparison with the time-invariant case, an online updating of matrices $H_{os}(k)$, $H_{us}(k)$, $H_{ds}(k)$ and $H_{fs}(k)$ at every time step k is necessary for LDTV systems. This will result in high real-time computation cost especially with the increase of parity space order s. To handle this difficulty, a recursive updating algorithm for these matrices has been developed by Zhong et al. [20] as summarized in Algorithm 3.3.1;
- Parity space-based optimal FD in the sense of H_2/H_2- and H_i/H_∞-optimization for LTI systems has been well established, see [28] for more details. It has shown that a larger parity space order s generally means better FD performance, but higher computation cost, while a smaller s would delivers a lower computation burden

Algorithm 3.3.1 Recursive computation of $H_{os}(k)$, $H_{us}(k)$, $H_{ds}(k)$ and $H_{fs}(k)$

1: At time step k, construct

$$B_{ds}(k) = \text{diag}\{B_d(k-s+1),\ \ldots,\ B_d(k)\},$$
$$\Gamma_1(k) = [\Gamma_1(k,1)\ \ \Gamma_1(k,2)\ \ \ldots\ \ \Gamma_1(k,s-1)],$$
$$\Gamma_1(k,i) = \prod_{j=k}^{k-s+i+1} A(j),\ i = 1,\ 2,\ \ldots,\ s-1;$$

2: Get $H_{ds,2}(k-1)$ from $H_{ds}(k-1)$ using

$$H_{ds,2}(k-1) = \begin{bmatrix} 0 & 0 \\ 0 & I \end{bmatrix} \begin{bmatrix} D_d(k-s-1) & 0 \\ * & H_{ds}(k-1) \end{bmatrix} \begin{bmatrix} 0 & 0 \\ 0 & I \end{bmatrix};$$

3: Get $\Gamma_1(k-1)$ from $\Gamma(k-1)$ according to $\Gamma(k-1) = [\Gamma_1(k-1)\ \ I]$;
4: Compute $\Gamma(k)$ using $\Gamma(k) = [A(k)\Gamma_1(k-1)\ \ I]$.;
5: Compute $\bar{H}_{ds,2}(k)$ by using

$$\bar{H}_{ds,2}(k) = C(k)\Gamma(k)B_{ds}(k-1);$$

6: Get $H_{ds}(k)$ by

$$H_{ds}(k) = \begin{bmatrix} H_{ds,2}(k-1) & 0 \\ \bar{H}_{ds,2}(k) & D_d(k) \end{bmatrix};$$

7: Get $H_{us}(k)$ and $H_{fs}(k)$ by performing Steps 1–6 with the $\{B_d(k),\ D_d(k)\}$ being replaced by $\{B(k),\ D(k)\}$ and $\{B_f(k),\ D_f(k)\}$, respectively;
8: Let $C_s(k) = \text{diag}\{C(k-s),\ C(k-s+1),\ \ldots,\ C(k)\}$ and

$$\Xi(k) = \begin{bmatrix} I \\ \Gamma_1(k-s+1, s-1) \\ \Gamma_1(k-s+2, s-2) \\ \vdots \\ \Gamma_1(k-1, 1) \\ A(k)\Gamma_1(k-1, 1) \end{bmatrix}.$$

Compute the $H_{os}(k)$ by $H_{os}(k) = C_s(k)\Xi(k-1)$.

---

but possible poor FD performance. A compromise between the FD performance and computation load remains a practically nontrivial and challenging topic in parity space-based FD study. To handle this issue, a Krein space based solution will be given in Chap. 14 and a wavelet transform aided parity space method will be introduced in Chap. 17. Extension of the results in event-triggered systems will be given in Chap. 16. Besides, parity space-based fault estimation issues will be discussed in Chap. 15.

3.4 Fault Estimation for LDTV Systems

Fault estimation aims to identify more fault information such as size, type and location, etc. Roughly speaking, there are two strategies for fault estimation in LDTV systems, that are: *(i)* modeling the fault as system parameter changes and then using parameter estimation schemes for fault estimation; *(ii)* augmenting the fault as a state and then estimating the augmented states by means of observer-based state estimation schemes in the sense of LMS, LS, or H_∞-optimization. In this section, a brief introduction of the second strategy is given.

3.4.1 Observer-Based State Augmentation Method

Consider LDTV system (3.3) with $f(k+1) = F_f(k)f(k)$. Define an augmented state

$$x_a(k) = \left[x^T(k) \;\; f^T(k)\right]^T. \tag{3.41}$$

The augmented system can be represented by

$$\begin{cases} x_a(k+1) = A_a(k)x_a(k) + B_a(k)u(k) + B_{da}(k)d(k), \\ y(k) = C_a(k)x_a(k) + D(k)u(k) + D_d(k)d(k), \\ z(k) = H(k)x_a(k), \end{cases} \tag{3.42}$$

where

$$A_a(k) = \begin{bmatrix} A(k) & B_f(k) \\ 0 & F_f(k) \end{bmatrix}, \; B_a(k) = \begin{bmatrix} B(k) \\ 0 \end{bmatrix}, \; B_{da}(k) = \begin{bmatrix} B_d(k) \\ 0 \end{bmatrix},$$
$$C_a(k) = \begin{bmatrix} C(k) & D_f(k) \end{bmatrix}, \; H(k) = [0 \;\; I].$$

Then a state observer based fault estimator can be constructed as

$$\begin{cases} \hat{x}_a(k+1) = A_a(k)\hat{x}_a(k) + B_a(k)u(k) + L_a(k)(y(k) - \hat{y}(k)), \\ \hat{y}(k) = C_a(k)\hat{x}_a(k) + D(k)u(k), \\ \hat{z}(k) = H(k)\hat{x}_a(k), \end{cases} \tag{3.43}$$

where $\hat{z}(k)$ is the estimate of $z(k)$, $L_a(k)$ is the gain matrix of the observer and should be designed stabilizing the state transfer matrix $(A_a(k) - L_a(k)C_a(k))$, $\forall k$ and, simultaneously, guarantees desired estimation performances that can be defined in the sense of like LMS, LS or H_∞-optimization, etc. For examples, when $d(k)$ is considered to be noises, the well-established Kalman filter can be applied that gives a recursive implementation of $L_a(k)$ and delivers an estimate of fault with minimum estimation errors in the context of LMS estimation.

Remark 3.5 In [29], fault estimation has been embedded in the fault detection solution and a unified design framework of fault detection and estimation in LS sense has been established both for the static and dynamic, stochastic and deterministic, linear and nonlinear systems. Moreover, based on operator theory, an operator based solution and interpretation to fault detection and estimation for LDTV systems have been derived in [30].

3.4.2 H_∞ Fault Estimation

The H_∞-optimization approach to robust control was introduced in the earlier 1980s with initial motivation rooted in handling uncertainties. The early work of using H_∞-optimization techniques for robust fault diagnosis can be found in the early 1990s. Elegant solutions to H_∞-optimal design for fault diagnosis have been developed by means of LMI technique or with respect to solving algebraic Riccati equations [5, 11, 13, 23]. Different from the objective of H_∞ robust control, H_∞-optimization based fault diagnosis require not only disturbance attenuation but also satisfactory sensitivity to faults.

Given LDTV system (3.3) under Assumptions 3.1 and 3.2, we can construct a fault estimator (3.43) with the matrices $L_a(k)$ being designed in the sense of H_∞ optimization, i.e., find a matrix $L_a(k)$ solving

$$\sup_{x_0,\, \tilde{d}\neq 0} \frac{\sum\limits_{k=0}^{N} ||z(k) - \hat{z}(k)||^2}{x_0^T P_0^{-1} x_0 + \sum\limits_{k=0}^{N} ||\tilde{d}(k)||^2} \leq \gamma, \tag{3.44}$$

where $\gamma > 0$ is a given scalar, $N > 0$ is a known integer, $\tilde{d}(k) = [d^T(k)\ f^T(k)]^T$, and P_0 is a weighting matrix.

In LTI cases, an H_∞-optimization problem is generally reduced into a model-matching problem and then solved by means of LMI techniques. For LDTV systems, thanks to the interesting connections of H_∞ estimation with indefinite quadratic form minimization in Krein spaces [2], a finite horizon H_∞ estimation problem can be cast into a problem of calculating the minimum point of a certain quadratic form and then, by virtue of Krein space projection, one can calculate the minimum point recursively via Riccati equations. Inspired by this, Krein space aided H_∞-filtering approach to fault estimation in LDTV systems will be presented in Chap. 11. Also, along this line, a design scheme of H_∞-FDFs in LDTV systems will be discussed in Chap. 12. The application of the results in delayed state LDTV systems is demonstrated in Chap. 13.

Remark 3.6 Aside from the above observer-based fault estimator (3.43), in Chap. 15 of this book we will also demonstrate a parity space-based fault estimation scheme in the sense of H_2-optimization.

3.5 Conclusion

This chapter has been dedicated to the preliminaries of LDTV system modeling, typical schemes of fault detection and estimation for LDTV systems. On the basis of establishing state space models and an alternative input-output model of LDTV systems both in nominal and faulty cases, we have introduced classical FD methods by using Kalman filter, UIO, FDF, and the parity space based approach, with main attention being paid to residual generation issues. Moreover, observer-based H_∞-optimal fault estimation has also be reviewed.

References

1. Khan, A., Abid, Q., & Ding, S. X. (2014). Fault detection filter design for discrete-time nonlinear systems: a mixed H_-/H_∞ optimization. *Systems and Control Letters, 67*(1), 46–54.
2. Hassibi, B., Sayed, A. H., & Kailath, T. (1999). *Indefinite-quadratic estimation and control: A unified approach to H_2 and H_∞ theories*. Philadelphia: SIAM.
3. Shen, B., Ding, S. X., & Wang, Z. (2013). Finite-horizon H_-/H_∞ fault estimation for uncertain linear discrete time-varying systems with known inputs. *IEEE Transactions on Circuits and Systems II: Express Briefs, 60*(12), 902–906.
4. Shen, B., Ding, S. X., & Wang, Z. D. (2013). Finite-horizon H_- fault estimation for linear discrete time-varying systems with delayed measurements. *Automatica, 49*(1), 293–296.
5. Chu, D., & Van Dooren, P. (2006). A novel numerical method for exact model matching problem with stability. *Automatica, 42*(10), 1697–1704.
6. Guo, D., Zhong, M., Ji, H., et al. (2018). A hybrid feature model and deep learning based fault diagnosis for unmanned aerial vehicle sensors. *Neurocomputing, 319*, 155–163.
7. Zhao, D., Ding, S. X., Karimi, H. R., et al. (2019). On robust Kalman filter for two-dimensional uncertain linear discrete time-varying systems: A least squares method. *Automatica, 99*, 203–212.
8. Zhao, D., Ding, S. X., Wang, Y., et al. (2018). Krein-space based robust H_∞ fault estimation for two-dimensional uncertain linear discrete time-varying systems. *Systems & Control Letters, 115*, 41–47.
9. Zhao, D., Wang, Y., Li, Y., et al. (2017). H_∞ fault estimation for 2-D linear discrete time-varying systems based on krein space method. *IEEE Transactions on Systems, Man, and Cybernetics: Systems, 48*(12), 2070–2079.
10. Estrada, F. L., Ponsart, J. C., Theilliol, D., et al. (2015). Robust H_-/H_∞ fault detection observer design for descriptor-LPV systems with unmeasurable gain scheduling functions. *International Journal of Control, 88*(11), 2380–2391.
11. Wang, H., Wang, J., & Lam, J. (2007). Worst-case fault detection observer design: Optimization approach. *Journal of Optimization Theory and Applications, 132*(3), 475–491.
12. Zhang, H., Duan, G., & Xie, L. (2005). Linear quadratic regulation for linear time-varying systems with multiple input delays part I: Discrete-time case. In *Proceedings of the 5th International Conference on Control and Automation*, June 26–29, Budapest, Hungary (Vol. 2, pp. 948–953).

13. Chen, J., & Patton, J. R. (2000). Standard H_∞ filtering formulation of robust fault detection. *IFAC-PaperOnline, 33*(11), 261–266.
14. Hu, L., Wang, Z., Han, Q.-L., et al. (2018). Event-based input and state estimation for linear discrete time-varying systems. *International Journal of Control, 91*(1), 101–113.
15. Chadli, M., Abdo, A., & Ding, S. X. (2013). H_-/H_∞ fault detection filter design for discrete-time Takagi-Sugeno fuzzy system. *Automatica, 49*(7), 1996–2005.
16. Zhong, M., Ding, S. X., & Ding, E. L. (2010). Optimal fault detection for linear discrete time-varying systems. *Automatica, 46*(8), 1395–1400.
17. Zhong, M., Ding, S. X., & Zhou, D. (2016). A new scheme for fault detection of linear discrete time-varying systems. *IEEE Transactions on Automatic Control, 61*(9), 2597–2602.
18. Zhang, M., Zhao, H., & Zhong, M. (2010). H_∞ fault detection for linear discrete time-varying systems with delayed state. *IET Control Theory & Applications, 4*(11), 2303–2314.
19. Zhong, M., Chen, J., & Geng, Y. (2017). Scheme of optimal fault detection for linear discrete time-varying systems with delayed state. *Journal of Systems Engineering and Electronics, 28*(5), 979–985.
20. Zhong, M., Ding, Q., & Shi, P. (2009). Parity space-based fault detection for Markovian jump systems. *International Journal of Systems Science, 40*(4), 421–428.
21. Zhong, M., Ding, S. X., Han, Q.-L., et al. (2010). Parity space-based fault estimation for linear discrete time-varying systems. *IEEE Transactions on Automatic Control, 55*(7), 1726–1731.
22. Zhong, M., Ding, S. X., Han, Q.-L., et al. (2022). A Krein space-based approach to event-triggered H_∞ filtering for linear discrete time-varying systems. *Automatica, 135,* 110001.
23. Zhong, M., Ding, S. X., Lam, J., et al. (2003). An LMI approach to design robust fault detection filter for uncertain LTI systems. *Automatica, 39*(3), 543–550.
24. Zhong, M., Guo, D., & Zhou, D. (2014). A Krein space approach to H_∞ filtering of discrete-time nonlinear systems. *IEEE Transactions on Circuits and Systems I: Regular Papers, 61*(9), 2644–2652.
25. Zhong, M., Liu, S., & Zhao, H. (2008). Krein space-based H_∞ fault estimation for linear discrete time-varying systems. *Acta Automatica Sinica, 34*(12), 1529–1533.
26. Wang, P., & Yu, C. (2021). Distributed actuator fault detection of ldtv systems using relative output measurements. In *Proceedings of the 40th Chinese Control Conference (CCC)*, Shanghai, China, July 26–28 (pp. 4449–4454).
27. Wang, P., Zou, P., Yu, C., et al. (2021). Distributed fault detection and isolation for uncertain linear discrete time-varying heterogeneous multi-agent systems. *Information Sciences, 579,* 483–507.
28. Ding, S. X. (2013). *Model-based fault diagnosis techniques: Design schemes, algorithms, and tools* (2nd ed.). London, U.K.: Springer.
29. Ding, S. X. (2018). Optimal fault detection and estimation: A unified scheme and least squares solutions. *IFAC-PapersOnLine, 51*(24), 465–472.
30. Ding, S. X. (2020). *Advanced methods for fault diagnosis and fault-tolerant control.* Berlin, Germany: Springer.
31. Varrier, S., Koenig, D., & Martinez, J. J. (2012). A parity space-based fault detection on LPV systems: approach for vehicle lateral dynamics control system. *IFAC Proceedings Volumes, 45*(20), 1191–1196.
32. Varrier, S., Koenig, D., & Martinez, J. J. (2014). Robust fault detection for uncertain unknown inputs LPV system. *Control Engineering Practice, 22,* 125–134.
33. Zhang, T., Deng, F., Sun, Y., et al. (2021). Fault estimation and fault-tolerant control for linear discrete time-varying stochastic systems. *SCIENCE CHINA Information Sciences, 64*(10), 1–16.
34. Wu, Y., Zhao, D., Liu, S., et al. (2022). Fault detection for linear discrete time-varying systems with multiplicative noise based on parity space method. *ISA Transactions, 121,* 156–170.

Chapter 4
Krein Space and Krein Space Based Optimization

In this chapter, we introduce basic knowledge of Krein space and Krein space based optimization issues. The related techniques will serve as important mathematical tools in coping with optimal fault detection and estimation problems in LDTV systems in the later chapters. The basic concepts in Krein space and Krein space projection are first introduced. Then the minimization issues of indefinite quadratic form and partially deterministic quadratic form are reviewed, followed by the derivation of recursive computation algorithms in Krein spaces.

4.1 Introduction

Krein space theory has emerged in the early 1980s and its applications in control engineering, signal processing, as well as fault diagnosis have been extensively studied over the past forty years [1–4, 6, 13, 19, 30]. In control engineering, the pioneer works on linear estimation in indefinite-metric space (i.e., the so-called Krein space) were given by Hassibi et al. in 1996 [1, 2]. Later, in the monograph [3] Hassibi et al. have proposed new arguments of classical Kalman filtering and H_∞-filtering and, more importantly, suggested a unified Krein space formulation of linear estimation under both H_2 and H_∞ criteria. In such a unified framework, the problems of Kalman filtering and H_∞ estimation can be converted into minimization issues of a certain indefinite quadratic form and then, by applying innovation analysis and projections in Krein space, the necessary and sufficient conditions for the existence of the minimum can be derived and computed recursively. Such a handling can reduce significantly the conservativeness of the results derived in Hilbert space and, simultaneously, leads to some computation attractiveness. Inspired by these remarkable achievements, a great number of literature on Krein space aided Kalman filtering [26, 27, 31] and H_∞-filtering (including estimation and smoothing) [8, 10, 25] has been increasingly

M. Zhong et al., *Fault Diagnosis for Linear Discrete Time-Varying Systems and Its Applications*, https://doi.org/10.1007/978-981-19-5438-2_4

reported for various different types of control systems, see e.g., [7, 12, 14, 18, 22, 24, 28]. Even today, study on the linear estimation in Krein space is still an active topic [11, 15, 17, 23, 29].

Early works on Krein space based fault detection and estimation for LDTV systems can be found in 2008 [22] and 2010 [9, 21]. In [22], the concerned fault vector was augmented as a state and then an H_∞ fault estimator was constructed by finding a minimum of a quadratic form function in Krein space. In [21], the design of an FDF-based FD system was formulated as a finite horizon H_∞ smoothing problem and solved by means of Krein space projection. In [16], a Krein space aided design of H_i/H_∞-FDF for FD purpose was given for LDTV systems. Krein space based solution to parity space based optimal FD can also be found in [20]. Very recently, an increasing trend is to apply Krein space techniques to cope with fault detection and estimation issues for more complex systems such as the system with delayed states [5, 9], nonlinear systems [18], event-triggered systems [17] and systems with various uncertainties [6, 7, 30], etc.

Thanks to the merits of Krein space techniques in addressing H_2- and H_∞-optimization issues, in this book Krein space based approaches to fault diagnosis for LDTV systems will be illustrated in Chaps. 5, 7, 8, 10–15 and 18. To facilitate subsequent studies, this chapter is devoted to introducing the preliminaries of Krein space and Krein space based optimization issues. On the other hand, we would like to mention that, much expositions of this chapter can be found in the book [3] and this chapter is given mainly for the sake of completeness.

In what follows, notations and basic concepts involved in Krein space are first reviewed. On this basis, Krein space projection, minimization issues of indefinite and deterministic quadratic forms, and recursive computation of Krein space projections are demonstrated. Finally, some notes are given in the conclusion.

4.2 Krein Spaces and Projections in Krein Space

Even Krein space shares many common properties with Hilbert space, it has some specific characteristics that contribute the advantages of Krein space based optimization. In this section we introduce the basic concepts of Krein spaces and the projections in Krein space.

Remark 4.1 For ease of description, in this and subsequent chapters, the elements in a Krein space are denoted by **boldface** letters, and elements in the Euclidean space are denoted by normal letters. With a little abuse of notation, in this chapter we denote a Krein space by $\mathcal{K}$, which is different from the projection operator $\mathcal{K}$ introduced in (2.2) for residual generation purpose.

4.2.1 Krein Spaces

We start with the definition of Krein spaces given in [3].

Definition 4.1 ([3]) An abstract vector space $\{\mathcal{K}, \langle\cdot,\cdot\rangle\}$ that satisfies the following requirements is called a Krein space:

(i) $\mathcal{K}$ is a linear space over $\mathbb{C}$, the field of complex numbers ;
(ii) There exists a bilinear form $\langle\cdot,\cdot\rangle \in \mathbb{C}$ on $\mathcal{K}$ such that for any $\mathbf{x}, \mathbf{y}, \mathbf{z} \in \mathcal{K}$, $a, b \in \mathbb{C}$,

$$\langle \mathbf{x}, \mathbf{y}\rangle^* = \langle \mathbf{y}, \mathbf{x}\rangle\,,\ \ \langle a\mathbf{x} + b\mathbf{y}, \mathbf{z}\rangle = a\,\langle \mathbf{x}, \mathbf{z}\rangle + b\,\langle \mathbf{y}, \mathbf{z}\rangle\,, \tag{4.1}$$

where $*$ denotes complex conjugation;
(iii) The vector space $\mathcal{K}$ admits a direct orthogonal sum decomposition

$$\mathcal{K} = \mathcal{K}_+ \oplus \mathcal{K}_-$$

such that $\{\mathcal{K}_+, \langle\cdot,\cdot\rangle\}$ and $\{\mathcal{K}_-, \langle\cdot,\cdot\rangle\}$ are Hilbert spaces, and

$$\langle \mathbf{x}, \mathbf{y}\rangle = 0$$

for any $\mathbf{x} \in \mathcal{K}_+$ and $\mathbf{y} \in \mathcal{K}_-$.

It is noted that, differ from Hilbert space where $\langle x, x\rangle > 0,\ \forall x \neq 0$ should hold except for the conditions (i) and (ii) in Definition 4.1, in Krein space the self-inner-product $\langle \mathbf{x}, \mathbf{x}\rangle$ for any vector $\mathbf{x} \in \mathcal{K}$ is indefinite but not necessarily positive as required in Hilbert space. We say a vector $\mathbf{x} \in \mathcal{K}$ to be *positive* if $\langle \mathbf{x}, \mathbf{x}\rangle > 0$; *neutral* if $\langle \mathbf{x}, \mathbf{x}\rangle = 0$, and *negative* if $\langle \mathbf{x}, \mathbf{x}\rangle < 0$. Correspondingly, a subspace $\mathcal{S}_{\mathcal{K}} \subseteq \mathcal{K}$ can be positive, neutral, or negative, if all its elements are so, respectively.

Let $\{\mathbf{y}(i)\}_{i=0}^N$ be the collection of $\{\mathbf{y}(0),\ \mathbf{y}(1),\ \ldots,\ \mathbf{y}(N)\}$ and $\mathcal{L}\left\{\{\mathbf{y}(i)\}_{i=0}^N\right\}$ be a linear subspace of $\mathcal{K}$ spanned by $\{\mathbf{y}(i)\}_{i=0}^N$ for $\mathbf{y}(i) \in \mathcal{K}$. The definitions of Gramian and cross-Gramian are given as follows.

Definition 4.2 ([3]) The Gramian of the collection of elements in $\{\mathbf{y}(i)\}_{i=0}^N$ is defined as the $(N+1) \times (N+1)$ block matrix

$$R_{\mathbf{y}_N} = [\langle \mathbf{y}(i), \mathbf{y}(j)\rangle]_{i,j=0:N}.$$

Let $\mathbf{y}_N = [\mathbf{y}^T(0)\ \mathbf{y}^T(1)\ \cdots\ \mathbf{y}^T(N)\ ]^T$. The Gramian $R_{\mathbf{y}_N}$ can be rewritten as

$$R_{\mathbf{y}_N} = \left\langle \mathbf{y}_N, \mathbf{y}_N \right\rangle.$$

Definition 4.3 Given $\mathbf{y}_N = [\mathbf{y}^T(0)\ \mathbf{y}^T(1)\ \cdots\ \mathbf{y}^T(N)]^T$ and $\mathbf{x}_M = [\mathbf{x}^T(0)\,\mathbf{x}^T(1)\ \ldots\ \mathbf{x}^T(M)]^T$, the $(M+1) \times (N+1)$ cross-Gramian of $\mathbf{x}_M$ and $\mathbf{y}_N$ is defined as

$$R_{\mathbf{x}_M\mathbf{y}_N} = [\langle \mathbf{x}(i), \mathbf{y}(j)\rangle]_{i=0:M, j=0:N} = \left\langle \mathbf{x}_M, \mathbf{y}_N \right\rangle.$$

Referring to Definitions 4.1 and 4.2, we would like to point out that

(i) the Gramian $R_{\mathbf{y}_N}$ is a Hermitian matrix due to $\langle \mathbf{y}(i), \mathbf{y}(j)\rangle^* = \langle \mathbf{y}(j), \mathbf{y}(i)\rangle$. Similarly, it holds for the cross-Gramian $R_{\mathbf{x}_M\mathbf{y}_N}$ that $R_{\mathbf{x}_M\mathbf{y}_N} = R^*_{\mathbf{y}_N\mathbf{x}_M}$;
(ii) The sequences $\{\mathbf{y}(0), \mathbf{y}(1), \ldots, \mathbf{y}(N)\}$ can be regarded as a random variable and the Gramian $R_{\mathbf{y}_N}$ can be understood as their "covarinace matrix", i.e.,

$$R_{\mathbf{y}_N} = [\mathbb{E}[\mathbf{y}(i)\mathbf{y}^*(j)]]_{i,j=0:N} = \mathbb{E}[\mathbf{y}_N\mathbf{y}_N^*].$$

Such a covariance matrix is indefinite. This is quite different from Hilbert space.

Moreover, in terms of Gramian, a Krein space can be determined either positive or negative by using the following lemma.

Lemma 4.1 ([3]) *Suppose* $\{y(0), y(1), \ldots, y(N)\}$ *are linearly independent elements in Krein space* $\mathcal{K}$. *Then* $\mathcal{L}\left\{\{y(i)\}_{i=0}^N\right\}$ *is a positive (negative) subspace of* $\mathcal{K}$ *if and only if* $R_{y_N} > 0$ ($R_{y_N} < 0$).

Remark 4.2 The key difference between Krein space and Hilbert space is the existence of neutral and isotropic vectors. A neutral vector in Krein space is a nonzero vector that has zero length. An isotropic vector is a nonzero vector lying in a linear subspace of $\mathcal{K}$ that is orthogonal to every element in that linear subspace. Note that such vectors do not exist in Euclidean or Hilbert spaces. The interested reader can refer to [3] for more detailed geometric interpretation of Krein space.

4.2.2 Projections in Krein Space

The definition and basic properties of Krein space projections are reviewed below.

Definition 4.4 Given elements $\mathbf{x}$ and $\{\mathbf{y}(i)\}_{i=0}^N$ in Krein space $\mathcal{K}$, $\hat{\mathbf{x}}$ is defined as the projection of $\mathbf{x}$ onto $\mathcal{L}\left\{\{\mathbf{y}(i)\}_{i=0}^N\right\}$ if

$$\mathbf{x} = \hat{\mathbf{x}} + \tilde{\mathbf{x}},$$

where $\hat{\mathbf{x}} \in \mathcal{L}\left\{\{\mathbf{y}(i)\}_{i=0}^N\right\}$ and $\tilde{\mathbf{x}}$ satisfies the orthogonality condition $\tilde{\mathbf{x}} \perp \mathcal{L}\left\{\{\mathbf{y}(i)\}_{i=0}^N\right\}$ in the sense of

$$\langle \tilde{\mathbf{x}}, \mathbf{y}(i)\rangle = 0,\ i = 0, 1, \ldots, N.$$

Different from the Hilbert space where projections always exist and are unique, the existence condition of projections in Krein space is given in the following lemma.

Lemma 4.2 ([3]) *If the Gramian matrix* $R_{y_N} = \left\langle y_N, y_N \right\rangle$ *is nonsingular, then the projection of* $\boldsymbol{x}$ *onto* $\mathcal{L}\left\{\{y(i)\}_{i=0}^N\right\}$ *exists, is unique, and is given by*

$$\hat{x} = \langle x, y_N \rangle \langle y_N, y_N \rangle^{-1} y = R_{xy_N} R_{y_N}^{-1} y_N. \tag{4.2}$$

Remark 4.3 Due to the significance of the existence and uniqueness of a Krein space projection for our later discussion, in the rest of this book we make the standing assumption that the Gramian $R_{\mathbf{y}_N}$ is nonsingular without loss of generality.

4.3 Minimization Issues in Krein Spaces

When it comes to minimization issues of a certain quadratic form, the arguments in Krein space and Hilbert space are different. To show this clearly, we consider a simple LMS estimation problem in Krein space.

Let $\mathbf{y}_N = [\mathbf{y}^T(0)\ \mathbf{y}^T(1)\ \cdots,\ \mathbf{y}^T(N)]^T \in \mathbb{R}^{(N+1)q}$ and $\hat{\mathbf{x}} \in \mathbb{R}^m$ be a linear combination of the elements in $\mathbf{y}_N$, i.e., $\hat{\mathbf{x}} = L\mathbf{y}_N$ with $L \in \mathbb{R}^{m\times(N+1)q}$. The linear mapping matrix L in LMS estimation sense is found such that the error Gramian $P(L)$ is minimized, i.e.,

$$\min_L P(L) := \langle \mathbf{x} - L\mathbf{y}_N, \mathbf{x} - L\mathbf{y}_N \rangle. \tag{4.3}$$

In Hilbert space, an optimal solution of L to problem (4.3) can be easily found by letting

$$P(L) = \langle \mathbf{x} - L\mathbf{y}_N, \mathbf{x} - L\mathbf{y}_N \rangle = \|\mathbf{x} - L\mathbf{y}_N\|_2^2 = 0,$$

because the inner product in Hilbert space is always nonnegative. However, in Krein space the inner produce $\langle \mathbf{x} - L\mathbf{y}_N, \mathbf{x} - L\mathbf{y}_N \rangle$ can be negative and, $P(L)$ may be not at its minimum even if $\|\mathbf{x} - L\mathbf{y}_N\|_2^2 = 0$ due to the existence of isotropic vectors.

Observing this difference, the existence and minimization conditions of certain quadratic form in Krein spaces should be studied.

4.3.1 *Minimization Conditions*

We begin with the definition of the so-called stationary point.

Definition 4.5 ([3]) The matrix $L_o \in \mathbb{R}^{m\times k_v}$ is said to be a stationary point of $J(L)$ if and only if Lv is a stationary point of the scalar quadratic form $v^T J(L)v$ for all vectors $v \in \mathbb{R}^{k_v}$, i.e., if and only if

$$\left.\frac{\partial(v^T J(L)v)}{\partial(Lv)}\right|_{L=L_o} = 0.$$

Due to Lemma 2.4.1 in [3], the existence condition for the minimum of a quadratic form $J(L)$ is given in the following lemma.

Lemma 4.3 ([3]) *A stationary point of* $J(L)$ *is a minimum if and only if for all* $v \in \mathbb{R}^{k_v}$

$$\left.\frac{\partial^2(v^T J(L)v)}{\partial(Lv)\partial(Lv)^T}\right|_{L=L_o} \geq 0.$$

Moreover, it is a unique minimum if and only if

$$\left.\frac{\partial^2(v^T J(L)v)}{\partial(Lv)\partial(Lv)^T}\right|_{L=L_o} > 0.$$

Remark 4.4 In the initial definition of stationary point given in [3], the vector v involved in the quadratic form can be complex, and Lemma 4.3 also holds for this situation. In this book, all variables are considered to be real. So in Definition 4.5 and Lemma 4.3 we consider the vector v and the matrix L_o are in real number spaces.

Rewrite the error Gramian $P(L)$ in (4.3) as the following quadratic form

$$\begin{aligned} P(L) &= \langle \mathbf{x}, \mathbf{x} \rangle - \langle \mathbf{x}, \mathbf{y}_N \rangle L - L^T \langle \mathbf{y}_N, \mathbf{x} \rangle - L^T \langle \mathbf{y}_N, \mathbf{y}_N \rangle L \\ &= \begin{bmatrix} I & -L^T \end{bmatrix} \begin{bmatrix} R_{\mathbf{x}} & R_{\mathbf{x}\mathbf{y}_N} \\ R_{\mathbf{y}_N\mathbf{x}} & R_{\mathbf{y}_N} \end{bmatrix} \begin{bmatrix} I \\ -L \end{bmatrix}. \end{aligned}$$

The stationary point of error Gramian $P(L)$ is then given in the following theorem.

Theorem 4.1 ([3]) *If* $R_{\mathbf{y}_N}$ *is nonsingular, the unique coefficient matrix* $L_o = R_{\mathbf{y}_N\mathbf{x}} R_{\mathbf{y}_N}^{-1}$ *in the projection of* $\mathbf{x}$ *onto* $\mathcal{L}\{\{\mathbf{y}(i)\}_{i=0}^N\}$*, i.e.,* $\hat{\mathbf{x}} = L_o \mathbf{y}_N$*, yields the unique stationary point of the error Gramian*

$$P(L) := \langle \mathbf{x} - L\mathbf{y}_N, \mathbf{x} - L\mathbf{y}_N \rangle = \begin{bmatrix} I & -L^T \end{bmatrix} \begin{bmatrix} R_{\mathbf{x}} & R_{\mathbf{x}\mathbf{y}_N} \\ R_{\mathbf{y}_N\mathbf{x}} & R_{\mathbf{y}_N} \end{bmatrix} \begin{bmatrix} I \\ -L \end{bmatrix}.$$

Moreover, the value of $P(L)$ *at the stationary point is given by*

$$P(L_o) = R_{\mathbf{x}} - R_{\mathbf{x}\mathbf{y}_N} R_{\mathbf{y}_N}^{-1} R_{\mathbf{y}_N\mathbf{x}}.$$

A detailed proof of Theorem 4.1 is referred to [3]. As a byproduct of Theorem 4.1 and Lemma 4.3, the minimum and uniqueness condition of the error Gramian is given in the following corollary.

Corollary 4.1 ([3]) *In Theorem 4.1, a unique minimum of* $P(L)$ *is achievable with respect to* L_o *if and only if* $R_{\mathbf{y}_N} > 0$*, i.e., the Gramian matrix* $R_{\mathbf{y}_N}$ *is required non-singular and positive definite.*

It is worth emphasizing that, for any vectors $\mathbf{x}, \mathbf{y} \in \mathcal{K}$, a projection matrix L_o (of $\mathbf{x}$ onto subspace $\mathcal{L}\{\{\mathbf{y}(i)\}_{i=0}^N\}$) that minimizes the error Gramian matrix $P(L)$

exists and is unique on condition that the Gramian matrix $R_{\mathbf{y}_N}$ is nonsingular and positive definite. Note that this conclusion holds for vectors in Krein spaces. One coming question is: is the result applicable in the case that the involved variables are in Hilbert spaces? Such a problem is referred to as the "*partially equivalent deterministic problem*" according to [3]. It has been proven in [3] that, the stationary point of a certain scalar quadratic form in this deterministic problem is same with the one in Krein space, while the condition for a minimum is different, as demonstrated in the following theorem and corollary.

Theorem 4.2 ([3]) *Given* $x \in \mathbb{R}^n$, $y \in \mathbb{R}^q$, $y_N = [y^T(0),\ y^T(1),\ \ldots,\ y^T(N)]^T$ *and the following scalar quadratic form*

$$J(x, y_N) = \begin{bmatrix} x^T & y_N^T \end{bmatrix} \begin{bmatrix} R_x & R_{xy_N} \\ R_{y_N x} & R_{y_N} \end{bmatrix}^{-1} \begin{bmatrix} x \\ y_N \end{bmatrix} \tag{4.4}$$

with $\begin{bmatrix} R_x & R_{xy_N} \\ R_{y_N x} & R_{y_N} \end{bmatrix}$ *and* $R_{\mathbf{y}_N}$ *being nonsingular, then the stationary point* x_o *of* $J(x, y_N)$ *over* x *is given by*

$$x_o = R_{xy_N} R_{\mathbf{y}_N}^{-1} y_N. \tag{4.5}$$

The value of $J(x, y_N)$ *at the stationary point is given by*

$$J(x_o, y_N) = y_N^T R_{\mathbf{y}_N}^{-1} y_N.$$

Corollary 4.2 ([3]) *In Theorem 4.2,* x_o *yields a unique minimum of* $J(x, y_N)$ *if and only if*

$$R_x - R_{xy_N} R_{\mathbf{y}_N}^{-1} R_{y_N x} > 0.$$

4.3.2 Some Remarks and Notes

Theorems 4.1 and 4.2 are of great importance for our later studies on Krein space based fault detection and estimation. To understand the results better, the following points are highlighted.

Firstly, in Theorem 4.2 vectors x and y are not in Krein space. y_N can be any vector in space $\mathbb{R}^{(N+1)q}$ and the stationary point x_o in (4.5) is not the projection of x onto subspace $\mathcal{L}\{\{y(i)\}_{i=0}^{N}\}$. Simultaneously, matrices $R_{\mathbf{x}}$, $R_{\mathbf{xy}_N}$, $R_{\mathbf{y}_N\mathbf{x}}$, and $R_{\mathbf{y}_N}$ are arbitrary matrices while can be obtained from Gramian and cross-Gramians of the vectors in Krein spaces. This observation allows us to handle minimization problems of a certain deterministic quadratic form (involving vectors in general Hilbert spaces) by establishing a partially equivalent model in Krein space and then addressing the equivalent minimum problem of quadratic form by means of Krein space projections. In such a way, the targeting deterministic minimization problem can be addressed in

Krein space with more computation efficiency and less conservativeness. In fact, this builds the core of Krein space based approaches to fault detection and estimation for LDTV systems in this book.

Secondly, according to Theorem 4.2 and Corollary 4.2, the existence and uniqueness of a minimum $J(x, y_N)$ can be checked by verifying whether in Krein space the Gramian $R_{\mathbf{y}_N}$ in $J(x, y_N)$ is nonsingular and the matrix $R_{\mathbf{x}} - R_{\mathbf{x}\mathbf{y}_N} R_{\mathbf{y}_N}^{-1} R_{\mathbf{y}_N\mathbf{x}}$ is positive or not. As direct checking such conditions is complicate, the following inertia conditions can be alternatively used for our purpose.

Lemma 4.4 ([3]) *If R_{y_N} and R_x are nonsingular, then the deterministic problem in Theorem 4.2 will have a minimizing solution if and only if*

$$I_{-}[R_{y_N}] = I_{-}[R_x] + I_{-}[R_x - R_{xy_N} R_{y_N}^{-1} R_{y_N x}],$$

where $I_{-}[A]$ denotes the negative inertia (i.e., the number of negative eigenvalues) of A. Especially, when $R_x > 0$, then the deterministic problem of Theorem 4.2 will have a minimizing solution if and only if

$$I_{-}[R_{y_N}] = I_{-}[R_x - R_{xy_N} R_{y_N}^{-1} R_{y_N x}],$$

i.e., if and only if R_{y_N} and $R_x - R_{xy_N} R_{y_N}^{-1} R_{y_N x}$ have the same negative inertia.

Thirdly, with x and y given in Hilbert space, $R_{\mathbf{x}} - R_{\mathbf{x}\mathbf{y}_N} R_{\mathbf{y}_N}^{-1} R_{\mathbf{y}_N\mathbf{x}} > 0$ implies $R_{\mathbf{y}_N} > 0$ due to Schur complement, which means that the minimums of quadratic forms $P(L)$ and $J(x, y_N)$ can be achieved simultaneously. However, in Krein space these two quadratic forms are not necessarily at their minimums at the same time.

4.4 Recursive Computation of Krein Space Projections

It is known from Sect. 4.2.2 that a projection of vector $\mathbf{x}$ onto subspace $\mathcal{L}\{\{\mathbf{y}(i)\}_{i=0}^{N}\}$ can be obtained using (4.2), where the inverse of Gramian $R_{\mathbf{y}_N}$ should be computed. In many applications $\mathbf{y}(i)$ arises sequentially and brings a series of projections of $\mathbf{x}$. To improve the computation efficiency, a recursive computation of Krein space projections is necessary.

Let $\mathbf{y}_k = [\mathbf{y}^T(0)\ \mathbf{y}^T(1),\ \ldots,\ \mathbf{y}^T(k)]^T$, $k = 0, 1, \ldots$, and $\hat{\mathbf{y}}(k) = \hat{\mathbf{y}}(k|k-1)$ be the projection of $\mathbf{y}(k)$ onto subspace $\mathcal{L}\left\{\{\mathbf{y}(i)\}_{i=0}^{k-1}\right\}$. The so-called innovation is defined as

$$\mathbf{e}(k) = \mathbf{y}(k) - \hat{\mathbf{y}}(k).$$

It is known from Definition 4.4 that innovations $\{\mathbf{e}(0), \mathbf{e}(1), \ldots, \mathbf{e}(k)\}$ form an orthogonal basis of subspace $\mathcal{L}\{\{\mathbf{y}(i)\}_{i=0}^{k}\}$. That means the projection of $\mathbf{y}(k)$ onto the subspace $\mathcal{L}\left\{\{\mathbf{y}(i)\}_{i=0}^{k-1}\right\}$ equals to the projection of it onto the subspace $\mathcal{L}\{\{\mathbf{e}(i)\}_{i=0}^{k-1}\}$. Following this observation, we have

$$\begin{aligned}\hat{\mathbf{y}}(k) &= \langle \mathbf{y}(k), \mathbf{e}(0)\rangle\, R_{\mathbf{e},0}^{-1}\mathbf{e}(0) + \cdots + \langle \mathbf{y}(k), \mathbf{e}(k-1)\rangle\, R_{\mathbf{e},k-1}^{-1}\mathbf{e}(k-1)\\ &= \sum_{i=0}^{k-1} \langle \mathbf{y}(i), \mathbf{e}(i)\rangle\, R_{\mathbf{e},i}^{-1}\mathbf{e}(i),\end{aligned}$$

where $R_{\mathbf{e},i} = \langle \mathbf{e}(i), \mathbf{e}(i)\rangle$. Together with $\mathbf{y}(k) = \hat{\mathbf{y}}(k) + \mathbf{e}(k)$, it holds, given $k = N$,

$$\underbrace{\begin{bmatrix} \mathbf{y}(0) \\ \mathbf{y}(1) \\ \vdots \\ \mathbf{y}(N) \end{bmatrix}}_{\mathbf{y}_N} = \underbrace{\begin{bmatrix} I & & & \\ \langle \mathbf{y}(1), \mathbf{e}(0)\rangle\, R_{\mathbf{e},0}^{-1} & I & & \\ \vdots & \vdots & & \\ \langle \mathbf{y}(N), \mathbf{e}(0)\rangle\, R_{\mathbf{e},0}^{-1} & \langle \mathbf{y}(N), \mathbf{e}(1)\rangle\, R_{\mathbf{e},1}^{-1} & \cdots & I \end{bmatrix}}_{H} \underbrace{\begin{bmatrix} \mathbf{e}(0) \\ \mathbf{e}(1) \\ \vdots \\ \mathbf{e}(N) \end{bmatrix}}_{\mathbf{e}_N},$$

and then

$$R_{\mathbf{y}_N} = \langle \mathbf{y}_N, \mathbf{y}_N\rangle = H R_{\mathbf{e}_N} H^T, \quad R_{\mathbf{e}_N} = \langle \mathbf{e}_N, \mathbf{e}_N\rangle\,. \tag{4.6}$$

If all submatrices of Gramian $R_{\mathbf{y}_N}$ are nonsingular, $R_{\mathbf{y}_N}$ and these submatrices are called strongly regular. Obviously, such strongly regular condition is required for a recursive computation of Krein space projection. Note from (4.6) that $R_{\mathbf{y}_N}$ and $R_{\mathbf{e}_N}$ have the same inertia. Referring to Lemma 4.2, the following lemma is thus obtained.

Lemma 4.5 ([3]) *The Gramian R_{y_N} of $\boldsymbol{y}_N$ has the same inertia as the Gramian of the innovations $\boldsymbol{e}_N$. The strong regularity of R_{y_N} implies the nonsingularity of $R_{e,i}$, $i \in [0, N]$. In particular, $R_{y_N} > 0$ if and only if*

$$R_{e,i} > 0, \ \forall i = 0, 1, \ldots, N.$$

Now we consider a projection of $\mathbf{x}$ onto the subspace $\mathcal{L}\left\{\{\mathbf{y}(i)\}_{i=0}^{k}\right\}$ (or equally, the subspace $\mathcal{L}\left\{\{\mathbf{e}(i)\}_{i=0}^{k}\right\}$) and denote it by $\hat{\mathbf{x}}(k)$ with $k = 1, 2, \ldots$. It is easy to obtain

$$\hat{\mathbf{x}}(k) = \sum_{i=0}^{k} \langle \mathbf{x}, \mathbf{e}(i)\rangle\, R_{\mathbf{e},i}^{-1}\mathbf{e}(i),$$

which can be recursively computated by

$$\hat{\mathbf{x}}(k) = \hat{\mathbf{x}}(k-1) + \langle \mathbf{x}, \mathbf{e}(k)\rangle\, R_{\mathbf{e},k}^{-1}\mathbf{e}(k).$$

Let $y_k = [y^T(0)\ y^T(1),\ \ldots,\ y^T(k)]^T, k = 0, 1, \ldots$. According to the connection between Krein space projection and the minimization of deterministic quadratic form, given

$$J(x, y_k) = \begin{bmatrix} x^T & y_k^T \end{bmatrix} \begin{bmatrix} R_{\mathbf{x}} & R_{\mathbf{x}\mathbf{y}_k} \\ R_{\mathbf{y}_k\mathbf{x}} & R_{\mathbf{y}_k} \end{bmatrix}^{-1} \begin{bmatrix} x \\ y_k \end{bmatrix},$$

the stationary point $x_o(k)$ of $J(x, y_k)$ over x can also be obtained as

$$x_o(k) = \sum_{i=0}^{k} \langle \mathbf{x}, \mathbf{e}(i) \rangle \, R_{\mathbf{e},i}^{-1} e(i)$$

that can be computed recursively by

$$x_o(k) = x_o(k-1) + \langle \mathbf{x}, \mathbf{e}(k) \rangle \, R_{\mathbf{e},k}^{-1} e(k).$$

If the minimization condition (stated in Corollary 4.2) can be satisfied at each time step k, the minimum of $J(x_o(k), y_k) = y_k^T R_{\mathbf{y}_k}^{-1} y_k$ can also be computed recursively, i.e.,

$$J(x_o(k), y_k) = J(x_o(k-1), y_{k-1}) + e^T(k) R_{\mathbf{e},k}^{-1} e(k).$$

4.5 Conclusion

In this chapter preliminaries of Krein space and optimization in Krein space have been introduced, that build important mathematical fundamentals of our studies on Krein space based fault detection and estimation for LDTV systems in later chapters. By introducing Gramian, cross-Gramian and Krein space projections, the minimization issue of stochastic quadratic form in Krein space has been investigated. The existence and uniqueness conditions of the minimum have been derived, which, due to the existence of neutral and isotropic vectors in Krein space, are different from the results in Hilbert space. Moreover, connections between the indefinite quadratic form minimization in Krein space and the deterministic quadratic form minimization in Hilbert space have been established, which allows us to handle deterministic quadratic form minimization issues in Krein space, so as to achieve less conservative results with more computation efficiency.

References

1. Hassibi, B., Sayed, A. H., & Kailath, T. (1996). Linear estimation in Krein spaces-Part I: Theory. *IEEE Transactions on Automatic Control, 41*(1), 18–33.
2. Hassibi, B., Sayed, A. H., & Kailath, T. (1996). Linear estimation in Krein spaces-Part II: Applications. *IEEE Transactions on Automatic Control, 41*(1), 34–49.
3. Hassibi, B., Sayed, A. H., & Kailath, T. (1999). *Indefinite-quadratic estimation and control: A unified approach to* H_2 *and* H_∞ *theories*. Philadelphia: SIAM.

4. Shen, B. (2014). A survey on the applications of the Krein-space theory in signal estimation. *Systems Science & Control Engineering: An Open Access Journal, 2*(1), 143–149.
5. Shen, B., Ding, S. X., & Wang, Z. (2013). Finite-horizon H_∞ fault estimation for linear discrete time-varying systems with delayed measurements. *Automatica, 49*(1), 293–296.
6. Zhao, D., Ding, S. X., Wang, Y., et al. (2018). Krein-space based robust H_∞ fault estimation for two-dimensional uncertain linear discrete time-varying systems. *Systems & Control Letters, 115,* 41–47.
7. Zhao, D., Wang, Y., Li, Y., et al. (2017). H_∞ fault estimation for 2-D linear discrete time-varying systems based on Krein space method. *IEEE Transactions on Systems, Man, and Cybernetics: Systems, 48*(12), 2070–2079.
8. Yu, F., Lv, C., & Dong, Q. (2016). A novel robust H_∞ filter based on Krein space theory in the SINS/CNS attitude reference system. *Sensors, 16*(3), 396.
9. Zhao, H., Zhong, M., & Zhang, M. (2010). H_∞ fault detection for linear discrete time-varying systems with delayed state. *IET Control Theory & Applications, 4*(11), 2303–2314.
10. Zhang, H., Xie, L., & Soh, Y. C. (2000). H_∞ deconvolution filtering, prediction, and smoothing: A Krein space polynomial approach. *IEEE Transactions on Signal Processing, 48*(3), 888–892.
11. Zhao, J., & Mili, L. (2018). A decentralized H-infinity unscented Kalman filter for dynamic state estimation against uncertainties. *IEEE Transactions on Smart Grid, 10*(5), 4870–4880.
12. Zhao, J., & Mili, L. (2019). A theoretical framework of robust H_∞ unscented Kalman filter and its application to power system dynamic state estimation. *IEEE Transactions on Signal Processing, 67*(10), 2734–2746.
13. Bognar, J. (1974). *Indefinite inner product spaces*. New York: Springer.
14. Feng, J., Yu, F., Yang, N., et al. (2012). Krein space approach to robust H_∞ filtering for linear uncertain systems. *Journal of Systems Engineering and Electronics, 23*(4), 596–602.
15. Vollmering, M. (2018). *Damage localization of mechanical structures by subspace identification and Krein space based H_∞ estimation*. Ph.D. thesis, Bauhaus-University Weimar.
16. Zhong, M., Ding, S. X., & Zhou, D. (2016). A new scheme of fault detection for linear discrete time-varying systems. *IEEE Transactions on Automatic Control, 61*(9), 2597–2602.
17. Zhong, M., Ding, S. X., Han, Q.-L., et al. (2022). A Krein space-based approach to event-triggered H_∞ filtering for linear discrete time-varying systems. *Automatica, 135,* 110001.
18. Zhong, M., Guo, D., & Zhou, D. (2014). A Krein space approach to H_∞ filtering of discrete-time nonlinear systems. *IEEE Transactions on Circuits and Systems I: Regular Papers, 61*(9), 2644–2652.
19. Zhong, M., Li, S., & Zhao, Y. (2013). Robust H_∞ fault detection for uncertain LDTV systems using Krein space approach. *International Journal of Innovative Computing, Information and Control, 9*(4), 1637–1649.
20. Zhong, M., Song, Y., & Ding, S. X. (2015). Parity space-based fault detection for linear discrete time-varying systems with unknown input. *Automatica, 59,* 120–126.
21. Zhong, M., Zhou, D., & Ding, S. X. (2010). On designing H_∞ fault detection filter for linear discrete time-varying systems. *IEEE Transactions on Automatic Control, 55*(7), 1689–1695.
22. Zhong, M., Liu, S., & Zhao, H. (2008). Krein space-based H_∞ fault estimation for linear discrete time-varying systems. *Acta Automatica Sinica, 34*(12), 1529–1533.
23. Zhang, Q., Li, Y., Li, Y., et al. (2020). Fault estimation for a class of nonlinear time-variant systems through a Krein space-based approach. *Measurement and Control, 53*(3–4), 541–550.
24. Ding, S. X., Shen, B., Wang, Z., et al. (2014). A fault detection scheme for linear discrete-time systems with an integrated online performance evaluation. *International Journal of Control, 87*(12), 2511–2521.
25. Jin, S., Park, J., Kim, K., et al. (2001). Krein space approach to decentralised H_∞ state estimation. *IEE Proceedings-Control Theory and Applications, 148*(6), 502–508.
26. Lee, T. H., Ra, W. S., Yoon, T., et al. (2004). Robust Kalman filtering via Krein space estimation. *IEE Proceedings-Control Theory and Applications, 151*(1), 59–63.
27. Lee, T. H., Ra, W. S., Yoon, T. S., et al. (2003). Robust filtering for linear discrete-time systems with parametric uncertainties: a Krein space estimation approach. In *Proceedings of the 42nd IEEE International Conference on Decision and Control*, Maui, Hawaii USA (Vol. 2, pp. 1285–1290).

28. Feng, X., Wen, C., & Park, J. H. (2017). Sequential fusion H_∞ filtering for multi-rate multi-sensor time-varying systems-a Krein-space approach. *IET Control Theory & Applications, 11*(3), 369–381.
29. Li, Y., Liu, S., Zhao, D., et al. (2021). Event-triggered fault estimation for discrete time-varying systems subject to sector-bounded nonlinearity: A Krein space based approach. *International Journal of Robust and Nonlinear Control, 31*(11), 5360–5380.
30. Li, Y., Song, X., Zhang, Z., et al. (2019). H_∞ deconvolution filter design for uncertain linear discrete time-variant systems: A Krein space approach. *Applied Mathematics and Computation, 361,* 131–143.
31. Zhu, Y., & Shi, X. (2006). Adaptive robust Kalman filtering via Krein space estimation. In *Proceedings of the 6th World Congress on Intelligent Control and Automation*, June 27–29, Dalian China (Vol. 1, pp. 1818–1822).

Part II
H_2- and H_∞-Optimization Based Fault Diagnosis for LDTV Systems

Chapter 5
H_2-Optimization-Based Fault Detection for LDTV Systems

H_2-optimization is a prevalent concept in control theory and is adopted, in its initial form, for optimization issues related to H_2-norm of transfer functions. Thanks to its optimization interpretation in the context of the well-known linear quadratic Gaussian (LQG) control problem, H_2-optimization is widely accepted as an established term for optimal control and filtering of linear systems under a quadratic cost function. In this regard, we will, in this chapter, introduce fault detection schemes for LDTV systems. Apart from presenting the standard algorithms, we will focus on gaining deep insights into optimal estimation, residual generation and optimal fault detection issues behind the optimization, and learning alternative interpretations. To this end, some basic estimation and fault detection problems will be, at first, addressed. Although these problems are formulated in the context of static processes, they are helpful to understand the centerpiece of conceptual solutions of any fault detection problems, independent of the system type under consideration. The subsequent study on fault detection in LDTV systems is in fact a natural extension and applications of these solutions, with the aid of advanced system theoretical handlings.

5.1 Basic Estimation and Fault Detection Problems

In this section, we will firstly review a very basic estimation problem, the so-called least mean squares (LMS) estimation, and then illustrate how to apply an LMS estimate for an optimal fault detection in static processes. We will demonstrate that the resulted detection approach is indeed the well-established canonical correlation analysis (CCA) algorithm, a popular multivariate analysis (MVA) method. Finally, we will extend the estimation problem to a more general formulation and introduce the concept of innovation, a special form of residual signals. On this basis, recursive algorithms for LMS estimation computations can be realized, which allow efficient and reliable fault detection.

M. Zhong et al., *Fault Diagnosis for Linear Discrete Time-Varying Systems and Its Applications*, https://doi.org/10.1007/978-981-19-5438-2_5

5.1.1 LMS Estimation

Consider two random vectors, $x \in \mathbb{R}^n$ and $y \in \mathbb{R}^q$,

$$x \sim \mathcal{N}(0, \Sigma_x), y \sim \mathcal{N}(0, \Sigma_y), \Sigma_y > 0, \tag{5.1}$$

$$\mathbb{E}\left[xy^T\right] = \Sigma_{xy} \neq 0. \tag{5.2}$$

Definition 5.1 Given x and y satisfying (5.1)–(5.2), an estimate of x by means of a linear mapping of y,

$$\hat{x} = Ly \in \mathbb{R}^n,$$

is called least mean squares and denoted by $\hat{x}_{LMS}$, when

$$\forall \hat{x}, \mathbb{E}\left[\left(x - \hat{x}_{LMS}\right)^T \left(x - \hat{x}_{LMS}\right)\right] \leq \mathbb{E}\left[\left(x - \hat{x}\right)^T \left(x - \hat{x}\right)\right]. \tag{5.3}$$

For our purpose, we introduce the projection scheme for the determination of $\hat{x}_{LMS}$. A projection of x onto y is a linear mapping of y, $Py \in \mathbb{R}^n$, and satisfies

$$\mathbb{E}\left[(x - Py)^T y\right] = 0. \tag{5.4}$$

Since for $P = \mathbb{E}\left[xy^T\right]\left(\mathbb{E}\left[yy^T\right]\right)^{-1}$ (5.4) holds, it is clear that $\mathbb{E}\left[xy^T\right]\left(\mathbb{E}\left[yy^T\right]\right)^{-1} y$ is a projection of x onto y. Moreover, it is straightforward that $\forall \hat{x} = Ly$,

$$\begin{aligned} \mathbb{E}\left[\left(x - \hat{x}\right)^T \left(x - \hat{x}\right)\right] &= \mathbb{E}\left[(x - Py - (L - P)y)^T (x - Py - (L - P)y)\right] \\ &= \mathbb{E}\left[(x - Py)^T (x - Py)\right] + \mathbb{E}\left[\left(\hat{x} - Py\right)^T \left(\hat{x} - Py\right)\right]. \end{aligned}$$

As a result,

$$\hat{x}_{LMS} = \mathbb{E}\left[xy^T\right]\left(\mathbb{E}\left[yy^T\right]\right)^{-1} y = \Sigma_{xy}\Sigma_y^{-1} y. \tag{5.5}$$

In other words, the LMS estimate of x is the projection of x onto y. It is noteworthy that $\forall \hat{x} = Ly$,

$$\begin{aligned} &\mathbb{E}\left[\left(x - \hat{x}\right)\left(x - \hat{x}\right)^T\right] \\ &= \mathbb{E}\left[\left(x - \hat{x}_{LMS} - (L - P)y\right)\left(x - \hat{x}_{LMS} - (L - P)y\right)^T\right] \\ &= \mathbb{E}\left[\left(x - \hat{x}_{LMS}\right)\left(x - \hat{x}_{LMS}\right)^T\right] + \mathbb{E}\left[\left(\hat{x} - \hat{x}_{LMS}\right)\left(\hat{x} - \hat{x}_{LMS}\right)^T\right] \\ &\geq \mathbb{E}\left[\left(x - \hat{x}_{LMS}\right)\left(x - \hat{x}_{LMS}\right)^T\right]. \end{aligned} \tag{5.6}$$

The last equality is the result of (5.4) that implies independence of $x - \hat{x}_{LMS}$ and y. Relation (5.6) means that the LMS estimate of x yields the minimum covariance matrix of estimation error, which is given by

$$\mathbb{E}\left[\left(x - \hat{x}_{LMS}\right)\left(x - \hat{x}_{LMS}\right)^T\right] = \mathbb{E}\left[xx^T\right] - \mathbb{E}\left[xy^T\right]P^T = \Sigma_x - \Sigma_{xy}\Sigma_y^{-1}\Sigma_{xy}^T. \tag{5.7}$$

Results (5.6) and (5.7) are of importance for our subsequent fault detection study.

Next, we will study the application of LMS estimation to optimal fault detection issue and its relation to the well-known MVA-method CCA.

5.1.2 A Basic Fault Detection Problem, LMS and CCA

Consider a measurement (process) model

$$x \sim \mathcal{N}\left(f, \Sigma_x\right), x \in \mathbb{R}^n, f = \begin{cases} 0, & \text{fault-free} \\ \text{an unknown constant vector different from zero.} \end{cases} \tag{5.8}$$

Give a measurement vector x, detecting the fault vector is a basic fault detection problem, whose optimal solution is well-known and summarized in the following algorithm [6]:

- form the test statistic

$$J = x^T \Sigma_x^{-1} x \sim \chi^2\left(n\right), \tag{5.9}$$

  where $\chi^2\left(n\right)$ denotes χ^2 distribution with n degree of freedom;
- given upper-bound of false alarm rate (FAR) α, set the threshold

$$J_{th} = \chi^2_{\alpha}\left(n\right);$$

- run online detection logic

$$\begin{cases} J \leq J_{th} \Longrightarrow \text{fault-free,} \\ J > J_{th} \Longrightarrow \text{faulty.} \end{cases}$$

This detection algorithm delivers the best fault detectability with a guarantee that FAR$\leq \alpha$. It is apparent from (5.9) that the fault detectability depends on the strength of uncertainty which is represented by Σ_x. Specifically, weaker uncertainty implies a higher fault detectability.

Now, assume that an additional measurement vector y, as given in (5.1), is available, and

$$\mathbb{E}\left[\left(x - \mathbb{E}x\right)y^T\right] = \Sigma_{xy}, rank\left(\Sigma_{xy}\right) = \kappa > 0$$

with known matrix Σ_{xy}. It is well-known that the following CCA-based detection algorithm delivers the best detectability [6, 7, 9]:

- form the matrix
$$K = \Sigma_x^{-1/2}\Sigma_{xy}\Sigma_y^{-1/2}, rank\,(K) = \kappa;$$
- run an SVD of K
$$K = R\Sigma V^T, R^T R = I, V^T V = I,$$
$$\Sigma = \begin{bmatrix} \text{diag}(\sigma_1, \ldots, \sigma_\kappa) & 0 \\ 0 & 0 \end{bmatrix}, 1 > \sigma_1 \geq \cdots \geq \sigma_\kappa > 0;$$
- set
$$P = \Sigma_x^{-1/2}R, L = \Sigma_y^{-1/2}V;$$
- construct residual vector
$$r = P^T x - \Sigma L^T y; \tag{5.10}$$
- define the test statistic
$$J = r^T \text{diag}((1-\sigma_1)^{-1}, \ldots, (1-\sigma_\kappa)^{-1}, 1, \ldots, 1)r;$$
- set threshold for a given FAR upper-bound α,
$$J_{th} = \chi_\alpha^2(n);$$
- run online detection logic
$$\begin{cases} J \leq J_{th} \Longrightarrow \text{fault-free}, \\ J > J_{th} \Longrightarrow \text{faulty}. \end{cases}$$

Remark 5.1 CCA is a standard MVA-method [8]. Its application to fault detection is generally implemented in a data-driven form, in which Σ_x, Σ_{xy}, Σ_y are unknown, but estimated using collected process data [6, 9].

In [7], it is proven that the residual vector (5.10) is equivalent to the difference of the normalized measurement vector x and its LMS estimate using y. Below, we schematically illustrate this claim. Note that

$$P^T\Sigma_x P = I, L^T\Sigma_y L = I, P^T\Sigma_{xy}L = \Sigma.$$

It yields

$$r = P^T x - \Sigma L^T y = P^T\left(x - \Sigma_{xy}LL^T y\right) = R\left(\Sigma_x^{-1/2}x - \Sigma_x^{-1/2}\Sigma_{xy}\Sigma_y^{-1}y\right). \tag{5.11}$$

Comparing with (5.7) makes it clear that

$$\hat{x}_{N,LMS} = \Sigma_x^{-1/2} \Sigma_{xy} \Sigma_y^{-1} y$$

is an LMS estimate of $x_N = \Sigma_x^{-1/2} x$, the normalized measurement vector x. Since R is a unitary matrix and has no influence on the fault detectability, we can conclude that the residual vector is constructed by the normalized measurement vector and its LMS estimate using y.

From the fault detection point of view, the expression (5.11) can be well explained. Recall our discussion about the fault detection solution and the influence of uncertainty (expressed in form of covariance matrix of the measurement noise) on the detection performance. It is obvious that integrating the additional measurement vector y leads to reduction of uncertainty. In particular, when the information embedded in y about x is optimally used, namely in form of an LMS estimate, the uncertainty is at minimum. This can be clearly seen by comparing the covariance matrix of x_N, which is a unity matrix due to the normalization, and the residual signal, $r = R\left(x_N - \hat{x}_{N,LMS}\right)$,

$$\begin{aligned} \operatorname{cov}(r) &= \mathbb{E}\left[\left(x_N - \hat{x}_{N,LMS}\right) RR^T \left(x_N - \hat{x}_{N,LMS}\right)^T\right] \\ &= \mathbb{E}\left[\left(x_N - \hat{x}_{N,LMS}\right) \left(x_N - \hat{x}_{N,LMS}\right)^T\right] \\ &= \operatorname{diag}((1-\sigma_1), \ldots, (1-\sigma_\kappa), 1, \ldots, 1) \le I = \mathbb{E}\left[x_N x_N^T\right]. \end{aligned} \quad (5.12)$$

It is of interest to notice that (5.12) tells us, in the measurement subspace, uncertainty will be reduced in the directions (subspaces), where x and y are correlated. In a nutshell, a residual vector generated by the measurement vector and its estimate constructed on the basis of additional measurements can reduce uncertainty and hence enhance the fault detectability. When the LMS estimate is applied for this purpose, the uncertainty is reduced to a minimum, and consequently the fault performance reaches the optimum.

5.1.3 Innovation, Residual and Recursive Residual Generation

We now extend our study to a more practical problem setting. Suppose that the process model (5.8) is under consideration. To detect the fault vector f, additional measurement data, $y(k), k = 0, 1, \ldots,$ are available, which satisfy

$$y(k) \sim \mathcal{N}\left(0, \Sigma_y(k)\right), y(k) \in \mathbb{R}^q, \Sigma_y(k) > 0,$$

and $y(i), y(j), i, j = 0, 1, \ldots,$ may be correlated. For the sake of simplicity, we first use data $y(0), \ldots, y(N), N > 1$, for our detection purpose and define the data vector,

$$y_N = \begin{bmatrix} y(0) \\ \vdots \\ y(N) \end{bmatrix} \in \mathbb{R}^{q(N+1)}, \; y_N \sim \mathcal{N}\left(0, \Sigma_{y_N}\right).$$

On the assumption that $\Sigma_{y_N} > 0$,

$$\begin{aligned} \mathbb{E}\left[xy_N^T\right] &= \left[\mathbb{E}\left[xy^T(0)\right] \cdots \mathbb{E}\left[xy^T(N)\right]\right] = \left[\Sigma_{xy}(0) \cdots \Sigma_{xy}(N)\right] \\ &=: \Sigma_{xy_N} \neq 0, \end{aligned}$$

it is straightforward to build an LMS estimate of x and, on its basis, to generate the residual vector

$$\hat{x}_{LMS} = \Sigma_{xy_N} \Sigma_{y_N}^{-1} y_N, \, r = x - \hat{x}_{LMS}. \tag{5.13}$$

However, for large N, the computation of $\Sigma_{y_N}^{-1}$ could be problematic. Furthermore, in case of further data, $y(N+1), y(N+2), \ldots,$ being available, it is evidently inefficient to repeat the computation by each new data. Apparently, a recursive algorithm would be helpful. To this end, we introduce the concept innovation and, based on it, a recursive LMS estimation algorithm.

Let $\hat{y}(k\,|k-1)$ denote the projection of $y(k)$ onto $\{y(i), i = 0, \ldots, k-1\}$, and call it one-step ahead prediction of $y(k)$. Set the initial value $\hat{y}(0) = 0$. Remember that $\hat{y}(k\,|k-1)$ is an LMS estimate of $y(k)$ using available data $y(i), i = 0, \ldots, k-1$. Define

$$\hat{y}(i+1\,|i) = T_{i+1} y_i = \sum_{j=1}^{i+1} T_{(i+1)j} y(j-1),$$

$$T_{i+1} = \left[T_{(i+1)1} \cdots T_{(i+1)(i+1)}\right], \, y_i = \begin{bmatrix} y(0) \\ \vdots \\ y(i) \end{bmatrix} \in \mathbb{R}^{q(i+1)},$$

$$e(i+1) = y(i+1) - \hat{y}(i+1\,|i) = \left[-T_{i+1} \; I\right] \begin{bmatrix} y_i \\ y(i+1) \end{bmatrix}.$$

Matrix $T_{i+1} \in \mathbb{R}^{q \times q(i+1)}$ represents the linear mapping to build $\hat{y}(i+1\,|i)$ using $y(j), j = 0, \ldots, i$. It turns out

$$e_k = T_{0,k} y_k, \tag{5.14}$$

$$e_k = \begin{bmatrix} e(0) \\ \vdots \\ e(k) \end{bmatrix}, y_k = \begin{bmatrix} y(0) \\ \vdots \\ y(k) \end{bmatrix}, T_{0,k} = \begin{bmatrix} I & 0 & \cdots & 0 \\ T_{11} & I & \cdots & 0 \\ \vdots & \ddots & \ddots & 0 \\ T_{k1} & \cdots & T_{kk} & I \end{bmatrix} \in \mathbb{R}^{q(k+1))\times q(k+1)}.$$

Note that due to the projection,

$$\mathbb{E}\left[e(i)e^T(j)\right] = \Sigma_e(i)\delta_{ij}, \delta_{ij} = \begin{cases} 1, i = j, \\ 0, i \neq j, \end{cases} \tag{5.15}$$
$$\Sigma_e(i) = \mathbb{E}\left[e(i)e^T(i)\right], i, j = 0, \ldots, k.$$

That is, $e(k), k = 0, 1, \ldots,$ are white. $e(k)$ is called innovation series. Moreover, (5.14) and (5.15) lead to

$$\mathbb{E}\left[e_k e_k^T\right] = T_k \mathbb{E}\left[y_k y_k^T\right] T_k^T = T_k \Sigma_{y_k} T_k^T = \text{diag}\left(\Sigma_e(0), \ldots, \Sigma_e(k)\right) > 0.$$

Aided by these results related to the defined innovation, LMS estimate (5.13) can be re-written as

$$\hat{x}_{LMS} = \Sigma_{xy_N} \Sigma_{y_N}^{-1} y_N = \Sigma_{xe_N} \Sigma_{e_N}^{-1} e_N = \sum_{i=0}^{N} \Sigma_{xe}(i) \Sigma_e^{-1}(i) e(i), \tag{5.16}$$
$$\Sigma_{xe_N} = \mathbb{E}\left[x e_N^T\right] = \left[\mathbb{E}\left[x e^T(0)\right] \cdots \mathbb{E}\left[x e^T(N)\right]\right] =: \left[\Sigma_{xe}(0) \cdots \Sigma_{xe}(N)\right],$$
$$e_N = \begin{bmatrix} e(0) \\ \vdots \\ e(N) \end{bmatrix}, \Sigma_{e_N}^{-1} = \text{diag}\left(\Sigma_e^{-1}(0), \ldots, \Sigma_e^{-1}(N)\right).$$

Equation (5.16) enables to compute $\hat{x}_{LMS}$ using the following recursive algorithm:

- set the initial value

$$\hat{x}_{LMS}(0) = \Sigma_{xe}(0)\Sigma_e^{-1}(0)e(0) = \Sigma_{xy}(0)\Sigma_y^{-1}(0)y(0);$$

- recursively compute

$$\hat{x}_{LMS}(k) = \hat{x}_{LMS}(k-1) + \Sigma_{xe}(k)\Sigma_e^{-1}(k)e(k). \tag{5.17}$$

The step (5.17) reveals that the innovation series, $e(k)$, plays an essential role in the recursive computation. It contains the information about the vector x, which is not included in $\hat{x}_{LMS}(k-1)$. Thus, the term $\Sigma_{xe}(k)\Sigma_e^{-1}(k)e(k)$ can be interpreted as an LMS of x using signal $e(k)$. Together with $\hat{x}_{LMS}(k-1)$, it results in a new LMS estimate $\hat{x}_{LMS}(k)$ for x.

In the step (5.17), $\Sigma_{xe}(k)$ and $\Sigma_e^{-1}(k)$ are computed based on the relations

$$\Sigma_{xe}(k) = \mathbb{E}\left[xe^T(k)\right] = \mathbb{E}\left[xy^T(k)\right] - \mathbb{E}\left[xy_{k-1}^T\right]T_k^T = \Sigma_{xy}(k) - \Sigma_{xy_{k-1}}T_k^T,$$
$$\Sigma_e^{-1}(k) = \left(\mathbb{E}\left[e(k)e^T(k)\right]\right)^{-1} = \left(\begin{bmatrix} -T_k & I \end{bmatrix}\Sigma_{y_k}\begin{bmatrix} -T_k^T \\ I \end{bmatrix}\right)^{-1}, \Sigma_{y_k} = \mathbb{E}\left[y_k y_k^T\right],$$

which can be recursively realised on some assumptions. With the aid of this recursive algorithm, residual signal can be generated at each sampling instant,

$$r(k) = x - \hat{x}_{LMS}(k), k = 0, 1, \ldots,$$

which allows a reliable fault detection. Since our intention in this section is to introduce the basic schemes, we will not discuss about the realisation details. Exemplarily, similar computation steps are delineated in the following case study. Nevertheless, it is straightforward that

$$\mathbb{E}\left[r(k)r^T(k)\right] = \mathbb{E}\left[r(k-1)r^T(k-1)\right] - \Sigma_{xe}(k)\Sigma_e^{-1}(k)\Sigma_{xe}^T(k). \tag{5.18}$$

That means, the covariance matrix of the residual vector is reduced at each iteration, as far as $\Sigma_{xe}(k) \neq 0$. In other words, the fault detectability can be improved with each new data $y(k)$.

Remark 5.2 We would like to emphasize that a data set, e.g. $\{y(k), k = 0, \ldots, N-1\}$, could also be viewed as data collected from a set of N sensors, instead as a time series. In this context, the relation (5.18) implies that each additional sensor can result in improvement of the fault detectability, as far as the sensor signal is correlated with x. Moreover, this is independent of the measurement noise of the (added) sensor.

Example 5.1 While the above-mentioned detection scheme is, more or less, a conceptual procedure, we now study the solution for a more practical case. Consider the process model

$$y(k) = f + \varepsilon(k) \in \mathbb{R}^q, f = \begin{cases} 0, \text{ fault-free}, \\ \text{an unknown constant vector different from zero}, \end{cases} \tag{5.19}$$

$$\varepsilon(k) \sim \mathcal{N}(0, \Sigma_\varepsilon(k)), \Sigma_\varepsilon(k) > 0, \mathbb{E}\left[\varepsilon(i)\varepsilon^T(j)\right] = \begin{cases} 0, i > j+1, \\ \Phi_\varepsilon, i = j+1, \quad j \leq i = 0, \ldots k. \\ \Sigma_\varepsilon, i = j, \end{cases} \tag{5.20}$$

It is supposed that Σ_ε, Φ_ε are known. Our task is to optimally detect the fault vector using available data $y(k), k = 0, 1, \ldots$. Note that this is a special case of the above discussed fault detection problem, namely, $x = y$. Therefore, for our detection purpose, we determine the LMS estimate of $y(k)$, which is equal to $\hat{y}(k\,|k-1)$ and applied for building innovation and residual

$$e(k) = r(k) = y(k) - \hat{y}(k\,|k-1).$$

To this end, we first apply (5.5), which leads to, for $k > 0$,

$$\hat{y}(k\,|k-1) = \mathbb{E}\left[y(k)y_{k-1}^T\right]\Sigma_{y_{k-1}}^{-1}y_{k-1},$$
$$y_{k-1} = \begin{bmatrix} y(0) \\ \vdots \\ y(k-1) \end{bmatrix}, \Sigma_{y_{k-1}} = \mathbb{E}\left[y_{k-1}y_{k-1}^T\right].$$

Considering that

$$\mathbb{E}\left[y(k)y_{k-1}^T\right] = \left[0 \cdots 0 \; \mathbb{E}\left[\varepsilon(k)\varepsilon^T(k-1)\right]\right] = \left[0 \cdots 0 \; \Phi_\varepsilon\right],$$
$$\mathbb{E}\left[y_{k-1}y_{k-1}^T\right] = \begin{bmatrix} \mathbb{E}\left[y_{k-2}y_{k-2}^T\right] & \mathbb{E}\left[y_{k-2}y^T(k-1)\right] \\ \mathbb{E}\left[y(k-1)y_{k-2}^T\right] & \mathbb{E}\left[y(k-1)y^T(k-1)\right] \end{bmatrix} = \begin{bmatrix} \Sigma_{y_{k-2}} & \mathbb{I}^T \\ \mathbb{I} & \Sigma_\varepsilon \end{bmatrix},$$
$$\mathbb{I} = \left[0 \cdots 0 \; \Phi_\varepsilon\right],$$
$$\begin{bmatrix} \Sigma_{y_{k-2}} & \mathbb{I}^T \\ \mathbb{I} & \Sigma_\varepsilon \end{bmatrix}^{-1} = \begin{bmatrix} * & * \\ -\Delta\mathbb{I}\Sigma_{y_{k-2}}^{-1} & \Delta^{-1} \end{bmatrix}, \Delta = \Sigma_\varepsilon - \mathbb{I}\Sigma_{y_{k-2}}^{-1}\mathbb{I}^T,$$

with $*$ denoting some matrix of no interest, it holds,

$$\begin{aligned} &\mathbb{E}\left[y(k)y_{k-1}^T\right]\Sigma_{y_{k-1}}^{\;1} = \Phi_\varepsilon\left[-\Delta\mathbb{I}\Sigma_{y_{k-2}}^{-1} \; \Delta^{-1}\right] \Longrightarrow \\ &\hat{y}(k\,|k-1) = \Phi_\varepsilon\left[-\Delta\mathbb{I}\Sigma_{y_{k-2}}^{-1} \; \Delta^{-1}\right]y_{k-1}. \end{aligned} \tag{5.21}$$

Alternatively, by means of (5.16) we have

$$\hat{y}(k\,|k-1) = \sum_{i=0}^{k-1}\mathbb{E}\left[y(k)e^T(i)\right]\Sigma_e^{-1}(i)e(i).$$

Since

$$\forall i < k-1, \mathbb{E}\left[y(k)e^T(i)\right] = \mathbb{E}\left[y(k)\left(y(i) - \hat{y}(i\,|i-1)\right)^T\right] = 0,$$

it turns out

$$\begin{aligned} \hat{y}(k\,|k-1) &= \mathbb{E}\left[y(k)e^T(k-1)\right]\Sigma_e^{-1}(k-1)e(k-1) \\ &= \Phi_\varepsilon\Sigma_e^{-1}(k-1)e(k-1), \end{aligned} \tag{5.22}$$

which is obviously easier for online implementation than the algorithm (5.21), as far as $\Sigma_e(k-1)$ can be determined in a recursive manner. Note that $\Sigma_e^{-1}(k-1)$ is also necessary for building the χ^2 test statistic for the detection purpose. To this end, write

$$e(k) = y(k) - \hat{y}(k\,|k-1) = \varepsilon(k) - \Phi_\varepsilon \Sigma_e^{-1}(k-1)e(k-1),$$

which yields

$$\mathbb{E}\left[e(k)e^T(k)\right] = \Sigma_e(k) = \Sigma_\varepsilon - \Phi_\varepsilon \Sigma_e^{-1}(k-1)\Phi_\varepsilon^T. \tag{5.23}$$

As a result, the following algorithm is capable for an optimal fault detection:

- set the initial value, for $k = 0$:

$$\hat{y}(0) := 0 \Longrightarrow \Sigma_e(0) := \mathbb{E}\left[e(0)e^T(0)\right] = \Sigma_\varepsilon;$$

- recursively compute, for $k > 0$:

$$\begin{aligned} \Sigma_e(k) &= \Sigma_\varepsilon - \Phi_\varepsilon \Sigma_e^{-1}(k-1)\Phi_\varepsilon^T, \\ \hat{y}(k+1\,|k) &= \Phi_\varepsilon \Sigma_e^{-1}(k)e(k); \end{aligned}$$

- online compute

$$J = e(k)^T \Sigma_e^{-1}(k)e(k) \sim \chi^2(q);$$

- make decision, for given FAR α,

$$\begin{cases} J \le J_{th} \Longrightarrow \text{fault-free,} \\ J > J_{th} \Longrightarrow \text{faulty,} \end{cases} \quad J_{th} = \chi_\alpha^2(q).$$

This example clearly illustrates the role of the innovation series in the recursive implementation of LMS estimation and residual generation.

5.1.4 *Analogous Estimation Problems in Krein Space*

As a basis for our study in Sect. 5.3, the LMS estimation and basic fault detection problems presented in the previous subsections will be analogously formulated in Krein space. With the aid of the results introduced in Chap. 4 and knowledge of Krein space [3], serving as a tool for the necessary mathematical handling, we will present the analogous solutions shortly and without mathematical details.

Consider vectors $\mathbf{x}$ and $\mathbf{y}$ in Krein space with Gramian matrices

$$\langle \mathbf{x}, \mathbf{x} \rangle = R_{\mathbf{x}} > 0, \langle \mathbf{y}, \mathbf{y} \rangle = R_{\mathbf{y}} > 0, \langle \mathbf{x}, \mathbf{y} \rangle = R_{\mathbf{xy}} \neq 0.$$

A least squares (LS) estimate of $\mathbf{x}$ by means of $\mathbf{y}$, denoted by $\hat{\mathbf{x}}_{LS}$, is given by

$$\hat{\mathbf{x}}_{LS} = R_{\mathbf{xy}} R_{\mathbf{y}}^{-1} \mathbf{y}, \tag{5.24}$$

and the estimation error Gramian is

$$\left\langle \mathbf{x} - \hat{\mathbf{x}}_{LS}, \mathbf{x} - \hat{\mathbf{x}}_{LS} \right\rangle = R_{\mathbf{x}} - R_{\mathbf{xy}} R_{\mathbf{y}}^{-1} R_{\mathbf{xy}}^T > 0. \tag{5.25}$$

Substituting vector $\mathbf{y}$ by $\mathbf{y}_N$,

$$\mathbf{y}_N = \begin{bmatrix} \mathbf{y}(0) \\ \vdots \\ \mathbf{y}(N) \end{bmatrix}$$

with

$$\left\langle \mathbf{y}_N, \mathbf{y}_N \right\rangle = R_{\mathbf{y}_N} = \left[\langle \mathbf{y}(i), \mathbf{y}(j) \rangle \right]_{i,j=0,\ldots,N} > 0, \left\langle \mathbf{x}, \mathbf{y}_N \right\rangle = R_{\mathbf{xy}_N} \neq 0,$$

the LS estimate of $\mathbf{x}$ by means of $\mathbf{y}_N$,

$$\hat{\mathbf{x}}_{LS} = R_{\mathbf{xy}_N} R_{\mathbf{y}_N}^{-1} \mathbf{y}_N,$$

can be equivalently written as

$$\hat{\mathbf{x}}_{LS} = \sum_{i=0}^{N} R_{\mathbf{xe}}(i) R_{\mathbf{e}}^{-1}(i) \mathbf{e}(i). \tag{5.26}$$

Here, $\mathbf{e}(i), i = 0, \ldots, N$, are an innovation series in Krein space with Gramian

$$\langle \mathbf{e}(i), \mathbf{e}(j) \rangle = R_{\mathbf{e}}(i) \delta_{ij}, R_{\mathbf{e}}(i) > 0, i, j = 0, \ldots, N,$$

and $R_{\mathbf{xe}}(i)$ is the cross-Gramian

$$R_{\mathbf{xe}}(i) = \langle \mathbf{x}, \mathbf{e}(i) \rangle, i = 0, \ldots, N.$$

As a result, we are able to estimate $\mathbf{x}$ recursively and optimally according to

$$\hat{\mathbf{x}}_{LS}(k) = \hat{\mathbf{x}}_{LS}(k-1) + R_{\mathbf{xe}}(k) R_{\mathbf{e}}^{-1}(k) \mathbf{e}(k), \hat{\mathbf{x}}_{LS}(0) = 0,$$

when $\mathbf{y}(k), i = 0, 1, \ldots,$ are available. Define the residual vector in Krein space,

$$\mathbf{r}(k) = \mathbf{x}(k) - \hat{\mathbf{x}}_{LS}(k).$$

It holds

$$R_{\mathbf{r}}(k) := \langle \mathbf{x}(k), \mathbf{x}(k) \rangle = R_{\mathbf{r}}(k-1) - R_{\mathbf{xe}}(k) R_{\mathbf{e}}^{-1}(k) R_{\mathbf{xe}}^{T}(k). \tag{5.27}$$

In Sect. 5.3, the applications of Krein space LS estimation to LDTV systems and further to solutions of optimal fault detection issues will be delineated.

At the end of this section, we would like to highlight the major results. There is no doubt that LMS estimation is in the focus of this section. By means of this concept, we are able to give new interpretations of established fault detection schemes for static processes, for instance, the CCA method. In the subsequent sections, LMS estimation concept will be applied to dynamic systems, the major attention of our investigation. Associated with LMS estimation, we have introduced the concept innovation, which is essential for realizing recursive computations of an LMS estimate. On the other hand, the information aspect of innovation series should be emphasized. In this context, an innovation series is an optimal residual signal, based on which best fault detection performance can be achieved.

5.2 Optimal Detection of Faults in Stochastic Systems

In this section, we would like to investigate optimal fault detection issues for LDTV systems with process and measurement noises. We will address the issues along the lines introduced in the previous section with the concepts LMS estimation and innovation in focus.

5.2.1 *Problem Formulation*

Consider following LDTV systems,

$$\begin{cases} x(k+1) = A(k)x(k) + B(k)u(k) + w(k), \\ y(k) = C(k)x(k) + D(k)u(k) + v(k), \\ x(0) = x_0, \end{cases} \tag{5.28}$$

where $x(k) \in \mathbb{R}^n$, $u(k) \in \mathbb{R}^p$, $y(k) \in \mathbb{R}^q$ are the state, control input, measurement output vectors, respectively. $\omega(k) \in \mathbb{R}^n$, $v(k) \in \mathbb{R}^q$, and $x(0)$ are (unknown) process noise, measurement noise and initial state vectors, and satisfy

$$w(k) \sim \mathcal{N}(0, \Sigma_w(k)), v(k) \sim \mathcal{N}(0, \Sigma_v(k)), x(0) \sim \mathcal{N}(0, \Pi_0), \tag{5.29}$$

$$\mathbb{E}\left[\begin{bmatrix} w(i) \\ v(i) \\ x(0) \end{bmatrix} \begin{bmatrix} w(j) \\ v(j) \\ x(0) \end{bmatrix}^T\right] = \begin{bmatrix} \begin{bmatrix} \Sigma_w(i) & S(i) \\ S^T(i) & \Sigma_v(i) \end{bmatrix} \delta_{ij} & 0 \\ 0 & \Pi_0 \end{bmatrix}. \tag{5.30}$$

It is further assumed that $u(k) \in l_{2,[0,N]}$ is a deterministic signal. Matrices $A(k)$, $B(k)$, $C(k)$, $D(k)$ as well as $S(i)$, $\Sigma_w(i)$, $\Sigma_v(i)$, Π_0 are known and of appropriate dimensions. The faults to be detected are modelled by an unknown vector $f(k) \in \mathbb{R}^l$, $f(k) \in l_{2,[0,N]}$, whose influence on the system dynamics is described by

$$\begin{cases} x(k+1) = A(k)x(k) + B(k)u(k) + B_f(k)f(k) + w(k), \\ y(k) = C(k)x(k) + D(k)u(k) + D_f(k)f(k) + v(k), \\ x(0) = x_0 \end{cases} \tag{5.31}$$

with known matrices $B_f(k)$ and $D_f(k)$. For our purpose, model (5.31) is further written into

$$\begin{cases} x(k+1) = A(k)x(k) + B(k)u(k) + w(k), \\ y(k) = C(k)x(k) + D(k)u(k) + v(k) + \bar{f}(k), \end{cases} \tag{5.32}$$

$$\begin{cases} x_f(k+1) = A(k)x_f(k) + B_f(k)f(k), x_f(0) = 0, \\ \bar{f}(k) = C(k)x_f(k) + D_f(k)f(k), \\ f(k) = 0, \text{ fault-free, otherwise, faulty.} \end{cases} \tag{5.33}$$

We assume that the output of the fault model (5.33) $\bar{f}(k)$ spans the whole q-dimensional measurement space.

Our optimal fault detection problem is formulated as: given $u(i), y(i), i = 0, 1, \ldots, k$, design an LDTV fault detection system that delivers the maximal fault detectability for a given upper-bound of FAR α. In order to achieve an early and reliable fault detection, it is required that

- a detection (running the detection logic) is to be performed at each sampling instant k with the new data $u(k), y(k), k = 0, 1, \ldots$, aiming at detecting $\bar{f}(k)$ as given in (5.32),
- the online computation cost should be as low as possible.

It is well-known that the Kalman filter-based fault detection system is a standard scheme for dealing with fault detection in stochastic systems [5, 7]. In this section, we will delineate that this scheme solves our optimal fault detection problem formulated above.

5.2.2 Kalman Filter-Based Optimal Solution

To begin with, we assume, for the sake of simplicity, that $u(k) = 0, k = 0, 1, \ldots$. This allows us to re-formulate our optimal fault detection problem in terms of the LMS estimation-based fault detection presented in the previous section as follows:

- find a (recursive) LMS estimate for $y(k)$ using the data up to $k-1$, i.e. using $y(i), i = 0, 1, \ldots, k-1$;

- build the residual (innovation series) vector and, based on it,
- form the test statistic and set threshold.

Note that, different from the detection problem addressed in the previous section, on the one hand the measurement vector $y(k)$ is a (linear) function of the (unknown) state vector $x(k)$, on the other hand the state space model is available. In the remaining part of this section, the following notations are adopted:

$$r(k) = y(k) - \hat{y}(k\,|k-1),\, e(k) = x(k) - \hat{x}(k\,|k-1)$$

represent the residual vector and state estimation error vector, respectively. Here, $\hat{y}(k\,|k-1)$ is the LMS estimate of $y(k)$ using $y(i), i = 0, 1, \ldots, k-1$. Hence, $\hat{y}(k\,|k-1)$ is a projection of $y(k)$ onto $\{y(i), i = 0, 1, \ldots, k-1\}$ and also called one-step ahead prediction. Consequently, $r(k)$ is an innovation series as well. Similarly, $\hat{x}(k\,|k-1)$ denotes the the LMS estimate of $x(k)$ by means of measurement data $y(i), i = 0, 1, \ldots, k-1$.

Applying the LMS estimation formulas (5.16)–(5.17) to our case yields

$$\begin{aligned}\hat{y}(k\,|k-1) &= \sum_{i=0}^{k-1} \mathbb{E}\left[y(k)\, r^T(i)\right] \Sigma_r^{-1}(i) r(i)\\ &= C(k) \sum_{i=0}^{k-1} \mathbb{E}\left[x(k) r^T(i)\right] \Sigma_r^{-1}(i) r(i) = C(k)\hat{x}(k\,|k-1),\\ \hat{x}(k+1\,|k) &= \sum_{i=0}^{k} \mathbb{E}\left[x(k+1)\, r^T(i)\right] \Sigma_r^{-1}(i) r(i)\\ &= A(k) \sum_{i=0}^{k} \mathbb{E}\left[x(k)\, r^T(i)\right] \Sigma_r^{-1}(i) r(i) + \sum_{i=0}^{k} \mathbb{E}\left[w(k)\, r^T(i)\right] \Sigma_r^{-1}(i) r(i)\\ &= A(k)\hat{x}(k\,|k-1) + \left(\mathbb{E}\left[w(k)\, r(k)^T\right] + \mathbb{E}\left[x(k)\, r^T(k)\right]\right) \Sigma_r^{-1}(k) r(k).\end{aligned}$$

The results in the first and last steps are due to the whiteness of $w(k)$ and $v(k)$ so that they are uncorrelated with $r(i), i = 0, 1, \ldots, k-1$. Here,

$$\Sigma_r(i) = \mathbb{E}\left[r(i) r^T(i)\right].$$

Since

$$\begin{gathered}\mathbb{E}\left[w(k)\, r^T(k)\right] = \mathbb{E}\left[w(k)\, v^T(k)\right] = S(k),\\ r(k) = y(k) - \hat{y}(k\,|k-1) = C(k) e(k) + v(k) \Longrightarrow\\ \Sigma_r(k) = \mathbb{E}\left[r(k) r^T(k)\right] = C(k)\mathbb{E}\left[e(k) e^T(k)\right] C^T(k) + \Sigma_v(k),\\ \mathbb{E}\left[x(k)\, r^T(k)\right] = \mathbb{E}\left[x(k)\, e^T(k)\right] C^T(k) = \mathbb{E}\left[e(k)\, e^T(k)\right] C^T(k),\end{gathered}$$

the remaining task is to determine $\mathbb{E}\left[e(k)\,e(k)^T\right]$. Consider the state-space model of $e(k)$,

$$x(k+1)-\hat{x}(k+1\,|k)=A(k)\left(x(k)-\hat{x}(k\,|k-1)\right)+w(k)-L(k)r(k) \Longleftrightarrow$$
$$e(k+1) = A(k)e(k) + w(k) - L(k)C(k)e(k) - L(k)v(k), \tag{5.34}$$
$$L(k) = \left(\mathbb{E}\left[w(k)\,r^T(k)\right] + \mathbb{E}\left[x(k)\,r^T(k)\right]\right)\Sigma_r^{-1}(k)$$
$$= \left(S(k) + \mathbb{E}\left[e(k)\,e^T(k)\right]C^T(k)\right)\left(C(k)\mathbb{E}\left[e(k)e^T(k)\right]C^T(k) + \Sigma_v(k)\right)^{-1}.$$

It turns out,

$$\Sigma_e(k) := \mathbb{E}\left[e(k)\,e^T(k)\right],\, A_L(k) := A(k) - L(k)C(k),\, \bar{w}(k) := w(k) - L(k)v(k),$$
$$\Sigma_e(k+1) = \mathbb{E}\left[\left(A_L(k)e(k) + \bar{w}(k)\right)\left(A_L(k)e(k) + \bar{w}(k)\right)^T\right]$$
$$= A_L(k)\Sigma_e(k)A_L^T(k) + \Sigma_w(k) - L(k)\left(C(k)\Sigma_e(k)C^T(k) + \Sigma_v(k)\right)L^T(k),$$
$$L(k) = \left(S(k) + \Sigma_e(k)C^T(k)\right)\Sigma_r^{-1}(k),$$
$$\Sigma_r(k) = C(k)\Sigma_e(k)C^T(k) + \Sigma_v(k) \Longrightarrow \tag{5.35}$$
$$\Sigma_e(k+1) = A(k)\Sigma_e(k)A^T(k) + \Sigma_w(k) - L(k)\Sigma_r(k)L^T(k). \tag{5.36}$$

The last equation is the well-known Riccati recursion. In summary, we have the well-known Kalman filter algorithm:

- set the initial values, for $k = 0$:

$$\hat{x}(0) := 0,\, \hat{y}(0) = 0,\, r(0) = y(0),\, \Sigma_e(0) = \Pi_0;$$

- recursively compute, for $k > 0$, $\Sigma_r(k)$, $L(k)$ according to (5.35) and

$$\hat{y}(k\,|k-1) = C(k)\hat{x}(k\,|k-1),\, r(k) = y(k) - \hat{y}(k\,|k-1); \tag{5.37}$$

- update, for $k+1$, $\Sigma_e(k+1)$ according to (5.36) and

$$\hat{x}(k+1\,|k) = A(k)\hat{x}(k\,|k-1) + L(k)r(k).$$

In case of faulty operations, the dynamics of the residual vector, considering the system model (5.32)–(5.33), is governed by

$$r(k) = y(k) - \hat{y}(k\,|k-1) = C(k)e(k) + v(k) + \bar{f}(k).$$

Since $C(k)e(k) + v(k)$ is an innovation series, detecting fault vector $f(k)$ becomes the basic fault detection problem addressed in the last section, i.e. detecting $\bar{f}(k)$ in a static process described by $C(k)e(k) + v(k)$. Therefore, an optimal fault detection is achieved by computing

$$J = r(k)^T \Sigma_r^{-1}(k) r(k) \sim \chi^2(q)$$

and running the detection logic

$$\begin{cases} J \leq J_{th} \Longrightarrow \text{fault-free}, \\ J > J_{th} \Longrightarrow \text{faulty}, \end{cases} \quad J_{th} = \chi_\alpha^2(q).$$

Next, we consider the general case with $u(k) \neq 0$. It is clear that the Kalman filter-based residual generator is then extended to

$$\hat{x}(k+1|k) = A(k)\hat{x}(k|k-1) + B(k)u(k) + L(k)r(k), \tag{5.38}$$

$$\hat{y}(k|k-1) = C(k)\hat{x}(k|k-1) + D(k)u(k) \tag{5.39}$$

with $L(k)$ given in (5.35). Note that, in the fault-free case, the dynamics of the residual vector $r(k)$ is governed by

$$e(k+1) = A_L(k)e(k) + w(k) - L(k)v(k),$$

and thus independent of $u(k)$ and identical with (5.34). This implies that the fault detection system (5.38)–(5.39) is equivalent with Kalman filter-based solution (5.37).

5.2.3 An Alternative Interpretation of Kalman Filter-Based Solution

A Kalman filter is known for its optimal (in sense of LMS) estimate of the state vector, which results in the minimum variance of the residual (innovation) vector. Notice that the process noise $w(k)$ in the state equation in (5.28) is unknown. It is of interest to check the LMS estimate of $w(k)$ using the measurement data $y(i), i = 0, 1, \ldots, k$, denoted by $\hat{w}(k|k)$. It follows from the LMS estimation formulas (5.16)–(5.17) that

$$\begin{aligned} \hat{w}(k|k) &= \sum_{i=0}^{k} \mathbb{E}\left[w(k) r^T(i)\right] \Sigma_r^{-1}(i) r(i) = \mathbb{E}\left[w(k) r^T(k)\right] \Sigma_r^{-1}(k) r(k) \\ &= S(k)\Sigma_r^{-1}(k) r(k) =: L_w(k) r(k). \end{aligned} \tag{5.40}$$

This result allows us to re-write Kalman filter (5.38) as

$$\hat{x}(k+1|k) = A(k)\hat{x}(k|k-1) + B(k)u(k) + \hat{w}(k|k) + L_x(k)r(k), \tag{5.41}$$

$$L_x(k) = \Sigma_e(k)C^T(k)\Sigma_r^{-1}(k), \tag{5.42}$$

which can be interpreted as: a Kalman filter consists of

- an LMS estimate of $x(k)$ based on the process model without process noise $w(k)$, i.e.
$$\begin{cases} x(k+1) = A(k)x(k) + B(k)u(k), \\ y(k) = C(k)x(k) + D(k)u(k) + v(k), \end{cases}$$

- an LMS estimate of $w(k)$ based on a process model without input $u(k)$,
$$\begin{cases} x(k+1) = A(k)x(k) + w(k), \\ y(k) = C(k)x(k) + v(k). \end{cases}$$

In case that $S(k) = 0$, it is possible to achieve an LMS estimate of $w(k)$ by means of the measurement data $y(i), i = 0, 1, \ldots, k+1$, a smoothing problem and denoted by $\hat{w}(k\,|k+1)$. It holds

$$\begin{aligned} \hat{w}(k\,|k+1) &= \sum_{i=0}^{k+1} \mathbb{E}\left[w(k)\,r^T(i)\right] \Sigma_r^{-1}(i)r(i) \\ &= \mathbb{E}\left[w(k)\,r^T(k+1)\right] \Sigma_r^{-1}(k+1)r(k+1) \\ &= \Sigma_w(k)C^T(k+1)\Sigma_r^{-1}(k+1)r(k+1). \end{aligned} \tag{5.43}$$

For our purpose of fault diagnosis, this interpretation of Kalman filter is useful. For instance, when $w(k)$ represents a fault vector, Eq. (5.40) or (5.43) can be adopted as an optimal fault estimate.

5.3 H_2-optimal Detection of Faults in LDTV Systems: A Krein Space Solution

In our previous study, fault detection issues are addressed in the statistic and stochastic system settings with process and measurement noises. In an analogous manner, we will, in this section, investigate optimal fault detection issues for LDTV systems with unknown inputs. To this end, Krein space based methods [3] introduced in Chap. 4 and Sect. 5.1.4 are applied. Due to the similar mathematical handlings with the Kalman filter-based algorithms, we will briefly outline the major estimation and detection algorithms, and pay more attention to the interpretations. It is noteworthy that the main results presented in this section are known. The reader can, for instance, refer to [7], in which a different solution procedure is followed.

5.3.1 Problem Formulation

Analogous to the system model (5.31), we consider following LDTV systems,

$$\begin{cases} x(k+1) = A(k)x(k) + B_d(k)d(k) + B_f(k)f(k), \\ y(k) = C(k)x(k) + D_d d(k) + D_f(k)f(k), \end{cases} \tag{5.44}$$

$$B_d(k) = \begin{bmatrix} B_w(k) \ 0 \end{bmatrix}, D_d(k) = \begin{bmatrix} 0 \ I \end{bmatrix}, d(k) = \begin{bmatrix} w(k) \\ v(k) \end{bmatrix},$$

where $B_d(k)d(k)$ and $D_d(k)d(k)$ model the influences of unknown inputs on the system dynamics with known matrix $B_w(k)$ and $l_{2,[0,N]}$ bounded unknown input vectors $w(k) \in \mathbb{R}^m$, $v(k) \in \mathbb{R}^q$. Corresponding to it, we have the following Krein space system model,

$$\begin{cases} \mathbf{x}(k+1) = A(k)\mathbf{x}(k) + B_d(k)\mathbf{d}(k) + B_f(k)\mathbf{f}(k), \ \mathbf{x}(0) = \mathbf{0}, \\ \mathbf{y}(k) = C(k)\mathbf{x}(k) + D_d(k)\mathbf{d}(k) + D_f(k)\mathbf{f}(k) \end{cases} \tag{5.45}$$

with vectors $\mathbf{x}(k)$, $\mathbf{y}(k)$, $\mathbf{d}(k)$, $\mathbf{f}(k)$ in Krein space. It is assumed that the Gramian of $\mathbf{d}(k)$ is subject to

$$\langle \mathbf{d}(i), \mathbf{d}(j) \rangle = I\delta_{ij}, i, j = 0, 1, \ldots. \tag{5.46}$$

Remark 5.3 For the sake of simplicity and without loss of generality, the system input vector $u(k)$ and the initial uncertainty are not considered in our subsequent work.

Adopting the same notations used in the previous section for fault-free dynamics, we define the residual vector and LS estimate of the output as well as the state estimation error vectors in Krein space as follows:

$$\mathbf{r}(k) = \mathbf{y}(k) - \hat{\mathbf{y}}(k\,|k-1), \mathbf{e}(k) = \mathbf{x}(k) - \hat{\mathbf{x}}(k\,|k-1).$$

In order to deal with fault detection issues, the corresponding observer-based residual generator in the conventional data space is firstly determined with

$$r(k) = y(k) - \hat{y}(k\,|k-1) \in \mathbb{R}^q$$

as the residual vector. Different from the detection scheme for stochastic systems introduced in the last section, the residual evaluation function will be defined by

$$J(k) = \sum_{i=k}^{N+k} r^T(i)W_r(i)r(i), \tag{5.47}$$

where $N \geq 0$ is some integer and $W_r(i)$ is a weighting matrix to be determined. For the threshold setting, we follow the following widely adopted strategy for systems with deterministic disturbances [5, 7],

$$J_{th} = \sup_d J(k), \tag{5.48}$$

on the assumption

$$\forall k, \sup_d \sum_{i=k}^{N+k} d^T(i)d(i) \leq \delta_d. \tag{5.49}$$

5.3.2 An H_2-Optimal Observer-Based Residual Generator and Fault Detection Scheme

By means of LS estimation formula (5.26), we have

$$\begin{aligned}
\hat{\mathbf{y}}(k\,|k-1) &= \sum_{i=0}^{k-1} R_{\mathbf{yr}}(i)R_{\mathbf{r}}^{-1}(i)\mathbf{r}(i) = C(k)\sum_{i=0}^{k-1} R_{\mathbf{xr}}(i)R_{\mathbf{r}}^{-1}(i)\mathbf{r}(i) \\
&= C(k)\hat{\mathbf{x}}(k\,|k-1), \\
R_{\mathbf{r}}(i) &:= \langle \mathbf{r}(i), \mathbf{r}(i)\rangle, \; R_{\mathbf{xr}}(i) := \langle \mathbf{x}(i), \mathbf{r}(i)\rangle, \; i = 0, \ldots, k-1, \\
\hat{\mathbf{x}}(k+1\,|k) &= A(k)\hat{\mathbf{x}}(k\,|k-1) + \left(B_d(k)D_d^T(k) + R_{\mathbf{xr}}(k)\right) R_{\mathbf{r}}^{-1}(k)\mathbf{r}(k).
\end{aligned}$$

It turns out,

$$\begin{aligned}
\mathbf{e}(k+1) &= A(k)\mathbf{e}(k) + (B_d(k) - L(k)D_d(k))\,\mathbf{d}(k) - L(k)C(k)\mathbf{e}(k) \\
&= (A(k) - L(k)C(k))\,\mathbf{e}(k) + (B_d(k) - L(k)D_d(k))\,\mathbf{d}(k), \\
L(k) &= \left(B_d(k)D_d^T(k) + A(k)R_{\mathbf{e}}(k)C^T(k)\right) R_{\mathbf{r}}^{-1}(k) \\
&= A(k)R_{\mathbf{e}}(k)C^T(k)R_{\mathbf{r}}^{-1}(k),
\end{aligned} \tag{5.50}$$

and based on it,

$$\begin{aligned}
R_{\mathbf{e}}(k) &:= \langle \mathbf{e}(k), \mathbf{e}(k)\rangle, \\
R_{\mathbf{e}}(k+1) &= A_L(k)R_{\mathbf{e}}(k)A_L^T(k) + (B_d(k) - L(k)D_d(k))\,(B_d(k) - L(k)D_d(k))^T, \\
R_{\mathbf{r}}(k) &= C(k)R_{\mathbf{e}}(k)C^T(k) + I \Longrightarrow
\end{aligned} \tag{5.51}$$

$$R_{\mathbf{e}}(k+1) = A(k)R_{\mathbf{e}}(k)A^T(k) + B_w(k)B_w^T(k) - L(k)R_{\mathbf{r}}(k)L^T(k), \; R_{\mathbf{e}}(0) = 0. \tag{5.52}$$

By means of (5.50), (5.51) and (5.52), we are able to compute $L(k)$ as an observer gain matrix recursively, which is proven to be identical with the one for conventional LS observer [3] and thus adopted for the realization of the following LDTV observer-based residual generator,

$$\hat{x}(k+1\,|k) = A(k)\hat{x}(k\,|k-1) + L(k)r(k),\ \hat{x}(0) = 0, \tag{5.53}$$

$$r(k) = y(k) - \hat{y}(k\,|k-1),\ \hat{y}(k\,|k-1) = C(k)\hat{x}(k\,|k-1). \tag{5.54}$$

It is apparent that the influence of the unknown input vector $d(k)$ on the residual vector $r(k)$ should be found, in order to realize the detection algorithm described by (5.47)–(5.49). To this end, we introduce the results given in Theorem 3.4.2 by [3].

Theorem 5.1 *Given system model*

$$\begin{cases} x(k+1)=A(k)x(k)+B_d(k)d(k)=A(k)x(k)+B_w(k)w(k),\ x(0)=0, \\ y(k) = C(k)x(k) + D_d(k)d(k) = C(k)x(k) + v(k), \end{cases} \tag{5.55}$$

and residual generator (5.53)–(5.54), it holds

$$\sum_{i=0}^{N} \left(\hat{w}^T(i\,|N)\hat{w}(i\,|N) + \hat{v}^T(i\,|N)\hat{v}(i\,|N)\right) = \sum_{i=0}^{N} r^T(i)\,R_{\mathbf{r}}^{-1}(i)r(i), \tag{5.56}$$

where $\hat{w}(i\,|N)$ and $\hat{v}(i\,|N)$, $i = 0, \ldots, N$, are the LS estimates for $w(i)$ and $v(i)$ subject to (5.55) and satisfying

$$\hat{w}(i\,|N) = \hat{w}(i\,|N-1)+B_w^T(i)\Phi^T(N, i+1)C^T(N)R_{\mathbf{r}}^{-1}(N)r(N),\ \hat{w}(i\,|i) = 0, \tag{5.57}$$

$$\Phi(N, i+1) = \prod_{j=i+1}^{N-1} A_L(j),$$

$$\hat{v}(i\,|N) = R_{\mathbf{r}}^{-1}(i)r(i) + \sum_{j=i+1}^{N} R_{\mathbf{vr}}(j)R_{\mathbf{r}}^{-1}(j)r(j). \tag{5.58}$$

Remark 5.4 We would like to mention that the formulation in the above theorem is slightly different from the original one given in [3] to fit our problem setting. The LS estimation for $\hat{v}(i\,|N)$ is achieved according to (5.26) as follows,

$$\hat{\mathbf{v}}(i\,|N) = \sum_{j=i}^{N} R_{\mathbf{vr}}(j)R_{\mathbf{r}}^{-1}(j)\mathbf{r}(j) = R_{\mathbf{r}}^{-1}(i)\mathbf{r}(i) + \sum_{j=i+1}^{N} R_{\mathbf{vr}}(j)R_{\mathbf{r}}^{-1}(j)\mathbf{r}(j).$$

It follows from the above theorem that

$$\forall k \geq 0, \sum_{i=k}^{N+k} r^T(i) R_{\mathbf{r}}^{-1}(i) r(i) = \sum_{i=k}^{N+k} \hat{d}^T(i)\hat{d}(i), \hat{d}(i) = \begin{bmatrix} \hat{w}(i\,|N+k) \\ \hat{v}(i\,|N+k) \end{bmatrix}.$$

Since $\hat{d}(i)$ is an LS estimate of $d(i)$ subject to (5.55), we have

$$\sup_d \sum_{i=k}^{N+k} r^T(i) R_{\mathbf{r}}^{-1}(i) r(i) \leq \delta_d.$$

Consequently, the detection algorithm (5.47)–(5.49) can be realized as follows:

- run the observer-based residual generator (5.53)–(5.54);
- compute the residual evaluation function

$$J(k) = \sum_{i=k}^{N+k} r^T(i) W_r(i) r(i),\ W_r(i) = R_{\mathbf{r}}^{-1}(i);$$

- perform detection logic

$$\begin{cases} J \leq J_{th} \Longrightarrow \text{fault-free}, \\ J > J_{th} \Longrightarrow \text{faulty}, \end{cases} \quad J_{th} = \delta_d.$$

5.3.3 A Discussion

As the so-called unified solution [7], the optimal residual generator (5.53)–(5.54) has been alternatively derived by means of a system co-inner-outer factorization, which reveals an important aspect of optimal residual generation and threshold setting in an integrated framework. In the sequel, we will briefly discuss about our solution, in particular relation (5.56), in the context of a system co-inner-outer factorization.

At first, we are going to find an alternative expression for $\hat{w}(i\,|N),\ i = 0, \ldots, N$, given in (5.56). Note that (5.57) holds for any $N, i, N > i, i \geq 0$, and can be re-written as

$$\begin{aligned}
\hat{w}(i\,|N) &= \sum_{j=i+1}^{N} B_w^T(i) \Phi^T(j, i+1) C^T(j) R_{\mathbf{r}}^{-1}(j) r(j) \\
&= B_w^T(i) \left(C^T(i+1) R_{\mathbf{r}}^{-1}(i+1) r(i+1) + \sum_{j=i+2}^{N} \Phi^T(j, i+1) C^T(j) R_{\mathbf{r}}^{-1}(j) r(j) \right) \\
&= B_w^T(i) \Big(C^T(i+1) R_{\mathbf{r}}^{-1}(i+1) r(i+1) \\
&\qquad + A_L^T(i+1) \sum_{j=i+2}^{N} \Phi^T(j, i+2) C^T(j) R_{\mathbf{r}}^{-1}(j) r(j) \Big).
\end{aligned} \tag{5.59}$$

Define a vector $\xi(k) \in \mathbb{R}^n$, $k = 1, \ldots, N-1$,

$$\xi(k) = \sum_{j=k}^{N} \Phi^T(j,k) C^T(j) R_{\mathbf{r}}^{-1}(j) r(j).$$

Equation (5.59) is equivalent to

$$\begin{cases} \xi(k) = A_L^T(k)\xi(k+1) + \bar{C}^T(k)\bar{r}(k+1), \xi(N) = 0, \\ \bar{w}(k+1) = B_w^T(k)\xi(k+1), k = 0, \ldots, N-1, \end{cases} \tag{5.60}$$
$$\bar{C}(k) = R_{\mathbf{r}}^{-1/2}(k), \bar{r}(k+1) = R_{\mathbf{r}}^{-1/2}(k) r(k), \bar{w}(k+1) = \hat{w}(k\,|N).$$

Next, the LS estimation for $\hat{v}(i\,|N)$ given by (5.58) is analogously re-written into a space representation form as follows,

$$\begin{cases} \xi(k) = A_L^T(k)\xi(k+1) + \bar{C}^T(k)\bar{r}(k+1), \xi(N) = 0, \\ \bar{v}(k+1) = \hat{v}(k\,|N) = R_{\mathbf{r}}^{-1/2}(k)\bar{r}(k+1) - L^T(k)\xi(k+1), \\ k = 0, \ldots, N-1 \end{cases} \tag{5.61}$$

with $L(k)$ being the observer gain matrix (5.50). Summarizing (5.60) and (5.61) yields

$$\begin{cases} \xi(k) = A_L^T(k)\xi(k+1) + \bar{C}^T(k)\bar{r}(k+1), \xi(N) = 0, \\ \bar{d}(k+1) = \begin{bmatrix} \bar{w}(k+1) \\ \bar{v}(k+1) \end{bmatrix} = \begin{bmatrix} B_w^T(k) \\ -L^T(k) \end{bmatrix} \xi(k+1) + \begin{bmatrix} 0 \\ R_{\mathbf{r}}^{-1/2}(k) \end{bmatrix} \bar{r}(k+1), \\ k = 0, \ldots, N-1, \bar{d}(N+1) = \begin{bmatrix} 0 \\ \bar{v}(N+1) \end{bmatrix} = \begin{bmatrix} 0 \\ R_{\mathbf{r}}^{-1/2}(k) r(N) \end{bmatrix}. \end{cases} \tag{5.62}$$

Recall that the dynamics of the (normalized) residual generator is governed by

$$\begin{cases} e(k+1) = A_L(k)e(k) + (B_d(k) - L(k)D_d(k))\, d(k) \\ \qquad\qquad = A_L(k)e(k) + B_w(k)w(k) - L(k)v(k), \\ r_N(k) = R_{\mathbf{r}}^{-1/2}(k) r(k) = R_{\mathbf{r}}^{-1/2}(k)\left(C(k)e(k) + D_d(k)d(k)\right) \\ \qquad\quad = \bar{C}(k)e(k) + R_{\mathbf{r}}^{-1/2}(k)v(k), \\ d(k) = \begin{bmatrix} w(k) \\ v(k) \end{bmatrix}. \end{cases} \tag{5.63}$$

It is known that the LS estimator (5.62) is a dual system of the residual generator (5.63) [4]. Furthermore, following Theorem 5.1 it holds

$$
\begin{aligned}
\left\|\bar{d}(k+1)\right\|_{2,[0,N]}^{2} &= \sum_{i=0}^{N} \bar{d}^{T}(i+1)\bar{d}(i+1) \\
&= \sum_{i=0}^{N} \left(\hat{w}^{T}(i\,|N)\hat{w}(i\,|N) + \hat{v}^{T}(i\,|N)\hat{v}(i\,|N)\right) \\
&= \|r_N(k)\|_{2,[0,N]}^{2} = \sum_{i=0}^{N} r^{T}(i)\, R_{\mathbf{r}}^{-1}(i) r(i).
\end{aligned} \tag{5.64}
$$

In other words, the $l_{2,[0,N]}$ norm of the (normalized) residual is equal to the one of the output (estimate for d) of the dual system. This interesting result motivates us to delineate another aspect of the LS estimation-based fault detection scheme.

To simplify our discussion, we introduce some notations and definitions. Let $\mathcal{H}_{yu,[0.N]}$ be a linear operator that maps $l_{2,[0,N]}$ input signal u to $l_{2,[0,N]}$ output signal y, $\mathcal{H}_{yu,[0.N]} : l_{2,[0,N]} \to l_{2,[0,N]}$. The adjoint of $\mathcal{H}_{yu,[0.N]}$ is denoted by $\mathcal{H}_{yu,[0.N]}^{*}$, $\mathcal{H}_{yu,[0.N]}^{*} : l_{2,[0,N]} \to l_{2,[0,N]}$. As described in [7], the residual generator (5.63) can be represented by an operator $\mathcal{H}_{r_N d,[0.N]}$, whose adjoint is the dual system (5.62), $\mathcal{H}_{\bar{d}r_N,[0.N]}^{*}$. Now, consider relation (5.64), which can be, in the framework of operator theory [1, 2, 4], equivalently written as

$$
\left\langle \mathcal{H}_{\bar{d}r_N,[0.N]}^{*} r_N, \bar{d} \right\rangle = \left\langle r_N, \mathcal{H}_{r_N d,[0.N]} d \right\rangle, \tag{5.65}
$$

where $\langle \cdot, \cdot \rangle$ denotes the inner product of two vectors in the $l_{2,[0,N]}$ Hilbert space. Equation (5.65) implies that $\mathcal{H}_{r_N d,[0.N]}$ is a co-inner [7] and it holds

$$
\mathcal{H}_{r_N d,[0.N]} \circ \mathcal{H}_{\bar{d}r_N,[0.N]}^{*} = \mathcal{I} \tag{5.66}
$$

with $\mathcal{I}$ being a unit operator. As a conclusion of the above discussion, we claim, without a rigorous mathematical proof, that the LS (or LMS) estimation problem and its solution can be described as follows: given a linear operator $\mathcal{S}_{yu}$ defined in a Hilbert space endowed with an inner product $\langle \cdot, \cdot \rangle$ and representing a co-inner system

$$
y = \mathcal{S}_{yu} u,
$$

an LS/LMS estimate of u by y is given by

$$
\hat{u} - \mathcal{S}_{yu}^{*} y.
$$

Here, $\mathcal{S}_{yu}^{*}$ is the adjoint of $\mathcal{S}_{yu}$ and satisfies

$$
\left\langle \mathcal{S}_{yu}^{*} y, \hat{u} \right\rangle = \left\langle y, \mathcal{S}_{yu} u \right\rangle.
$$

In order to understand the result that the estimator (5.62) is a co-inner, we would like to outline a schematic procedure for finding the co-inner $\mathcal{H}_{\bar{d}r_N,[0.N]}$, which gives a system theoretical interpretation of the LS estimation. Suppose that the system model (5.44) is represented by

$$y(k) = \mathcal{S}_{yd,[0.N]}d(k)$$

with a linear operator $\mathcal{S}_{yd,[0.N]} : l_{2,[0,N]} \to l_{2,[0,N]}$. By means of a co-inner-outer factorization, $\mathcal{S}_{yd,[0.N]}$ can be factorized into

$$\mathcal{S}_{yd,[0.N]} = \mathcal{H}_{co,[0.N]} \circ \mathcal{H}_{ci,[0.N]},$$

where $\mathcal{H}_{co,[0.N]}$ is a co-outer whose inverse $\mathcal{H}_{co,[0.N]}^{-1}$ is a stable and causal system and $\mathcal{H}_{ci,[0.N]}$ is co-inner [7]. Let

$$\mathcal{R}_{r_N y,[0.N]} = \mathcal{H}_{co,[0.N]}^{-1}$$

be a post-filter and used for residual generation as follows,

$$r_N(k) = \mathcal{R}_{r_N y,[0.N]}y(k).$$

As a result, we have

$$r_N(k) = \mathcal{R}_{r_N y,[0.N]}y(k) = \mathcal{H}_{ci,[0.N]}d(k),$$

for which it holds, due to the property of a co-inner,

$$\left\|\bar{d}(k+1)\right\|_{2,[0,N]}^2 = \left\langle \mathcal{H}_{ci,[0.N]}^* r_N, \bar{d}\right\rangle = \left\langle r_N, \mathcal{H}_{ci,[0.N]}d\right\rangle = \|r_N(k)\|_{2,[0,N]}^2 .$$

That is the result (5.56) given in Theorem 5.1. In a nutshell, the LS estimation problem can be alternatively solved by finding the inverse of the system co-outer [7].

5.4 Concluding Remarks

Although we have presented the major algorithms of H_2-optimal fault detection schemes for LDTV systems with stochastic noises and deterministic unknown disturbances, our focus in this chapter is on introducing the ideas, concepts and some mathematical and control theoretical tools for deriving the solutions. To this end, we have begun with a basic and simple fault detection problem, detection of a fault in a static process corrupted with noise. It is demonstrated that the fault detection performance can be improved when the uncertainty in the process caused by the noise is reduced. For this purpose, additional measurements are utilized for an LMS estimate of the process variable under consideration, and the estimation error, whose variance is at minimum, is then used to build the test statistic. This procedure leads to the optimal detection performance. The idea behind this handling is the use of correlations between the process variables and measurement to suppress the uncertainty, a well-known method in MVA, namely the CCA algorithm.

An extension of this idea to dynamic processes is straightforward, when the consecutive measurement data are appropriately utilized. On the one hand, new measurement data imply additional information useful for improving the LMS estimation. On the other hand, repeated computations of the LMS estimate are computationally demanding, in particular when the data are collected over a large time interval. To solve this problem, the concept of innovation is introduced, which is the estimation error of the process measurement variables and their LMS estimate. Innovations are white noise series. From the information point of view, innovation series contains the exclusive information in the current process measurement variables, which does not exist in the past measurement data. This property enables a recursive computation of LMS estimation of the (unknown) process variables. For the realization, the projection method serves as an efficient tool. The application of this idea to an LDTV system with process and measurement noises leads to the well-known Kalman filter algorithm with the innovation as the residual signal. Thanks to its whiteness property, innovation is then adopted to build the test statistic, based on which the detection decision is made.

In order to extend the LMS estimation-based design scheme to LDTV systems with (deterministic) unknown disturbances, the well-established Krein space framework and its applications to estimation issues [3] are applied as a mathematical tool. The centerpiece of this extension is the interpretation of Gramian of a Krein space vector (signal) as uncertainty in the sense of variance for stochastic systems. In this framework, both the concepts of LMS (LS for deterministic systems) estimation and innovation series as well as the projection method can be fully adopted. A further important result is Theorem 5.1 proven in [3], which establishes the relation between the $l_{2,[0,N]}$-norm of the innovation series and the one of the LS estimation of the unknown inputs. On this basis, the residual evaluation function, i.e. the normalized $l_{2,[0,N]}$-norm of the innovation, and the upper-bound of the unknown inputs as the threshold can be determined.

At the end of this chapter, we have revealed the equivalence between the LS-based optimal fault detection solution and the known solution achieved in the co-inner-outer framework [7]. It is delineated that

- the dual system of the LS/LMS estimation-based residual generator delivers the LS/LMS estimate of the process uncertainty like the unknown inputs,
- the LS/LMS estimation-based residual generator is a co-inner, which explains the optimal setting of the threshold equal to the upper-bound of the unknown inputs,
- the LS/LMS estimation-based residual generator and the LS/LMS estimator for the process uncertainty can be systematically designed in the co-inner-outer factorization framework, where the LS/LMS estimator is the adjoint of the co-inner, namely the LS/LMS estimation-based residual generator.

These results unify the LS/LMS estimation-based optimal fault detection schemes and the unified solution of optimal fault detection filters [5, 7] and give deep insights of the H_2-optimal fault detection schemes from two important aspects: the LS/LMS estimation and the system factorization views.

References

1. Feintuch, A. (1998). *Robust control theory in Hilbert space*. New York, USA: Springer.
2. Francis, B. A. (1987). *A course in H-infinity control theory*. Berlin-New York: Springer.
3. Hassibi, B. Sayed, A. H., & Kailath, T. (1999). *Indefinite-quadratic estimation and control: A unified approach to H_2 and H_∞ theories*. SIAM.
4. Callier, F. M., & Desoer, C. A. (1991). *Linear system theory*. New York, USA: Springer.
5. Ding, S. X. (2013). *Model-based fault diagnosis techniques: Design schemes, algorithms, and tools* (2nd ed.). London, U.K.: Springer.
6. Ding, S. X. (2014). *Data-driven design of fault diagnosis and fault-tolerant control systems*. London, U.K.: Springer.
7. Ding, S. X. (2020). *Advanced methods for fault diagnosis and fault-tolerant control*. Berlin, Germany: Springer.
8. Härdle, W. K., & Simar, L. (2012). *Applied multivariate statistical analysis* (3rd ed.). Berlin Heidelberg: Springer.
9. Chen, Z., Ding, S. X., Zhang, K., et al. (2016). Canonical correlation analysis-based fault detection methods with application to alumina evaporation process. *Control Engineering Practice, 46*, 51–58.

Chapter 6
Optimal Fault Detection for LDTV Systems

In this chapter, we deal with observer-based optimal FD issues for LDTV systems in the sense of finite horizon H_∞/H_∞ or H_-/H_∞-optimization. By generating a residual using an FDF, the design of the FDF is expressed as maximizing the finite horizon H_∞/H_∞ or H_-/H_∞ index. A unified optimal solution is developed by solving a discrete-time Riccati equation. It is proven that the presented FD approach can be applied for LDTV systems subject to l_2-norm bounded unknown inputs or stochastic noise sequences.

6.1 Introduction

It is well-known that the core of observer based FD is the generation of residual signals which should be robust to unknown inputs and, at the same time, sensitive to faults. If a complete unknown input decoupling is impossible, one can evaluate the worst case robustness of residual to unknown inputs with H_∞ index, while measure the fault sensitivity in best case with H_∞ index and the H_- index in worst case. In this way, a residual generator can be designed maximizing the H_∞/H_∞ or H_-/H_∞ performance index, towards a trade-off between the robustness and fault sensitivity indices. For LTI systems with l_2-norm bounded unknown inputs and faults, a unified solution to the H_∞/H_∞- and H_-/H_∞-optimal FD has been given in [9]. The results were further extended to linear continuous time-varying systems [5], for example, in [10] Li and Zhou proposed a unified optimal FDF to robust FD under different performance indices H_-/H_∞, H_2/H_∞ and H_∞/H_∞.

In this chapter, an optimal FDF-based FD method for LDTV systems is developed in the sense of H_∞/H_∞- and H_-/H_∞-optimization. A unified solution to H_∞/H_∞- and H_-/H_∞-FDFs is developed by recursively computing discrete-time

M. Zhong et al., *Fault Diagnosis for Linear Discrete Time-Varying Systems and Its Applications*, https://doi.org/10.1007/978-981-19-5438-2_6

Riccati equations. Furthermore, we will show that the optimal solution is not unique and it is applicable for LDTV systems subject to either l_2-norm bounded unknown inputs or stochastic noise sequences.

6.2 Problem Formulation

Consider the following LDTV system

$$\begin{cases} x(k+1) = A(k)x(k) + B(k)u(k) + B_d(k)d(k) + B_f(k)f(k), \; x(0) = x_0, \\ y(k) = C(k)x(k) + D_d(k)d(k) + D_f(k)f(k), \end{cases} \tag{6.1}$$

where $x(k) \in \mathbb{R}^n$, $u(k) \in \mathbb{R}^p$, $y(k) \in \mathbb{R}^q$, $d(k) \in \mathbb{R}^m$, and $f(k) \in \mathbb{R}^l$ are the state, system input, measurement output, unknown input vectors, and fault vector to be detected, respectively, $A(k)$, $B(k)$, $B_d(k)$, $B_f(k)$, $C(k)$, $D_d(k)$ and $D_f(k)$ are known matrices with appropriate dimensions. Assumptions 3.1 and 3.2 are supposed to be satisfied.

Remembering the core role of residual generation in FD, the following FDF is constructed for system (6.1)

$$\begin{cases} \hat{x}(k+1) = A(k)\hat{x}(k) + B(k)u(k) + L(k)(y(k) - C(k)\hat{x}(k)), \\ \varepsilon(k) = y(k) - C(k)\hat{x}(k), \; \hat{x}(0) = \hat{x}_0, \\ r(k) = W(k)\varepsilon(k), \end{cases} \tag{6.2}$$

where $\hat{x}(k)$ is the estimate of $x(k)$, $r(k)$ is the residual, $\hat{x}_0$ is a guess of initial state, $L(k)$ is the observer gain matrix and $W(k)$ is the so-called post-filter.

Let $e(k) = x(k) - \hat{x}(k)$. It follows from (6.1)–(6.2) that

$$\begin{cases} e(k+1) = (A(k) - L(k)C(k))e(k) + (B_d(k) - L(k)D_d(k))d(k), \\ \qquad\qquad + (B_f(k) - L(k)D_f(k))f(k), \; e(0) = x_0 - \hat{x}_0, \\ \varepsilon(k) = C(k)e(k) + D_d(k)d(k) + D_f(k)f(k), \\ r(k) = W(k)\varepsilon(k). \end{cases} \tag{6.3}$$

Refer to (3.27) for the state transition matrix $\Phi_L(k, i)$ of system (6.3), i.e.,

$$\Phi_L(k, i) = \prod_{j=i}^{k-1}(A(j) - L(j)C(j)), \; \Phi_L(k, k) = I, \; i \le k - 1.$$

which under Assumption 3.1, guarantees the stability of the FDF (6.3). Recall the notations

$$
\begin{aligned}
d_k &= \left[e^T(0),\, d^T(0),\, d^T(1),\, \ldots,\, d^T(k)\right]^T, \\
f_k &= \left[f^T(0),\, f^T(1),\, \ldots,\, f^T(k)\right]^T, \\
g_{\varepsilon d}(k) &= \left[\, g_e(k,0)\; g_d(k,0)\; \cdots\; g_d(k,k-1)\; D_d(k)\,\right], \\
g_{\varepsilon f}(k) &= \left[\, g_f(k,0)\; \cdots\; g_f(k,k-1)\; D_f(k)\,\right], \\
g_e(k,0) &= C(k)\Phi_L(k,0), \\
g_d(k,i) &= C(k)\Phi_L(k,i+1)(B_d(i) - L(i)D_d(i)), \\
g_f(k,i) &= C(k)\Phi_L(k,i+1)(B_f(i) - L(i)D_f(i)).
\end{aligned}
$$

The dynamics of residual generator (6.2) is obtained as

$$
\begin{aligned}
r(k) &= r_d(k) + r_f(k), \\
r_d(k) &= W(k)g_{\varepsilon d}(k)d_k, \\
r_f(k) &= W(k)g_{\varepsilon f}(k)f_k.
\end{aligned}
$$

The design of an FDF-based residual generator thus lies in the design of the gain matrix $L(k)$ and the post-filter $W(k)$.

To evaluate the robustness of residual to unknown input and the fault sensitivity, we define

$$||G_{rd}||_{\infty,[0,k]} = \sup_{d \in l_{2,[0,k]}} \frac{\sum_{i=0}^{k} ||r_d(i)||^2}{||e(0)||^2 + \sum_{i=0}^{k} ||d(i)||^2}, \tag{6.4}$$

$$||G_{rf}||_{\infty,[0,k]} = \sup_{f \in l_{2,[0,k]}} \frac{\sum_{i=0}^{k} ||r_f(i)||^2}{\sum_{i=0}^{k} ||f(i)||^2}, \tag{6.5}$$

$$||G_{rf}||_{-,[0,k]} = \inf_{f \in l_{2,[0,k]}} \frac{\sum_{i=0}^{k} ||r_f(i)||^2}{\sum_{i=0}^{k} ||f(i)||^2}. \tag{6.6}$$

The index $||G_{rd}||_{\infty,[0,k]}$ indicates the worst-case robustness of residual to unknown initial state and unknown input, while the sensitivity of residual to fault is measured by the index $||G_{rf}||_{\infty,[0,k]}$ in best case or $||G_{rd}||_{-,[0,k]}$ in worst case. The matrices $L(k)$ and $W(k)$ can then be designed towards maximizing the sensitivity/robustness ratio, i.e., with respect to solving the following maximization problem $\forall k = 1, 2, \ldots, N$

$$\max_{L(k),W(k)} \frac{||G_{rf}||_{\infty,[0,k]}}{||G_{rd}||_{\infty,[0,k]}} \quad \text{or} \quad \max_{L(k),W(k)} \frac{||G_{rf}||_{-,[0,k]}}{||G_{rd}||_{\infty,[0,k]}}.$$

Moreover, let

$$r_{dk} = \begin{bmatrix} r_d(0) \\ r_d(1) \\ \vdots \\ r_d(k) \end{bmatrix}, \; r_{fk} = \begin{bmatrix} r_f(0) \\ r_f(1) \\ \vdots \\ r_f(k) \end{bmatrix}.$$

We have

$$r_{dk} = H_{rd}(k)d_k,\ r_{fk} = H_{rf}(k)f_k,$$

where

$$H_{rd}(k) = \begin{bmatrix} \bar{g}_e(0,0) & \bar{D}_d(0) & 0 & \cdots & \cdots & 0 \\ \bar{g}_e(1,0) & \bar{g}_d(1,0) & \bar{D}_d(1) & 0 & \cdots & 0 \\ \vdots & \vdots & \ddots & \ddots & \ddots & 0 \\ \bar{g}_e(k,0) & \bar{g}_d(k,0) & \bar{g}_d(k,1) & \cdots & \bar{g}_d(k,k-1) & \bar{D}_d(k) \end{bmatrix},$$

$$H_{rf}(k) = \begin{bmatrix} \bar{D}_f(0) & 0 & \cdots & \cdots & 0 \\ \bar{g}_f(1,0) & \bar{D}_f(1) & 0 & \cdots & 0 \\ \vdots & \ddots & \ddots & \ddots & 0 \\ \bar{g}_f(k,0) & \bar{g}_f(k,1) & \cdots & \bar{g}_f(k,k-1) & \bar{D}_f(k) \end{bmatrix},$$

$$\begin{aligned} \bar{g}_e(i,0) &= W(i)g_e(i,0),\ \bar{g}_d(i,j) = W(i)g_d(i,j), \\ \bar{D}_d(i) &= W(i)D_d(i),\ \bar{g}_f(i,j) = W(i)g_f(i,j), \\ \bar{D}_f(i) &= W(i)D_f(i),\ i \le k,\ j \le i-1. \end{aligned}$$

It follows from (6.4)–(6.6) that

$$\begin{aligned} &||G_{rd}||_{\infty,[0,k]} = ||H_{rd}(k)||_2,\ ||G_{rf}||_{\infty,[0,k]} = ||H_{rf}(k)||_2, \\ &||G_{rf}||_{-,[0,k]} = ||H_{rf}(k)||_- = \underline{\sigma}(H_{rf}(k)). \end{aligned}$$

where $\underline{\sigma}(H_{rf}(k))$ denotes the minimum singular value of matrix $H_{rf}(k)$.

Now we are in the position to formulate the underlying problems as follows.

Problem 6.1 Given residual generator (6.2) for system (6.1), find $L(k)$ ensuring the stability of FDF (6.2) and $W(k)$ such that

$$\max_{L(k),W(k)} \frac{\|H_{rf}(k)\|_2}{\|H_{rd}(k)\|_2}. \tag{6.7}$$

Problem 6.2 Given residual generator (6.2) for system (6.1), find $L(k)$ ensuring the stability of FDF (6.2) and $W(k)$ such that

$$\max_{L(k),W(k)} \frac{\|H_{rf}(k)\|_-}{\|H_{rd}(k)\|_2}. \tag{6.8}$$

Remark 6.1 In case that $d(k)$ and $f(k)$ are stochastic sequences, we have

$$\begin{aligned}
\|r_{dk}\|_E &\leq ||H_{rd}(k)||_2 \cdot \|d_k\|_E, \\
\|r_{fk}\|_E &\leq ||H_{rf}(k)||_2 \cdot \|f_k\|_E, \\
\|r_{fk}\|_E &\geq ||H_{rf}(k)||_- \cdot \|f_k\|_E,
\end{aligned}$$

where $\|(\cdot)\|_E$ denotes the mean value of $\|(\cdot)\|$. So, $\|H_{rd}(k)\|_2$ and $\|H_{rf}(k)\|_2$ (or $\|H_{rd}(k)\|_-$) are still reasonable indices to evaluate the robustness of residual to unknown input and the sensitivity of residual to fault, respectively. That means the H_∞/H_∞- and H_-/H_∞-FDF problem formulations in Problems 6.1 and 6.2 can be applied to FD for LDTV systems subject to either l_2-norm bounded unknown inputs or stochastic noise sequences.

Remark 6.2 It is easy to see that the defined H_∞/H_∞ and H_-/H_∞ FD performance indices are equivalent to the ones in [11], but the problem formulations are different. Especially, if system (6.1) is LTI, then for $k \to \infty$, the $||G_{rd}||_{\infty,[0,k]}$ with $e(0) = 0$ is the H_∞ norm of transfer function matrix from d to r, the $||G_{rf}||_{\infty,[0,k]}$ and $||G_{rf}||_{-,[0,k]}$ are the H_∞ norm and smallest singular value of transfer function matrix from f to r, respectively.

6.3 Optimal FDFs for Fault Detection

The following Lemma 6.1 plays an important role in deriving the major solution.

Lemma 6.1 *Consider the following residual generators*

$$\begin{cases} \hat{x}^i(k+1) = A(k)\hat{x}^i(k) + L_i(k)(y(k) - C(k)\hat{x}^i(k)) + B(k)u(k), \\ \varepsilon^i(k) = y(k) - C(k)\hat{x}^i(k), \;\; \hat{x}^i(0) = \hat{x}_0, \end{cases} \tag{6.9}$$

where $L_i(k)$ $(i = 1, 2)$ are observer gain matrices such that $A(k) - L_i(k)C(k)$ are exponentially stable. Then

$$\varepsilon^2(k) = \mathcal{Q}(k)\varepsilon^1(k), \tag{6.10}$$

where $\mathcal{Q}(k) : \nu(k) \to \varepsilon^{\mathcal{Q}}(k)$ is an operator described by

$$\begin{cases} \eta(k+1) = (A(k) - L_2(k)C(k))\eta(k) + (L_1(k) - L_2(k))\nu(k), \\ \varepsilon^{\mathcal{Q}}(k) = C(k)\eta(k) + \nu(k), \;\; \eta(0) = 0. \end{cases} \tag{6.11}$$

Proof Define $e^i(k) = x(k) - \hat{x}^i(k)$. For any given $L_i(k)$ ensuring the exponential stability of $A(k) - L_i(k)C(k)$ $(i = 1, 2)$, it follows from (6.1) and (6.9) that

$$\begin{cases} e^i(k+1) = (A(k) - L_i(k)C(k))e^i(k) + (B_d(k) - L_i(k)D_d(k))d(k), \\ \qquad\qquad +(B_f(k) - L_i(k)D_f(k))f(k), \\ \varepsilon^i(k) = C(k)e^i(k) + D_d(k)d(k) + D_f(k)f(k),\ e^i(0) = x_0 - \hat{x}_0. \end{cases} \tag{6.12}$$

We now study $\mathcal{Q}(k)\varepsilon^1(k)$. Let

$$A_\xi(k) = \begin{bmatrix} A(k) - L_1(k)C(k) & 0 \\ (L_1(k) - L_2(k))C(k) & A(k) - L_2(k)C(k) \end{bmatrix},$$

$$B_{\xi d}(k) = \begin{bmatrix} B_d(k) - L_1(k)D_d(k) \\ (L_1(k) - L_2(k))D_d(k) \end{bmatrix},\ C_\xi(k) = \begin{bmatrix} C^T(k) \\ C^T(k) \end{bmatrix}^T,$$

$$B_{\xi f}(k) = \begin{bmatrix} B_f(k) - L(k)D_f(k) \\ (L_1(k) - L_2(k))D_f(k) \end{bmatrix},\ \xi(k) = \begin{bmatrix} e^1(k) \\ \eta^(k) \end{bmatrix}.$$

It holds

$$\begin{cases} \xi(k+1) = A_\xi(k)\xi(k) + B_{\xi d}(k)d(k) + B_{\xi f}(k)f(k), \\ \mathcal{Q}(k)\varepsilon^1(k) = C_\xi(k)\xi(k) + D_d(k)d(k) + D_f(k)f(k). \end{cases} \tag{6.13}$$

Thus, the state transition matrix of (6.13) is

$$\Phi_\xi(k,i) = A_\xi(k-1)A_\xi(k-2)\cdots A_\xi(i),\ \Phi_\xi(k,k) = I, \tag{6.14}$$

where $i \le k-1$. Furthermore, rewrite $\mathcal{Q}(k)\varepsilon^1(k)$ into

$$\begin{aligned} \mathcal{Q}(k)\varepsilon^1(k) &= \bar{g}_{\xi d}(k)d_k + \bar{g}_{\xi f}(k)f_k\bar{g}_{\xi d}(k) \\ &= \begin{bmatrix} g_\xi(k,0)\ g_{\xi d}(k,0) \cdots g_{\xi d}(k,k-1)\ D_d(k) \end{bmatrix}, \\ \bar{g}_{\xi f}(k) &= \begin{bmatrix} g_{\xi f}(k,0) \cdots g_{\xi f}(k,k-1)\ D_f(k) \end{bmatrix}, \\ g_\xi(k,0) &= C_\xi(k)\Phi_\xi(k,0)\begin{bmatrix} I\ 0 \end{bmatrix}^T, \\ g_{\xi d}(k,i) &= C_\xi(k)\Phi_\xi(k,i+1)B_{\xi d}(i), \\ g_{\xi f}(k,i) &= C_\xi(k)\Phi_\xi(k,i+1)B_{\xi f}(i). \end{aligned}$$

Note that, for $0 \le i \le k-1$,

$$\begin{aligned} &C_\xi(k)\Phi_\xi(k,i) = C(k)\Phi_2(k,i)\begin{bmatrix} I & I \end{bmatrix}, \\ &\Phi_2(k,i) := \Phi_L(k,i)|_{L(i)=L_2(i)}, \\ &\begin{bmatrix} I & I \end{bmatrix} B_{\xi d}(i) = B_d(i) - L_2(i)D_d(i). \end{aligned}$$

Hence,

$$\begin{aligned} &C_\xi(k)\Phi_\xi(k,0)\begin{bmatrix} I \\ 0 \end{bmatrix} = C(k)\Phi_2(k,0), \\ &C_\xi(k)\Phi_\xi(k,i)B_{\xi d}(i) = C(k)\Phi_2(k,i)(B_d(i) - L_2(i)D_d(i)), \\ &C_\xi(k)\Phi_\xi(k,i)B_{\xi f}(i) = C(k)\Phi_2(k,i)(B_f(i) - L_2(i)D_f(i)), \end{aligned}$$

which lead to

$$
\begin{aligned}
g_e^2(k,0) &:= g_e(k,0)|_{L(0)=L_2(0)} = g_\xi(k,0),\\
g_d^2(k,i) &:= g_d(k,i)|_{L(i)=L_2(i)} = g_{\xi d}(k,i),\\
g_f^2(k,i) &:= g_f(k,i)|_{L(i)=L_2(i)} = g_{\xi f}(k,i).
\end{aligned}
$$

Therefore,

$$
g_{\varepsilon d}^2(k) = \bar{g}_{\xi d}(k),\ \ g_{\varepsilon f}^2(k) = \bar{g}_{\xi f}(k),\ \ \varepsilon^2(k) = g_{\varepsilon d}^2(k)d_k + g_{\varepsilon f}^2(k) f_k.
$$

From the above analysis, we have (6.10)–(6.11). □

Next, we study $g_{\varepsilon d}(k)g_{\varepsilon d}^T(k)$, which is needed for the computation of $\|H_{rd}(k)\|_2$. Suppose that $L(k)$ ensures the exponential stability of $A(k) - L(k)C(k)$ and $P(k) \geq 0$ is a solution of the following Riccati equation

$$
\begin{cases}
P(k+1) = (A(k) - L(k)C(k))P(k)(A(k) - L(k)C(k))^T \\
\qquad\qquad +(B_d(k) - L(k)D_d(k))(B_d(k) - L(k)D_d(k))^T, \\
P(0) = I.
\end{cases}
\tag{6.15}
$$

Note that

$$
g_{\varepsilon d}(k)g_{\varepsilon d}^T(k) = g_e(k,0)g_e^T(k,0) + D_d(k)D_d^T(k) + \sum_{i=0}^{k-1} g_d(k,i)g_d^T(k,i).
$$

Moreover, $\Phi_L(k,i) = \Phi_L(k,j)\Phi_L(j,i),\ i \leq j \leq k$. Then we have

$$
\begin{aligned}
P(0) &= \Phi_L(0,0)\Phi_L^T(0,0) = I,\\
g_{\varepsilon d}(0)g_{\varepsilon d}^T(0) &= C(0)P(0)C^T(0) + D_d(0)D_d^T(0),\\
P(1) &= \Phi_L(1,0)\Phi_L^T(1,0) + \Phi_L(1,1)(B_d(0) - L(0)D_d(0))\\
&\quad \times (B_d(0) - L(0)D_d(0))^T\Phi_L^T(1,1),\\
g_{\varepsilon d}(1)g_{\varepsilon d}^T(1) &= C(1)P(1)C^T(1) + D_d(1)D_d^T(1).
\end{aligned}
$$

Suppose that

$$
\begin{aligned}
P(j) &= \Phi_L(j,0)\Phi_L^T(j,0)\\
&\quad + \sum_{i=0}^{j-1} \Phi_L(j,i+1)(B_d(i) - L(i)D_d(i))(B_d(i) - L(i)D_d(i))^T\Phi_L^T(j,i+1),\\
g_{\varepsilon d}(j)g_{\varepsilon d}^T(j) &= C(j)P(j)C^T(j) + D_d(j)D_d^T(j).
\end{aligned}
$$

It is easy to show that

$$
\begin{aligned}
P(j+1) &= \Phi_L(j+1,0)\Phi_L^T(j+1,0) + \sum_{i=0}^{j}\Phi_L(j+1,i+1)(B_d(i)-L(i)D_d(i)) \\
&\quad \times (B_d(i)-L(i)D_d(i))^T\Phi_L^T(j+1,i+1), \\
g_{\varepsilon d}(j+1)g_{\varepsilon d}^T(j+1) &= C(j+1)P(j+1)C^T(j+1) + D_d(j+1)D_d^T(j+1).
\end{aligned}
$$

Using the induction method, we are able to derive

$$
g_{\varepsilon d}(k)g_{\varepsilon d}^T(k) = C(k)P(k)C^T(k) + D_d(k)D_d^T(k).
$$

We now present solutions to optimization problems (6.7) and (6.8). First one is the case of static post-filter $W(k)$.

Theorem 6.1 *Under the assumptions of* $(C(k), A(k))$ *being uniformly detectable and* $(A(k), B_d(k))$ *being uniformly stabilizable,* $L_o(k)$ *and* $W_{p,o}(k)$ *given by*

$$
L_o(k) = (A(k)P_o(k)C^T(k) + B_d(k)D_d^T(k))W_{p,o}^2(k), \tag{6.16}
$$

$$
W_{p,o}(k) = R_d^{-1/2}(k) \tag{6.17}
$$

with

$$
R_d(k) = C(k)P_o(k)C^T(k) + D_d(k)D_d^T(k) > 0
$$

deliver an optimal solution to both Problems 6.1 and 6.2, where $P_o(k) \geq 0$ *is the solution of the following Riccati equation*

$$
\begin{cases}
P_o(k+1) = A(k)P_o(k)A(k) - L_o(k)W_{p,o}^{-2}(k)L_o^T(k) + B_d(k)B_d^T(k), \\
P_o(0) = I.
\end{cases}
$$

Proof Let

$$
\begin{aligned}
L_o(k) &= (A(k)P(k)C^T(k) + B_d(k)D_d^T(k))(C(k)P(k)C^T(k) + D_d(k)D_d^T(k))^{-1}, \\
H_{rd,o}(k) &= H_{rd}(k)|_{L(k)=L_o(k)}, \quad H_{rf,o}(k) = H_{rf}(k)|_{L(k)=L_o(k)}.
\end{aligned}
$$

It follows from (6.15) and (6.16) that

$$
\begin{aligned}
P_o(k+1) &= (A(k)-L_o(k)C(k))\,P_o(k)\,(A(k)-L_o(k)C(k))^T \\
&\quad + (B_d(k)-L_o(k)D_d(k))(B_d(k)-L_o(k)D_d(k))^T \\
&= A(k)P_o(k)A^T(k) - L_o(k)W_{p,o}^{-2}(k)L_o^T(k) + B_d(k)B_d^T(k), \\
g_{\varepsilon d,o}(k)g_{\varepsilon d,o}^T(k) &= C(k)P_o(k)C^T(k) + D_d(k)D_d^T(k).
\end{aligned}
$$

It is well known that, under the assumption of $(C(k), A(k))$ being detectable and $(A(k), B_d(k))$ being stabilizable, $L_o(k)$ is a Kalman filter gain matrix and $A(k) - L_o(k)C(k)$ is exponentially stable (see [1]).

Let

$$\Psi(k) = (A(k) - L_o(k)C(k))P_o(k)C^T(k) + (B_d(k) - L_o(k)D_d(k))D_d^T(k).$$

In light of (6.16)–(6.17), we have $\Psi(k) = 0$. Moreover,

$$\begin{aligned}
W_{p,o}(i)g_{\varepsilon d,o}(i)g_{\varepsilon d,o}^T(i)W_{p,o}^T(i) &= I, \;\; i \le k, \; \forall k = 1, 2, \ldots, N, \\
\begin{bmatrix} g_{\varepsilon d,o}(0) & 0 \end{bmatrix} g_{\varepsilon d,o}^T(j) &= C(0)\Phi_o^T(j,0)\Phi_o^T(j,1)C^T(j) \\
&\quad + D_d(0)(B_d(0) - L_o(0)D_d(0))^T\Phi_o^T(j,1)C^T(j) \\
&= \Psi(0)^T\Phi_o^T(j,1)C^T(j) = 0, \;\; j > 0, \\
\begin{bmatrix} g_{\varepsilon d,o}(i) & 0 \end{bmatrix} g_{\varepsilon d,o}^T(j) &= C(i)\Phi_o(i,0)\Phi_o^T(j,0)\Phi_o^T(j,i+1)C^T(j) \\
&\quad + D_d(i)(B_d(i) - L_o(i)D_d(i))^T\Phi_o^T(j,i+1)C^T(j) \\
&\quad + \sum_{t=0}^{i-1} C(i)\Phi_o(i,t+1)(B_d(t) - L_o(t)D_d(t)) \\
&\quad \times (B_d(t) - L_o(t)D_d(t))^T\Phi_o^T(j,t+1)C^T(j) \\
&= \Psi(i)^T\Phi_o^T(j,i+1)C^T(j) = 0, \; i \le k, \;\; j > i.
\end{aligned}$$

Thus it is easy to get

$$H_{rd,o}(k)H_{rd,o}^T(k) = I, \tag{6.18}$$

which leads to

$$\frac{\|H_{rf,o}(k)\|_2}{\|H_{rd,o}(k)\|_2} = \|H_{rf,o}(k)\|_2, \;\; \frac{\|H_{rf,o}(k)\|_-}{\|H_{rd,o}(k)\|_2} = \|H_{rf,o}(k)\|_-. \tag{6.19}$$

On the other hand, it follows from Lemma 6.1 that for any $L(k)$ ensuring the exponential stability of $A(k) - L(k)C(k)$ and $\varepsilon(k)$ given in (6.3), we have

$$\varepsilon(k) = \mathcal{Q}_o(k)r_o(k) \;\text{ with }\; r_o(k) = W_{p,o}(k)\varepsilon_o(k),$$

where $\mathcal{Q}_o(k) : r_o(k) \mapsto \varepsilon(k)$ denotes the system

$$\begin{cases} \eta(k+1) = (A(k) - L(k)C(k))\eta(k) + (L_o(k) - L(k))W_{p,o}^{-1}(k)r_o(k), \\ \varepsilon(k) = C(k)\eta(k) + W_{p,o}^{-1}(k)r_o(k), \;\; \eta(0) = 0, \end{cases} \tag{6.20}$$

and $\varepsilon_o(k)$ satisfying

$$\begin{cases} e_o(k+1) = (A(k) - L_o(k)C(k))e_o(k) + (B_d(k) \\ \qquad\qquad -L_o(k)D_d(k))d(k) + (B_f(k) - L_o(k)D_f(k))f(k), \\ \varepsilon_o(k) = C(k)e_o(k) + D_d(k)d(k) + D_f(k)f(k), \\ e_o(0) = x_0 - \hat{x}_0. \end{cases} \tag{6.21}$$

Let

$$r_k = \left[r^T(0),\ r^T(1),\ \ldots,\ r^T(k)\right]^T,\ r_{k,o} = r_k|_{r(k)=r_o(k)}.$$

For any given $L(k)$ and $W(k)$ with appropriate dimensions, it is obtained from (6.3), (6.20) and (6.21) that

$$r_k = H_{rr_o}(k)(H_{rd,o}(k)d_k + H_{rf,o}(k)f_k),$$

where

$$H_{rr_o}(k) = \begin{bmatrix} \bar{W}_{p,o}^{-1}(0) & 0 & \cdots & \cdots & 0 \\ \bar{g}_{r_o}(1,0) & \bar{W}_{p,o}^{-1}(1) & 0 & \cdots & 0 \\ \vdots & \ddots & \ddots & \ddots & 0 \\ \bar{g}_{r_o}(k,0) & \bar{g}_{r_o}(k,1) & \cdots & \bar{g}_{r_o}(k,k-1) & \bar{W}_{p,o}^{-1}(k) \end{bmatrix},$$

$$\bar{g}_{r_o}(i,j) = W(i)C(i)\Phi(i,j+1)(L_o(i) - L(i))W_{p,o}^{-1}(i),$$
$$\bar{W}_{p,o}^{-1}(i) = W(i)W_{p,o}^{-1}(i),\ i \le k,\ j \le i-1.$$

In view of (6.18), we have

$$\begin{aligned} \frac{\|H_{rf}(k)\|_2}{\|H_{rd}(k)\|_2} &= \frac{\|H_{rr_o}(k)H_{rf,o}(k)\|_2}{\|H_{rr_o}(k)H_{rd,o}(k)\|_2} \\ &\le \frac{\|H_{rr_o}(k)\|_2 \cdot \|H_{rf,o}(k)\|_2}{\|H_{rr_o}(k)\|_2} \\ &= \|H_{rf,o}(k)\|_2. \end{aligned} \tag{6.22}$$

Similarly, one can further prove that

$$\begin{aligned} \frac{\|H_{rf}(k)\|_-}{\|H_{rd}(k)\|_2} &= \frac{\|H_{rr_o}(k)H_{rf,o}(k)\|_-}{\|H_{rr_o}(k)H_{rd,o}(k)\|_2} \\ &\le \frac{\|H_{rr_o}(k)\|_2 \cdot \|H_{rf,o}(k)\|_-}{\|H_{rr_o}(k)\|_2} \\ &= \|H_{rf,o}(k)\|_-. \end{aligned} \tag{6.23}$$

In light of (6.19) and (6.22)–(6.23), $\{L_o(k),\ W_{p,o}(k)\}$ is a solution to both (6.7) and (6.8). □

Remark 6.3 The optimal solution given in Theorem 6.1 can also be derived and interrupted based on co-inner-outer factorization technique. The interested reader can refer to [8] for more details.

It should be pointed out that the optimal solution $\{L_o(k), W_{p,o}(k)\}$ to problems (6.7) and (6.8) is not unique and, for any unitary matrix $U(k)$ with appropriate dimensions, the $\{L_o(k), U(k)W_{p,o}(k)\}$ is still an optimal solution to both (6.7) and (6.8). Moreover, in Theorem 6.1, we only focused on the case of static post-filter $W(k)$. As a more generalized extension, the result to the dynamic post-filter case is obtained in the following Theorem 6.2.

Theorem 6.2 *For any given $L(k)$ ensuring the exponential stability of $A(k) - L(k)C(k)$, then $\mathcal{W}_d(k) : \varepsilon(k) \to r(k)$ denotes a corresponding dynamic optimal post-filter*

$$\begin{cases} \eta_o(k+1) = (A(k) - L_o(k)C(k))\eta_o(k) + (L(k) - L_o(k))\varepsilon(k), \\ r(k) = W_{p,o}(k)(C(k)\eta_o(k) + \varepsilon(k)), \ \ \eta_o(0) = 0, \end{cases} \tag{6.24}$$

where $L_o(k)$ and $W_{p,o}(k))$ are given in (6.16)–(6.17).

Proof For any given $L(k)$ ensuring the exponential stability of $A(k) - L(k)C(k)$, it is obtained by using Lemma 6.1 that $\varepsilon_o(k) = \mathcal{Q}_L(k)\varepsilon(k)$, where $\mathcal{Q}_L(k) : \varepsilon(k) \to \varepsilon_o(k)$ denotes the following LDTV system

$$\begin{cases} \eta_o(k+1) = (A(k) - L_o(k)C(k))\eta_o(k) + (L(k) - L_o(k))\varepsilon(k), \\ \varepsilon_o(k) = C(k)\eta_o(k) + \varepsilon(k), \ \ \eta_o(0) = 0. \end{cases}$$

Hence, if a dynamic post-filter is chosen as $\mathcal{W}_d(k) = W_{p,o}(k)\mathcal{Q}_L(k)$, then the generated residual is $r(k) = r_o(k)$, which implies that the $L(k)$ and the dynamic post-filter $\mathcal{W}_d(k)$ in (6.24) lead to an optimal FDF satisfying both (6.7) and (6.8). □

Remark 6.4 The optimal solution solves both the finite horizon H_∞/H_∞- and H_-/H_∞-FDF problems and the optimal value of (6.7) or (6.8) is independent of the choice of observer gain matrix $L(k)$. The optimal static post-filter is a special case of the optimal dynamic post-filter, that is, the one with $L(k) = L_o(k)$. Moreover, it is easy to see that

- if system (6.1) is LTI, then for $k \to \infty$, the optimal solution given in Theorem 6.1 is identical with the one given in [4] for optimal FDF design in discrete LTI systems;
- if system (6.1) is linear time periodic, then the optimal solution given in Theorem 6.1 is identical with the one given by [6] for the periodic FDF design;
- for LDTV system (6.1) with full row rank $D_d(k)$, the optimal FDF with dynamic post-filter $\mathcal{W}_d(k)$ in Theorem 6.2 is the same as the one using the time domain coprime factorization approach in [11].

By generating residual signal using (6.2) with $L(k)$ and $W(k)$ respectively given in (6.16) and (6.17), we now turn to residual evaluation. The following evaluation function is applied to this end

$$J_{2,[0,N]}(r) = \|r(k)\|_{2,[0,N]}^2 = \sum_{j=0}^{N} r^T(j)r(j). \tag{6.25}$$

Given $||d(k)||_{2,[0,N]} \leq \delta_d$, $||e(0)||_2 \leq \delta_e$, we have $||d_k||_{2,[0,N]}^2 \leq \delta_d^2 + \delta_e^2$. Due to (6.18), the threshold with a zero FAR can be set as

$$J_{th} = \delta_d^2 + \delta_e^2. \tag{6.26}$$

It has been proven in [8] that such $\{J_{2,[0,N]}(r),\ J_{th}\}$ solves the optimal FD problem defined in Definition 2.4. For the sake of completeness, the following theorem is given.

Theorem 6.3 (Theorem 7.1 in [8]) *Given LDTV system (6.1), the FDF (6.2) and evaluation function (6.25), then $L_o(k)$ and $W_{p,o}(k)$) given in (6.16)–(6.17) and the threshold in (6.25) solve the optimal FD problem in Definition 2.4.*

Also, we can apply the following evaluation function

$$J_e(k) = \left(\frac{1}{k+1}\sum_{i=0}^{k} \|r(i)\|^2\right)^{\frac{1}{2}} \tag{6.27}$$

which, together with a threshold $J_{th} = \delta_d + \delta_e$, leads to the optimal FD towards maximal fault detectability with a zero false fault alarm, as defined in Definition 2.4.

6.4 A Numerical Example

To illustrate the effectiveness of the above developed optimal FD method, we consider LDTV system (6.1) with the parameter matrices

$$A(k) = \begin{bmatrix} 0.2e^{-\frac{k}{100}+1} & 0.6 & 0 \\ 0 & 0.5 & \sin(k) \\ 0 & 0 & 0.7 \end{bmatrix},\ B_d(k) = \begin{bmatrix} 1.3 \\ 0.5 \\ 0.9^k \end{bmatrix},\ B_f(k) = \begin{bmatrix} 0.7 \\ 0.4 \\ 0.5 \end{bmatrix},$$
$$C(k) = \begin{bmatrix} -0.5 & 1.5 & 0 \end{bmatrix},\ D_d(k) = 0.5.$$

Optimal FDFs for the following two cases are first designed by applying Theorem 6.1 and choose residual evaluation function $J_e(k)$ in (6.27), i.e.,

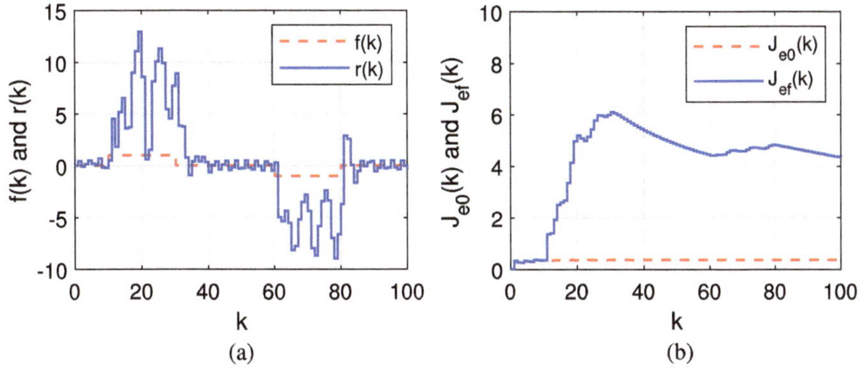

Fig. 6.1 Case 1: **a** the $f(k)$ and $r(k)$; **b** $J_{e0}(k)$ and $J_{ef}(k)$

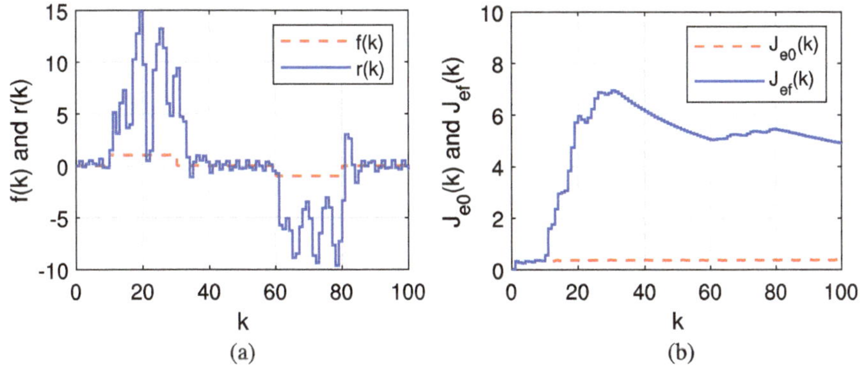

Fig. 6.2 Case 2: **a** the $f(k)$ and $r(k)$; **b** $J_{e0}(k)$ and $J_{ef}(k)$

Case 1: $D_f(k) = 0.5,\ d(k) = 0.5\sin(2k),\ J_{th} = 0.36$;

Case 2: $D_f(k) = 0$, $d(k)$ is white noise, $J_{th} = 1.23$.

Figs. 6.1 to 6.2 show the simulated fault $f(k)$ and the generated residual $r(k)$, the fault-free case residual evaluation function $J_{e0}(k)$ and the faulty case residual evaluation function $J_{ef}(k)$, respectively. It is seen that fault alarms are delivered at $k = 10$ in Case 1 and $k = 11$ in Case 2, respectively.

Next, we use the fault estimation in [3] as a residual and consider the following case, i.e.,

Case 3: $D_f(k) = 0.5,\ d(k) = 0.5\sin(2k),\ J_{th} = 0.005$.

When a fault shown in Fig. 6.1 occurs, Fig. 6.3 show the fault estimation and residual evaluation function, respectively. It is seen that a fault alarm is delivered at $k = 11$. When $D_f(k) = 0$, however, the fault has no influence on residual.

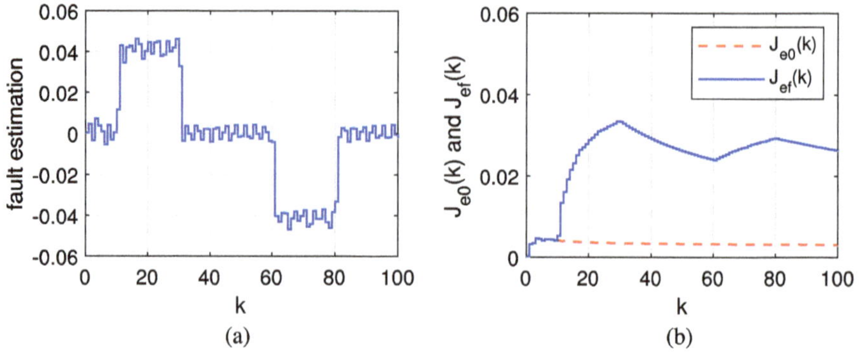

Fig. 6.3 Case 3: **a** the fault estimation; **b** $J_{e0}(k)$ and $J_{ef}(k)$

These simulation results show that the fault can be detected quickly after its occurrence in the three cases. The influence of fault on residual of Case 1 is larger than that of Case 3, while the previous fault estimation result in [3] is not applicable when $D_f(k) = 0$.

6.5 Conclusion

In this chapter, FDF-based FD issues have been addressed for LDTV systems in the context of H_∞/H_∞- and H_-/H_∞-optimization. By introduction finite horizon H_∞ index to measure the robustness of residual to unknown input and the H_∞ or H_- criterion to indicate fault sensitivity, the design of an FDF-based residual generator has been formulated as an H_∞/H_∞- or H_-/H_∞-optimization problem, that concerned with finding an observer gain matrix and a post-filter so as to maximize the sensitivity/robustness ratio criterion. It has shown that the optimal solution to FDF is not unique and a unified solution to the H_∞/H_∞- and H_-/H_∞-FDFs has been obtained by recursively computing Riccati difference equation. The results can be applied to the design of FDFs for LDTV systems subject to either l_2-norm bounded unknown inputs or stochastic noise sequences. A numerical example is finally demonstrated to show the effectiveness of the achieved results.

References

1. Anderson, B., & Moore, J. (1981). Detectability and stabilizability of time-varying discrete-time linear systems. *SIAM Journal of Control and Optimization, 19*, 20–32.
2. Wang, J., Yang, G., & Liu, J. (2007). An LMI approach to H_- index and mixed H_-/H_∞ fault detection observer design. *Automatica, 43*, 1656–1665.

3. Zhong, M., Liu, S., & Zhao, H. (2008). Krein space-based H_∞ fault estimation for linear discrete time-varying systems. *ACTA Automatica Sinica, 34*, 1529–1533.
4. Liu, N., & Zhou, K. (2007). Optimal robust fault detection for linear discrete time systems. In *Proceedings of the 46th IEEE Conference on Decision and Control*, pp. 989–994. New Orleans, USA.
5. Zhang, P., & Ding, S. X. (2004). Observer-based fault detection of linear time-varying systems. *Automatisierungstechniik, 52*, 370–376. (in German).
6. Zhang, P., Ding, S. X., Wang, G., et al. (2005). Fault detection of linear discrete-time periodic systems. *IEEE Transactions on Automatic Control, 50*, 239–244.
7. Zhang, P., Ding, S. X., Wang, G., et al. (2007). Disturbance decoupling in fault detection of linear periodic systems. *Automatica, 43*, 1410–1417.
8. Ding, S. X. (2020). *Advanced methods for fault diagnosis and fault-tolerant control.* Berlin Germany: Springer.
9. Ding, S. X., Jeinsch, T., Frank, P. M., et al. (2000). A unified approach to the optimization of fault detection systems. *International Journal of Adaptive Control and Signal Processing, 14*, 725–745.
10. Li, X., & Zhou, K. (2009). A time domain approach to robust fault detection of linear time-varying systems. *Automatica, 45*, 94–102.
11. Li, X. (2009). *Fault detection filter design for linear systems*, Ph.D. dissertation. Louisiana State University, USA.

Chapter 7
A Projection-Based Method of Fault Detection for LDTV Systems

This chapter confines to a projection-based method for FD in LDTV systems in the sense of maximizing the sensitivity/robustness ratio. The core of the method lies in the construction of a residual generator by making use of an arbitrary linear combination of measurement output estimation error sequences. By proposing Gramian matrix based criteria to measure the influences of an unknown input and a fault on residual, the residual generator design is formulated into a sensitivity/robustness ratio maximization problem. It is shown that the optimal solution is not unique and one of them can be derived by directly applying an orthogonal projection of the measurement output. Moreover, with the aid of innovation analysis, a more generalized residual generator is obtained, whose state space description provides us with an optimal FDF for LDTV systems subject to l_2-norm bounded unknown inputs or stochastic sequences. A numerical example is given at last to show the effectiveness of the method.

7.1 Introduction

For LDTV systems subject to either l_2-norm bounded unknown inputs or stochastic noises, finite horizon H_∞/H_∞- and H_-/H_∞-optimization based FD issues have been widely studied, see e.g., [1–5]. In Chap. 6, we have also demonstrated a unified optimal solution to the H_∞/H_∞- and H_-/H_∞-optimal design of FDFs. We know that most of the existing residual generators for LDTV systems can be constructed by using a weighted output estimation errors. Especially, when an LDTV system with stochastic disturbances is concerned, the residual sequence is indeed a weighted innovation that can be derived via orthogonal projections, as discussed in Chap. 5.

Inspired by the idea of orthogonal projection and innovation analysis, this chapter focuses on a projection-based method for FD in LDTV systems. On the basis of introducing Gramian matrix based criteria to measure the influences of unknown

M. Zhong et al., *Fault Diagnosis for Linear Discrete Time-Varying Systems and Its Applications*, https://doi.org/10.1007/978-981-19-5438-2_7

input and fault on residual, our main attention is paid to addressing the following problems:

- design a residual generator by using an arbitrary linear combination of output estimation error sequences in the sense of maximizing the sensitivity/robustness ratio criterion;
- derive an optimal solution to the formulated maximization problem by means of orthogonal projection and innovation analysis;
- develop a state space description of the optimal solution and, further, give an H_∞/H_∞- and/or H_-/H_∞-FDF for FD purpose.

7.2 A Projection Approach to Residual Generation

Consider LDTV systems described by

$$\begin{cases} x(k+1) = A(k)x(k) + B(k)u(k) + B_d(k)d(k) + B_f(k)f(k), x(0) = x_0, \\ y(k) = C(k)x(k) + D(k)u(k) + D_d(k)d(k) + D_f(k)f(k), \end{cases} \tag{7.1}$$

which can be equally modeled in the following input-output form

$$y_k = H_{y0}(k)x_0 + H_{yu}(k)u_k + H_{yd}(k)d_k + H_{yf}(k)f_k, \tag{7.2}$$

where y_k, u_k, d_k, f_k are given in (3.6), $H_{y0}(k)$, $H_{yu}(k)$, $H_{yd}(k)$ and $H_{yf}(k)$ are referred to (3.8)–(3.11). Assumption 3.1 is supposed to be satisfied. Besides, the initial state $x_0 \in \mathbb{R}^n$, system input $u(k) \in \mathbb{R}^p$, unknown input $d(k) \in \mathbb{R}^m$ and fault $f(k) \in \mathbb{R}^l$ satisfy the following standard assumption.

Assumption 7.1 $u(k)$ is online known and

$$\left\langle \begin{bmatrix} x_0 \\ d(i) \\ f(i) \end{bmatrix}, \begin{bmatrix} x_0 \\ d(j) \\ f(j) \end{bmatrix} \right\rangle = \begin{bmatrix} \Pi_0 & 0 & 0 \\ 0 & R_d(i,j) & 0 \\ 0 & 0 & R_f(i,j) \end{bmatrix}.$$

where $R_d(i,j) = \langle d(i), d(j) \rangle$, $R_f(i,j) = \langle f(i), f(j) \rangle$.

Similar to Definitions 4.2 and 4.3 in Krein spaces, we first introduce the following definitions in Euclidean space for ease of subsequent study.

Definition 7.1 The Gramian of y_k is defined as the $(k+1) \times (k+1)$ block matrix

$$R_{y_k} = [\langle y(i), y(j) \rangle]_{i,j=0:k},$$

where $\langle y(i), y(j) \rangle$ stands for the Gramian of $y(i)$ and $y(j)$, $\langle \alpha(i)y(i), \beta(j)y(j) \rangle = \alpha(i)\langle y(i), y(j) \rangle \beta^T(j)$ for all $\alpha(i)$ and $\beta(j)$ with appropriate dimensions.

Definition 7.2 $\hat{x}$ is defined as the projection of x onto $\mathcal{L}\{\{y(i)\}_{i=0}^{k}\}$ if

$$x = \hat{x} + \tilde{x},$$

where $\hat{x} \in \mathcal{L}\{\{y(i)\}_{i=0}^{k}\}$ and $\tilde{x}$ satisfies the orthogonality condition $\tilde{x} \perp \mathcal{L}\{\{y(i)\}_{i=0}^{k}\}$ in the sense of $\langle \tilde{x}, y(i) \rangle = 0,\ i = 0, 1, \ldots, k$.

For FD purpose, we construct a residual generator of the following form

$$\begin{cases} \hat{y}_k = G(k)y_{k-1} + H(k)u_k, \\ r_k = W(k)(y_k - \hat{y}_k), \end{cases} \tag{7.3}$$

where $\hat{y}(k)$ is the estimate of $y(k)$ at time step k, $r(k)$ is the generated residual, and

$$\hat{y}_k = \left[\hat{y}^T(0)\ \hat{y}^T(1) \cdots \hat{y}^T(k)\right]^T,\ r_k = \left[r^T(0)\ r^T(1) \cdots r^T(k)\right]^T,$$
$$G(k) = \begin{bmatrix} g(0,0) & 0 & \cdots & 0 \\ g(1,0) & 0 & \cdots & 0 \\ \vdots & & \ddots & 0 \\ g(k,0) & \cdots & \cdots & g(k,k-1) \end{bmatrix},\ H(k) = \begin{bmatrix} h(0,0) & 0 & & \cdots & 0 \\ h(1,0) & h(1,1) & 0 & \cdots & 0 \\ \vdots & & & \ddots & 0 \\ h(k,0) & \cdots & & \cdots\cdots & h(k,k) \end{bmatrix},$$
$$W(k) = \begin{bmatrix} w(0,0) & 0 & & \cdots & 0 \\ w(1,0) & w(1,1) & 0 & \cdots & 0 \\ \vdots & & & \ddots & 0 \\ w(k,0) & \cdots & & \cdots\cdots & w(k,k) \end{bmatrix}.$$

From (7.2) and (7.3), we have

$$\begin{cases} \hat{y}_k = G(k)(H_{y0}(k-1)x_0 + H_{yd}(k-1)d_{k-1} + H_{yf}(k-1)f_{k-1}) \\ \qquad + (\left[G(k)H_{yu}(k-1)\ 0\right] + H(k))u_k, \\ r_k = W(k)\left(G_{y0}(k)x_0 + G_{yu}(k)u_k + G_{yd}(k)d_k + G_{yf}(k)f_k\right), \end{cases} \tag{7.4}$$

where

$$\begin{aligned} G_{y0}(k) &= H_{y0}(k) - G(k)H_{y0}(k-1), \\ G_{yu}(k) &= H_{yu}(k) - \left[G(k)H_{yu}(k-1)\ 0\right] - H(k), \\ G_{yd}(k) &= H_{yd}(k) - \left[G(k)H_{yd}(k-1)\ 0\right], \\ G_{yf}(k) &= H_{yf}(k) - \left[G(k)H_{yf}(k-1)\ 0\right]. \end{aligned}$$

Let

$$H(k) = H_{yu}(k) - \left[G(k)H_{yu}(k-1)\ 0\right]. \tag{7.5}$$

We have $G_{yu}(k) = 0$. Thus (7.4) can be rewritten as

$$\begin{cases} \hat{y}_k = G(k)(H_{y0}(k-1)x_0 + H_{yd}(k-1)d_{k-1} \\ \qquad +H_{yf}(k-1)f_{k-1}) + H_{yu}(k)u_k, \\ r_k = W(k)\left(G_{y0}(k)x_0 + G_{yd}(k)d_k + G_{yf}(k)f_k\right). \end{cases} \tag{7.6}$$

It is seen that complete decoupling of control input from the residual is achieved. From here, we shall only deal with the case of $u(k) \equiv 0$ for simplicity.

Introduce

$$\bar{d}_k = \left[x_0^T \;\; d_k^T\right]^T, \; G_{y\bar{d}}(k) = \left[G_{y0}(k) \; G_{yd}(k)\right],$$
$$r_{\bar{d}k} = W(k)G_{y\bar{d}}(k)\bar{d}_k, \; r_{fk} = W(k)G_{yf}(k)f_k.$$

With Assumption 7.1, this immediately gives

$$r_k = r_{\bar{d}k} + r_{fk}, \; R_{r_k} = \langle r_k, r_k\rangle = R_{r_{\bar{d}k}} + R_{r_{fk}}, \tag{7.7}$$

where

$$R_{r_{\bar{d}k}} = \langle r_{\bar{d}k}, r_{\bar{d}k}\rangle = W(k)G_{y_{\bar{d}}}(k)R_{\bar{d}k}G_{y_{\bar{d}}}^T(k)W^T(k), \;\; R_{\bar{d}k} = \langle \bar{d}_k, \bar{d}_k\rangle,$$
$$R_{r_{fk}} = \langle r_{fk}, r_{fk}\rangle = W(k)G_{y_f}(k)R_{fk}G_{y_f}^T(k)W^T(k), \;\; R_{fk} = \langle f_k, f_k\rangle.$$

Thus, the influences of $\bar{d}_k$ and f_k on residual r_k can be represented by $R_{r_{\bar{d}k}}$ and $R_{r_f k}$, respectively. Define

$$J_d(k) = R_{r_{\bar{d}k}}|_{R_{\bar{d}k}=I} = W(k)G_{y_{\bar{d}}}(k)G_{y_{\bar{d}}}^T(k)W^T(k), \tag{7.8}$$
$$J_f(k) = R_{r_{fk}}|_{R_{fk}=I} = W(k)G_{y_f}(k)G_{y_f}^T(k)W^T(k). \tag{7.9}$$

Without loss of generality, we propose to use $\lambda_{max}(J_d(k))$ as a robustness index of the residual to an unknown input, while use $\lambda_{max}(J_f(k))$ or $\lambda_{min}(J_f(k))$ as a sensitivity index of the residual to a fault.

The design of the residual generator (7.3) can then be formulated into a sensitivity/robustness ratio maximization problem, i.e.,

$$\max_{G(k),W(k)} \frac{\lambda_{max}(J_f(k))}{\lambda_{max}(J_d(k))} \tag{7.10}$$

or

$$\max_{G(k),W(k)} \frac{\lambda_{min}(J_f(k))}{\lambda_{max}(J_d(k))}. \tag{7.11}$$

Remark 7.1 Similar to the definition of the Gramian matrix in Krein space, a useful mnemonic device for recalling Definition 7.1 is to think of the $\{y(0), y(1), \ldots, y(k)\}$ as "random variables" and R_{y_k} as the "covariance matrix". With the introduction of the Gramian matrix as a measure of a vector, $J_d(k)$ and $J_f(k)$ can be viewed as a measure of residual generated by unit $\bar{d}_k$ and f_k, respectively.

Remark 7.2 The design objectives in (7.10) and (7.11) are independent of the exact calculation of the Gramian matrix.

Remark 7.3 It is noted that the residual generator (7.3) is constructed by an arbitrary linear combination of output estimation error sequence, which implies that

$$\begin{cases} \hat{y}(k) = \sum\limits_{i=0}^{k-1} g(k,i)y(i) + \sum\limits_{i=0}^{k} h(k,i)u(i), \\ r(k) = \sum\limits_{i=0}^{k} w(k,i)(y(i) - \hat{y}(i)). \end{cases}$$

The residual generator (7.3) with the coefficient structures is not loss of generality. Moreover, such a residual generator is realizable directly by using input/output data. When $d(k)$ and $f(k)$ are l_2-norm bounded and $w(k,i) = 0$ for $i \neq k$, the design objectives (7.10) and (7.11) are equivalent to the H_∞/H_∞ and H_-/H_∞ criteria given in (6.7) and (6.8).

7.3 Innovation Analysis Aided Optimal Solutions

In this section, optimal solutions of $W(k)$ and $\hat{y}(k)$ in residual generator (7.3) are investigated via innovation analysis. To this aim, We shall first discuss here how to conveniently find $W(k)$ and $\hat{y}(k)$ satisfying (7.10) or (7.11), when the innovations $\{\tilde{y}(0), \tilde{y}(1), \ldots, \tilde{y}(k-1)\}$ are given.

Denote by $\mathcal{L}\{\{\theta(k)\}_{k=0}^{N}\}$ the linear subspace spanned by $\{\theta(k)\}_{k=0}^{N}$ for a vector $\theta \in \mathbb{R}^{k_\theta}$. Using the induction method, we are able to derive the following results:

- For $k = 0$, let $\hat{y}(0) = 0$ and $\tilde{y}(0) = y(0)$. We have $\mathcal{L}\{y(0)\} = \mathcal{L}\{\tilde{y}(0)\}$.
- For $k = 1$, let $\hat{y}(1) = g(1,0)y(0) = g(1,0)\tilde{y}(0)$. We have

$$\tilde{y}(1) = y(1) - g(1,0)y(0),\ \ \tilde{y}(1) \in \mathcal{L}\{\{y(i)\}_{i=0}^{1}\},$$
$$y(1) = g(1,0)\tilde{y}(0) + \tilde{y}(1),\ \ y(1) \in \mathcal{L}\{\{y(i)\}_{i=0}^{1}\}.$$

  which implies $\mathcal{L}\{\{y(i)\}_{i=0}^{1}\} = \mathcal{L}\{\{\tilde{y}(i)\}_{i=0}^{1}\}$.
- For $k = j$, it is assumed that $\mathcal{L}\{\{y(i)\}_{i=0}^{j}\} = \mathcal{L}\{\{\tilde{y}(i)\}_{i=0}^{j}\}$.
- For $k = j + 1$, it holds

$$\hat{y}(k) = \sum_{i=0}^{k-1} g(k, i) y(i), \quad \tilde{y}(k) = y(k) - \sum_{i=0}^{k-1} g(k, i) y(i),$$

$$\tilde{y}(j + 1) \in \mathcal{L}\{\{y(i)\}_{i=0}^{j+1}\},$$

$$y(k) = (g(k, 0) + g(k, 1) g(1, 0) + \cdots + g(k, k-1) g(k-1, k-2) \cdots g(1, 0)) \tilde{y}(0),$$
$$+ \cdots + (g(k, k-2) + g(k, k-1) g(k-1, k-2)) \tilde{y}(k-2)$$
$$+ g(k, k-1) \tilde{y}(k-1) + \tilde{y}(k),$$

$$y(j + 1) \in \mathcal{L}\{\{\tilde{y}(i)\}_{i=0}^{j+1}\},$$

$$\mathcal{L}\{\{y(i)\}_{i=0}^{j+1}\} = \mathcal{L}\{\{\tilde{y}(i)\}_{i=0}^{j+1}\}.$$

As a result, we have

$$\mathcal{L}\{\{y(i)\}_{i=0}^{k}\} = \mathcal{L}\{\{\tilde{y}(i)\}_{i=0}^{k}\}, \tag{7.12}$$

$$y_k = G_{y\tilde{y}}(k) \tilde{y}_k, \quad \tilde{y}_k = G_{\tilde{y}y}(k) y_k, \tag{7.13}$$

where

$$G_{y\tilde{y}}(k) = \begin{bmatrix} I & 0 & 0 & \cdots & 0 \\ g_{y\tilde{y}}(1, 0) & I & 0 & \cdots & 0 \\ g_{y\tilde{y}}(2, 0) & g_{y\tilde{y}}(2, 1) & I & \ddots & 0 \\ \vdots & & \ddots & \ddots & 0 \\ g_{y\tilde{y}}(k, 0) & \cdots & \cdots & g_{y\tilde{y}}(k, k-1) & I \end{bmatrix}, \tag{7.14}$$

$$g_{y\tilde{y}}(i, j) = g(i, j) + g(i, j+1) g(j+1, j) + \cdots$$
$$+ g(i, i-1) g(i-1, i-2) \cdots g(j+1, j), \tag{7.15}$$

$$G_{\tilde{y}y}(k) = \begin{bmatrix} I & 0 & 0 & \cdots & 0 \\ -g(1, 0) & I & 0 & \cdots & 0 \\ -g(2, 0) & -g(2, 1) & I & \ddots & 0 \\ \vdots & & \ddots & \ddots & 0 \\ -g(k, 0) & \cdots & \cdots & -g(k, k-1) & I \end{bmatrix}. \tag{7.16}$$

Thus, the residual generator (7.3) can be given in the following form

$$\begin{cases} \hat{y}(k) = \sum_{i=0}^{k-1} \tilde{g}(k, i) \tilde{y}(i), \\ \tilde{y}(i) = y(i) - \hat{y}(i), \quad i = 0, 1, \ldots, k, \\ r_k = W(k) \tilde{y}_k. \end{cases} \tag{7.17}$$

We are now in the position to reformulate the problem of FD as to find $W(k)$ and $\hat{y}(k)$ in terms of innovations $\{\tilde{y}(0), \tilde{y}(1), \ldots, \tilde{y}(k-1)\}$ solving

$$\max_{\hat{y}(k),W(k)} \frac{\lambda_{max}(J_f(k))}{\lambda_{max}(J_d(k))} \tag{7.18}$$

or

$$\max_{\hat{y}(k),W(k)} \frac{\lambda_{min}(J_f(k))}{\lambda_{max}(J_d(k))}. \tag{7.19}$$

We now consider the construction of innovations by using an orthogonal projection. Let

$$\begin{aligned}
y_d(k) &= y(k)|_{f(k)=0}, \quad y_f(k) = y(k)|_{\bar{d}(k)=0}, \\
\hat{y}_{do}(k) &= \text{Proj}\{y_d(k)|y_d(0), y_d(1), \ldots, y_d(k-1)\}, \\
\tilde{y}_{do}(k) &= y_d(k) - \hat{y}_{do}(k), \quad R_{\tilde{y}_{do}}(i,j) = \langle \tilde{y}_{do}(i), \tilde{y}_{do}(j) \rangle.
\end{aligned}$$

where $\hat{y}_{do}(k)$ is the orthogonal projection of $y_d(k)$ on linear space $\mathcal{L}\{\{y_d(i)\}_{i=0}^{k-1}\}$.

It is known from the famous Gram-Schmidit orthogonalization procedure that the innovations $\{\tilde{y}_{do}(0), \tilde{y}_{do}(1), \ldots, \tilde{y}_{do}(k-1)\}$ form an orthogonal basis for $\mathcal{L}\{\{y_d(i)\}_{i=0}^{k-1}\}$ with respect to the inner product defined by the Gramian matrix. We thus have

$$\begin{aligned}
\hat{y}_{do}(k) &= \sum_{i=0}^{k-1} \tilde{g}_o(k,i)\tilde{y}_{do}(i), \\
\tilde{g}_o(k,i) &= \langle y_d(k), \tilde{y}_{do}(i) \rangle \langle \tilde{y}_{do}(i), \tilde{y}_{do}(i) \rangle^{-1}, \\
R_{\tilde{y}_{do}}(i,j) &= \langle \tilde{y}_{do}(i), \tilde{y}_{do}(j) \rangle = 0, \ i \neq j,
\end{aligned}$$

which implies that the Gramian of $\tilde{y}_{dk,o}$, i.e., $R_{\tilde{y}_{dk,o}} = [R_{\tilde{y}_{do}}(i,j)]_{i,j=0:k}$, is a diagonal block matrix. Now an optimal residual generator constructed by innovations $\{\tilde{y}_{do}(0), \tilde{y}_{do}(1), \ldots, \tilde{y}_{do}(k)\}$ is obtained in the following theorem.

Theorem 7.1 *Suppose that $\hat{y}_{do}(k)$ is the fault-free case orthogonal projection of $y(k)$ on linear space $\mathcal{L}\{\{y(i)\}_{i=0}^{k}\}$ and when $R_{\bar{d}k} = I$, it holds*

$$R_{\tilde{y}_{dk,o}} > 0. \tag{7.20}$$

Then

$$\begin{cases}
\hat{y}_o(k) = \sum_{i=0}^{k-1} \tilde{g}_o(k,i)\tilde{y}_o(i) \\
\tilde{y}_o(k) = y(k) - \hat{y}_o(k) \\
r_o(k) = w_o(k)\tilde{y}_o(k)
\end{cases} \tag{7.21}$$

provides an optimal residual generator satisfying both (7.18) and (7.19) with

$$w_o(k) = [R_{\tilde{y}_{do}}(k,k)|_{R_{\bar{d}k}=I}]^{-\frac{1}{2}}. \tag{7.22}$$

Proof Define

$$\tilde{y}_{ko} = \left[\, \tilde{y}_o^T(0)\ \tilde{y}_o^T(1) \cdots \tilde{y}_o^T(k) \,\right]^T,\ \tilde{y}_{dk,o} = \left[\, \tilde{y}_{do}^T(0)\ \tilde{y}_{do}^T(1) \cdots \tilde{y}_{do}^T(k) \,\right]^T,$$
$$\tilde{y}_{fk,o} = \tilde{y}_{ko} - \tilde{y}_{dk,o},\ r_{ko} = \left[\, r_o^T(0)\ r_o^T(1) \cdots r_o^T(k) \,\right]^T.$$

In light of (7.21)–(7.22), we have

$$r_{ko} = r_{\bar{d}k,o} + r_{fk,o}, \tag{7.23}$$

where $r_{\bar{d}k,o} = W_o(k)\tilde{y}_{dk,o},\ r_{fk,o} = W_o(k)\tilde{y}_{fk,o}$ and

$$W_o(k) = \mathrm{diag}(w_o(0), w_o(1), \ldots, w_o(k)). \tag{7.24}$$

Thus, we get

$$R_{r_{\bar{d}k},o} = \langle r_{\bar{d}k,o}, r_{\bar{d}k,o}\rangle = W_o(k)R_{\tilde{y}_{dk},o}W_o^T(k),$$
$$R_{r_{fk},o} = \langle r_{fk,o}, r_{fk,o}\rangle = W_o(k)R_{\tilde{y}_{fk},o}W_o^T(k).$$

Observing (7.20) and (7.22), we have

$$J_{do}(k) = W_o(k)R_{\tilde{y}_{dk},o}W_o^T(k)|_{R_{\bar{d}k}=I} = I,$$
$$J_{fo}(k) = W_o(k)R_{\tilde{y}_{fk},o}W_o^T(k)|_{R_{fk}=I}.$$

Moreover, it is obtained from (7.21) that

$$y(k) = \sum_{i=0}^{k-1} \tilde{g}_o(k,i)\tilde{y}_o(i) + \tilde{y}_o(k).$$

Thus we have

$$y_k = G_{y\tilde{y}_o}(k)\tilde{y}_{ko}, \tag{7.25}$$

where $G_{y\tilde{y}_o}(k) = G_{y\tilde{y}}(k)|_{g_{y\tilde{y}}(i,j)=\tilde{g}_o(i,j)}$. Using (7.13), (7.21), and (7.23)–(7.25), we further have

$$\begin{aligned}
\tilde{y}_k &= G_{\tilde{y}y}(k)y_k = G_{\tilde{y}y}(k)G_{y\tilde{y}_o}(k)\tilde{y}_{ko},\\
r_k &= W(k)\tilde{y}_k = W(k)G_{\tilde{y}y}(k)G_{y\tilde{y}_o}(k)\tilde{y}_{ko}\\
&= W(k)G_{\tilde{y}y}(k)G_{y\tilde{y}_o}(k)W_o^{-1}(k)r_{ko}\\
&= W(k)G_{\tilde{y}y}(k)G_{y\tilde{y}_o}(k)W_o^{-1}(k)(r_{\bar{d}k,o} + r_{fk,o}).
\end{aligned} \tag{7.26}$$

Therefore, the Gramian matrices $R_{r_{\bar{d}k}}$ and $R_{r_{fk}}$ are given by

$$R_{r_{\bar{d}k}} = W(k)G_{\tilde{y}y}(k)G_{y\tilde{y}_o}(k)W_o^{-1}(k)R_{r_{\bar{d}k},o}W_o^{-1}(k)G_{y\tilde{y}_o}^T(k)G_{\tilde{y}y}^T(k)W^T(k),$$
$$R_{r_{fk}} = W(k)G_{\tilde{y}y}(k)G_{y\tilde{y}_o}(k)W_o^{-1}(k)R_{r_{fk},o}W_o^{-1}(k)G_{y\tilde{y}_o}^T(k)G_{\tilde{y}y}^T(k)W^T(k).$$

In this case, the $J_d(k)$ and $J_f(k)$ are

$$J_d(k) = W(k)G_{\tilde{y}y}(k)G_{y\tilde{y}_o}(k)W_o^{-1}(k)W_o^{-1}(k)G_{y\tilde{y}_o}^T(k)G_{\tilde{y}y}^T(k)W^T(k),$$
$$J_f(k) = W(k)G_{\tilde{y}y}(k)G_{y\tilde{y}_o}(k)W_o^{-1}(k)J_{fo}(k)W_o^{-1}(k)G_{y\tilde{y}_o}^T(k)G_{\tilde{y}y}^T(k)W^T(k).$$

Note also that

$$\lambda_{max}(J_d(k)) = \bar{\sigma}(W(k)G_{\tilde{y}y}(k)G_{y\tilde{y}_o}(k)W_o^{-1}(k)),$$
$$\lambda_{max}(J_f(k)) \leq \bar{\sigma}(W(k)G_{\tilde{y}y}(k)G_{y\tilde{y}_o}(k)W_o^{-1}(k))\lambda_{max}(J_{fo}(k)),$$
$$\lambda_{min}(J_f(k)) \leq \bar{\sigma}(W(k)G_{\tilde{y}y}(k)G_{y\tilde{y}_o}(k)W_o^{-1}(k))\lambda_{min}(J_{fo}(k)),$$

where $\bar{\sigma}(\cdot)$ represents the maximal singular value of $(\cdot)$. Therefore,

$$\frac{\lambda_{max}(J_f(k))}{\lambda_{max}(J_d(k))} \leq \lambda_{max}(J_{fo}(k)) = \frac{\lambda_{max}(J_{fo}(k))}{\lambda_{max}(J_{do}(k))},$$
$$\frac{\lambda_{min}(J_f(k))}{\lambda_{max}(J_d(k))} \leq \lambda_{min}(J_{fo}(k)) = \frac{\lambda_{min}(J_{fo}(k))}{\lambda_{max}(J_{do}(k))},$$

which implies that (7.21)–(7.22) provide an optimal residual generator satisfying both (7.18) and (7.19). This completes the proof. □

It is seen from Theorem 7.1 that the fault-free case output orthogonal projection leads to an optimal solution to both (7.18) and (7.19). For $\hat{y}(k)$ given by an arbitrary linear combination of measurement output sequence, it will be shown in the following Theorem 7.2 that there exists a corresponding $W(k)$ such that both (7.18) and (7.19) are satisfied also.

Theorem 7.2 *For any $\hat{y}(k)$ given by an arbitrary combination of $\{y(0), y(1), \ldots y(k-1)\}$, there exists a corresponding $W(k)$ such that both (7.18) and (7.19) are satisfied. In this case, the $W(k)$ is given by*

$$W(k) = W_o(k)G_{\tilde{y}_o y}(k)G_{y\tilde{y}}(k). \tag{7.27}$$

Proof Considering residual generator (7.3), it follows from (7.13) and (7.25) that

$$r_k = W(k)\tilde{y}_k = W(k)G_{\tilde{y}y}(k)y_k = W(k)G_{\tilde{y}y}(k)G_{y\tilde{y}_o}(k)\tilde{y}_{ko}.$$

Let $W(k) = W_o(k)G_{\tilde{y}_o y}(k)G_{y\tilde{y}}(k)$. Using (7.15)–(7.16), it is easy to verify that

$$G_{y\tilde{y}}(k)G_{\tilde{y}y}(k) = I, \;\; G_{\tilde{y}y}(k)G_{y\tilde{y}}(k) = I.$$

Similarly, it is easy to show that $G_{\tilde{y}_o y}(k)G_{y\tilde{y}_o}(k) = I$. Therefore,

$$r_k = W_o(k)G_{\tilde{y}_o y}(k)G_{y\tilde{y}}(k)G_{\tilde{y}y}(k)G_{y\tilde{y}_o}(k)\tilde{y}_{ko} = W_o(k)\tilde{y}_{ko} = r_{ko},$$

which implies that $\hat{y}(k)$ and the $W(k)$) in (7.27) deliver an optimal residual generator satisfying both (7.18) and (7.19). This completes the proof. □

Since the optimal solution in Theorem 7.1 is obtained by using an orthogonal projection with respect to the inner product in the framework of Gramian matrix, Theorems 7.1 and 7.2 are applicable to LDTV systems described by the state space model (7.1) or an input-output model (7.2). In the next section, we focus on the derivation of state space recursions of $\hat{y}(k)$ and $W(k)$ for LDTV system (7.1).

7.4 Recursive Implementation in State Space

To obtain $\hat{y}(k)$ and $W(k)$ recursively, we first consider the recursive calculation of $\hat{y}_{do}(k)$. Let $x_d(k) = x(k)|_{f(k)=0}$. In fault-free case, the LDTV system (7.1) becomes

$$\begin{cases} x_d(k+1) = A(k)x_d(k) + B_d(k)d(k), \\ y_d(k) = C(k)x_d(k) + D_d(k)d(k), \\ x_d(0) = x_0. \end{cases} \tag{7.28}$$

Denote $\hat{x}_{do}(k) = \text{Proj}\{x_d(k)|y_d(0), y_d(1), \ldots, y_d(k-1)\}$. Similar to (7.12), it is easy to get

$$\mathcal{L}\{\{y_d(i)\}_{i=0}^k\} = \mathcal{L}\{\{\tilde{y}_{do}(i)\}_{i=0}^k\}.$$

We thus have

$$\begin{aligned} \hat{y}_{do}(k) &= \text{Proj}\{y_d(k)|\tilde{y}_{do}(0), \tilde{y}_{do}(1), \ldots, \tilde{y}_{do}(k-1)\}, \\ &= C(k)\hat{x}_{do}(k), \qquad (7.29) \\ \hat{x}_{do}(k+1) &= \text{Proj}\{x_d(k+1)|\tilde{y}_{do}(0), \tilde{y}_{do}(1), \ldots, \tilde{y}_{do}(k)\} \\ &= \text{Proj}\{x_d(k+1)|\tilde{y}_{do}(0), \tilde{y}_{do}(1), \ldots, \tilde{y}_{do}(k-1)\} \\ &\quad +\text{Proj}\{x_d(k+1)|\tilde{y}_{do}(k)\} \\ &= A(k)\hat{x}_{do}(k) + K_o(k)\tilde{y}_{do}(k), \qquad (7.30) \end{aligned}$$

where

$$K_o(k) = R_{x_d\tilde{y}_{do}}(k+1, k)R_{\tilde{y}_{do}}^{-1}(k, k). \tag{7.31}$$

Introduce innovations

$$\tilde{x}_{do}(k) = x_d(k) - \hat{x}_{do}(k).$$

It follows from (7.28)–(7.30) that

$$\begin{cases} \tilde{x}_{do}(k+1) = A(k)\tilde{x}_{do}(k) - K(k)\tilde{y}_{do}(k) + B_d(k)d(k), \\ \tilde{y}_{do}(k) = C(k)\tilde{x}_{do}(k) + D_d(k)d(k). \end{cases}$$

Let

$$P(k) = \langle \tilde{x}_{do}(k), \tilde{x}_{do}(k) \rangle, \quad P(0) = I.$$

Then we have

$$\begin{aligned} P(k+1) = &< A(k)\tilde{x}_{do}(k) - K_o(k)\tilde{y}_{do}(k) + B_d(k)d(k), \\ &A(k)\tilde{x}_{do}(k) - K_o(k)\tilde{y}_{do}(k) + B_d(k)d(k) > \\ = &\, A(k)P(k)A^T(k) - A(k)R_{\tilde{x}_{do}\tilde{y}_{do}}(k,k)K_o^T(k) \\ &- K_o(k)R_{\tilde{x}_{do}\tilde{y}_{do}}^T(k,k)A^T(k) + K_o(k)R_{\tilde{y}_{do}}(k,k)K_o^T(k) \\ &- K_o(k)R_{\tilde{y}_{do}d}(k,k)B_d^T(k) - B_d(k)R_{\tilde{y}_{do}d}^T(k,k)K_o^T(k) \\ &+ B_d(k)R_d(k,k)B_d^T(k). \end{aligned} \tag{7.32}$$

Note also that

$$R_{\tilde{y}_{do}}(k,k) = C(k)P(k)C^T(k) + D_d(k)R_d(k,k)D_d^T(k), \tag{7.33}$$

$$R_{\tilde{x}_{do}\tilde{y}_{do}}(k,k) = P(k)C^T(k), \quad R_{\tilde{y}_{do}d}(k,k) = D_d(k)R_d(k,k), \tag{7.34}$$

$$R_{x_{do}\tilde{y}_{do}}(k+1,k) = A(k)P(k)C^T(k) + B_d(k)R_d(k,k)D_d^T(k). \tag{7.35}$$

Substituting (7.33)–(7.35) into (7.22), (7.31), (7.32) and setting $R_d(k,k) - I$ yield

$$w_o(k) = (C(k)P(k)C^T(k) + D_d(k)D_d^T(k))^{-\frac{1}{2}}, \tag{7.36}$$

$$K_o(k) = (A(k)P(k)C^T(k) + B_d(k)D_d^T(k))w_o^2(k), \tag{7.37}$$

$$\begin{aligned} P(k+1) = &\, A(k)P(k)A^T(k) - A(k)P(k)C^T(k)K_o^T(k) \\ &- K_o(k)C(k)P(k)A^T(k) + K_o(k)w_o^2(k)K_o^T(k) \\ &- K_o(k)D_d(k)B_d^T(k) - B_d(k)D_d^T(k)K_o^T(k) + B_d(k)B_d^T(k) \\ = &\, (A(k) - K_o(k)C(k))P(k)(A(k) - K_o(k)C(k))^T \\ &+ (B_d(k) - K_o(k)D_d(k))(B_d(k) - K_o(k)D_d(k))^T \\ = &\, A(k)P(k)A^T(k) - K_o(k)w_o^2(k)K_o^T(k) + B_d(k)B_d^T(k). \end{aligned} \tag{7.38}$$

Introduce

$$\begin{cases} \hat{x}_o(k+1) = A(k)\hat{x}_o(k) + K_o(k)\tilde{y}_o(k), \\ \hat{y}_o(k) = C(k)\hat{x}_o(k), \\ \tilde{y}_o(k) = y(k) - \hat{y}_o(k), \\ r_o(k) = w_o(k)\tilde{y}_o(k), \\ \hat{x}_o(0) = 0. \end{cases} \tag{7.39}$$

Using Theorem 7.1, it is easy to see that (7.39) provides an optimal residual generator satisfying both (7.18) and (7.19).

On the other hand, with Assumption 7.1, the $K_o(k)$ in (7.37) is a Kalman filter gain matrix and $A(k) - K_o(k)C(k)$ is exponentially stable.

Based on the above analysis, the following Theorem 7.3 is now concluded.

Theorem 7.3 *Under Assumptions 7.1 and 3.1, the $\hat{y}_o(k)$ given by (7.39) and the $W_o(k)$ given by (7.36) and (7.22) provide us with an exponentially stable optimal residual generator satisfying both (7.18) and (7.19).*

Remark 7.4 Theorem 7.3 gives a state space recursion of $\hat{y}_o(k)$ and $W_o(k)$ in Theorem 7.1 in the framework of Gramian matrix. The analysis technique is similar to the innovation derivation of the Kalman filter and $K_o(k)$ is the same with the Kalman filter. However, the obtained results can be used to handle all the cases of unknown input and fault satisfying Assumption 7.1, including the white noise case and the norm-bounded case. In particular, for the LDTV system (7.1) with l_2-norm bounded unknown input and fault, the resulting residual generator (7.39) provides also an optimal solution to the H_∞/H_∞- and H_-/H_∞-FDFs in Problems 6.1 and 6.2, while the $w_o(k)$ is the so-called static post-filter.

Next, we consider an arbitrary $\hat{y}(k)$ given by

$$\begin{cases} \hat{x}(k+1) = A(k)\hat{x}(k) + K(k)(y(k) - \hat{y}(k)), \\ \hat{y}(k) = C(k)\hat{x}(k), \end{cases} \tag{7.40}$$

where $K(k)$ is an observer gain matrix ensuring the exponential stability of $A(k) - K(k)C(k)$. A state space recursion of the corresponding optimal $W(k)$ in Theorem 7.2 is shown in the following Theorem 7.4.

Theorem 7.4 *For any given $K(k)$ ensuring the exponential stability of $A(k) - K(k)C(k)$, there exists a corresponding optimal $W(k)$ described by*

$$\begin{cases} \eta(k+1) = (A(k) - K_o(k)C(k))\eta(k) + (K(k) - K_o(k))\tilde{y}(k), \\ r(k) = w_o(k)(C(k)\eta(k) + \tilde{y}(k)), \\ \eta(0) = 0, \end{cases} \tag{7.41}$$

such that both (7.18) and (7.19) are satisfied, where $w_o(k)$ and $K_o(k)$ are given by (7.36)–(7.37).

Proof Applying Theorem 7.2, the $\hat{y}(k)$ given by an arbitrary combination of $y(i)$ $(i = 0, 1, \ldots, k-1)$ and the corresponding $W(k) = W_o(k)G_{\tilde{y}_o y}(k)G_{y\tilde{y}}(k)$ provide an optimal residual generator satisfying both (7.18) and (7.19), i.e.,

$$r_k = W(k)\tilde{y}_k = W_o(k)G_{\tilde{y}_o y}(k)G_{y\tilde{y}}(k)\tilde{y}_k = W_o(k)\tilde{y}_{ko} = r_{ko}.$$

Consider the $\hat{y}(k)$ in (7.40). Note that $y(k) = \hat{y}(k) + \tilde{y}(k)$. Then a state space description of $y_k = G_{y\tilde{y}}(k)\tilde{y}_k$ can be given by

$$\begin{cases} \hat{x}(k+1) = A(k)\hat{x}(k) + K(k)\tilde{y}(k), \\ y(k) = C(k)\hat{x}(k) + \tilde{y}(k). \end{cases} \tag{7.42}$$

Recall that a state space description of $\tilde{y}_{ko} = G_{\tilde{y}_o y}(k)y_k$ can be given by

$$\begin{cases} \hat{x}_o(k+1) = A(k)\hat{x}_o(k) + K_o(k)(y(k) - \hat{y}_o(k)), \\ \hat{y}_o(k) = C(k)\hat{x}_o(k), \\ \tilde{y}_o(k) = y(k) - \hat{y}_o(k). \end{cases} \tag{7.43}$$

Let $\eta(k) = \hat{x}(k) - \hat{x}_o(k)$. Using (7.42)–(7.43), a state space description of $\tilde{y}_{ko} = G_{\tilde{y}_o y}(k)G_{y\tilde{y}}(k)\tilde{y}_k$ is given by

$$\begin{cases} \eta(k+1) = (A(k) - K_o(k))\eta(k) + (K(k) - K_o(k))\tilde{y}(k), \\ \tilde{y}_o(k) = C(k)\eta(k) + \tilde{y}(k). \end{cases}$$

It is seen from the above analysis that the $r(k)$ given by (7.41) satisfies $r(k) = r_o(k)$. Therefore, (7.41) is a residual generator satisfying both (7.18) and (7.19). □

Remark 7.5 For LDTV system (7.1), Theorem 7.4 also provides us with a generalized optimal solution to the H_∞/H_∞- and H_-/H_∞-FDFs in Problems 6.1 and 6.2, i.e.,

$$\begin{cases} \hat{x}(k+1) = A(k)\hat{x}(k) + K(k)\tilde{y}(k), \\ \hat{y}(k) = C(k)\hat{x}(k), \\ \tilde{y}(k) = y(k) - \hat{y}(k), \\ \eta(k+1) = (A(k) - K_o(k)C(k))\eta(k) + (K(k) - K_o(k))\tilde{y}(k), \\ r(k) = w_o(k)(C(k)\eta(k) + \tilde{y}(k)), \\ \hat{x}(0) = 0,\ \eta(0) = 0. \end{cases} \tag{7.44}$$

As a summary, residual generator (7.3) is constructed by using a linear combination of input/output data. The results in Theorems 7.1 and 7.2 have potential applications to LDTV systems described by input-output model. The results illustrated in Chap. 6 can be regarded as the special cases of the above results with l_2-norm bounded unknown input and fault. In addition, the applied analysis and derivation techniques are different in these two chapters. In view of (7.17) and (7.44), the relationship between $K(i)$ and $\tilde{g}(k, i)$ $(i = 0, 1, \ldots, k-1)$ is

$$\begin{cases} \tilde{g}(k, k-1) = C(k)K(k-1), \\ \tilde{g}(k, i) = C(k)A(k-1)\cdots A(i+1)K(i),\ i = 0, 1, \ldots, k-2. \end{cases}$$

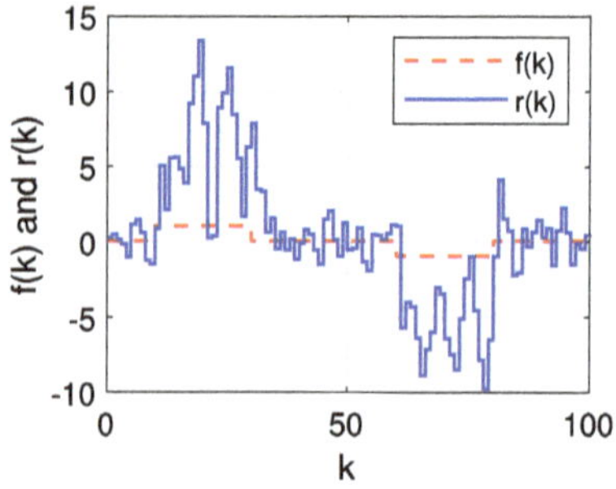

Fig. 7.1 The fault $f(k)$ and the residual $r(k)$ in Case 1

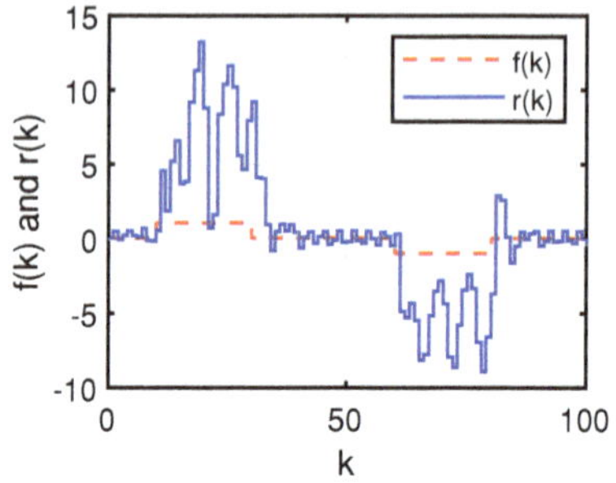

Fig. 7.2 The fault $f(k)$ and the residual $r(k)$ in Case 2

7.5 A Numerical Example

A numerical example is given below to verify the above results. Consider LDTV system (7.1) with the following parameter matrices

$$A(k) = \begin{bmatrix} 0.2e^{-\frac{k}{100}+1} & 0.6 & 0 \\ 0 & 0.5 & \sin(k) \\ 0 & 0 & 0.7 \end{bmatrix}, \; B_d(k) = \begin{bmatrix} 1.3 \\ 0.5 \\ 0.9^k \end{bmatrix}, \; B_f(k) = \begin{bmatrix} 0.2 \\ 1.8 \\ 0.3 \end{bmatrix},$$
$$C(k) = \begin{bmatrix} -0.5 & 1.5 & 0 \end{bmatrix}, \; D(k) = 0, \; D_d(k) = 0.5, \; D_f(k) = 0.5.$$

We first calculate $w_o(k)$ and $K_o(k)$ by using of (7.36)–(7.38). Set $P_0 = I$, $x_0 = 0$. By simulating two different unknown input cases, i.e., the white Gaussian noise $d(k)$ with power 0.01 (Case 1) and the sine wave unknown input $d(k) = 0.5\sin(2k)$ (Case 2), we calculate the residuals by applying (7.39) and the results are shown in Figs. 7.1 and 7.2, respectively.

7.6 Conclusion

This chapter aims at the development of a residual generator for LDTV systems by making use of an arbitrary linear combination of output estimation error sequence. Gramian matrix-based criteria have been used to measure the influences of an unknown input and a fault on the residual and, based on this, the problem of FD has been formulated in the sense of maximizing a sensitivity/robustness ratio. It has been shown that the optimal solution is not unique and an optimal residual generator can be obtained by applying orthogonal projection and innovation analysis. Furthermore, state space descriptions of the optimal residual generators have been given in terms of Riccati recursions and such optimal residual generators also provide us with optimal H_∞/H_∞- and H_-/H_∞-FDFs. Moreover, such a projection approach has potential application in dealing with data-driven FD problems for LDTV systems.

References

1. Zhong, M., Liu, S., & Zhao, H. (2008). Krein space-based H_∞ fault estimation for linear discrete time-varying systems. *Acta Automatica Sinica, 34*(12), 1529–1533.
2. Zhong, M., Ding, S. X., & Ding, E. L. (2010). Optimal fault detection for linear discrete time-varying systems. *Automatica, 46*(8), 1395–1400.
3. Zhong, M., Zhou, D., & Ding, S. X. (2010). On designing H_∞ fault detection filter for linear discrete time-varying systems. *IEEE Transactions on Automatic Control, 55*(7), 1689–1695.
4. Zhang, P., & Ding, S. X. (2004). Observer-based fault detection of linear time-varying systems. *Automatisierungstechniik, 52*, 370–376. (in German).
5. Li, X. (2009). *Fault detection filter design for linear systems*, Ph.D. dissertation. Louisiana State University, USA.

Chapter 8
An H_i/H_∞-Optimization Scheme of Fault Detection for LDTV Systems

In this chapter, an H_i/H_∞-optimization scheme of FD is developed for LDTV systems via Krein space projection. This leads to a natural design of an FD system with the $l_{2,[0,N]}$-norm boundedness of unknown inputs as a threshold and the $l_{2,[0,N]}$-norm of the unknown input estimate as the evaluation function. To avoid heavy computation burden, the projection technique in Krein space is applied that allows a recursive computation of the evaluation function. The achieved FD system satisfies both the worst-case and best-case sensitivity/robustness ratio criteria and, thus, provides us with an alternative design of H_∞/H_∞- and/or H_-/H_∞-FDFs.

8.1 Introduction

For LDTV systems subject to $l_{2,[0,N]}$-norm bounded unknown inputs, the widely accepted performance indices are H_∞/H_∞ and H_-/H_∞ with H_- denoting the fault sensitivity index and the H_∞-norm indicating the influence of unknown inputs or faults (in case that H_∞/H_∞ index is applied) on residual vector, as demonstrated in Chap. 6. Once an FDF is designed and an $l_{2,[0,N]}$-norm based evaluation function is applied for residual evaluation purpose, the threshold is generally determined depending on the l_2-norm boundedness of the unknown inputs [4, 8]. With recent progress of H_∞-filtering based fault diagnosis and the studies of H_∞ estimation in Krein spaces, Krein space based techniques have provided us with a promising way to address estimation issues with high computation efficiency [1–3, 5, 6].

In this chapter, we aim to develop an H_i/H_∞-optimal design scheme of FD systems for LDTV systems subject to $l_{2,[0,N]}$-norm bounded unknown inputs. It serves as an alternative solution to the well-established H_∞/H_∞, H_-/H_∞ and H_∞-filtering based FD system design. The basic idea is to compare an (optimal) estimation of the $l_{2,[0,N]}$-norm of the unknown inputs and their $l_{2,[0,N]}$-norm boundedness. In other words, the estimated $l_{2,[0,N]}$-norm of the unknown inputs builds the evaluation func-

M. Zhong et al., *Fault Diagnosis for Linear Discrete Time-Varying Systems and Its Applications*, https://doi.org/10.1007/978-981-19-5438-2_8

tion, while the $l_{2,[0,N]}$-norm boundedness serves as the threshold. In this way, a separate step for the threshold setting, as needed in the existing H_∞-filtering, H_-/H_∞- and H_∞/H_∞-optimization based FD schemes, and the associated (complicated) computation like H_∞-norm computation, becomes unnecessary. To this end, the following two problems should be addressed:

- finding a right estimation of the $l_{2,[0,N]}$-norm of unknown inputs, which plays a central role to our aim. In this chapter we formulate such a problem as finding a minimum of a quadratic form and solve it by converting the finite horizon linear estimation problem into a minimum problem of certain quadratic form in Krein space;
- developing a recursive solution to our estimation problem so as to reduce the online realization cost. The techniques of projection and innovation analysis in Krein space will be applied for this purpose.

8.2 Basic Ideas and Problem Formulation

Consider LDTV systems described by the following input-output model

$$y_k = H_{y0}(k)x_0 + H_{yd}(k)d_k + H_{yf}(k)f_k, \tag{8.1}$$

where $x_0 = x(0) \in \mathbb{R}^n$ is the system state at time step $k = 0$, $y_k,\ d_k,\ f_k$ are given in (3.6), $H_{y0}(k)$, $H_{yd}(k)$ and $H_{yf}(k)$ are referred to (3.8), (3.9) and (3.11), respectively, with $y(k) \in \mathbb{R}^q$, $d(k) \in \mathbb{R}^m$ and $f(k) \in \mathbb{R}^l$ are the measurement output, unknown input and fault vectors, respectively. Given $k = N$, let

$$\Omega = H_{y0}(N),\ \ \Gamma_d = H_{yd}(N),\ \ \Gamma_f = H_{yf}(N).$$

It yields from (8.1) that

$$y_N = \Omega x_0 + \Gamma_d d_N + \Gamma_f f_N. \tag{8.2}$$

Under the assumption of $\mathrm{rank}(D_d(k)) = \dim(y(k)) = q$, Γ_d is of full row rank.

Recall an FDF-based residual generator (6.2), i.e.,

$$\begin{cases} \hat{x}(k+1) = A(k)\hat{x}(k) + L(k)(y(k) - C(k)\hat{x}(k)),\ \hat{x}(0) = 0, \\ r(k) = W(k)(y(k) - C(k)\hat{x}(k)). \end{cases} \tag{8.3}$$

where $L(k)$ is the observer gain matrix and $W(k)$ is the post-filter. As given in Chap. 6, optimal solutions of $L(k)$ and $W(k)$ can be obtained in the sense of H_∞/H_∞- and/or H_-/H_∞-optimization by solving the Problems 6.1 and 6.2. In the stage of residual evaluation, the evaluation function $J_{2,[0,N]}(r)$ in (6.25) and the threshold J_{th} in (6.26), i.e.,

$$J_{2,[0,N]}(r) = ||r(k)||^2_{2,[0,N]} = \sum_{j=0}^{N} r^T(j)r(j),$$

$$J_{th} = \delta_d^2 + \delta_e^2 := \delta^2$$

can be applied to an optimal FD defined in Definition 2.4, due to

$$x_0^T x_0 + \sum_{k=0}^{N} d^T(k)d(k) \leq \delta^2 \tag{8.4}$$

for $||d(k)||_{2,[0,N]} \leq \delta_d$ and $||e(0)||_2 = ||x(0) - \hat{x}(0)||_2 \leq \delta_e$.

Alternatively, since $J_{2,[0,N]}(r) \leq \gamma^2\delta^2$ for $f(k) = 0$ with

$$\gamma^2 = \sup_{x_0, d(k)} \frac{\sum_{k=0}^{N} r_d^T(k)r_d(k)}{x_0^T x_0 + \sum_{k=0}^{N} d^T(k)d(k)}, \quad r_d(k) = r(k)|_{f(k)=0},$$

a threshold with a zero FAR can be set as $J_{th,r} = \gamma^2\delta^2$ by using detection logic

$$\begin{cases} J_{2,[0,N]}(r) < J_{th,r} \Rightarrow \text{no fault,} \\ J_{2,[0,N]}(r) \geq J_{th,r} \Rightarrow \text{fault alarm.} \end{cases}$$

Different from the above strategy, the core of our alternative scheme is to estimate the l_2-norm of the unknown inputs. It is known from [7] that the following regularized least-squares problem provides a minimum estimation for the unknown input d_N and initial value x_0, i.e.,

$$J_0(x_0) = x_0^T x_0 + ||y_N - \Omega x_0||^2_{(\Gamma_d \Gamma_d^T)^{-1}},$$

where the weighting matrix is introduced due to the relations

$$y_N = \Omega x_0 + \Gamma_d d_N$$
$$\Longleftrightarrow \begin{cases} (\Gamma_d\Gamma_d^T)^{-1/2}\Gamma_d d_N = (\Gamma_d\Gamma_d^T)^{-1/2}(y_N - \Omega x_0) \\ \| (\Gamma_d\Gamma_d^T)^{-1/2}\Gamma_d d_N \| \leq \| d_N \| . \end{cases} \tag{8.5}$$

With given measurement vector y_N, it follows from [7] that $J_0(x_0)$ has a minimum at

$$\begin{aligned} \hat{x}_0 &= (I + \Omega^T(\Gamma_d\Gamma_d^T)^{-1}\Omega)^{-1}\Omega^T(\Gamma_d\Gamma_d^T)^{-1}y_N \\ &= \Omega^T(\Omega\Omega^T + \Gamma_d\Gamma_d^T)^{-1}y_N \end{aligned} \tag{8.6}$$

and its value at the minimum is

$$J_0(\hat{x}_0) = y_N^T(\Omega\Omega^T + \Gamma_d\Gamma_d^T)^{-1}y_N. \tag{8.7}$$

Let $J_0(\hat{x}_0)$ be the evaluation function J_N. It holds

$$\begin{aligned} J_N &= J_0(\hat{x}_0) = y_N^T(\Omega\Omega^T + \Gamma_d\Gamma_d^T)^{-1}y_N \\ &= \begin{bmatrix} x_0^T & d_N^T \end{bmatrix}\begin{bmatrix} \Omega^T \\ \Gamma_d^T \end{bmatrix}(\Omega\Omega^T + \Gamma_d\Gamma_d^T)^{-1}\begin{bmatrix} \Omega & \Gamma_d \end{bmatrix}\begin{bmatrix} x_0 \\ d_N \end{bmatrix} \\ &\le \begin{bmatrix} x_0^T & d_N^T \end{bmatrix}\begin{bmatrix} x_0 \\ d_N \end{bmatrix} \le \delta^2. \end{aligned}$$

As a result, we can apply the following decision logic for FD

$$\begin{cases} J_N > J_{th} = \delta^2, & \text{fault alarm}, \\ J_N \le J_{th} = \delta^2, & \text{no alarm}. \end{cases}$$

Although the derivation of the above FD scheme sounds natural and reasonable, the following problems should be solved in order to have a successful realization.

At first, note that $\Omega \in \mathbb{R}^{(N+1)q\times n}$, $\Gamma_d \in \mathbb{R}^{(N+1)q\times(N+1)m}$. It is a heavy task to update Ω, Γ_d and calculate $[\Omega\Omega^T + \Gamma_d\Gamma_d^T]^{-1}$ online for LDTV systems, especially with the increase of N. To avoid complicated matrix computation, an H_∞ optimization approach in Krein space will be applied and a recursive computation of J_N will be derived, by establishing a relationship between the minimum problem of quadratic form and projection in Krein space.

Secondly, to demonstrate the FD performance of the developed scheme, we define

$$\begin{aligned} \gamma_d^2 &= \sup_{x_0, d_N} \frac{J_N|_{f_N=0}}{x_0^T x_0 + d_N^T d_N}, \\ \gamma_f^2 &= \sup_{f_N} \frac{J_N|_{x_0=0,\ d_N=0}}{f_N^T f_N}, \\ \gamma_{f-}^2 &= \inf_{f_N} \frac{J_N|_{x_0=0,\ d_N=0}}{f_N^T f_N}, \end{aligned}$$

where γ_d denotes the robustness index of the FD system against unknown inputs and initial state, γ_f and γ_{f-} represent the best-case and worst-case fault sensitivity criteria, respectively. The J_N is called an optimal evaluation function if the following best-case and/or worst-case sensitivity/robustness ratio criteria are satisfied

$$\max_{J_N} \frac{\gamma_f}{\gamma_d} \text{ and/or } \max_{J_N} \frac{\gamma_{f-}}{\gamma_d}. \tag{8.8}$$

In summary, in the sequel we are going to

- derive a recursive algorithm for computing J_N by means of Krein space projection, and

- analyze the feasibility of the optimization problems in (8.8).

As a byproduct, we will finally apply the achieved results to the design of H_-/H_∞- and H_∞/H_∞-FDFs.

8.3 A Krein Space Based Recursive Solution

We know that H_∞-filtering in Hilbert space can be cast into a minimum problem of certain indefinite quadratic forms and, via introducing a corresponding system in Krein space, the minimum problem can be solved by applying Krein space projection. A solution of the minimum problem can be obtained in terms of Riccati recursions and, therefore, it is computational attractive. Inspired by these, in this section we derive a recursive computation of J_N by applying Krein space projections.

8.3.1 Recursive Computation of J_N

Introduce the following Krein space system

$$\begin{cases} \mathbf{x}(k+1) = A(k)\mathbf{x}(k) + B_d(k)\mathbf{d}(k) + B_f(k)\mathbf{f}(k),\ \mathbf{x}(0) = \mathbf{x}_0, \\ \mathbf{y}(k) = C(k)\mathbf{x}(k) + D_d(k)\mathbf{d}(k) + D_f(k)\mathbf{f}(k), \end{cases} \tag{8.9}$$

where $\mathbf{x}_0$, $\mathbf{d}(k)$ and $\mathbf{f}(k)$ are vectors in Krein space with Gramian

$$\left\langle \begin{bmatrix} \mathbf{x}_0 \\ \mathbf{d}(i) \\ \mathbf{f}(i) \end{bmatrix}, \begin{bmatrix} \mathbf{x}_0 \\ \mathbf{d}(j) \\ \mathbf{f}(j) \end{bmatrix} \right\rangle = \begin{bmatrix} I & 0 & 0 \\ 0 & I\delta_{ij} & 0 \\ 0 & 0 & I\delta_{ij} \end{bmatrix} \tag{8.10}$$

and $\delta_{ii} = 1,\ \delta_{ij} = 0\ (i \neq j)$. Analogue to (8.2), we can get

$$\mathbf{y}_N = \Omega\mathbf{x}_0 + \Gamma_d\mathbf{d}_N + \Gamma_f\mathbf{f}_N.$$

Denote by $R_{\mathbf{y}_N}$ and $R_{\mathbf{y}_N\mathbf{x}_0}$ the Gramian of $\mathbf{y}_N$ and the cross-Gramian of $\mathbf{y}_N$ and $\mathbf{x}_0$, respectively. We have

$$R_{\mathbf{y}_N} = \Omega\Omega^T + \Gamma_d\Gamma_d^T,\ R_{\mathbf{y}_N\mathbf{x}_0} = \Omega = R_{\mathbf{x}_0\mathbf{y}_N}^T.$$

Note that Γ_d being full row rank implies

$$R_{\mathbf{y}_N} > 0,\ I - \Omega^T(\Omega\Omega^T + \Gamma_d\Gamma_d^T)^{-1}\Omega > 0. \tag{8.11}$$

For ease of later discussion, we introduce

$$J(x_0, d_N) = x_0^T x_0 + d_N^T d_N.$$

It is known from Theorem 4.2 that the minimization of $J(x_0, d_N)$ subject to (8.5) is partially equivalent to finding an orthogonal projection of $\mathbf{x}_0$, $\mathbf{d}_N$ onto subspace $\mathcal{L}\{\{\mathbf{y}(i)\}_{i=0}^N\}$. When $R_{\mathbf{y}_N} > 0$, according to Lemma 4.2, the orthogonal projections $\hat{\mathbf{x}}_0$, $\hat{\mathbf{d}}_N$ exist and are unique, which are given by

$$\hat{\mathbf{x}}_0 = \Omega^T R_{\mathbf{y}_N}^{-1} \mathbf{y}_N, \ \hat{\mathbf{d}}_N = \Gamma_d^T R_{\mathbf{y}_N}^{-1} \mathbf{y}_N.$$

Moreover, $J(x_0, d_N)$ subject to (8.5) has a minimum at

$$\check{x}_0 = \Omega^T R_{\mathbf{y}_N}^{-1} y_N, \ \check{d}_N = \Gamma_d^T R_{\mathbf{y}_N}^{-1} y_N$$

if and only if the conditions in (8.11) are satisfied and the value of $J(x_0, d_N)$ at the minimum is

$$J(\check{x}_0, \check{d}_N) = y_N^T R_{\mathbf{y}_N}^{-1} y_N.$$

In view of (8.6) and (8.7), the $\check{x}_0$ and $J(\check{x}_0, \check{d}_N)$ are the same with $\hat{x}_0$ and $J_0(\hat{x}_0)$, respectively. This means that J_N can be obtained alternatively by the minimum estimation of $J(x_0, d_N)$ and be solved by applying projection in Krein space.

On the other hand, as mentioned earlier, the online calculation of Ω, Γ_d and $R_{\mathbf{y}_N}^{-1}$ is a heavy task for LDTV systems, especially with large N. To avoid complex inverse matrix computation, an algorithm for recursively computing J_N will be derived by applying the Krein space projection.

Denote by $\hat{\mathbf{x}}(k)$ the projection of $\mathbf{x}(k)$ onto subspace $\mathcal{L}\{\{\mathbf{y}(k)\}_{k=0}^N\}$. In the case of $\mathbf{f}(k) = 0$, we have

$$\begin{cases} \hat{\mathbf{x}}(k+1) = A(k)\hat{\mathbf{x}}(k) + K_p(k)(\mathbf{y}(k) - C(k)\hat{\mathbf{x}}(k)), \ \hat{\mathbf{x}}(0) = 0, \\ \tilde{\mathbf{y}}(k) = \mathbf{y}(k) - C(k)\hat{\mathbf{x}}(k), \end{cases} \tag{8.12}$$

where

$$\begin{aligned} K_p(k) &= < \mathbf{x}(k+1), \tilde{\mathbf{y}}(k) > R_{\tilde{\mathbf{y}}}^{-1}(k) \\ &= (A(k)P(k)C^T(k) + B_d(k)D_d^T(k))R_{\tilde{\mathbf{y}}}^{-1}(k), \qquad (8.13) \\ R_{\tilde{\mathbf{y}}}(k) &= C(k)P(k)C^T(k) + D_d(k)D_d^T(k), \qquad (8.14) \\ P(k+1) &= A(k)P(k)A^T(k) - K_p(k)R_{\tilde{\mathbf{y}}}(k)K_p^T(k) \\ &\quad + B_d(k)B_d^T(k), \ P(0) = I. \qquad (8.15) \end{aligned}$$

It is known from Sect. 4.4 that $\{\tilde{\mathbf{y}}(k)\}_{i=0}^N$ forms an orthogonal basis of $\mathcal{L}\{\{\mathbf{y}(k)\}_{k=0}^N\}$ and $R_{\mathbf{y}_N}$ has the same inertia as diag$(R_{\tilde{\mathbf{y}}}(0), R_{\tilde{\mathbf{y}}}(1), \ldots, R_{\tilde{\mathbf{y}}}(k))$. As a result, $J_0(\hat{x}_0)$ can be rewritten as

Algorithm 8.3.2 Calculation of J_N

1: Update $K_p(k)$, $R_{\tilde{\mathbf{y}}}(k)$ and $P(k)$ by using (8.13)–(8.15);
2: If $R_{\tilde{\mathbf{y}}}(k) > 0$ is satisfied, then we calculate innovation $\tilde{y}(k)$ by using (8.16);
3: Calculate J_N recursively by using (8.17).

$$J_0(\hat{x}_0) = J(\check{x}_0, \check{d}_N) = \sum_{k=0}^{N} \tilde{y}^T(k) R_{\tilde{\mathbf{y}}}^{-1}(k) \tilde{y}(k),$$

where $\tilde{y}(k)$ is computed using the formulae of $\tilde{\mathbf{y}}(k)$, i.e.,

$$\begin{cases} \hat{x}(k+1) = A(k)\hat{x}(k) + K_p(k)(y(k) - C(k)\hat{x}(k)),\ \hat{x}(0) = 0, \\ \tilde{y}(k) = y(k) - C(k)\hat{x}(k). \end{cases} \tag{8.16}$$

Therefore, we can compute J_N using (8.13)–(8.15), (8.16) and

$$J_N = \sum_{k=0}^{N} \tilde{y}^T(k) R_{\tilde{\mathbf{y}}}^{-1}(k) \tilde{y}(k). \tag{8.17}$$

Now the following proposition is readily concluded.

Proposition 8.1 *$J(x_0, d_N)$ subject to (8.5) has a unique minimum if and only if $R_{\tilde{\mathbf{y}}}(k) > 0 for k = 0, 1, \ldots, N$. Moreover, an evaluation function J_N can be obtained via recursions (8.13)–(8.17).*

We now summarize the calculation of J_N into the Algorithm 8.3.2. Comparing Algorithm 8.3.2 with the direct computation of J_N makes it clear that a reduction in the computation effort in finding matrix inverse of $R_{\mathbf{y}_N} \in \mathbb{R}^{(N+1)q \times (N+1)q}$ and $R_{\tilde{\mathbf{y}}}(k) \in \mathbb{R}^{q \times q}$ is achieved, since additions are much faster than matrix inverse. The main purpose of applying Krein space projection is to derive the recursive formulae for computing J_N.

Remark 8.1 In the FD study, it is helpful to introduce the concept of "direction" of the disturbance subspace with the aid of Krein space. The projection $\hat{\mathbf{d}}_N = \Gamma_d^T R_{\mathbf{y}_N}^{-1} \mathbf{y}_N$ represents the "direction" of unknown input d_N with minimizing "size", while the minimum value of $J(x_0, d_N)$ represents its "size". The core of the presented scheme is to find such a "direction" and, when a fault is taken into account, to define its "size" as an evaluation function.

8.3.2 Analysis of Sensitivity/Robustness Performance

After computing J_N, the remaining task is to check if the sensitivity/robustness ratio criteria in (8.8) are satisfied. Define $\mathbf{e}(k) = \mathbf{x}(k) - \hat{\mathbf{x}}(k)$. Subtracting (8.12) from

(8.9) yields

$$\begin{cases} \mathbf{e}(k+1) = A_K(k)\mathbf{e}(k) + B_{Kd}(k)\mathbf{d}(k) + B_{Kf}(k)\mathbf{f}(k), \\ \tilde{\mathbf{y}}(k) = C(k)\mathbf{e}(k) + D_d(k)\mathbf{d}(k) + D_f(k)\mathbf{f}(k), \end{cases}$$

where

$$\begin{aligned} A_K(k) &= A(k) - K_p(k)C(k), \\ B_{Kd}(k) &= B_d(k) - K_p(k)D_d(k), \\ B_{Kf}(k) &= B_f(k) - K_p(k)D_f(k). \end{aligned}$$

Let

$$\begin{aligned} &\tilde{\mathbf{y}}_{\mathbf{d}}(k) = \tilde{\mathbf{y}}(k)|_{\mathbf{f}(k)=0}, \ \tilde{\mathbf{y}}_{\mathbf{f}}(k) = \tilde{\mathbf{y}}(k)|_{\mathbf{d}(k)=0, \mathbf{x}_0=0}, \\ &\mathbf{r}_{\mathbf{d}}(k) = R_{\tilde{\mathbf{y}}}^{-1/2}(k)\tilde{\mathbf{y}}_{\mathbf{d}}(k), \ \mathbf{r}_{\mathbf{f}}(k) = R_{\tilde{\mathbf{y}}}^{-1/2}(k)\tilde{\mathbf{y}}_{\mathbf{f}}(k). \end{aligned}$$

We then have

$$\begin{gathered} \mathbf{r}_{\mathbf{d}N} = G_{\mathbf{r}\bar{\mathbf{d}}_N}\bar{\mathbf{d}}_N, \ \mathbf{r}_{\mathbf{f}N} = G_{\mathbf{r}\mathbf{f}_N}\mathbf{f}_N, \\ G_{\mathbf{r}\bar{\mathbf{d}}_N} = R_{\tilde{\mathbf{y}}_N} \quad \left[\tilde{\Omega} \ \tilde{\Gamma}_d \right], \ G_{\mathbf{r}\mathbf{f}_N} = R_{\tilde{\mathbf{y}}_N}\tilde{\Gamma}_f, \ \bar{\mathbf{d}}_N = \left[\mathbf{x}_0^T \ \mathbf{d}_N^T \right]^T, \end{gathered}$$

where $\tilde{\Omega}$, $\tilde{\Gamma}_d$ and $\tilde{\Gamma}_f$ are constructed by replacing $A(k)$, $B_d(k)$, $B_f(k)$ in Ω, Γ_d and Γ_f with $A_K(k)$, $B_{Kd}(k)$, $B_{Kf}(k)$, respectively. Recall that $\{\tilde{\mathbf{y}}(k)\}_{i=0}^N$ forms an orthogonal basis of $\mathcal{L}\{\{\mathbf{y}(k)\}_{k=0}^N\}$. It is immediate to get

$$\begin{aligned} R_{\tilde{\mathbf{y}}_N} &= \text{diag}(R_{\tilde{\mathbf{y}}}(0), R_{\tilde{\mathbf{y}}}(1), \ldots, R_{\tilde{\mathbf{y}}}(N)), \\ R_{\mathbf{r}_{\mathbf{d}N}} &= \text{diag}(R_{\mathbf{r}_{\mathbf{d}}}(0), R_{\mathbf{r}_{\mathbf{d}}}(1), \ldots, R_{\mathbf{r}_{\mathbf{d}}}(N)) = I. \end{aligned}$$

In view of (8.10), we have

$$\begin{aligned} R_{\mathbf{r}_{\mathbf{d}_N}} &= G_{\mathbf{r}\bar{\mathbf{d}}_N} R_{\bar{\mathbf{d}}_N} G_{\mathbf{r}\bar{\mathbf{d}}_N}^T = G_{\mathbf{r}\bar{\mathbf{d}}_N} G_{\mathbf{r}\bar{\mathbf{d}}_N}^T, \\ R_{\mathbf{r}_{\mathbf{f}_N}} &= G_{\mathbf{r}\mathbf{f}_N} R_{\mathbf{f}_N} G_{\mathbf{r}\mathbf{f}_N}^T = G_{\mathbf{r}\mathbf{f}_N} G_{\mathbf{r}\mathbf{f}_N}^T. \end{aligned}$$

Consequently, we obtain

$$\begin{aligned} \gamma_d^2 &= \sup_{x_0, d(k)} \frac{J_N|_{f_N=0}}{x_0^T x_0 + \sum_{k=0}^N d^T(k)d(k)} = \lambda_{max}(\mathcal{R}_{\mathbf{r}_{dN}}) = 1, \\ \gamma_f^2 &= \sup_{f(k)} \frac{J_N|_{x_0=0, \ d_N=0}}{\sum_{k=0}^N f^T(k)f(k)} = \lambda_{max}(\mathcal{R}_{\mathbf{r}_{fN}}), \\ \gamma_{f-}^2 &= \inf_{f(k)} \frac{J_N|_{x_0=0, \ d_N=0}}{\sum_{k=0}^N f^T(k)f(k)} = \lambda_{-}(\mathcal{R}_{\mathbf{r}_{fN}}), \end{aligned}$$

where $\lambda_{max}(\mathcal{R}_{\mathbf{r}_{fN}})$ and $\lambda_{max}(\mathcal{R}_{\mathbf{r}_{dN}})$ denote the maximum eigenvalues of $\mathcal{R}_{\mathbf{r}_{fN}}$ and $\mathcal{R}_{\mathbf{r}_{dN}}$, respectively, $\lambda_{-}(\mathcal{R}_{\mathbf{r}_{fN}})$ is the nonzero minimum eigenvalue of $\mathcal{R}_{\mathbf{r}_{fN}}$. It turns out that

$$\frac{\gamma_f^2}{\gamma_d^2} = \lambda_{max}(\mathcal{R}_{\mathbf{r}_{\mathbf{f}N}}), \quad \frac{\gamma_{f-}^2}{\gamma_d^2} = \lambda_{-}(\mathcal{R}_{\mathbf{r}_{\mathbf{f}N}}).$$

Next, we consider the case of arbitrary sequence $\{\tilde{y}_1(k)\}_{k=0}^{N}$ given by

$$\begin{cases} \hat{x}_1(k+1) = A(k)\hat{x}_1(k) + L_1(k)(y(k) - C(k)\hat{x}_1(k)), \\ \tilde{y}_1(k) = y(k) - C(k)\hat{x}_1(k), \ \hat{x}_1(0) = 0, \end{cases} \tag{8.18}$$

where $L_1(k)$ is a given gain matrix. It is known from Theorem 4.2 that there exists a mapping $\mathcal{W}: \ \tilde{y}(k) \mapsto \tilde{y}_1(k)$, i.e.,

$$\begin{cases} \eta(k+1) = (A(k) - L_1(k)C(k))\eta(k) + (K_p(k) - L_1(k))\tilde{y}(k), \\ \tilde{y}_1(k) = C(k)\eta(k) + \tilde{y}(k), \ \eta(0) = 0, \end{cases}$$

such that $\tilde{y}_{1N} = W_r(N)\tilde{y}_N$, where $K_p(k), \tilde{y}(k)$ are given in (8.13) and (8.16), $W_r(N)$ is constructed by replacing $A(k)$, $B_d(k)$ and $D_d(k)$ in Γ_d with $A(k) - L_1(k)C(k)$, $K_p(k) - L_1(k)$ and I, respectively. Define

$$\tilde{\mathbf{y}}_{1N} := W_r(N)\tilde{\mathbf{y}}_N, \ \mathbf{r}_{1dN} := R_{\tilde{\mathbf{y}}_{1N}}^{-1/2}\tilde{\mathbf{y}}_{1N}|_{\mathbf{f}_N=0}, \ \mathbf{r}_{1fN} := R_{\tilde{\mathbf{y}}_{1N}}^{-1/2}\tilde{\mathbf{y}}_{1N}|_{\bar{\mathbf{d}}_N=0}.$$

We have

$$\begin{aligned} \mathbf{r}_{1dN} &= \Gamma_r(N)\mathbf{r}_{dN}, \ \mathcal{R}_{\mathbf{r}_{1dN}} = \Gamma_r(N)\mathcal{R}_{\mathbf{r}_{dN}}\Gamma_r^T(N), \\ \mathbf{r}_{1fN} &= \Gamma_r(N)\mathbf{r}_{fN}, \ \mathcal{R}_{\mathbf{r}_{1fN}} = \Gamma_r(N)\mathcal{R}_{\mathbf{r}_{fN}}\Gamma_r^T(N), \end{aligned}$$

where

$$\Gamma_r(N) = R_{\tilde{\mathbf{y}}_{1N}}^{-1/2} W_r(N) R_{\tilde{\mathbf{y}}_N}^{1/2}, \ R_{\tilde{\mathbf{y}}_{1N}} = W_r(N) R_{\tilde{\mathbf{y}}_N} W_r^T(N).$$

Let the evaluation function be calculated by J_{1N}, which equals to $\tilde{y}_{1N}^T R_{\tilde{\mathbf{y}}_{1N}}^{-1} \tilde{y}_{1N}$. We can get

$$\begin{aligned} \gamma_{1d}^2 = \lambda_{max}(\mathcal{R}_{\mathbf{r}_{1dN}}) = \lambda_{max}(\Gamma_r(N)\mathcal{R}_{\mathbf{r}_{\mathbf{d}N}}\Gamma_r^T(N)) &= \lambda_{max}(\Gamma_r(N)\Gamma_r^T(N)), \\ \gamma_{1f}^2 = \lambda_{max}(\mathcal{R}_{\mathbf{r}_{1fN}}) &= \lambda_{max}(\Gamma_r(N)\mathcal{R}_{\mathbf{r}_{\mathbf{f}N}}\Gamma_r^T(N)), \\ \gamma_{1f_-}^2 = \lambda_{-}(\mathcal{R}_{\mathbf{r}_{1fN}}) &= \lambda_{-}(\Gamma_r(N)\mathcal{R}_{\mathbf{r}_{\mathbf{f}N}}\Gamma_r^T(N)). \end{aligned}$$

Note that

$$\frac{\lambda_{max}(\Gamma_r(N)\mathcal{R}_{\mathbf{r}_{\mathbf{f}N}}\Gamma_r^T(N))}{\lambda_{max}(\Gamma_r(N)\Gamma_r^T(N))} \leq \lambda_{max}(\mathcal{R}_{\mathbf{r}_{\mathbf{f}N}}),$$
$$\frac{\lambda_{-}(\Gamma_r(N)\mathcal{R}_{\mathbf{r}_{\mathbf{f}N}}\Gamma_r^T(N))}{\lambda_{max}(\Gamma_r(N)\Gamma_r^T(N))} \leq \lambda_{-}(\mathcal{R}_{\mathbf{r}_{\mathbf{f}N}}).$$

We then have

$$\frac{\gamma_{1f}^2}{\gamma_{1d}^2} \leq \frac{\gamma_f^2}{\gamma_d^2}, \ \frac{\gamma_{1f-}^2}{\gamma_{1d}^2} \leq \frac{\gamma_{f-}^2}{\gamma_d^2},$$

which leads to the following proposition.

Proposition 8.2 *The evaluation function J_N given by (8.16)–(8.17) with (8.13)–(8.15) is an optimal solution satisfying the sensitivity/robustness ratio criteria (8.8).*

Remark 8.2 Although evaluation function J_N in (8.17) is determined by using the minimal estimate of $J(x_0, d_N)$, the sensitivity/robustness ratio criteria in (8.8) are satisfied. As expected, the developed FD scheme presents a new way to generate evaluation function.

8.4 Applications to Observer-Based Fault Detection

In this section we apply the developed scheme to observer-based FD and compare it with the existing H_∞/H_∞ and H_-/H_∞-optimal FD methods.

Let

$$L(k) = K_p(k), \ W(k) = R_{\tilde{\mathbf{y}}}^{-\frac{1}{2}}(k). \tag{8.19}$$

The residual generator (8.3) becomes

$$\begin{cases} \hat{x}(k+1) = A(k)\hat{x}(k) + K_p(k)(y(k) - C(k)\hat{x}(k)), \ \hat{x}(0) = 0, \\ \tilde{y}(k) = y(k) - C(k)\hat{x}(k), \\ r(k) = R_{\tilde{\mathbf{y}}}^{-\frac{1}{2}}(k)\tilde{y}(k). \end{cases} \tag{8.20}$$

Then the evaluation function J_N can be rewritten as

$$J_N = \sum_{k=0}^{N} r^T(k)r(k) = J_{2,[0,N]}(r).$$

As a result, for indices $\|G_{rd}\|_{\infty,[0,N]}$, $\|G_{rf}\|_{\infty,[0,N]}$ and $\|G_{rf}\|_{-,[0,N]}$ given respectively in (6.4)–(6.6), we have

$$\|G_{rd}\|_{\infty,[0,N]} = \gamma_d^2, \ \|G_{rf}\|_{\infty,[0,N]} = \gamma_f^2, \ \|G_{rf}\|_{-,[0,N]} = \gamma_{f-}^2.$$

In view of (8.8), we can conclude that residual generator (8.3) with $L(k)$, $W(k)$ given by (8.19) delivers an observer-based FD solving the Problems 6.1 and 6.2. The residual generator (8.3) with (8.19) is thus the optimal H_∞/H_∞- and/or H_-/H_∞-FDF, while $W(k) = R_{\tilde{\mathbf{y}}}^{-\frac{1}{2}}(k)$ is a post-filter corresponding to $L(k)$.

Similarly, for an arbitrary sequence $\{\tilde{y}_1(k)\}_{k=0}^N$ in (8.18), there exists a mapping $\mathcal{W}_I : \tilde{y}_1(k) \mapsto \tilde{y}(k)$ such that

$$\begin{cases} \mu(k+1) = (A(k) - K_p(k)C(k))\mu(k) + (L_1(k) - K_p(k))\tilde{y}_1(k), \\ \tilde{y}(k) = C(k)\mu(k) + \tilde{y}_1(k), \\ \mu(0) = 0. \end{cases}$$

Suppose that the post-filter $W(k)$ is described by

$$\begin{cases} \mu(k+1) = (A(k) - K_p(k)C(k))\mu(k) + (L_1(k) - K_p(k))\tilde{y}_1(k), \\ r(k) = R_{\tilde{\mathbf{y}}}^{-1/2}(k)(C(k)\mu(k) + \tilde{y}_1(k)), \\ \mu(0) = 0. \end{cases} \tag{8.21}$$

It is known from Theorem 6.2 that, the residual generator (8.3) with $L(k) = L_1(k)$ and (8.21) delivers also an optimal H_∞/H_∞- and/or H_-/H_∞-FDF, while $W(k)$ described by (8.21) is the so-called dynamic post-filter. Therefore, the dynamic post-filter $W(k)$ with arbitrary innovation sequences $\{\tilde{y}_1(k)\}_{k=0}^N$ described by (8.18) delivers a so-called dynamic optimal H_∞/H_∞- and/or H_-/H_∞-FDF. The optimal H_∞/H_∞- and H_-/H_∞-FDFs with respect to solving (6.7) and (6.8) can be obtained by using (8.13)–(8.15) and (8.20). The corresponding residual evaluation function $J_{2,[0,N]}(r)$ is the same as the new generated J_N. From this point of view, the developed FD method provides an alternative realization form of optimal H_∞/H_∞- and H_-/H_∞ FDFs.

8.5 Conclusion

In this chapter, we have presented an H_i/H_∞-optimal design scheme of FD for LDTV systems subject to $l_{2,[0,N]}$-norm bounded unknown inputs, which serves as an alternative solution to the well-established H_∞/H_∞, H_-/H_∞ and H_∞-filtering based FD system design. The basic idea is to compare an optimal estimation of the $l_{2,[0,N]}$-norm of the unknown inputs and their $l_{2,[0,N]}$-norm boundedness. Different from the H_-/H_∞ or H_∞/H_∞ trade-off optimization in Chap. 6 and the H_∞ filtering based FD in later Chap. 11, such a design method is based on available knowledge of the unknown inputs, i.e. their $l_{2,[0,N]}$-norm boundedness. A successful FD can be achieved if, in the fault-free case, the estimated $l_{2,[0,N]}$-norm is bounded by the known $l_{2,[0,N]}$-norm boundedness of the unknown inputs and by a fault the estimated $l_{2,[0,N]}$-norm is larger than the $l_{2,[0,N]}$-norm boundedness. By using such an $l_{2,[0,N]}$-norm estimate as evaluation function, a separate step for the threshold setting and the associated computation needed in H_∞-filtering, H_-/H_∞- and H_∞/H_∞-optimization based FD schemes are unnecessary any more. To obtain a right estimate of the $l_{2,[0,N]}$-norm of the unknown inputs, we have formulated the problem as finding a minimum of quadratic form usually applied in linear quadratic regulation, and then projection and innovation analysis in Krein space have been applied aiming at avoiding complicated online computation.

References

1. Hassibi, B., Sayed, A. H., & Kailath, T. (1999). *Indefinite-quadratic estimation and control: A unified approach to H_2 and H_∞ theories*. SIAM.
2. Shen, B., Ding, S. X., & Wang, Z. (2013). Finite-horizon H_∞ fault estimation for linear discrete time-varying systems with delayed measurements. *Automatica, 49*(1), 293–296.
3. Zhong, M., Liu, S., & Zhao, H. (2008). Krein space-based H_∞ fault estimation for linear discrete time-varying systems. *Acta Automatica Sinica, 34*(12), 1529–1533.
4. Zhong, M., Ding, S. X., & Ding, E. L. (2010). Optimal fault detection for linear discrete time-varying systems. *Automatica, 46*(8), 1395–1400.
5. Zhong, M., Zhou, D., & Ding, S. X. (2010). On designing H_∞ fault detection filter for linear discrete time-varying systems. *IEEE Transactions on Automatic Control, 55*(7), 1689–1695.
6. Zhong, M., Li, S., & Zhao, Y. (2013). Robust H_∞ fault detection for uncertain LDTV systems using Krein space approach. *International Journal of Innovative Computing, Information and Control, 9*(4), 1637–1649.
7. Kailath, T., Sayed, A. H., & Hassibi, B. (1999). *Linear estimation*. New Jersey: Prentice Hall.
8. Li, X. (2009). *Fault detection filter design for linear systems*. Ph.D. dissertation. Louisiana State University, USA.

Chapter 9
An H_i/H_∞ Approach to Event-Triggered Optimal Fault Detection

Event-triggered communication has been increasingly utilized in the networked control systems for saving communication resources. As a response to this surge, this chapter is devoted to an event-triggered H_i/H_∞-optimization scheme of FD for linear discrete time systems. By describing the dynamics of an event-triggered system as a linear discrete time system with varying sampling periods, an observer-based FDF is constructed as an event-triggered residual generator in the framework of H_i/H_∞-optimization. An optimal solution can then be obtained by recursive calculation of Riccati equations. In this way, the generated residual signal is completely decoupled from the event-triggered transmission error such that the design of an FDF and the event generator can be carried out independently. Also, a satisfactory trade-off between the robustness of residual to disturbances and sensitivity to faults can be guaranteed. A vehicle lateral dynamic system is used as a simulation example to demonstrate the effectiveness of the obtained results.

9.1 Introduction

With the rapid development of communication technologies, more and more practical systems are implemented via networks. Although the measurement outputs of a control system are sampled periodically and transmitting all the sampled-data packets is beneficial to system analysis and synthesis, it is sometimes undesirable due to the limitation of network resources, such as the communication bandwidth and power energy. When a system works in steady state, it may have a very small measurement outputs fluctuation over a certain time interval. For the purpose of efficient network resource utilization, event-triggered communication has gained increasing interests in the fields of control, state estimation, and filtering of complex systems [4, 5, 9, 10, 12]. In event-triggering schemes, necessary data transmissions are determined by an "event generator" and an event is occurred only when a predefined triggering condition is achieved. Because of the loss of some system information as well

M. Zhong et al., *Fault Diagnosis for Linear Discrete Time-Varying Systems and Its Applications*, https://doi.org/10.1007/978-981-19-5438-2_9

as the introduction of time-varying dynamics by nonuniform sampling pattern, the residual generated for an event-triggered FD system is affected not only by disturbances and faults, but also by the event-triggered transmission errors. Due to this, the problem of FD becomes more complicated for a control system with event-triggered communication.

In recent years, numerous efforts have been made to solve the problem of event-triggered FD, see for instance, [2, 3, 6–9, 11, 13–16], to mention only some of them. In [11], the problem of event-triggered FD filtering was considered for networked switched systems subject to repeated scalar nonlinearities and stochastic disturbance. In [3], the problem of event-triggered FD was studied for networked control systems with communication delay and nonlinear perturbation. In [13], the coordinated design of event-triggered FDF and controller was dealt with for a continuous-time networked control system with biased sensor faults. In [14], the adaptively adjusted event-triggering based FD was applied to a networked aircraft control system. In [7], the problem of event-based H_∞-FDF design was investigated for complex systems over communication networks subject to Markovian jump parameters. In [16], an event-triggered FDF was designed for fuzzy stochastic systems with missing measurements. In [2, 9, 15], the event-triggered H_∞-FDF was designed for Takagi-Sugeno Fuzzy systems in a network environment. In [8], an event-triggered non-fragile H_∞-FDF was designed for discrete time-delayed nonlinear systems with channel fadings. In [6], a multi-objective formulation of the event-triggered FD was presented for discrete time linear systems based on l_1, H_- and H_∞ performance indices.

It is noted that most of the above mentioned studies dealt with the problem of event-triggered FD in the framework of H_∞ filtering. While, only sufficient conditions for the design of H_∞-FDF were developed, which lead to the solutions of being conservative. More recently, the H_∞/H_∞-optimization based FD was applied to event-triggered FD in [1], and a solution to the optimization problem was obtained by solving discrete time Riccati equation. It is worth pointing out that the aforementioned event-triggered FD can only attenuate the effects associated with the event-triggered transmission errors to a certain extent. Achieve a full-decoupling of residual from the event-triggered transmission errors while guaranteeing a satisfactory FD performance remains a challenging task.

Motivated by these observations, we investigate an H_i/H_∞-optimization approach to FD for event-triggered linear discrete time systems in this chapter. To this end, a linear discrete time system with nonuniform sampling periods is established to describe the time-varying system dynamics caused by event-triggering data transmission mechanism. Concerning the influence of event-triggered transmission errors, our main attention will be paid to

- construct an FDF for achieving complete decoupling of residual from event-triggered transmission error, such that the event-triggered H_i/H_∞-FDF is independent of the configuration of the event generator;
- find an optimal solution to the event-triggered H_i/H_∞-optimization problem;
- and then give an evaluation function and a threshold towards optimal FD.

9.2 Preliminaries of an Event-Triggered FD System

The event-trigged FD under consideration is illustrated in Fig. 9.1, in which the plant is a linear discrete time system described by

$$\begin{cases} x(k+1) = Ax(k) + B_u u(k) + B_d d(k) + B_f f(k), \\ y(k) = Cx(k) + D_u u(k) + D_d d(k) + D_f f(k), \end{cases} \tag{9.1}$$

where $x(k) \in \mathbb{R}^n$, $y(k) \in \mathbb{R}^q$, $u(k) \in \mathbb{R}^p$, $d(k) \in \mathbb{R}^m$ and $f(k) \in \mathbb{R}^l$ are the state, measurement output, control input, unknown input and fault vectors, respectively, A, B_u, B_d, B_f, C, D_u, D_d and D_f are known constant matrices with appropriate dimensions. It is assumed that the control input $u(k)$ is time-triggered and available for FD and the measurement output $y(k)$ is transmitted to the FD module based on an event-triggering mechanism.

Let k_i $(i = 0, 1, 2, \ldots)$ be the release instant time step of the event-triggered FD and $k_{i+1} > k_i$. The event generator is used to check whether or not the current measurement output $y(k)$ satisfies the following event condition

$$||\Omega[y(k) - y(k_i)]|| < \epsilon||\Omega y(k)||, \tag{9.2}$$

where $\Omega \in \mathbb{R}^{q\times q}$ is a weighting matrix, $\epsilon > 0$ is a threshold for event generator. Once condition (9.2) is violated, the current measurement output is transmitted to the FD module. Otherwise, the data packet is discarded. Therefore, when the measurement output $y(k_i)$ is released by the event generator, the next release instant time step k_{i+1} is determined by the following condition

$$k_{i+1} = k_i + \min_{1\le j\le \tau_M} \{j| \; ||\Omega e_y(k_i)]|| \ge \epsilon||\Omega y(k)||\}, \tag{9.3}$$

where $e_y(k_i) = y(k_i + j) - y(k_i)$, τ_M is a given maximum event interval. Hence, the updating of input data $\bar{y}(k)$ of FD module can be described by

$$\bar{y}(k) = y(k_i), \; \forall k \in [k_i, k_{i+1}).$$

The main tasks of event-triggered FD consist of residual generation and residual evaluation by using the available $u(k)$ and $\bar{y}(k)$. However, because the data

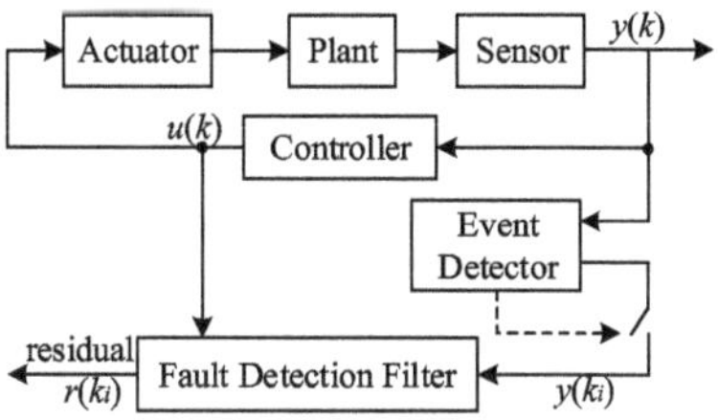

Fig. 9.1 The structure of an event-triggered residual generator

transmission is triggered by the event, the error between $\bar{y}(k)$ and $y(k)$ exists for $k \in (k_i,\ k_{i+1})$, i.e., the so-called event-triggered transmission error $e_y(k) = y(k) - \bar{y}(k)$. As is known from the existing schemes reported in the literature, one of the major challenges encountered in event-triggered FD comes from the influences of event-triggered transmission error on the residual signal. Despite some efforts have been made to attenuate their effects on both the residual generation and residual evaluation, achieving complete decoupling remains a challenging task. In what follows, we will focus on solving this problem in the framework of H_i/H_∞-optimization.

Remark 9.1 The time-based assumption on control input may not be available for a practical event-triggered control system. For example, the control input is event-triggered and $u(k) = u_c(k_j),\ \text{for}\ k \in [k_j, k_{j+1})$, where $u_c(k_j)$ is the released control input from controller at time step k_j. If this is the case, the control input to the FD unit is chosen as $u(k)$. For the sake of simplicity, the control input is assumed to be time-based in this book. Especially, this approach is meaningful when the sensors are far from the actuators and communicate with the FD unit through a network, while the FD unit is implemented near the actuator.

9.3 Event-Triggered H_i/H_∞-Optimal Fault Detection

Aiming to achieve complete decoupling of event-triggered transmission error on residual, this section focuses on the development of an event-triggered FD system in the framework of H_i/H_∞-optimization.

In observing (9.1), the system state at event-triggering instant k_{i+1} can be expressed as

$$\begin{aligned} x(k_{i+1}) = {} & A^{k_{i+1}-k_i}x(k_i) + \sum_{j=k_i}^{k_{i+1}-1} A^{k_{i+1}-j-1}B_u u(j) \\ & + \sum_{j=k_i}^{k_{i+1}-1} A^{k_{i+1}-j-1}B_d d(j) + \sum_{j=k_i}^{k_{i+1}-1} A^{k_{i+1}-j-1}B_f f(j). \end{aligned} \tag{9.4}$$

Let

$$\begin{cases} \underline{A}(k_i) = A^{k_{i+1}-k_i}, \\ \underline{B}_\vartheta(k_i) = [\, A^{k_{i+1}-k_i-1}B_\vartheta \cdots AB_\vartheta\ B_\vartheta \,], \\ \underline{\vartheta}(k_i) = [\, \vartheta^T(k_i)\ \vartheta^T(k_i+1) \cdots \vartheta^T(k_{i+1}-1) \,]^T, \end{cases} \tag{9.5}$$

where ϑ stands for u, d and f in the sequel. Rewrite (9.4) as

$$x(k_{i+1}) = \underline{A}(k_i)x(k_i) + \underline{B}_u(k_i)\underline{u}(k_i) + \underline{B}_d(k_i)\underline{d}(k_i) + \underline{B}_f(k_i)\underline{f}(k_i).$$

Then the event-triggered system dynamics can be described by the following linear discrete time system model with varying sampling periods

$$\begin{cases} x(k_{i+1}) = \underline{A}(k_i)x(k_i) + \underline{B}_u(k_i)\underline{u}(k_i) + \underline{B}_d(k_i)\underline{d}(k_i) + \underline{B}_f(k_i)\underline{f}(k_i), \\ y(k_i) = Cx(k_i) + \underline{D}_u(k_i)\underline{u}(k_i) + \underline{D}_d(k_i)\underline{d}(k_i) + \underline{D}_f(k_i)\underline{f}(k_i), \end{cases} \tag{9.6}$$

where

$$\underline{D}_\vartheta(k_i) = \begin{bmatrix} D_\vartheta\ 0 \cdots 0 \end{bmatrix}. \tag{9.7}$$

For the purpose of FD, we use the following FDF as an event-triggered residual generator

$$\begin{cases} \hat{x}(k_{i+1}) = \underline{A}(k_i)\hat{x}(k_i) + \underline{B}_u(k_i)\underline{u}(k_i) + L(k_i)\tilde{y}(k_i), \\ \tilde{y}(k_i) = y(k_i) - C\hat{x}(k_i), \\ r(k_i) = W(k_i)\tilde{y}(k_i), \end{cases} \tag{9.8}$$

where $r(k_i)$ is the generated residual when data $y(k_i)$ is transmitted to the FD module, $L(k_i)$ and $W(k_i)$ denote respectively the observer gain matrix and weighting matrix.

Let $e(k_i) = x(k_i) - \hat{x}(k_i)$. It follows from (9.6) and (9.8) that

$$\begin{cases} e(k_{i+1}) = \underline{A}_L(k_i)e(k_i) + \underline{B}_{dL}(k_i)\underline{d}(k_i) + \underline{B}_{fL}(k_i)\underline{f}(k_i), \\ r(k_i) = W(k_i)(Ce(k_i) + \underline{D}_d(k_i)\underline{d}(k_i) + \underline{D}_f(k_i)\underline{f}(k_i)), \end{cases} \tag{9.9}$$

where

$$\begin{aligned} \underline{A}_L(k_i) &= \underline{A}(k_i) - L(k_i)C, \\ \underline{B}_{dL}(k_i) &= \underline{B}_d(k_i) - L(k_i)\underline{D}_d(k_i), \\ \underline{B}_{fL}(k_i) &= \underline{B}_f(k_i) - L(k_i)\underline{D}_f(k_i). \end{aligned}$$

It is easily seen that, by using the FDF (9.8), complete decoupling of residual $r(k_i)$ from the event-triggered transmission error $e_y(k)$ is realized. Thus, the major issue of event-triggered residual generation is to find $L(k_i)$ and $W(k_i)$ such that the generated residual is simultaneously robust to unknown input and sensitive to fault.

For a given positive integer N, define

$$||\mathcal{G}_{rd}||_{2,[0,k_N]} = \sup_{f(k)=0} \frac{\sum_{i=0}^{N} ||r(k_i)||^2}{||x_0||^2 + \sum_{k=0}^{k_N} ||d(k)||^2}, \tag{9.10}$$

$$||\mathcal{G}_{rf}||_{2,[0,k_N]} = \sup_{x_0, d(k)=0} \frac{\sum_{i=0}^{N} ||r(k_i)||^2}{\sum_{k=k_0}^{k_N} ||f(k)||^2}, \tag{9.11}$$

$$||\mathcal{G}_{rf}||_{-,[0,k_N]} = \inf_{x_0, d(k)=0} \frac{\sum_{i=0}^{N} ||r(k_i)||^2}{\sum_{k=k_0}^{k_N} ||f(k)||^2}. \tag{9.12}$$

Similar to the H_i/H_∞-optimization based FD for LDTV systems, $||\mathcal{G}_{rd}||_{2,[0,k_N]}$ is the worst-case robustness index of residual against unknown input, $||\mathcal{G}_{rf}||_{2,[0,k_N]}$ and $||\mathcal{G}_{rf}||_{-,[0,k_N]}$ are defined as the best-case and worst-case fault sensitivity indices, respectively. Then the design of the event-triggered FDF can be formulated into the optimization problem towards a trade-off between the robustness and sensitivity indices

$$\max_{L(k_i),W(k_i)} \frac{||\mathcal{G}_{rf}||_{2,[0,k_N]}}{||\mathcal{G}_{rd}||_{2,[0,k_N]}}, \text{ subject to (9.9)} \tag{9.13}$$

and

$$\max_{L(k_i),W(k_i)} \frac{||\mathcal{G}_{rf}||_{-,[0,k_N]}}{||\mathcal{G}_{rd}||_{2,[0,k_N]}}, \text{ subject to (9.9)}. \tag{9.14}$$

The residual generator (9.8) satisfying (9.13) is the so-called event-triggered H_∞/H_∞-FDF, while the one satisfying (9.14) is the so-called event-triggered H_-/H_∞-FDF.

By choose an evaluation function $J(r(k_i))$ and an appropriate threshold J_{th}, the detection of fault can be carried out by using the following decision logic

$$\begin{cases} J(r(k_i)) > J_{th}, & \text{fault alarm,} \\ J(r(k_i)) \leq J_{th}, & \text{no alarm.} \end{cases} \tag{9.15}$$

We now summarize the basic ideas of the above event-triggered FD scheme as follows:

- Based on the established system model (9.6), an event-triggered FDF (9.8) is constructed such that the residual $r(k_i)$ is completely decoupled from the event-triggered transmission error $e_y(k)$;
- The design of an H_i/H_∞-FDF is formulated as to find $L(k_i)$ and $W(k_i)$ such that a trade-off between the robustness of residual to unknown input and the sensitivity to fault is achieved in the sense of (9.13) and (9.14);
- A novel residual evaluation strategy is presented unifiedly with the event-triggered residual generator (9.8), i.e., the determination of $J(r(k_i))$ and J_{th}.

Remark 9.2 The complete decoupling of residual $r(k_i)$ from the event-triggered transmission error $e_y(k)$ allows to improve the performance of event-triggered FD. Thanks to this feature, residual generator (9.8) and event detector satisfying (9.3) can be designed independently. Therefore, the presented FD scheme can be applied to the other kinds of event-triggering mechanisms, such as the event-triggering strategy in [9, 12] and the adaptively adjusted ones in [14, 16]. In addition, it is easy to see from (9.8)–(9.9) that the network-induced delay via $y(k_i)$ has no influence on $r(k_i)$, which means that this approach is applicable even if the network-induced delay being taken into account.

9.3.1 Design of an Event-Triggered H_i/H_∞-FDF

In this subsection, we design an event-triggered H_i/H_∞-FDF by using the optimal approach demonstrated in Chap. 8 for LDTV systems.

First, we introduce the following LDTV system

$$\begin{cases} z(i+1) = (\underline{A}(k_i) - L(k_i)C)z(i) + (\underline{B}_d(k_i) - L(k_i)\underline{D}_d(k_i))v(i) \\ \qquad\qquad + (\underline{B}_f(k_i) - L(k_i)\underline{D}_f(k_i))f_z(i), \\ r_z(i) = W(k_i)(Cz(i) + \underline{D}_d(k_i)v(i) + \underline{D}_f(k_i)f_z(i)), \\ z(0) = x_0, \end{cases} \tag{9.16}$$

where $z(i) = e(k_i)$, $r_z(i) = r(k_i)$, and

$$v(i) = \begin{cases} \underline{d}(k_i), \ i \le N-1 \\ d(k_N), \ i = N \end{cases}, \ f_z(i) = \begin{cases} \underline{f}(k_i), \ i \le N-1 \\ f(k_N), \ i = N \end{cases}.$$

It is clear that

$$\begin{aligned} \sum_{i=0}^{N} ||r_z(i)||^2 &= \sum_{i=0}^{N} ||r(k_i)||^2, \\ \sum_{i=0}^{N} ||v(i)||^2 &= \sum_{i=0}^{N} ||\underline{d}(k_i)||^2 = \sum_{k=k_0}^{k_N} ||d(k)||^2, \\ \sum_{i=0}^{N} ||f_z(i)||^2 &= \sum_{i=0}^{N} ||\underline{f}(k_i)||^2 = \sum_{k=k_0}^{k_N} ||f(k)||^2. \end{aligned}$$

Define

$$\begin{aligned} \gamma_v^2 &= \sup_{f_z=0} \frac{\sum_{i=0}^{N} ||r_z(i)||^2}{||x_0||^2 + \sum_{i=0}^{N} ||v(i)||^2}, \\ \gamma_f^2 &= \sup_{x_0, v=0} \frac{\sum_{i=0}^{N} ||r_z(i)||^2}{\sum_{i=0}^{N} ||f_z(i)||^2}, \\ \gamma_{f-}^2 &= \inf_{x_0, v=0} \frac{\sum_{i=0}^{N} ||r_z(i)||^2}{\sum_{i=0}^{N} ||f_z(i)||^2}. \end{aligned}$$

Thus, the optimization problem (9.13) and (9.14) can be equivalent respectively to

$$\max_{L(k_i),\, W(k_i)} \frac{\gamma_f}{\gamma_v}, \text{ subject to (9.16)} \tag{9.17}$$

and

$$\max_{L(k_i),\,W(k_i)} \frac{\gamma_{f-}}{\gamma_v}, \text{ subject to (9.16).} \tag{9.18}$$

By applying the H_i/H_∞-optimization approach in Chap. 8, the optimization problem (9.17) and (9.18) can be solved based on the following lemma.

Lemma 9.1 *Considering LDTV system (9.16) with norm bounded $v(i)$ and $f_z(i)$, the observer gain $L(k_i)$ and post-filter matrix $W(k_i)$ given by*

$$L(k_i) = (\underline{A}(k_i)P_z(i)C^T + \underline{B}_d(k_i)\underline{D}_d^T(k_i))W^2(k_i), \tag{9.19}$$
$$W(k_i) = (CP_z(i)C^T + \underline{D}_d(k_i)\underline{D}_d^T(k_i))^{-1/2}. \tag{9.20}$$

provide a solution to both the optimization problems (9.17) and (9.18), where $P_z(i) > 0$ is computed recursively by

$$\begin{cases} P_z(i+1) = \underline{A}(k_i)P_z(i)\underline{A}^T(k_i) - L(k_i)W^{-2}(k_i)L^T(k_i) + \underline{B}_d(k_i)\underline{B}_d^T(k_i), \\ P_z(0) = I. \end{cases} \tag{9.21}$$

Moreover, we have $\gamma_v^* = 1$ and

$$\max_{L(k_i),\,W(k_i)} \frac{\gamma_f}{\gamma_v} = \gamma_f^*, \quad \max_{L(k_i),\,W(k_i)} \frac{\gamma_{f-}}{\gamma_v} = \gamma_{f-}^*.$$

By the definitions of $\underline{A}(k_i)$, $\underline{B}_d(k_i)$ and $\underline{D}_d(k_i)$, we have

$$\underline{A}(k_i)P_z(i)C^T = A^{\tau_i}P_z(i)C^T, \tag{9.22}$$
$$\underline{B}_d(k_i)\underline{B}_d^T(k_i) = A^{\tau_i-1}B_dD_d^T, \tag{9.23}$$
$$\underline{D}_d(k_i)\underline{D}_d^T(k_i) = D_dD_d^T, \tag{9.24}$$
$$\underline{A}(k_i)P_z(i)\underline{A}^T(k_i) = A^{\tau_i}P_z(i)(A^{\tau_i})^T, \tag{9.25}$$
$$\underline{B}_d(k_i)\underline{B}_d^T(k_i) = \sum_{j=0}^{\tau_i-1} A^jB_dB_d^T(A^j)^T, \tag{9.26}$$

where $\tau_i = k_{i+1} - k_i$. Substituting (9.22)–(9.26) into (9.19)–(9.20) and (9.21), and letting $P(k_i) = P_z(i)$, yields

$$L(k_i) = (A^{\tau_i}P(k_i)C^T + A^{\tau_i-1}B_dD_d^T)R_{\tilde{y}}^{-1}(k_i), \tag{9.27}$$
$$P(k_{i+1}) = A^{\tau_i}P(k_i)(A^{\tau_i})^T - L(k_i)R_{\tilde{y}}(k_i)L^T(k_i)$$
$$+\sum_{j=0}^{\tau_i-1} A^jB_dB_d^T(A^j)^T, \ P(k_0) = I, \tag{9.28}$$
$$R_{\tilde{y}}(k_i) = CP(k_i)C^T + D_dD_d^T, \tag{9.29}$$
$$W(k_i) = R_{\tilde{y}}^{\frac{1}{2}}(k_i). \tag{9.30}$$

Thus, the following proposition can be concluded and the proof is omitted here for simplicity.

Proposition 9.1 *The residual generator (9.8) with* $L(k_i),\ W(k_i)$ *given recursively by (9.27)–(9.30) delivers a unified solution to the optimization problems (9.13) and (9.14). Moreover, the robustness index corresponding to the given* $L(k_i),\ W(k_i)$ *is* $||\mathcal{G}^*_{rd}||_{2,[0,k_N]} = 1$. *Furthermore, we have*

$$\max_{L(k_i),W(k_i)} \frac{||\mathcal{G}_{rf}||_{2,[0,k_N]}}{||\mathcal{G}_{rd}||_{2,[0,k_N]}} = ||\mathcal{G}^*_{rf}||_{2,[0,k_N]},$$
$$\max_{L(k_i),W(k_i)} \frac{||\mathcal{G}_{rf}||_{-,[0,k_N]}}{||\mathcal{G}_{rd}||_{2,[0,k_N]}} = ||\mathcal{G}^*_{rf}||_{-,[0,k_N]}.$$

9.3.2 Residual Evaluation

Similar to the traditional time-triggered FD scheme, the stage of residual evaluation concerns with the determination of a residual evaluation function $J(r(k_i))$ and a threshold J_{th}. For a given $N > 0$, define an evaluation function as

$$J(r(k_i)) = \frac{1}{k_N - k_0 + 1}\sum_{i=0}^{N} r^T(k_i)r(k_i), \tag{9.31}$$

while a threshold can be chosen based on

$$J_{th} \le \sup_{f(k)=0} J(r(k_i)). \tag{9.32}$$

It is known from Proposition 9.1 that the robustness index corresponding to the $L(k_i),\ W(k_i)$ given by (9.27)–(9.30) is $||\mathcal{G}^*_{rd}||_{2,[0,k_N]} = 1$. In observing (9.10), the fault -free case residual $r(k_i)$ satisfies

$$\frac{\sum_{i=0}^{N}||r(k_i)||^2}{||x_0||^2 + \sum_{k=0}^{k_N}||d(k)||^2} \le 1.$$

We further have

$$J(r(k_i)) \le \frac{1}{k_N - k_0 + 1}(||x_0||^2 + \sum_{k=0}^{k_N}||d(k)||^2). \tag{9.33}$$

Under the assumption of $d(k)$ being norm-bounded and

Algorithm 9.3.3 Implementation of an event-triggered FD system

1: Set N, τ_M, $x(k_0)$, k_0, $P(k_0) = I$, J_{th} and $\hat{x}(k_0) = 0$;
2: Calculate $W(k_0)$, $r(k_0)$ by using (9.30) and (9.8);
3: If an event is generated and the FD module received the transmitted measurement output $y(k_{i+1})$, then calculate $L(k_i)$, $P(k_{i+1})$, $W(k_{i+1})$ by using (9.27)–(9.30);
4: Calculate residual $r(k_{i+1})$ and evaluate function $J(r(k_{i+1}))$ by using (9.8) and (9.31), respectively;
5: Perform the detection of fault by using (9.34);
6: Let $i = i + 1$, go to step (iii), till the end of the simulation.

$$\frac{1}{k+1}(||x_0||^2 + \sum_{j=0}^{k} ||d(j)||^2) \leq \delta_d^2, \ \forall k \in [N, N\tau_M],$$

where $\delta_d > 0$ is a given constant. Then a threshold can be set as $J_{th} = \delta_d^2$. The occurrence of a fault can be detected by

$$\begin{cases} J(r(k_i)) > J_{th}, & \text{fault alarm,} \\ J(r(k_i)) \leq J_{th}, & \text{no alarm.} \end{cases} \tag{9.34}$$

It is clear that such an evaluation mechanism may lead to a zero FAR, but the FDR is poor. To achieve an admissible trade-off between FAR and FDR, a threshold can be reset as $J_{th} = \beta\delta_d^2$ with $\beta \in (0, 1)$. The online implementation of event-triggered FD is summarized as the Algorithm 9.3.3.

To gain a deeper insight into the merits of the developed event-triggered FD system, here we would like to give a brief comparison of it with the method in [1] and remark the following points:

- In [1], the generated residual is not only affected by disturbance and fault, but also the control input as well as reference input, due to the coupling of event-triggered transmission errors. Regarding the maximum influence of event-triggered transmission errors $e_y(k)$, the worst-case possible effect of $d(k)$ on residual is considered to evaluate the robustness of the residual generator, while the sensitivity to fault is evaluated by the best-case possible effect of $f(k)$. In comparison, the robustness and sensitivity performance indices defined by (9.10)–(9.12) are less conservative comparing with [1] and better FD performance can be obtained by using the results in Proposition 9.1.
- In the stage of residual evaluation, root mean square (RMS) type evaluation function is considered in [1] and, due to the effect of event-triggered transmission error, the threshold need to be adjusted online according to the actual reference input. So, the threshold with zero FAR is not a constant and is computed online based on the l_2-norm bound of $d(k)$, the RMS of reference input over the considered moving time window, and the H_∞-norm from both the disturbance and reference input to the residual. As a comparison, the defined evaluation function in (9.31) is event-triggered, while the zero FAR threshold is a constant. Especially, because the

fault-free case residual is affected only by $d(k)$ and initial state, a less conservative threshold can be chosen based on (9.33), which can improve the FD performance in a certain extent;

- It is seen from Algorithm 9.3.3 that the presented event-triggered FD scheme increases the online computation burden compared with the approach in [1]. In fact, the major computational cost concerns with the calculation of $L(k_i)$ and $W(k_i)$, which is performed recursively with Riccati recursion (9.28) for $P(k_i)$ and thus similar to the well known Kalman filter. From this point of view, the increased amount of calculation is acceptable for most of the event-trigged control systems. On the other hand, this scheme results in a constant threshold that is equal to the l_2-norm bound of unknown inputs. This property is of reasonable interests in practical applications, for instance, it allows us to optimize the threshold according to the process operation conditions or in the probabilistic framework.

9.4 A Simulation Example

To show the effectiveness of the developed method, a benchmark of vehicle lateral dynamic system in [1] is considered as the plant of Fig. 9.1. The parameter matrices of the discretized linear system are given as

$$A = \begin{bmatrix} 0.6333 & -0.0672 \\ 2.0570 & 0.6082 \end{bmatrix}, \ B_u = B_f = \begin{bmatrix} -0.0653 \\ 3.4462 \end{bmatrix}, \ C = \begin{bmatrix} -156.7658 & 1.2493 \\ 0 & 1 \end{bmatrix},$$

$$D_u = D_f = \begin{bmatrix} 56 \\ 0 \end{bmatrix}, \ B_d = \begin{bmatrix} 0.1571 & 0.2395 & 0 & 0 \\ 0.3977 & 0.5156 & 0 & 0 \end{bmatrix}, \ D_d = \begin{bmatrix} 0 & 0 & 1 & 0 \\ 0 & 0 & 0 & 1 \end{bmatrix}.$$

The triggering parameters are set as the same with the ones in [1], i.e., $\Omega = I, \ \epsilon = 0.1, \tau_M = 6$. The moving time window length is chosen as $N = 20$.

Firstly, we model the event-triggered system and construct an event-triggered residual generator by using (9.6) and (9.8), respectively. Next, we design an event-triggered H_i/H_∞-FDF by applying Proposition 9.1. The residual evaluation function is computed by

$$J(r(k_i)) = \begin{cases} \frac{1}{k_i - k_{i-N} + 1} \sum\limits_{j=i-N}^{i} r^T(k_j) r(k_j), \ i \geq N, \\ \frac{1}{k_i - k_0 + 1} \sum\limits_{j=0}^{i} r^T(k_j) r(k_j), \ i < N. \end{cases}$$

The disturbance $d(k)$ and fault $f(k)$ are simulated as

$$d(k) = [0.5 \ \ 0.3 \ \ 0.3 \ \ 0.00001]^T \omega(k), \ \ f(k) = \begin{cases} \frac{\pi}{12}, \ k \geq 40, \\ 0, \ \text{elsewhere}, \end{cases}$$

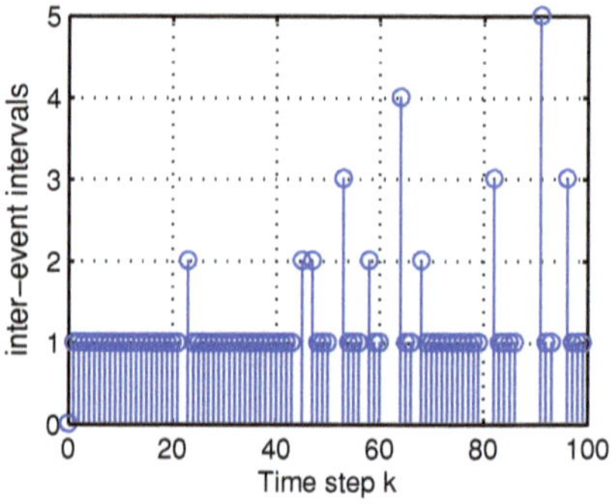

Fig. 9.2 The inter-event intervals $k_{i+1} - k_i$ of case 1

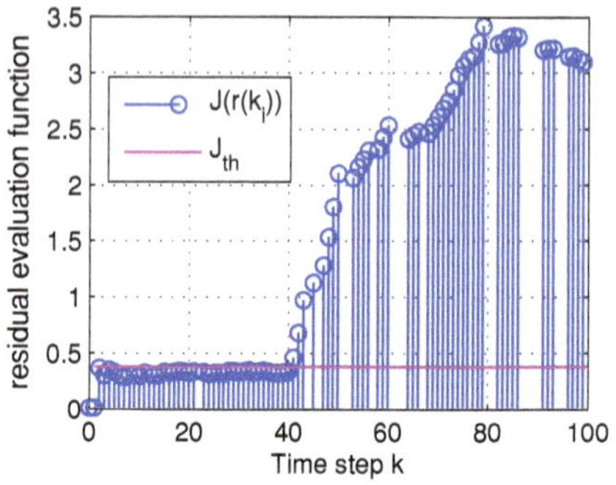

Fig. 9.3 The residual evaluation function $J(r(k_i))$ of case 1

respectively, where $\omega(k) \in \mathcal{U}[-1, 1]$ denotes the uniformly distributed random signal over $[-1, 1]$. Based on (9.32), we set the threshold as $J_{th} = 0.37$.

The first case we considered is $u(k) = 0$. The inter-event intervals $k_{i+1} - k_i$ are depicted in Fig. 9.2. The online event-triggered FD is performed by using Algorithm 9.3.3. The evolution of $J(r(k_i))$ and J_{th} are depicted in Fig. 9.3, which shows that the fault can be detected immediately and the number of false fault alarm is zero.

In order to compare the results of this study with the existing ones, the optimal event-triggered FD approach in [1] is also considered. We use the designed residual generator in [1] to generate residual signals, i.e.,

$$\begin{cases} \hat{x}(k+1) = A\hat{x}(k) + B_u u(k) + H(\bar{y}(k) - \hat{y}(k)), \\ \hat{y}(k) = C\hat{x}(k) + D_u u(k), \\ r_c(k) = V(\bar{y}(k) - \hat{y}(k)), \end{cases}$$

where $r_c(k)$ denotes the residual signal and

$$H = \begin{bmatrix} -0.0031 & -0.0001 \\ -0.0220 & 0.0008 \end{bmatrix}, V = \begin{bmatrix} 0.0227 & 0 \\ 0.0016 & 0.1106 \end{bmatrix}.$$

The residual evaluation function $J_{r_c}(k)$ is computed via

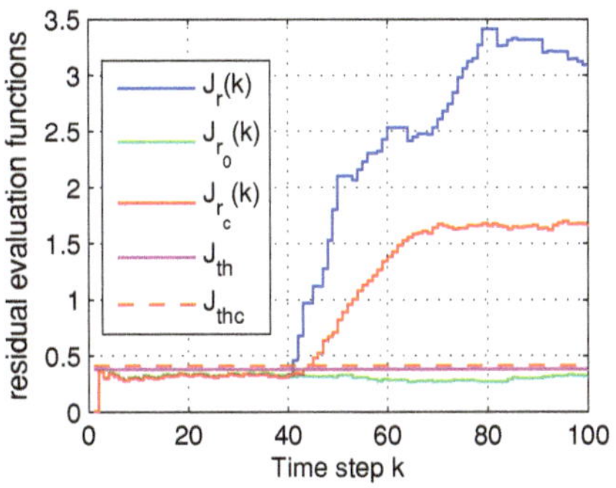

Fig. 9.4 The comparison results of the proposed method with [1] in case 1

$$J_{r_c}(k) = \begin{cases} \frac{1}{N+1} \sum_{j=k-N}^{k} r^T(j)r(j), \ k \geq N, \\ \frac{1}{k+1} \sum_{j=0}^{k} r^T(j)r(j), \ k < N. \end{cases} \tag{9.35}$$

In the case of $u(k) = 0$, the threshold in [1] is specified as $J_{thc} = \frac{\gamma_1}{\sqrt{N}}||d(k)||_2$ with $\gamma_1 = 1.4145$, which is set as $J_{thc} = 0.4031$. The evolution of $J_{r_c}(k)$ and J_{thc} are depicted in Fig. 9.4. It is clear that the fault can be detected at $k = 45$ and the number of false fault alarm is zero. For comparison purpose, the results of the developed event-triggered FD scheme are also presented in Fig. 9.4, where

$$J_r(k) = J(r(k_i)), \ k \in [k_i, k_{i+1}),$$

$J_{r_0}(k)$ denotes the fault-free case of $J_r(k)$. It is clear that the residual evaluation function $J_r(k)$ is more sensitive to fault than $J_{r_c}(k)$, and the threshold J_{th} is a little less than J_{thc}, which implies better FD performance of the developed FD scheme in this chapter.

Next, we consider the case of $u(k) = \frac{\pi}{15}\sin(0.2k)$. In case 2, the inter-event intervals $k_{i+1} - k_i$ are depicted in Fig. 9.5. The evolution of residual evaluation function $J(r(k_i))$ and threshold $J_{th} = 0.37$ are depicted in Fig. 9.6, which shows that the fault can also be detected immediately. The comparison results with the scheme in [1] are depicted in Fig. 9.7. As can be observed, the residual evaluation function $J_r(k)$ remains more sensitive to fault than $J_{r_c}(k)$. Moreover, because of the coupling effect of event-triggered transmission errors, the residual in [1] is affected not only by $d(k)$ and $f(k)$, but also by $u(k)$. For $u(k) \neq 0$, the threshold in [1] is determined by $J_{thc} = 0.4031 + \gamma_2||u(k)||_{rms}$ with $\gamma_2 = 2.2365$ and

$$||u(k)||_{rms} = \sqrt{\frac{1}{N} \sum_{i=k-N}^{k} u^T(i)u(i)}.$$

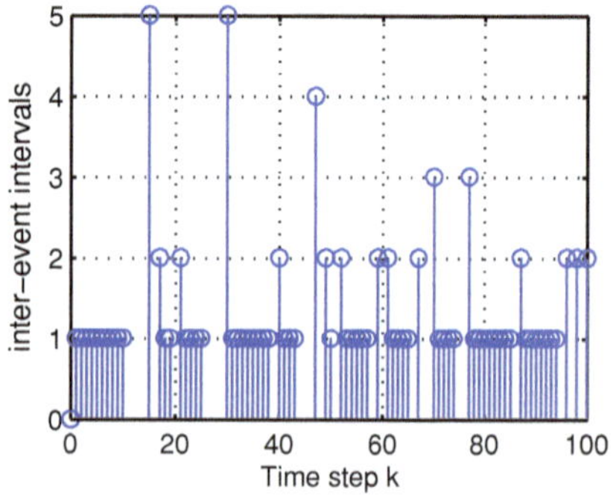

Fig. 9.5 The inter-event intervals $k_{i+1} - k_i$ of case 2

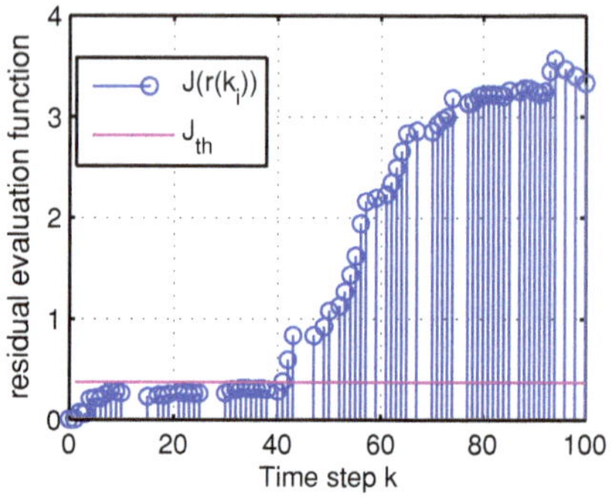

Fig. 9.6 The residual evaluation function $J(r(k_i))$ of case 2

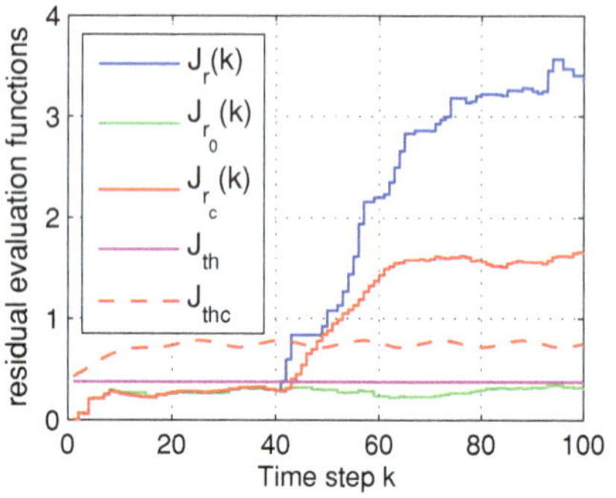

Fig. 9.7 The comparison results of the proposed method with [1] in case 2

Clearly, the J_{thc} in case 2 is greater than the one in case 1 and time-varying, that needs to be computed online. It is seen from Fig. 9.7 that the fault can be detected at $k = 48$ by using the method in [1]. Especially, due to the influence of event-triggered transmission errors, the zero-FAR threshold J_{thc} is obvious higher than J_{th}. It is definite that, in case 2, the developed FD scheme provides better FD performance than the one in [1].

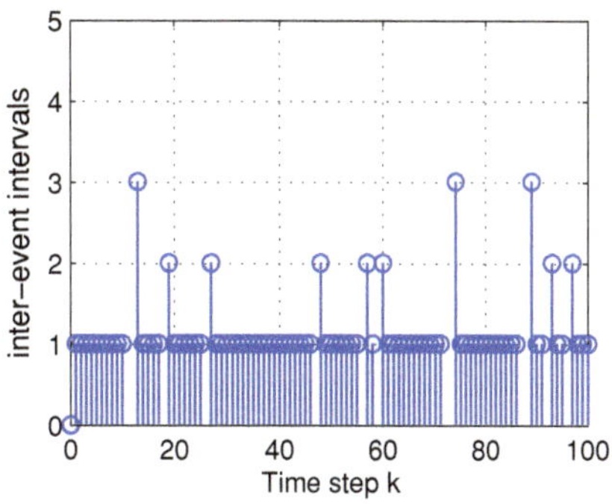

Fig. 9.8 The inter-event intervals $k_{i+1} - k_i$ of case 3

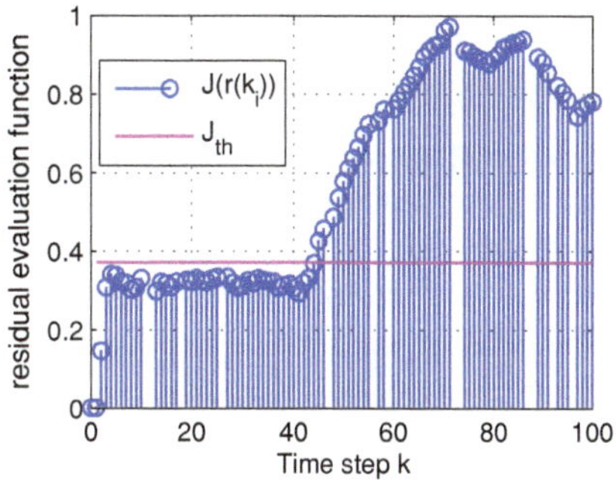

Fig. 9.9 The residual evaluation function $J(r(k_i))$ of case 3

To compare the fault detectability more clearly, the following case 3 is now taken into account, i.e.,

$$u(k) = \frac{\pi}{15}\sin(0.2k), \quad f(k) = \begin{cases} \frac{\pi}{60}, & k \geq 40, \\ 0, & \text{elsewhere.} \end{cases}$$

In this case, the corresponding inter-event intervals $k_{i+1} - k_i$, the $J(r(k_i))$, and the comparison results with [1] are shown in Figs. 9.8, 9.9 and 9.10, respectively. It is seen from Fig. 9.9 that the fault can be detected at time instant $k_i = 45$ by using our scheme and the number of false fault alarms remains zero. In Fig. 9.10, however, the fault cannot be detected by using the method of [1].

Now we can conclude from the above simulation results that better FD performance can be achieved by using the FD scheme developed in this chapter. Of course, as mentioned in Sect. 9.3.2, a certain acceptable increasing of online computational cost is needed comparing with the scheme in [1].

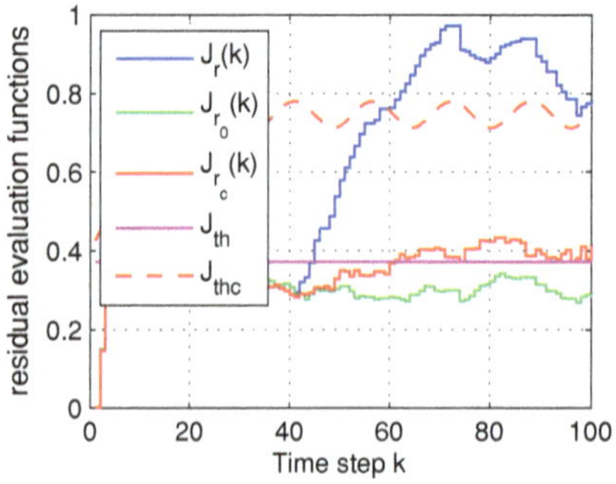

Fig. 9.10 The comparison results of the proposed method with [1] in case 3

9.5 Conclusion

In this chapter, an H_i/H_∞-optimization approach to event-triggered FD have been investigated. An event-triggered FDF has been constructed as a residual generator. In the framework of H_i/H_∞-optimization, an event-triggering parameter-independent optimal solution to FDF has been obtained that can be realized in terms of Riccati recursions. In addition, a residual evaluation unit has been designed involving the definition of a novel event-triggered residual evaluation function and a corresponding zero-FAR threshold. In comparison with the existing results, the developed event-triggered FD scheme can obtain a residual signal completely decoupled from the event-triggered transmission error, while a trade-off between the robustness of residual to unknown input and the sensitivity to fault can be achieved in the sense of H_i/H_∞-optimization. Moreover, the optimal solution of FDF-based residual generator is event-triggering parameter-independent. This allows the engineer to design an H_i/H_∞-FDF and the event generator independently. It should be pointed out that, the demonstrated event-triggered FD scheme can improve the FD performance significantly, but it brings moderate increases of online computational cost. And the solution of $L(k_i)$, $W(k_i)$ depends on the even-triggering parameters Ω and ϵ. As future works, the FD performance related event-triggering condition, the trade-off design between FD performance and data transmission reduction, etc., remain to be further investigated.

References

1. Qiu, A., Al-Dabbagh, A. W., & Chen, T. (2019). A tradeoff approach for optimal event-triggered fault detection. *IEEE Transactions on Industrial Electronics, 66*(3), 2111–2121.
2. Li, H., Chen, Z., Wu, L., et al. (2017). Event-triggered fault detection of nonlinear networked systems. *IEEE Transactions on Cybernetics, 47*(4), 298–311.
3. Liu, J., & Yue, D. (2013). Event-based fault detection for networked systems with communication delay and nonlinear perturbation. *Journal of the Franklin Institute, 350*(9), 2791–2807.
4. Zou, L., Wang, Z., & Zhou, D. (2017). Event-based control and filtering of networked systems: A survey. *International Journal of Automation and Computing, 14*(3), 239–253.

5. Li, Q., Shen, B., Wang, Z., et al. (2018). Event-triggered H_∞ state estimation for state-saturated complex networks subject to quantization effects and distributed delays. *Journal of the Franklin Institute, 355*(5), 2874–2891.
6. Hajshirmohamadi, S., Davoodi, M., Meskin, N., et al. (2016). Event-triggered fault detection and isolation for discrete time linear systems. *IET Control Theory and Applications, 10*(5), 526–533.
7. Wu, T., Li, F., Yang, C., et al. (2018). Event-based fault detection filtering for complexed networked jump systems. *IEEE/ASME Transactions on Mechatronics, 47*(8), 2299–2311.
8. Ren, W., Sun, S., Hou, N., et al. (2018). Event-triggered non-fragile H_∞ fault detection for discrete time-delayed nonlinear systems with channel fadings. *Journal of the Franklin Institute, 355*, 436–457.
9. Su, X., Xia, F., Wu, L., et al. (2018). Event-triggered fault detector and controller coordinated design of fuzzy systems. *IEEE Transactions on Fuzzy Systems, 26*(4), 2004–2016.
10. Zhang, X., Han, Q.-L., & Zhang, B. (2017). An overview and deep investigation on sampled-data-based event-triggered control and filtering for networked systems. *IEEE Transactions on Industrial Informatics, 13*(1), 4–16.
11. Liu, X., Su, X., Shi, P., et al. (2019). Fault detection filtering for nonlinear switched systems via event-triggered communication approach. *Automatica, 101*, 365–376.
12. Zhang, X., & Han, Q.-L. (2015). Event-based H_∞ filtering for sampled-data systems. *Automatica, 51*, 55–69.
13. Wang, Y., Shi, P., & Lim, C. (2016). Event-triggering fault detection filter design for a continuous-time networked control system. *IEEE Transactions on Cybernetics, 46*(12), 3414–3426.
14. Wang, Y., Lim, C., & Shi, P. (2017). Adaptively adjusted event-triggering mechanism on fault detection for networked control systems. *IEEE Transactions on Cybernetics, 47*(8), 2299–2311.
15. Pan, Y., & Yang, G. (2018). Event-triggered fault detection filter design for nonlinear networked systems. *IEEE Transactions on Systems, Man, Cybernetics: Systems, 48*(11), 1851–1862.
16. Ning, Z., Yu, J., Pan, Y., et al. (2018). Adaptive event-triggered fault detection for fuzzy stochastic systems with missing measurements. *IEEE Transactions on Fuzzy Systems, 26*(4), 2201–2212.

Chapter 10
Scheme of Optimal Fault Detection for LDTV Systems with Delayed State

This chapter presents an application of H_i/H_∞-optimal FD scheme to LDTV systems with delayed state, focusing on the designing of an evaluation function. The core idea is to directly construct an evaluation function by using a weighted l_2-norm of the measurement output, which, simultaneously, achieves an optimal trade-off between the fault sensitivity and the robustness to l_2-norm bounded unknown input. By means of orthogonal projection, a feasible solution is obtained via recursive computation. Such an evaluation function provides a unified scheme for both the cases of unknown input being l_2-norm bounded and jointly normal distributed, while a threshold may be chosen based on a priori knowledge of unknown input.

10.1 Introduction

Time-delay is frequently encountered in many fields such as networked control systems, manufacturing systems, economy, biology and other areas. In the past years, research on model-based FD for time-delay systems has been widely studied, see e.g. [3, 8–10]. Considering that unknown input is usually inevitable in practice and unknown input completely decoupling is almost impossible for time-delay systems, an FD system, without loss of generality, should be designed to be robust to unknown input and sensitive to fault. As well known, H_∞-optimization is accepted as a powerful tool to robust FD for linear time-delay systems subject to l_2-norm bounded unknown input. In [2, 4–7, 11] the problem of robust FD was formulated as to find an FDF such that the l_2-induced gain from unknown input to the error between residual and fault is minimized, i.e., the H_∞-filtering based FD. Literature [8–10, 12, 14] extend the H_∞-filtering based FD to networked control systems with communication time-delays. In [13], the H_-/H_∞-optimal FD method was applied to a class of nonlinear stochastic systems with Markovian switching and mixed time-delays.

M. Zhong et al., *Fault Diagnosis for Linear Discrete Time-Varying Systems and Its Applications*, https://doi.org/10.1007/978-981-19-5438-2_10

It should be pointed out that most of the above mentioned results were obtained by applying LMI techniques, while the solutions in terms of LMIs are conservative and not applicable to time-varying systems. A feasible solution to robust FD for linear time-varying systems with time-delay remains to be difficult to get. In [3], the problem of H_∞-filtering based FD was investigated by applying Krein space projection for LDTV systems with delayed state and a solution was obtained in terms of Riccati recursions. Under the assumption of unknown input being jointly normal distribution, Liu et al. [1] proposed a χ^2-testing approach to FD for LDTV systems with delayed state. Among the existing results of robust FD for time-delay systems, however, the trade-off between the robustness of residual to unknown input and sensitivity to fault was not considered in H_∞ filtering based FD, while the LMI techniques cannot be applied to H_∞/H_∞- and/or H_-/H_∞-optimal FD for LDTV systems. Besides, the χ^2-testing approach in [1] was not applicable to the case of unknown input being l_2-norm bounded.

Concerning LDTV systems with delayed state and l_2-norm bounded unknown input, in this chapter we directly formulate the FD problem as the design of an evaluation function towards an optimal trade-off between the robustness index and fault sensitivity criterion. The orthogonal projection technique will be applied to derive a feasible solution via recursive computation. The threshold determination issue is also addressed. A numerical example is given in Sect. 10.4 to demonstrate the effectiveness of the obtained results.

10.2 Optimal Fault Detection Under Delayed State

Consider the following LDTV systems with delayed state

$$\begin{cases} x(k+1) = A(k)x(k) + A_\tau(k)x(k-\tau) + B_d(k)d(k) + B_f(k)f(k), \\ y(k) = C(k)x(k) + v(k) + D_f(k)f(k), \\ x(0) = x_0, \end{cases} \tag{10.1}$$

where $x(k) \in \mathbb{R}^n$, $u(k) \in \mathbb{R}^p$, $y(k) \in \mathbb{R}^q$, $d(k) \in \mathbb{R}^m$, $v(k) \in \mathbb{R}^q$ and $f(k) \in \mathbb{R}^l$ are the state, system input, output, process noise, measurement noise and fault vectors to be detected, respectively, $x(k) = 0$ for $k < 0$, integer $\tau > 0$ is the state delay, $A(k)$, $A_\tau(k)$, $B_d(k)$, $B_f(k)$, $C(k)$ and $D_f(k)$ are known matrices with appropriate dimensions.

Define the state transition matrix of system (10.1) as

$$\begin{cases} \Phi(i+1, j) = A(i)\Phi(i, j) + A_\tau(i)\Phi(i-\tau, j), \\ \Phi(m, j) = 0 (m < 0), \quad \Phi(m, m) = I, \; i = 0, 1, \ldots, N-1; \; j \geq 0. \end{cases} \tag{10.2}$$

The dynamics of system (10.1) from time instant 0 to N can be described by

$$y_N = H_{0N}x_0 + H_{dN}d_N + H_{fN}f_N + v_N, \tag{10.3}$$

where $\theta_N = [\,\theta^T(0), \theta^T(1), \ldots, \theta^T(N)\,]^T$ with $\theta = y, d, f, v$ in sequence, and

$$\begin{aligned} H_{0N} &= \left[\, C^T(0)\ C^T(1)\Phi^T(1,0)\ C^T(2)\Phi^T(2,0)\ \cdots\ C^T(N)\Phi^T(N,0) \,\right]^T, \\ H_{dN} &= [h_{dN}(i,j)]_{(N+1)\times(N+1)},\ H_{fN} = [h_{fN}(i,j)]_{(N+1)\times(N+1)}, \\ h_{dN}(i,i) &= D_d(i-1),\ h_{fN}(i,i) = 0,\ i = 1, 2, \ldots, N+1, \\ h_{fN}(i,j) &= C(i-1)\Phi(i-1,j)B_f(j-1),\ i < j = 2, 3, \ldots, N+1. \end{aligned}$$

For the purpose of FD, the following $l_{2,[0,N]}$-norm type evaluation function is considered

$$J_N = y_N^T W_N y_N, \tag{10.4}$$

where W_N is a positive matrix to be determined. Under the assumption of $d(k),\ v(k)$ being $l_{2,[0,N]}$-norm bounded, i.e., $x_0^T x_0 + d_N^T d_N + v_N^T v_N \le \delta^2$, we define

$$\gamma_d^2 = \sup_{x_0, d_N, v_N} \frac{J_N|_{f_N=0}}{x_0^T x_0 + d_N^T d_N + v_N^T v_N}, \tag{10.5}$$

$$\gamma_f^2 = \sup_{f_N} \frac{J_N|_{x_0=0, d_N=0, v_N=0}}{f_N^T f_N}, \tag{10.6}$$

$$\gamma_{f-}^2 = \inf_{f_N} \frac{J_N|_{x_0=0, d_N=0, v_N=0}}{f_N^T f_N} \tag{10.7}$$

to measure the the robustness of the FD system against unknown input and initial state (i.e., γ_d) and the best-case (i.e., γ_f) and worst-case (i.e., γ_{f-}) fault sensitivities, respectively. Then the underlying problem can be formulated as to find W_N such that the evaluation function J_N satisfies the following best-case and/or worst-case sensitivity/robustness ratio criteria

$$\max_{W_N} \frac{\gamma_f}{\gamma_d} \ \text{and/or}\ \max_{W_N} \frac{\gamma_{f-}}{\gamma_d}. \tag{10.8}$$

Obviously, the problems in (10.8) are non-convex optimization problems, which makes it difficult to get a feasible solution. Moreover, the updating of $H_{0N},\ H_{dN}$ and H_{fN} are necessary at every time step and, therefore, the online computation is a heavy task for LDTV systems with the delayed state. Inspired by the stochastic χ^2-testing based FD in [1], projection in subspace spanned by $\{y(k)\}_{k=0}^N$ will be applied. For the later development, the following lemma is useful.

Lemma 10.1 ([1]) *Under the assumption of $x_0,\ d(k),\ v(k)$ being jointly normal distributions with $\mathbb{E}[x_0 x_0^T] = P_0$, $\mathbb{E}[d(k)d(k)^T] = R(k)$ and $\mathbb{E}[v(k)v(k)^T] = Q(k)$, an orthogonal projection of $x(k)$ onto subspace $\{y(i)\}_{i=0}^k$ with $f(k) = 0$ is given by*

$$\begin{cases} \hat{x}(k+1) = A(k)\hat{x}(k) + A_\tau(k)\hat{x}(k-\tau) \\ \qquad\qquad + A_\tau(k)\sum_{i=k-\tau}^{k-1} K_\tau(i)\tilde{y}(i) + K_p(k)\tilde{y}(k), \\ \tilde{y}(i) = y(i) - \hat{y}(i), \\ \hat{y}(i) = C(i)\hat{x}(i),\ k-\tau \le i \le k, \end{cases} \tag{10.9}$$

where $P(k),\ K_p(k),\ K_\tau(i)\ (i = k-\tau, k-\tau+1, \ldots, k-1)$ *are given by*

$$K_p(k) = (A(k)P(k)C^T(k) + A_\tau(k)P(k-\tau, k)C^T(k))R_{\tilde{y}}^{-1}(k), \tag{10.10}$$

$$K_\tau(i) = P(k-\tau, i)C^T(i)R_{\tilde{y}}^{-1}(i), \tag{10.11}$$

$$\begin{aligned} P(k+1) =& P(k+1, k+1) \\ =& A(k)P(k, k+1) + A_\tau(k)P(k-\tau, k+1) \\ &+ B_d(k)R(k)B_d^T(k) - K_p(k)C(k)P(k, k+1) \\ &- A_\tau(k)\sum_{i=k-\tau}^{k-1} K_\tau(i)C(i)P(i, k+1),\ P(0) = P_0, \end{aligned} \tag{10.12}$$

and

$$\begin{aligned} P(i, k+1) =& P(i, k)A^T(k) + P(i, k-\tau)A_\tau^T(k) - P(i, k)C^T(k)K_p^T(k) \\ &- \sum_{j=k-\tau}^{k-1} P(i, j)C^T(i)K_\tau^T(i)A_\tau^T(k), \end{aligned} \tag{10.13}$$

$$P(k, k+1) = P(k)A^T(k) + P^T(k-\tau, k)A_\tau^T(k) - \Theta(k)K_p^T(k), \tag{10.14}$$

$$\begin{aligned} P(k-\tau, k+1) =& P(k-\tau, k)A^T(k) + P(k-\tau)A_\tau^T(k) - \Theta_\tau(k)K_p^T(k) \\ &- \sum_{i=k-\tau}^{k-1} \Theta_\tau(k, i)K_\tau^T(i)A_\tau^T(k), \end{aligned} \tag{10.15}$$

$$R_{\tilde{y}}(k) = C(k)P(k)C^T(k) + Q(k). \tag{10.16}$$

In this case the evaluation function $\check{J}_N = y_N^T R_{y_N}^{-1} y_N$ *can be recursively computed by*

$$\check{J}_N = \sum_{k=0}^{N} \tilde{y}^T(k)R_{\tilde{y}}^{-1}(k)\tilde{y}(k). \tag{10.17}$$

After constructing an evaluation function, the remaining task is to determine a threshold. Since for $f(k) = 0$ we have $J_N \le \gamma_d^2\delta^2$, then a threshold with zero FAR can be chosen as $J_{th} = \gamma_d^2\delta^2$. The presence of a fault can be detected by using

$$\begin{cases} J_N \le J_{th}, & \text{no alarm}, \\ J_N > J_{th}, & \text{fault alarm}. \end{cases} \tag{10.18}$$

10.3 A Robust Optimal Residual Evaluation Scheme

In this section, we address the estimation issues of evaluation function J_N with W_N satisfying (10.8).

10.3.1 Design of Evaluation Function J_N

We start with the analysis on projection based evaluation function $\check{J}_N$ in Lemma 10.1. For the case of x_0, $d(k)$ and $v(k)$ being jointly normal distributions, it is easy to have

$$y_N = L_N \tilde{y}_N,$$

where

$$\begin{aligned} &L_N = [l_N(i,j)]_{(N+1)\times(N+1)},\ l_N(i,i) = I,\ l_N(i,j) = 0 \text{ for } i < j, \\ &l_N(i,j) = R_{y\tilde{y}}(i, j-1)R_{\tilde{y}}^{-1}(j-1) \text{ for } j < i = 2, 3, \ldots, N+1 \end{aligned}$$

with $R_{y\tilde{y}}(i, j-1)$ being the co-variance matrix of $y(i)$ and $\tilde{y}(j-1)$, respectively.

Define $e(k) = x(k) - \hat{x}(k)$. Subtracting (10.9) from (10.1) yields

$$\begin{cases} e(k+1) = A_p(k)e(k) + A_\tau(k)e(k-\tau) \\ \qquad - A_\tau(k) \sum\limits_{i=k-\tau}^{k-1} K_\tau(i)C(i)e(i) + B_d(k)d(k) \\ \qquad + B_{pf}(k)f(k) - K_p(k)v(k) - A_\tau(k) \sum\limits_{i=k-\tau}^{k-1} K_\tau(i)v(i), \\ \tilde{y}(k) = C(k)e(k) + v(k), \end{cases} \tag{10.19}$$

where $A_p(k) = A(k) - K_p(k)C(k)$, $B_{pf}(k) = B_f(k) - K_p(k)D_f(k)$. Similar to (10.3), it follows from (10.19) the following redundant relations

$$\tilde{y}_N = \tilde{\mathcal{H}}_{0N}x_0 + \tilde{\mathcal{H}}_{dN}d_N + \tilde{\mathcal{H}}_{vN}v_N + \tilde{\mathcal{H}}_{fN}f_N.$$

Since the exact values of $\tilde{\mathcal{H}}_{0N}$, $\tilde{\mathcal{H}}_{dN}$, $\tilde{\mathcal{H}}_{vN}$, $\tilde{\mathcal{H}}_{fN}$ are unnecessary to the end, the calculating formulate are not listed here for simplicity.

Let

$$\begin{aligned} &w_N = \begin{bmatrix} x_0^T & d_N^T & v_N^T \end{bmatrix}^T,\ \tilde{H}_{wN} = \begin{bmatrix} \tilde{\mathcal{H}}_{0N} & \tilde{\mathcal{H}}_{dN} & \tilde{\mathcal{H}}_{vN} \end{bmatrix}, \\ &R_{\tilde{y}_N} = \operatorname{diag}(R_{\tilde{y}}(0),\ R_{\tilde{y}}(1),\ \ldots,\ R_{\tilde{y}}(N)). \end{aligned}$$

Substituting into (10.17) yields

$$\check{J}_{0N} = J_N|_{f_N=0} = w_N^T \tilde{H}_{wN}^T R_{\tilde{y}_N}^{-1} \tilde{H}_{wN} w_N,$$
$$\check{J}_{fN} = J_N|_{w_N=0} = f_N^T \tilde{H}_{fN}^T R_{\tilde{y}_N}^{-1} \tilde{H}_{fN} f_N.$$

In the case of fault free and $\mathbb{E}[w_N] = 0,\ \mathbb{E}[w_N w_N^T] = I$, it is easy to have

$$R_{\tilde{y}_N} = \tilde{H}_{wN} \tilde{H}_{wN}^T.$$

We now turn to the case of $l_{2,[0,N]}$-norm bounded unknown input. Let $W_N = (L_N R_{\tilde{y}_N} L_N^T)^{-1}$, where $R_{\tilde{y}_N}$ and innovation $\tilde{y}(k)$ are generated by the same formulate of Lemma 10.1 with $P_0 = I,\ R(k) = I$ and $Q(k) = I$. Substituting into (10.4) yields

$$J_N = y_N^T W_N y_N = \sum_{k=0}^{N} \tilde{y}^T(k) R_{\tilde{y}}^{-1}(k) \tilde{y}(k), \tag{10.20}$$

which means that $J_N = \check{J}_N$. To prove that $W_N = (L_N R_{\tilde{y}N} L_N^T)^{-1}$ is a solution to the optimization problem (10.8), we next turn to analyze the sensitivity and robustness indices corresponding to J_N in (10.20).

In view of (10.5)–(10.7), the robustness and sensitivity indices corresponding to $l_{2,[0,N]}$-norm bounded $w(k)$ and $f(k)$ become to

$$\gamma_d^2 = \sup_{w_N} \frac{J_{0N}}{w_N^T w_N} = \sup_{w_N} \frac{w_N^T \tilde{H}_{wN}^T R_{\tilde{y}_N}^{-1} \tilde{H}_{wN} w_N}{w_N^T w_N} = \bar{\sigma}^2[(\tilde{H}_{wN} \tilde{H}_{wN}^T)^{-\frac{1}{2}} \tilde{H}_{wN}] = 1,$$
$$\gamma_f^2 = \sup_{f_N} \frac{J_{fN}}{f_N^T f_N} = \sup_{f_N} \frac{f_N^T \tilde{H}_{fN}^T R_{\tilde{y}_N}^{-1} \tilde{H}_{fN} f_N}{f_N^T f_N} = \bar{\sigma}^2(R_{\tilde{y}_N}^{-1/2} \tilde{H}_{fN}),$$
$$\gamma_{f-}^2 = \inf_{f_N} \frac{J_{fN}}{f_N^T f_N} = \inf_{f_N} \frac{f_N^T \tilde{H}_{fN}^T R_{\tilde{y}_N}^{-1} \tilde{H}_{fN} f_N}{f_N^T f_N} = \underline{\sigma}^2(R_{\tilde{y}_N}^{-1/2} \tilde{H}_{fN}).$$

As a result, we have

$$\frac{\gamma_f^2}{\gamma_d^2} = \bar{\sigma}^2(R_{\tilde{y}_N}^{-\frac{1}{2}} \tilde{H}_{fN}), \quad \frac{\gamma_{f-}^2}{\gamma_d^2} = \underline{\sigma}^2(R_{\tilde{y}_N}^{-\frac{1}{2}} \tilde{H}_{fN}).$$

where $\bar{\sigma}(\cdot)$ and $\underline{\sigma}(\cdot)$ denote the maximal and minimal singular values of $(\cdot)$, respectively.

Moreover, for an arbitrary positive matrix W_{aN} which can be rewritten as $W_{aN} = Q_{aN}^T W_N Q_{aN}$, the corresponding evaluation function is

$$J_{aN} = y_N^T W_{aN} y_N = y_N^T Q_{aN}^T W_N Q_{aN} y_N = \tilde{y}_N^T L_N^T Q_{aN}^T W_N Q_{aN} L_N \tilde{y}_N.$$

We thus have

$$\gamma_{ad}^2 = \sup_{w_N} \frac{w_N^T \tilde{H}_{wN}^T L_N^T Q_{aN}^T W_N Q_{aN} L_N \tilde{H}_{wN} w_N}{w_N^T w_N}$$
$$= \bar{\sigma}^2 (R_{\tilde{y}_N}^{-\frac{1}{2}} L_N^{-1} Q_{aN} L_N \tilde{H}_{wN}) = \bar{\sigma}^2 (Q_{aN}).$$

Similarly, we can further have

$$\gamma_{af}^2 = \sup_{f_N} \frac{f_N^T \tilde{H}_{fN}^T L_N^T Q_{aN}^T W_N Q_{aN} L_N \tilde{H}_{fN} f_N}{f_N^T f_N} = \bar{\sigma}^2 (R_{\tilde{y}_N}^{-\frac{1}{2}} L_N^{-1} Q_{aN} L_N \tilde{H}_{fN}),$$
$$\gamma_{af-}^2 = \inf_{f_N} \frac{f_N^T \tilde{H}_{fN}^T L_N^T Q_{aN}^T W_N Q_{aN} L_N \tilde{H}_{fN} f_N}{f_N^T f_N} = \underline{\sigma}^2 (R_{\tilde{y}_N}^{-\frac{1}{2}} L_N^{-1} Q_{aN} L_N \tilde{H}_{fN}),$$
$$\frac{\gamma_{af}^2}{\gamma_{ad}^2} = \frac{\bar{\sigma}(R_{\tilde{y}_N}^{-\frac{1}{2}} L_N^{-1} Q_{aN} L_N \tilde{H}_{fN})}{\bar{\sigma}(Q_{aN})} \leq \bar{\sigma}^2 (R_{\tilde{y}_N}^{-\frac{1}{2}} \tilde{H}_{fN}) = \frac{\gamma_f^2}{\gamma_d^2},$$
$$\frac{\gamma_{af-}^2}{\gamma_{ad}^2} = \frac{\underline{\sigma}(R_{\tilde{y}_N}^{-\frac{1}{2}} L_N^{-1} Q_{aN} L_N \tilde{H}_{fN})}{\bar{\sigma}^2(Q_{aN})} \leq \underline{\sigma}^2 (R_{\tilde{y}_N}^{-\frac{1}{2}} \tilde{H}_{fN}) = \frac{\gamma_{f-}^2}{\gamma_d^2}.$$

On this basis, we can conclude that $W_N = (L_N R_{\tilde{y}_N} L_N^T)^{-1}$ is a solution satisfying (10.8). In other words, J_N given by (10.20) delivers an optimal evaluation function that achieves an optimal trade-off between the fault sensitivity and the robustness to unknown input. As a summary, the following proposition is readily obtained.

Proposition 10.1 *Under the assumption of $d(k)$, $v(k)$ being $l_{2,[0,N]}$-norm bounded, the evaluation function J_N given by (10.20) with (10.9)–(10.16) satisfies the sensitivity/robustness ratio criteria (10.8).*

Proposition 10.1 shows that $\check{J}_N$ given in Lemma 10.1 also delivers an optimal evaluation function for the case of $d(k)$, $v(k)$ being $l_{2,[0,N]}$-norm bounded. In this regard, J_N provides us with a unified design of evaluation function, no matter x_0, $d(k)$, $v(k)$ are jointly normally distributed or l_2-norm bounded. A threshold can be determined according to the characteristics of unknown initial state x_0 and disturbances $d(k)$, $v(k)$, such as the χ^2-testing based FD for the case of $x_0, d(k)$, $v(k)$ being jointly normal distribution or a threshold with zero FAR for $l_{2,[0,N]}$-norm bounded unknown input. Moreover, J_N is obtained via recursive computation and, therefore, it is easy to implement online. Indeed, J_N and $\check{J}_N$ are calculated by using the same formulation.

Remark 10.1 As a further application to observer-based FD, we can define a residual as $r(k) = R_{\tilde{y}}^{-1/2}(k)\tilde{y}(k)$. Then J_N can be rewritten as

$$J_N = \sum_{k=0}^{N} r^T(k) r(k). \tag{10.21}$$

It implies that, in the case of $d(k)$, $v(k)$ being $l_{2,[0,N]}$-norm bounded, the residual satisfies both (10.21) and sensitivity/robustness ratio criteria (10.8). Therefore, $r(k) = R_{\tilde{y}}^{-1/2}(k)\tilde{y}(k)$ with (10.9) delivers an optimal H_∞/H_∞- and H_-/H_∞-FDFs for system (10.1). Especially, when the delayed state is not taken into account, it becomes the optimal H_∞/H_∞- and H_-/H_∞-FDFs as demonstrated in Chap. 8. In this sense, the presented method is an application of the scheme in Chap. 8 to LDTV systems subject to delayed state.

10.3.2 Determination of the Threshold

For the case of $d(k),\ v(k) \in l_{2,[0,N]}$, we have $\gamma_d = 1$. In the fault-free case, it holds for J_N that

$$J_{0N} = \tilde{y}_N^T R_{\tilde{y}_N}^{-1} \tilde{y}_N \le w_N^T w_N = x_0^T P_0^{-1} x_0 + d_N^T R_N^{-1} d_N + v_N^T Q_N^{-1} v_N \le \delta^2.$$

Thus, we can choose a threshold with zero FAR as $J_{th} = \delta^2$ and detect the occurrence of a fault by applying (10.18).

As δ^2 is often concerned to be a conservative estimation of $w_N^T w_N$ that in practice will be approached only in some extreme situations and missed detection of fault is inevitable, we can alternatively set a threshold being less than δ^2 to improve the fault detection rate while at the cost of a non-zero FAR. In particular, if δ is not exactly known, the boundedness of δ^2 can be estimated as

$$\hat{\delta}^2 = \sup_{f(k)=0} \sum_{k=0}^{N} \tilde{y}^T(k) R_{\tilde{y}}^{-1}(k) \tilde{y}(k),$$

which leads to a less conservative choice of threshold, i.e., $\hat{J}_{th} = \hat{\delta}^2$.

10.4 A Numerical Example

For verification purpose, we consider system (10.1) with

$$A(k) = \begin{bmatrix} 1.5e^{-k} & 0.2\sin(k) \\ 0 & 0.8 \end{bmatrix},\ A_\tau(k) = \begin{bmatrix} 0.7 & 0.3\cos(k) \\ 0 & 0.9 \end{bmatrix},\ B_f(k) = \begin{bmatrix} 0.9 \\ 1.8 \end{bmatrix},$$
$$B_d(k) = [0.3\ \ 0.7]^T,\ C(k) = \begin{bmatrix} 1.4 & 2.8 \end{bmatrix},\ D_f(k) = 0,\ \tau = 2.$$

Set $N = 50$ and $x(k) = 0$ for $k \le 0$. We first consider the case of $l_{2,[0,50]}$-norm bounded unknown input, i.e., the first case. Let $d(k) = \sin(0.5k)$ and $v(k)$ be normally distributed with zero-mean and $Q(k) = 1$. The fault is simulated as

Fig. 10.1 The evaluation functions $J_{0N}(k)$ and $J_N(k)$ in first case

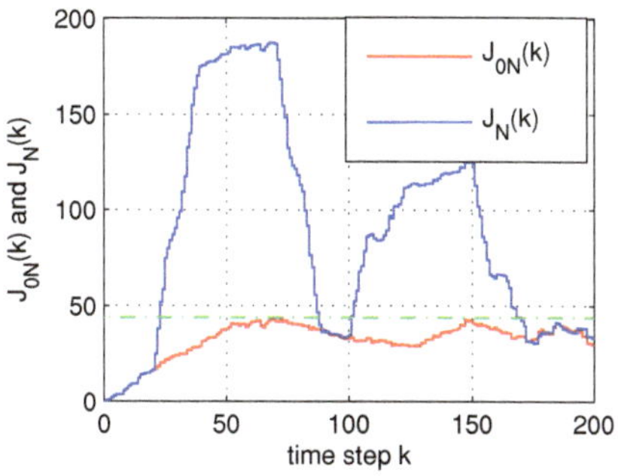

Fig. 10.2 The evaluation functions $J_{0N}(k)$ and $J_N(k)$ in second case

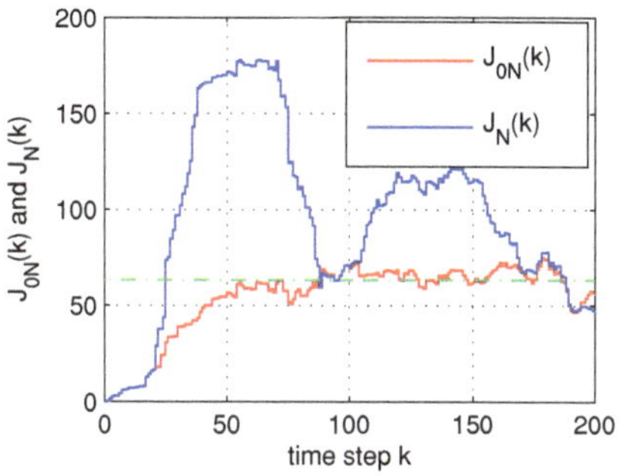

$$f(k) = \begin{cases} -1, & k \in [20, 40], \\ 1, & k \in [100, 120], \\ 0, & \text{otherwise}. \end{cases}$$

At every time step k, we calculate $J_N(k)$ and $J_{0N}(k)$ by applying Lemma 10.1 and

$$J_N(k) = \sum_{i=k-N}^{k} \tilde{y}^T(i) R_{\tilde{y}}^{-1}(i) \tilde{y}(i), \;\; J_{0N}(k) = J_N(k)|_{f(k)=0}. \tag{10.22}$$

The evolutions of $J_N(k)$ and $J_{0N}(k)$ are shown in Fig. 10.1. The threshold is determined by using an estimation of $\hat{\delta}^2$. We have $J_{th} = \max J_{0N} = 43.5579$, i.e., the green dashdot line in Fig. 10.1. It is seen that the simulated fault can be detected efficiently, while no false alarm exists.

To show that J_N provides a unified design of evaluation function, the case of jointly normally distributed x_0, $d(k)$ and $v(k)$ with $P_0 = I$, $P(0, i) = 0 (i = 1, 2, 3)$, $R(k) = 1$, $Q(k) = 1$ is also considered, i.e., the second case. The evolutions of $J_N(k)$ and $J_{0N}(k)$ are depicted in Fig. 10.2. With a given $FAR = 0.1$, a threshold is chosen as $Jth = \chi^2_{0.1}(51) = 64.2954$, i.e., the green dashdot line in Fig. 10.2. It is shown that the simulated fault can be detected, while false alarms exist at some time steps.

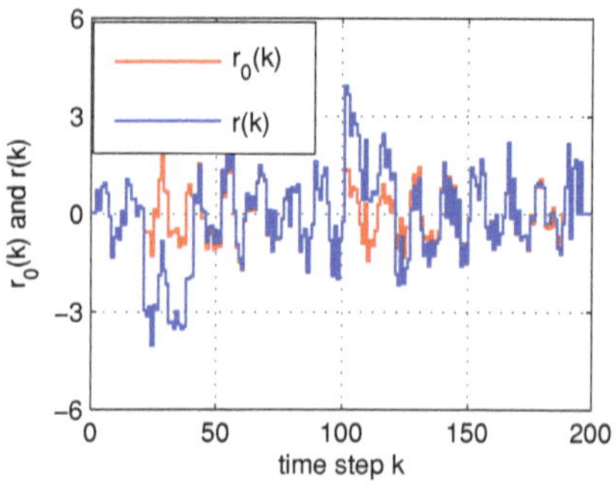

Fig. 10.3 The residual in first case

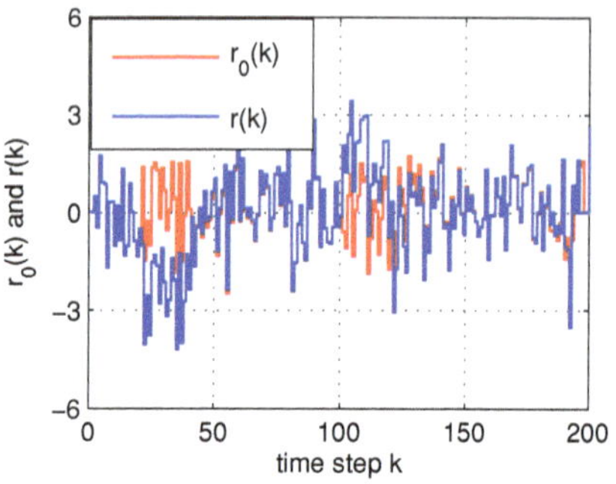

Fig. 10.4 The residual in second case

As an extended application to observer-based FD, the residual signals discussed in Remark 10.1 is also considered, i.e., $r(k) = R_{\tilde{y}}^{-1/2}(k)\tilde{y}(k)$. The generated residuals for the first case and second case are shown in Figs. 10.3 and 10.4, respectively. Since J_N remains an evaluation function for observer-based FD, we can detect the presence of the fault by using the results.

10.5 Conclusion

This chapter has confined to an H_i/H_∞-optimal FD scheme for LDTV systems subject to delayed state, along the research line of Chap. 8. The core idea lies in, in the presence of delayed state, constructing an estimate of residual evaluation function and, simultaneously, achieving a maximal sensitivity/robustness ratio criteria in the sense of finite horizon H_∞/H_∞- and H_-/H_∞-optimization. An optimal solution has been obtained via recursive computation and the resulted evaluation function follows the same formulae with the one for χ^2-testing based FD in [1] for jointly normal distribution. From this point of view, the presented scheme provides a unified design of evaluation function for LDTV systems with delayed state for both the cases of unknown input being l_2-norm bounded and jointly normal distributed. The threshold can be chosen according to priori knowledge of the unknown input.

References

1. Liu, B., & Ye, H. (2014). Statistical χ^2 testing based fault detection for linear discrete time-delay systems. *ACTA Automatica Sinica, 40*(7), 1278–1284.
2. Shen, B., Ding, S. X., & Wang, Z. (2013). Finite-horizon H_∞ fault estimation for linear discrete time-varying systems with delayed measurements. *Automatica, 49*(1), 293–296.
3. Zhao, H., Zhong, M., & Zhang, M. (2010). H_∞ fault detection for linear discrete time-varying systems with delayed state. *IET Control Theory and Applications, 4*(11), 2303–2314.
4. Yu, J., Liu, F., Yu, X., et al. (2013). Fault detection discrete-time switched systems with distributed delays: Input-output approach. *International Journal of Systems Science, 44*(12), 2255–2272.
5. Dong, Q., Zhong, M., & Ding, S. X. (2012). Active fault tolerant control for a class of linear time-delay systems in finite frequency domain. *International Journal of Systems Science, 43*(3), 543–551.
6. Ahmadizadeh, S., Zarei, J., Frank, P. M., et al. (2000). Robust unknown input observer design for linear uncertain time delay systems with application to fault detection. *Asian Journal of Control, 16*(4), 1006–1019.
7. Ding, S. X., Zhong, M., Tang, B., et al. (2001). An LMI approach to the design of fault detection filter for time-delay LTI systems with unknown inputs. In *Proceedings of the American Control Conference (Cat. No.01CH37148)*, June 25–27, Arlington, VA (Vol. 3, pp. 2137–2142).
8. Ding, W., Mao, Z., Jiang, B., et al. (2014). H_∞ fault detection for a class of T-S fuzzy model-based nonlinear networked control systems. *IFAC Proceedings Volumes, 47*(3), 11647–11652.
9. He, X., Wang, Z., Ji, Y., et al. (2008). Network-based fault detection for discrete-time state-delay systems: A new measurement model. *International Journal of Adaptive Control and Signal Processing, 22*(5), 510–528.
10. He, X., Wang, Z., & Zhou, D. (2009). Robust fault detection for networked systems with communication delay and data missing. *Automatica, 45*(11), 2634–2639.
11. Meng, X., & Yang, G. (2014). Simultaneous fault detection and control for stochastic time-delay systems. *International Journal of Systems Science, 45*(5), 1058–1069.
12. Wang, Y., Ding, S. X., Ye, H., et al. (2008). A new fault detection scheme for networked control systems subject to uncertain time-varying delay. *IEEE Transactions on Signal Processing, 56*(10), 5258–5268.
13. Long, Y., & Yang, G. (2014). Fault detection for a class of nonlinear stochastic systems with Markovian switching and mixed time-delay. *International Journal of Systems Science, 45*(3), 215–231.
14. Zhang, Y., & Fang, H. (2012). Fault detection for nonlinear networked control systems with stochastic interval delay characterization. *International Journal of Systems Science, 43*(5), 952–960.

Part III
H_∞-Filtering Based Fault Diagnosis for LDTV Systems

Chapter 11
A Krein Space Approach to H_∞ Fault Estimation for LDTV Systems

Issues of H_∞ fault estimation for LDTV systems are addressed in this chapter via Krein space techniques. To this end, the H_∞ fault estimation is first equated to a minimization problem of a quadratic function. By introducing a corresponding system of the monitored process in Krein space, a sufficient and necessary condition on the existence of an H_∞ fault estimator is derived and a solution to its parameter matrices is obtained in terms of matrix Riccati equation. The efficiency of the developed method is verified through numerical examples.

11.1 Introduction

H_∞ filtering is an important approach to robust FD, the original motivation of which lies in designing a residual generator such that the effects of disturbances and modeling errors to the residual is not exceeding the predefined level in worst-case setting. A standard formulation of H_∞-filtering for robust FD has been given in [4]. Over the past decades, a great number of achievements on H_∞ filtering based FD have been obtained and most of them are developed for LTI systems, see e.g., [4, 5, 8, 9].

With the development of Krein space theory, the gap between H_∞ filtering and Kalman filtering in Krein space has been filled [1–3, 6, 7, 10]. Especially, for LDTV systems, it has shown that a finite horizon linear estimation problem can be cast into a problem of calculating the minimum point of a certain quadratic form and, by means of Krein space projections, one can calculate the minimum point recursively. Comparing with the approaches in Hilbert space, linear estimation in Krein space is much more computationally attractive and, because of the existence of necessary and sufficient conditions, less conservative.

In this chapter we provide a Krein space approach to H_∞ fault estimation for LDTV systems with l_2-norm bounded unknown input. The main idea is to formulate

M. Zhong et al., *Fault Diagnosis for Linear Discrete Time-Varying Systems and Its Applications*, https://doi.org/10.1007/978-981-19-5438-2_11

the H_∞ fault estimation problem as a minimization problem of a certain quadratic form in Krein space and then address it via Riccati recursions. Bearing this in mind, we focus on coping with the following tasks:

- construct an H_∞ fault estimator for LDTV systems with l_2-norm bounded unknown input;
- establish an equivalent formulation of the H_∞ fault estimation problem in Krein space and,
- address the formulated problem and derive a recursive realization algorithm for online fault estimation.

11.2 Problem Statement

Consider the following LDTV system

$$\begin{cases} x(k+1) = A(k)x(k) + B_d(k)d(k) + B_f(k)f(k), \ x(0) = x_0, \\ y(k) = C(k)x(k) + v(k) + D_f(k)f(k), \end{cases} \tag{11.1}$$

where $x(k) \in \mathbb{R}^n$ and $y(k) \in \mathbb{R}^q$ are system state and output vectors, respectively, $d(k) \in \mathbb{R}^m$, $v(k) \in \mathbb{R}^q$, and $f(k) \in \mathbb{R}^l$ are the unknown input, measurement noise, and fault vectors, respectively, $d(k)$, $v(k)$, and $f(k)$ are considered to be $l_{2,[0,N]}$-norm bounded. Without loss of generality, it is assumed that $[A(k) \ B_d(k)]$ is of full row rank for all k.

Let $\hat{f}(k)$ be the estimate of $f(k)$ and

$$w(k) = \begin{cases} \begin{bmatrix} d(k) \\ f(k) \\ v(k) \end{bmatrix}, & \text{for } k \in [0, \ N-1], \\ \begin{bmatrix} f(k) \\ v(k) \end{bmatrix}, & \text{for } k = N, \end{cases}$$

$$r_f(k) = \hat{f}(k) - f(k), \ \ k = 0, \ 1, \ \ldots, \ N.$$

The H_∞ fault estimation problem under investigation can be stated as: given a scalar $\gamma > 0$, find a fault estimator such that

$$\frac{\sum_{i=0}^{N} r_f^T(k) r_f(k)}{x_0^T P_0^{-1} x_0 + \sum_{i=0}^{N} w^T(k) w(k)} < \gamma^2, \tag{11.2}$$

where P_0 is a given positive definite weighting matrix. Under zero initial condition, the fault estimation on the basis of (11.2) can be regarded as a finite horizon version of standard H_∞ filtering formulation, i.e., to find a fault estimate such that the $l_{2,[0,N]}$-induced gain from unknown input to fault estimate error is less than γ^2.

Define

$$\bar{x}(k) = \begin{bmatrix} x(k) \\ f(k) \end{bmatrix}, \ \bar{d}(k) = \begin{bmatrix} d(k) \\ f(k+1) \end{bmatrix}.$$

System (11.1) can be rewritten into the following augmented form

$$\begin{cases} \bar{x}(k+1) = \bar{A}(k)\bar{x}(k) + \bar{B}_d(k)\bar{d}(k), \\ y(k) = \bar{C}(k)\bar{x}(k) + v(k), \\ \hat{f}(k) = \bar{L}(k)\bar{x}(k) + r_f(k), \\ \bar{x}(0) = [x_0^T \ \ 0]^T. \end{cases} \tag{11.3}$$

where

$$\bar{A}(k) = \begin{bmatrix} A(k) & B_f(k) \\ 0 & 0 \end{bmatrix}, \ \bar{B}_d(k) = \begin{bmatrix} B_d(k) & 0 \\ 0 & I \end{bmatrix},$$
$$\bar{C}(k) = \begin{bmatrix} C(k) & D_f(k) \end{bmatrix}, \ \bar{L}(k) = \begin{bmatrix} 0 & I \end{bmatrix}.$$

Then, we consider the following fault estimator

$$\begin{cases} \hat{\bar{x}}(k+1) = F(k)\hat{\bar{x}}(k) + G(k)y(k), \ \hat{\bar{x}}(0) = 0, \\ \hat{f}(k) = M_1(k)\hat{\bar{x}}(k) + M_2(k)y(k). \end{cases} \tag{11.4}$$

The H_∞ fault estimation problem means solving $F(k)$, $G(k)$, $M_1(k)$, and $M_2(k)$ so as to guarantee (11.2).

Let

$$J_N = \begin{bmatrix} \bar{x}_0 \\ \bar{b}_{N-1} \\ v_N \\ r_{f,N} \end{bmatrix}^T \begin{bmatrix} \bar{P}_0 & 0 & 0 & 0 \\ 0 & I & 0 & 0 \\ 0 & 0 & I & 0 \\ 0 & 0 & 0 & -\gamma^2 I \end{bmatrix}^{-1} \begin{bmatrix} \bar{x}_0 \\ \bar{d}_{N-1} \\ v_N \\ r_{f,N} \end{bmatrix}, \tag{11.5}$$

where $\bar{P}_0 = \text{diag}\{P_0, \ I\}$ and

$$v_N = \begin{bmatrix} v(0) \\ v(1) \\ \vdots \\ v(N) \end{bmatrix}, \ \bar{d}_{N-1} = \begin{bmatrix} d(0) \\ d(1) \\ \vdots \\ d(N-1) \end{bmatrix}, \ r_{f,N} = \begin{bmatrix} r_f(0) \\ r_f(1) \\ \vdots \\ r_f(N) \end{bmatrix}.$$

We formulate the H_∞ fault estimation problem as follows.

Problem 11.1 Given system (11.1) and fault estimator (11.4), find $F(k)$, $G(k)$, $M_1(k)$, and $M_2(k)$ solving

$$\min_{\bar{x}_0, v_N, \bar{d}_{N-1}} J_N, \quad s.t.\ (11.4), (11.5)$$

such that the value of J_N at its minimum is positive.

That is to say, the H_∞ fault estimation problem is equivalent to solving the minimum problem of J_N with respect to $\bar{x}_0$, v_N and $\bar{d}_{N-1}$.

We know from Chaps. 4 and 5 that, the minimum of J_N over $\bar{x}_0$, v_N and $\bar{d}_{N-1}$ can be derived by using Kalman filtering theory in Krein space. In the following section, we will first define an associated Krein space stochastic system. Then, Kalman filtering theory in Krein space can be applied to solve the minimum problem of J_N. Finally, a solution of Problem 11.1 to parameter matrices $F(k)$, $G(k)$, $M_1(k)$, and $M_2(k)$ will be derived.

11.3 Design of an H_∞ Fault Estimator

Introduce the following Krein space stochastic system corresponding to (11.3)

$$\begin{cases} \mathbf{x}(k+1) = \bar{A}(k)\mathbf{x}(k) + \bar{B}_d(k)\mathbf{d}(k), \\ \mathbf{y}(k) = \bar{C}(k)\mathbf{x}(k) + \mathbf{v}(k), \\ \mathbf{z}(k) = \bar{L}(k)\mathbf{x} + \mathbf{r}_f(k), \\ \mathbf{x}(0) = \mathbf{x}_0, \end{cases} \tag{11.6}$$

where $\bar{A}(k)$, $\bar{B}_d(k)$, $\bar{C}(k)$, $\bar{L}(k)$ are the same as in (11.3), $\mathbf{x}(k)$ is a state vector, $\mathbf{d}(k)$, $\mathbf{v}(k)$, and $\mathbf{r}_f(k)$ are input vectors, $\mathbf{y}(k)$ and $\mathbf{z}(k)$ are output vectors, $\mathbf{x}(0)$, $\mathbf{d}(k)$, $\mathbf{v}(k)$, and $\mathbf{r}_f(k)$ are uncorrelated random vectors with zero means and the following covariance matrix

$$\left\langle \begin{bmatrix} \mathbf{x}(0) \\ \mathbf{d}(i) \\ \mathbf{v}(i) \\ \mathbf{r}_f(i) \end{bmatrix}, \begin{bmatrix} \mathbf{x}(0) \\ \mathbf{d}j) \\ \mathbf{v}(j) \\ \mathbf{r}_f(j) \end{bmatrix} \right\rangle = \begin{bmatrix} \bar{P}_0 & 0 & 0 & 0 \\ 0 & I\delta_{ij} & 0 & 0 \\ 0 & 0 & I\delta_{ij} & 0 \\ 0 & 0 & 0 & -\gamma^2 I\delta_{ij} \end{bmatrix}. \tag{11.7}$$

Let

$$\mathbf{y}_f(k) = \begin{bmatrix} \mathbf{y}(k) \\ \mathbf{z}(k) \end{bmatrix}, \mathbf{v}_{yf}(k) = \begin{bmatrix} \mathbf{v}(k) \\ \mathbf{r}_f(k) \end{bmatrix}.$$

We have

$$\langle \mathbf{v}_{yf}(i), \mathbf{v}_{yf}(j) \rangle = R_{yf}(i)\delta_{ij}, \quad R_{yf}(i) = \text{diag}\left(I, -\gamma^2 I\right). \tag{11.8}$$

It follows from (11.6) that

$$\begin{cases} \mathbf{x}(k+1) = \bar{A}(k)\mathbf{x}(k) + \bar{B}_d(k)\mathbf{d}(k), \ \mathbf{x}(0) = \mathbf{x}_0, \\ \mathbf{y}_f(k) = \bar{L}_C(k)\mathbf{x}(k) + \mathbf{v}_{yf}(k). \end{cases} \tag{11.9}$$

where $\bar{L}_C(k) = \left[\bar{C}^T(k) \ \ \bar{L}^T(k)\right]^T$. Denote

$$\hat{\mathbf{y}}_f(k|k-1) = \bar{L}_C(k)\hat{\mathbf{x}}(k|k-1) \tag{11.10}$$

and

$$\begin{aligned} &\mathbf{e}(k) = \mathbf{y}_f(k) - \hat{\mathbf{y}}_f(k|k-1), \ \mathbf{e}_k = \left[\mathbf{e}^T(0) \ \mathbf{e}^T(1) \cdots \mathbf{e}^T(k)\right]^T \\ &R_e(k) = \langle \mathbf{e}(k), \mathbf{e}(k) \rangle, \ R_{ek} = \langle \mathbf{e}_k, \mathbf{e}_k \rangle, \ \tilde{\mathbf{x}}(k|k-1) = \mathbf{x}(k) - \hat{\mathbf{x}}(k|k-1), \\ &P(k) = \langle \tilde{\mathbf{x}}(k|k-1), \tilde{\mathbf{x}}(k|k-1) \rangle, \ P(0) = \bar{P}_0. \end{aligned}$$

By applying Kalman filtering theory in Krein space, the following lemma can be obtained.

Lemma 11.1 *Given a scalar $\gamma > 0$, the minimum problem of J_N subject to (11.3) is solvable if and only if $R_e(k)$ and $R_{yf}(k)$ have the same inertia for all $k = 0, 1, \ldots, N$. If this is the case, the minimum of J_N is*

$$\min_{\bar{x}_0, \bar{d}(k)} J_N = e_N^T R_{eN}^{-1} e_N, \tag{11.11}$$

where

$$\begin{aligned} &e_N = \left[e^T(0) \ e^T(1) \cdots e^T(N)\right]^T, \ \ e(k) = \left[e_y^T(k) \ e_f^T(k)\right]^T, \\ &e_y(k) = y(k) - \bar{C}(k)\hat{\bar{x}}(k|k-1), \ \ e_f(k) = \breve{f}(k) - \bar{L}(k)\hat{\bar{x}}(k|k-1), \\ &\hat{\bar{x}}(0|-1) = 0, \ \ k = 0, 1, \ldots, N, \end{aligned}$$

$\hat{\bar{x}}(k|k-1)$ is given by

$$\hat{\bar{x}}(k+1|k) = \bar{A}(k)\hat{\bar{x}}(k|k-1) - K_p(k)\left(\begin{bmatrix} y(k) \\ \breve{f}(k) \end{bmatrix} - \bar{L}_C(k)\hat{\bar{x}}(k|k-1)\right), \tag{11.12}$$

$$K_p(k) = \bar{A}(k)P(k)\bar{L}_C^T(k)R_e^{-1}(k), \tag{11.13}$$

and $P(k)$ is calculated by the following Riccati equation

$$\begin{cases} P(k+1) = \bar{A}(k)P(k)\bar{A}^T(k) - \bar{A}(k)P(k)\bar{L}_C^T(k)\times \\ \qquad\qquad R_e^{-1}(k)\bar{L}_C(k)P(k)\bar{A}^T(k) + \bar{B}_d(k)\bar{B}_d^T(k), \\ P(0) = \bar{P}_0. \end{cases} \tag{11.14}$$

From the definition of $\mathbf{e}(k)$, (11.9) and (11.10), we have

$$\begin{aligned} \mathbf{e}(k) &= \bar{L}_C(k)\tilde{\mathbf{x}}(k|k-1) + \mathbf{v}_{yf}(k), \\ R_e(k) &= R_{yf}(k) + \bar{L}_C(k)P(k)\bar{L}_C^T(k) \\ &= \begin{bmatrix} I + \bar{C}(k)P(k)\bar{C}^T(k) & \bar{C}(k)P(k)\bar{L}^T(k) \\ \bar{L}(k)P(k)\bar{C}^T(k) & -\gamma^2 I + \bar{L}(k)P(k)\bar{L}^T(k) \end{bmatrix}. \end{aligned}$$

Define $\bar{e}_f(k) = \breve{f}(k) - \hat{f}(k|k)$ with

$$\hat{f}(k|k) = \hat{f}(k|k-1) + \bar{L}(k)P(k)\bar{C}^T(k)(\bar{C}(k)P(k)\bar{C}^T(k) + I)^{-1}e_y(k). \tag{11.15}$$

Then, (11.11) can be rewritten as

$$\begin{aligned} \min_{\bar{x}_0, \bar{d}(k)} J_N &= \sum_{i=0}^{N} \begin{bmatrix} e_y(k) \\ e_f(k) \end{bmatrix}^T R_e^{-1}(k) \begin{bmatrix} e_y(k) \\ e_f(k) \end{bmatrix} \\ &= \sum_{i=0}^{N} \begin{bmatrix} e_y(k) \\ \bar{e}_f(k) \end{bmatrix}^T \bar{R}_e^{-1}(k) \begin{bmatrix} e_y(k) \\ \bar{e}_f(k) \end{bmatrix}, \end{aligned} \tag{11.16}$$

where

$$\bar{R}_e(k) = \begin{bmatrix} \bar{C}(k)P(k)\bar{C}^T(k) + I & 0 \\ 0 & \bar{L}(k)(P^{-1}(k) + \bar{C}^T(k)\bar{C}(k))^{-1}\bar{L}^T(k) - \gamma^2 I \end{bmatrix}. \tag{11.17}$$

$R_e(k)$ and $\bar{R}_e(k)$ have the same inertia. Therefore, J_N is minimum if and only if $\bar{R}_e(k)$ and $R_{yf}(k)$ have the same inertia for all $k = 0, 1, \ldots, N$.

Moreover, from (11.14), we have

$$\begin{aligned} P(k+1) &= \begin{bmatrix} \bar{A}(k) & \bar{B}_d(k) \end{bmatrix} \begin{bmatrix} P(k)\phi(k) & 0 \\ 0 & I \end{bmatrix} \begin{bmatrix} \bar{A}^T(k) \\ \bar{B}_d^T(k) \end{bmatrix} \\ &= \begin{bmatrix} \bar{A}(k) & \bar{B}_d(k) \end{bmatrix} \begin{bmatrix} \varphi^{-1}(k) & 0 \\ 0 & I \end{bmatrix} \begin{bmatrix} \bar{A}^T(k) \\ \bar{B}_d^T(k) \end{bmatrix}, \end{aligned} \tag{11.18}$$

where

$$\begin{aligned} \phi(k) &= I - \bar{L}_C^T(k)R_e^{-1}(k)\bar{L}_C(k)P(k), \\ \varphi(k) &= P^{-1}(k) + \bar{L}_C^T(k)R_{yf}^{-1}(k)\bar{L}_C(k). \end{aligned}$$

Under the assumption that $\begin{bmatrix} A(k) & B_d(k) \end{bmatrix}$ is of full row rank, $\begin{bmatrix} \bar{A}(k) & \bar{B}_d(k) \end{bmatrix}$ is of full row rank too. Therefore, $P(k+1) > 0$ if and only if

$$P^{-1}(k) + \bar{L}_C^T(k) R_{yf}^{-1}(k) \bar{L}_C(k) > 0,$$

i.e.,

$$P^{-1}(k) + \bar{C}^T(k)\bar{C}(k) - \gamma^2 \bar{L}^T(k)\bar{L}(k) > 0. \tag{11.19}$$

In this case,

$$\bar{L}(k)(P^{-1}(k) + \bar{C}^T(k)\bar{C}(k))^{-1}\bar{L}^T(k) - \gamma^2 I < 0, \; \bar{C}(k)P(k)\bar{C}^T(k) + I > 0.$$

So, $\bar{R}_e(k)$ and $R_{yf}(k)$ have the same inertia if and only if (11.19) is satisfied.
Let $\bar{e}_f(k) = 0$, i.e.,

$$\breve{f}(k) = \hat{f}(k|k-1) + \bar{L}(k)P(k)\bar{C}^T(k)(\bar{C}(k)P(k)\bar{C}^T(k) + I)^{-1}e_y(k). \tag{11.20}$$

It follows from (11.16) that

$$\min_{\bar{x}_0, \bar{d}(k)} J_N = \sum_{i=0}^{N} e_y^T(k)(\bar{C}(k)P(k)\bar{C}^T(k) + I)^{-1}e_y(k) > 0,$$

which means that the H_∞ fault estimation problem is solvable.
Let $\hat{\bar{x}}(k) = \hat{\bar{x}}(k|k-1)$. Then, the parameter matrices $F(k)$, $G(k)$, $M_1(k)$, and $M_2(k)$ can be derived from (11.12) and (11.20) as

$$\begin{cases} G(k) = \bar{A}(k) - F(k)\bar{C}(k), \\ F(k) = \bar{A}(k)P(k)\bar{L}_C^T(k)R_e^{-1}(k)\begin{bmatrix} I \\ M_2(k) \end{bmatrix}, \\ M_1(k) = \bar{L}(k)(I + P(k)\bar{C}^T(k)\bar{C}(k))^{-1}, \\ M_2(k) = M_1(k)P(k)\bar{C}^T(k). \end{cases} \tag{11.21}$$

We now conclude the following theorem.

Theorem 11.1 *Given a scalar $\gamma > 0$, J_N has a minimum and its value at its minimum is positive if and only if (11.19) is satisfied. In this case, the fault estimator achieving (11.2) can be obtained from (11.4) and (11.21), where $P(k)$ is calculated recursively from matrix Riccati equation (11.14).*

The online realization algorithm of the designed H_∞ fault estimator is summarized in Algorithm 11.3.4.

Algorithm 11.3.4 Calculation of matrices $F(k)$, $G(k)$, $M_1(k)$, and $M_2(k)$

1: Set $\gamma > 0$, $P_0 > 0$ and $k = 1$;
2: Calculate $K_p(k)$ and $P(k)$ using (11.13) and (11.14), respectively;
3: Calculate $P(k+1)$ and $R_e(k)$ using (11.18) and (11.17), respectively;
4: Compute $F(k)$, $G(k)$, $M_1(k)$, and $M_2(k)$ using (11.21);
5: Let $k = k+1$, go to Step 2 till $k = N$.

11.4 Numerical Examples

In this section, we show the applicability of the presented H_∞ fault estimation method through two numerical examples.

Example 11.1 Consider system (11.1) with the following parameters

$$A(k) = \begin{bmatrix} 0.2 & 1+e^{-k} & -0.2\sin(10k) \\ 0 & 0.7 & 0.1 \\ 0 & 0 & 0.3 \end{bmatrix}, \quad B_d(k) = \begin{bmatrix} -\frac{1.6059}{k} \\ 0.57867 \\ -0.3569 \end{bmatrix},$$

$$B_f(k) = \begin{bmatrix} -\frac{1.6059}{k} \\ 0.57867 \\ -0.3569 \end{bmatrix}, \quad C(k) = \begin{bmatrix} -1.2193 & 1.8369 & 0.4253 \end{bmatrix}, \quad D_f(k) = 8.$$

Set $x_0 = \begin{bmatrix} 1 & -1 & 2 \end{bmatrix}^T$, $P_0 = I$ and $\gamma = 0.1$. An H_∞ fault estimator is designed using Algorithm 11.3.4 based on Theorem 11.1. Suppose that the unknown input $d(k)$ and the measurement noise $v(k)$ are simulated as in Figs. 11.1 and 11.2, respectively. A fault signal (dashed lines) and its estimation (solid lines) are shown in Fig. 11.3. It is seen that a satisfactory fault estimate can be obtained.

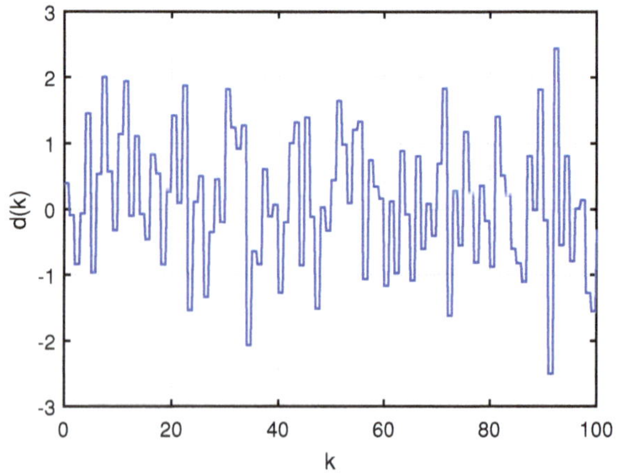

Fig. 11.1 The unknown input $d(k)$

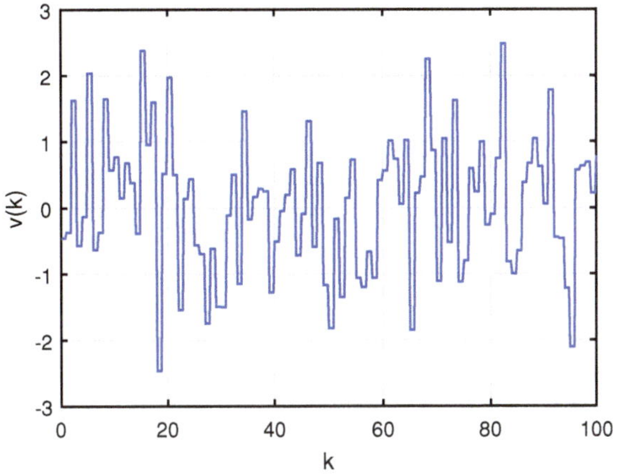

Fig. 11.2 The measurement noise $v(k)$

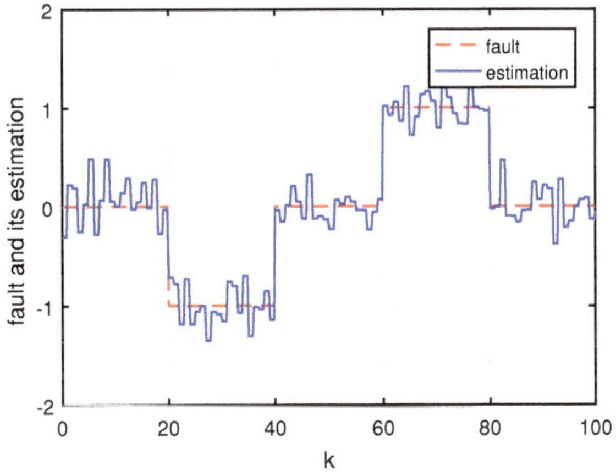

Fig. 11.3 The fault and its estimate

Example 11.2 Consider a linear periodic system in [10] that is described by (11.1) with period 2 and the following parameters

$$A(0) = \begin{bmatrix} 0.25 & 0.25 & 0.1 & -0.1 \\ 0.5 & 0.1 & 0.1 & 0.5 \\ 0.5 & -0.2 & 0.2 & 0.25 \\ 0.1 & 0 & 0.25 & 0.1 \end{bmatrix}, \; B_d(0) = \begin{bmatrix} 1.3 \\ 1.8 \\ 1.6 \\ 0.32 \end{bmatrix}, \; B_f(0) = \begin{bmatrix} 0.1 \\ -1 \\ 0.2 \\ 0.1 \end{bmatrix},$$

$$C(0) = \begin{bmatrix} 0.25 & 0.1 & 0.2 & 0.1 \\ -0.1 & 0.5 & 0.2 & 0.5 \\ 0.25 & 0.5 & -0.1 & 0.1 \end{bmatrix}, \; D_f(0) = \begin{bmatrix} 0.2 \\ 0.1 \\ 0.4 \end{bmatrix},$$

$$A(1) = \begin{bmatrix} 0.1 & 0.2 & 0.1 & -0.1 \\ -0.1 & 0.5 & 0 & 0.5 \\ 0.5 & 0.5 & 0.1 & 0.25 \\ 0 & 0.1 & 0.1 & 0.25 \end{bmatrix}, \quad B_d(1) = \begin{bmatrix} 3.2 \\ 2 \\ -1 \\ -2 \end{bmatrix}, \quad B_f(1) = \begin{bmatrix} 0.1 \\ -1 \\ 0.2 \\ 0.1 \end{bmatrix},$$

$$C(1) = \begin{bmatrix} 0.1 & 0.25 & 0.1 & -0.1 \\ 0.25 & 0.1 & 0.2 & 0.1 \\ 0.1 & 0.25 & -0.2 & 0.5 \end{bmatrix}, \quad D_f(1) = \begin{bmatrix} -0.2 \\ -0.1 \\ 0.3 \end{bmatrix}.$$

Set $x_0 = \begin{bmatrix} 0 & 0 & 0 & 0 \end{bmatrix}^T$, $P_0 = I$ and $\gamma = 0.3$. The unknown input $d(k) = \sin(0.01\pi k)$ is shown in Fig. 11.4. A fault and its estimation by applying Algorithm 11.3.4 are shown in Fig. 11.5. When the fault in Fig. 11.5 occurs, the residual signal generated by parity space-based scheme and observer-based approach in [10] are shown

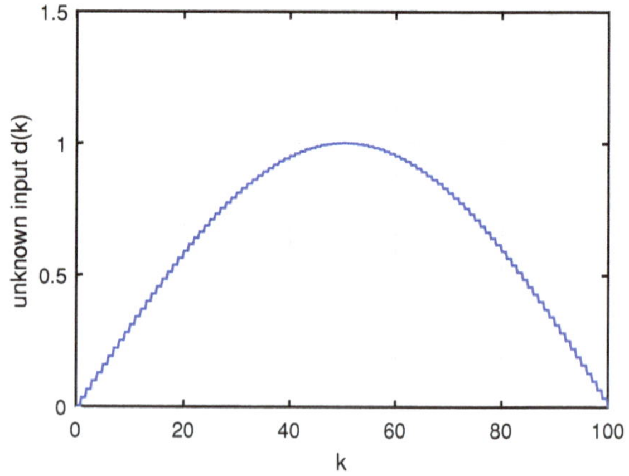

Fig. 11.4 The unknown input $d(k)$

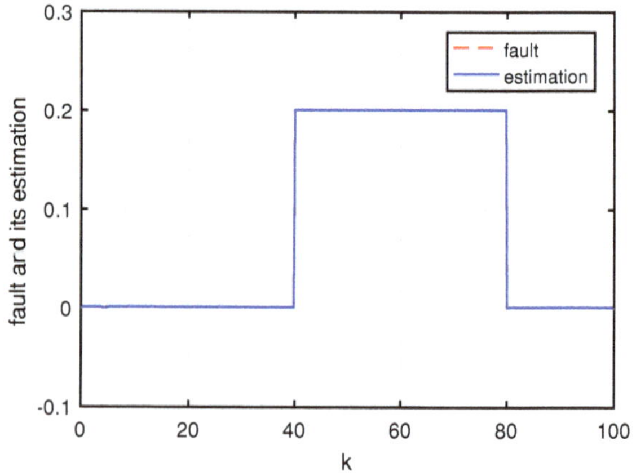

Fig. 11.5 The fault and its estimation

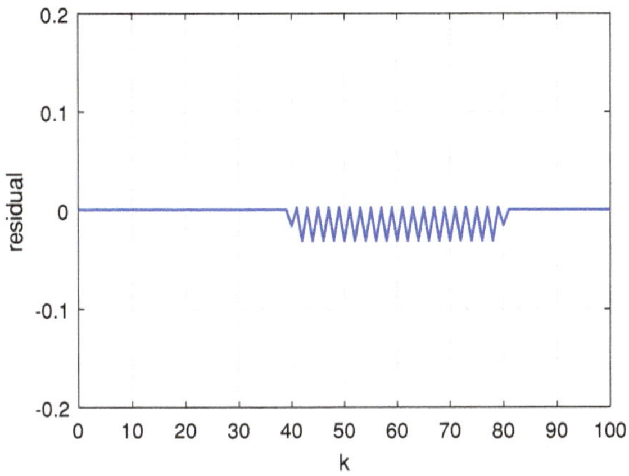

Fig. 11.6 The residual using parity space-based approach

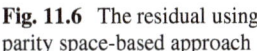

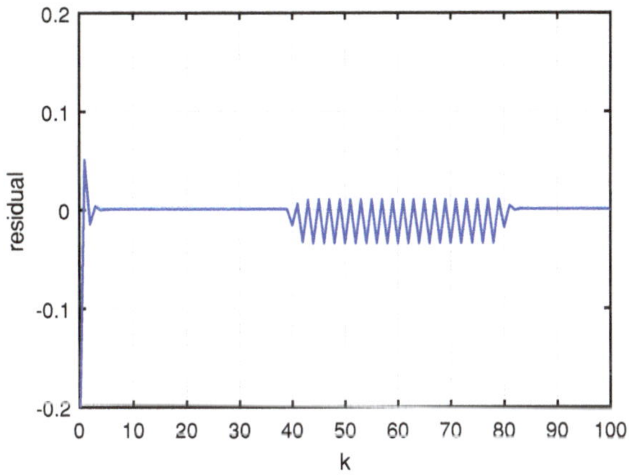

Fig. 11.7 The residual using observer-based approach

in Figs. 11.6 and 11.7, respectively. It is seen from the simulation results that fully unknown input decoupling in Fig. 11.5 is achieved and the obtained fault estimate is almost the same as the fault.

11.5 Conclusion

In this chapter, a Krein space approach to H_∞ fault estimation has been investigated for LDTV systems subject to l_2-norm bounded unknown input. On the basis of equating the H_∞ fault estimation problem to a minimization problem of a scalar quadratic form subject to an augmented system, a corresponding Krein space stochastic system has been introduced and a relationship between H_∞ fault estimation and Kalman filtering in Krein space has been built. More importantly, a sufficient and necessary solvable condition on the existence of the minimum has been derived. A recursive realization of the H_∞ fault estimator has been also developed in terms of Riccati equations.

References

1. Hassibi, B., Sayed, A. H., & Kailath, T. (1999). *Indefinite-quadratic estimation and control: A unified approach to H_2 and H_∞ theories*. Philadelphia: SIAM.
2. Hassibi, B., Sayed, A. H., & Kailath, T. (1996). Linear estimation in Krein spaces-Part I: Theory. *IEEE Transactions on Automatic Control, 41*(1), 18–33.
3. Hassibi, B., Sayed, A. H., & Kailath, T. (1996). Linear estimation in Krein spaces-Part II: Applications. *IEEE Transactions on Automatic Control, 41*(1), 34–49.
4. Chen, J., & Patton, R. J. (2000). Standard H_∞ filtering formulation of robust fault detection. *IFAC-PaperOnline, 33*(11), 261–266.
5. Chen, J., & Patton, R. J. (1999). *Robust model-based fault diagnosis for dynamic systems*. Boston, MA: Kluwer Academic Publishers.
6. Zhong, M., Liu, S., & Zhao, H. (2008). Krein space-based H_∞ fault estimation for linear discrete time-varying systems. *Acta Automatica Sinica, 34*(12), 1529–1533.
7. Zhong, M., Ding, S. X., Han, Q.-L., et al. (2022). A Krein space-based approach to event-triggered H_∞ filtering for linear discrete time-varying systems. *135*, 110001.
8. Ding, S. X., Zhong, M., Tang, B., et al. (2001) An LMI approach to the design of fault detection filter for time-delay LTI systems with unknown inputs. In *Proceedings of the American Control Conference (Cat. No.01CH37148)*, June 25–27, Arlington, VA (Vol. 3, pp. 2137–2142).
9. Ding, S. X. (2013). *Model-based fault diagnosis techniques: Design schemes, algorithms, and tools* (2nd ed.). London, U.K.: Springer.
10. Ding, S. X., & Zhang, P. (2007). Disturbance decoupling in fault detection of linear periodic systems. *Automatica, 43*(8), 1410–1417.

Chapter 12
On Designing H_∞ Fault Detection Filter for LDTV Systems

The objective of this chapter is to address the finite horizon H_∞-filtering based FD problem for LDTV systems. Using an observer-based FDF with residual feedback as a residual generator, the design of finite horizon H_∞-FDF is formulated as solving a minimum problem of certain quadratic form. By means of projection and innovation analysis in Krein space, a sufficient and necessary condition for the minimum is derived and a solution to the H_∞-FDF is obtained by recursively computing Riccati recursions.

12.1 Introduction

H_∞ optimization and H_∞ filtering are two widely used FDI approaches for linear dynamic systems with l_2-norm bounded unknown inputs. In Chap. 6, we have demonstrated a unified solution to the H_∞/H_∞- and H_-/H_∞-optimal FD for LDTV systems and, a Krein space based approach to H_i/H_∞-optimal FD is also given in Chap. 8. H_∞ filtering focuses on finding a residual generator such that the error between residual and fault is minimized under H_∞ criterion [6]. By noticing the research results on Krein space theory and the interesting connections of linear estimation problems (including filtering, prediction and smoothing) with projections in Krein spaces [1, 4, 5], we have also demonstrated a Krein space aided H_∞ filtering based fault estimation method for LDTV systems in Chap. 11, thanks to the computation attractiveness and the less conservativeness of linear estimation in Krein space.

In this chapter, we aim at handling FD issues for LDTV systems in the framework of finite horizon H_∞ filtering. An observer-based FDF with residual feedback is first proposed as a residual generator. On this basis, a modified finite horizon H_∞ smoothing formulation of FD is presented. It will be shown that such an H_∞-FDF problem can be converted into a minimization problem of an indefinite quadratic form, and

M. Zhong et al., *Fault Diagnosis for Linear Discrete Time-Varying Systems and Its Applications*, https://doi.org/10.1007/978-981-19-5438-2_12

a sufficient and necessary condition for the minimum is derived by using projection and innovation analysis in Krein space. An optimal solution to the observer-based H_∞-FDF will be derived recursively via Riccati recursions.

12.2 Problem Formulation

Consider LDTV system (11.1), i.e.,

$$\begin{cases} x(k+1) = A(k)x(k) + B_d(k)d(k) + B_f(k)f(k), x(0) = x_0, \\ y(k) = C(k)x(k) + v(k) + D_f(k)f(k), \end{cases} \tag{12.1}$$

where $d(k) \in l_{2,[0,N]}$ and $f(k) \in l_{2,[0,N]}$. For FD purpose, the following observer-based FDF is constructed as a residual generator

$$\begin{cases} \hat{x}(k+1|k) = A(k)\hat{x}(k) + H_1(k)\varepsilon(k), \\ \hat{x}(k+1) = \hat{x}(k+1|k) + H_2(k)\varepsilon(k) + H_3(k)r(k), \\ \varepsilon(k) = y(k) - C(k)\hat{x}(k), \\ r(k) = V_1(k)\varepsilon(k) + V_2(k+1)(y(k+1) - C(k+1)\hat{x}(k+1|k)), \\ \hat{x}(0) = 0, \end{cases} \tag{12.2}$$

where $\hat{x}(k+1|k)$ denotes the one-step state prediction, $\hat{x}(k)$ is the state estimation, $r(k)$ is the residual generated at time step $k+1$, $H_i(k)$ $(i = 1, 2, 3)$, $V_1(k)$ and $V_2(k+1)$ are parameter matrices to be designed.

It is remarkable that the term $V_1(k)\varepsilon(k)$ can be regarded as a standard residual and we call $r(k)$ an improvement of $V_1(k)\varepsilon(k)$. Note that $f(k)$ has no influence on $y(k)$ in case that $D_f(k) = 0$. Hence, when $y(k+1)$ is available, high fault detection performance is achievable through accommodating the residual with term $V_2(k+1)(y(k+1) - C(k+1)\hat{x}(k+1|k))$. On the other hand, in the state updating equation $\hat{x}(k+1)$, term $H_3(k)r(k)$ denotes the feedback of residual $r(k)$ and, from this view point, we call (12.2) a closed-loop residual generator. Term $H_2(k)\varepsilon(k)$ can be regarded as an improvement of $H_1(k)\varepsilon(k)$.

The design of a finite horizon H_∞-FDF is then formulated as the following problem.

Problem 12.1 *Given residual generator (12.2) for system (12.1), find $H_i(k)$ $(i = 1, 2, 3)$, $V_1(k)$ and $V_2(k+1)$ such that*

$$\sup_{(x_0, w) \neq 0} \frac{\sum_{k=0}^{N-1} \|r(k) - f(k)\|^2}{x_0^T P_0^{-1} x_0 + \sum_{k=0}^{N} \|w(k)\|^2} < \gamma^2, \tag{12.3}$$

where $\gamma > 0$, $P_0 > 0$ is a weighting matrix, $w(k) = \begin{bmatrix} d^T(k) & f^T(k) & v^T(k) \end{bmatrix}^T$.

For the sake of simplicity, let $\mathcal{S}$ be the set of $\{0, 1, \ldots, N\}$. Define the error between $r(k)$ and $f(k)$ as

$$v_f(k) = r(k) - f(k),\ \forall k \in \mathcal{S} \tag{12.4}$$

and introduce the following quadratic form

$$J_N = x_0^T P_0^{-1} x_0 + \sum_{k=0}^{N} \|w(k)\|^2 - \gamma^{-2} \sum_{k=0}^{N-1} \|v_f(k)\|^2.$$

Obviously, (12.3) is satisfied if and only if $J_N > 0$ for all $(x_0, w) \neq 0$. The H_∞-FDF problem in Problem 12.1 is then reformulated to find $H_i(k)$ $(i = 1, 2, 3)$, $V_1(k)$ and $V_2(k+1)$ such that $J_N > 0$.

Remark 12.1 When the parameter matrices in the monitored system (12.1) are constant and $H_2(k) = 0$, $H_3(k) = 0$, $V_2(k+1) = 0$, the presented FDF leads to the so-called standard H_∞ filtering formulation of FD in [6]. When $H_2(k) = 0$ and $H_3(k) = 0$, it leads to a one-step H_∞ smoothing of $f(k)$. In this regard, the presented FDF can be considered as a closed-loop H_∞ smoothing formulation.

It is worth noting that, even in the case of $H_2(k) = 0$ and $H_3(k) = 0$, the formulated FDF problem cannot be solved by simple extension of existing H_∞ smoothing results such as the Krein space approach in [4], the differential game approach in [3], and the lifting state augmentation method in [9]. Inspired by the results in Chap. 11, below we present a Krein space approach to the design of H_∞-FDF.

12.3 Krein Space Based Design of H_∞-FDF

In this section, an artificial stochastic system in Krein space is first introduced and a relationship of it with the minimum problem of J_N is built. Then a sufficient and necessary condition for the minimum of J_N is derived by using Krein space projection and innovation analysis. On this basis, a solution to H_∞-FDF concerned in Problem 12.1 is obtained.

12.3.1 An Equivalent Problem in Krein Space

To motivate the subsequent discussion, let us first define

$$y_s(i) = \begin{cases} y_f(i), & i \leq k-1 \\ y(k), & i = k \end{cases},\quad y_f(i) = \begin{bmatrix} y(i) \\ r(i) \end{bmatrix},$$

$$v_s(i) = \begin{cases} \begin{bmatrix} v(i) \\ v_f(i) \end{bmatrix}, & i \leq k-1 \\ v(k), & i = k \end{cases},\ \forall k \in \mathcal{S}. \tag{12.5}$$

Combining system (12.1) and (12.4) yields

$$\begin{cases} x(i+1) = A(i)x(i) + B_d(i)d(i) + B_f(i)f(i),\ x(0) = x_0, \\ y_s(i) = C_s(i)x(i) + v_s(i) + D_{sf}(i)f(i), \end{cases} \tag{12.6}$$

where

$$C_s(i) = \begin{cases} \bar{C}(i),\ i \le k-1 \\ C(k),\ i = k \end{cases},\quad \bar{C}(i) = \begin{bmatrix} C(i) \\ 0 \end{bmatrix},$$
$$D_{sf}(i) = \begin{cases} \bar{D}_f(i),\ i \le k-1 \\ D_f(k),\ i = k \end{cases},\quad \bar{D}_f(i) = \begin{bmatrix} D_f(i) \\ I \end{bmatrix}. \tag{12.7}$$

Let

$$f_N = \begin{bmatrix} f(0) \\ f(1) \\ \vdots \\ f(N) \end{bmatrix},\ d_N = \begin{bmatrix} d(0) \\ d(1) \\ \vdots \\ d(N) \end{bmatrix},\ v_{sN} = \begin{bmatrix} v_s(0) \\ v_s(1) \\ \vdots \\ v_s \end{bmatrix} y_{sN} = \begin{bmatrix} y_s(0) \\ y_s(1) \\ \vdots \\ y_s(N) \end{bmatrix},$$
$$Q_{v_sN} = \text{diag}(Q_{v_s}(0), Q_{v_s}(1), \ldots, Q_{v_s}(N)), \tag{12.8}$$

where

$$Q_{v_s}(k) = \begin{cases} \text{diag}(I, -\gamma^2 I),\ k \le N-1 \\ I,\qquad\qquad\quad k = N \end{cases}. \tag{12.9}$$

Rewrite J_N in (12.5) as

$$J_N = \begin{bmatrix} x_0 \\ d_N \\ f_N \\ v_{sN} \end{bmatrix}^T \begin{bmatrix} P_0 & 0 & 0 & 0 \\ 0 & I & 0 & 0 \\ 0 & 0 & I & 0 \\ 0 & 0 & 0 & Q_{v_sN} \end{bmatrix}^{-1} \begin{bmatrix} x_0 \\ d_N \\ f_N \\ v_{sN} \end{bmatrix}. \tag{12.10}$$

Note that v_{sN} can be expressed into a linear combination of y_{sN}, d_N and f_N. Hence, J_N is an indefinite quadratic form of $\{x_0, d_N, f_N, y_{sN}\}$.

Similar to Problem 11.1, we reformulate the problem of H_∞-FDF as follows.

Problem 12.2 *Given residual generator (12.2) for system (12.1) and J_N in (12.10), find matrices $H_i(k)$ $(i = 1, 2, 3)$, $V_1(k)$ and $V_2(k+1)$ solving*

$$\min_{x_0, d_N, f_N} J_N,\ s.t.\ (12.6)$$

such that J_N at its minimum is positive.

As well known, different from usual Hilbert space, the inner product defined in Krein space may be positive definite, negative definite, or indefinite. The basic idea of

H_∞ estimation in Krein space is to convert the estimation problem into the minimum of indefinite quadratic form J_N and construct an equivalent projection problem in Krein space. Next, we consider to construct an equivalent Krein space problem to the minimum for J_N. To do so, we introduce the following Krein space stochastic system corresponding to (12.6)

$$\begin{cases} \mathbf{x}(i+1) = A(i)\mathbf{x}(i) + B_d(i)\mathbf{d}(i) + B_f(i)\mathbf{f}(i),\ \mathbf{x}(0) = \mathbf{x}_0, \\ \mathbf{y}_s(i) = C_s(i)\mathbf{x}(i) + \mathbf{v}_s(i) + D_{sf}(i)\mathbf{f}(i), \end{cases} \tag{12.11}$$

where

$$\mathbf{y}_s(i) = \begin{cases} \mathbf{y}_f(i),\ i \le k-1 \\ \mathbf{y}(k),\ \ i = k \end{cases},\ \mathbf{y}_f(i) = \begin{bmatrix} \mathbf{y}(i) \\ \mathbf{r}(i) \end{bmatrix},$$

$$\mathbf{v}_s(i) = \begin{cases} \left[\mathbf{v}^T(i)\ \mathbf{v}_f^T(i)\right]^T,\ i \le k-1 \\ \mathbf{v}(i),\ i = k \end{cases},\ \ \mathbf{r}(i) = \mathbf{f}(i) + \mathbf{v}_f(i),$$

$$\left\langle \begin{bmatrix} \mathbf{x}_0 \\ \mathbf{d}(i) \\ \mathbf{f}(i) \\ \mathbf{v}_s(i) \end{bmatrix}, \begin{bmatrix} \mathbf{x}_0 \\ \mathbf{d}(j) \\ \mathbf{f}(j) \\ \mathbf{v}_s(j) \end{bmatrix} \right\rangle = \text{diag}(P_0, I\delta_{ij}, I\delta_{ij}, Q_{vs}(i)\delta_{ij}).$$

We now state and establish an existence condition for the minimum of J_N in the following lemma. Since the proof is similar to that of Lemma 3.2.4 in [1], it is omitted here.

Lemma 12.1 *The J_N subject to (12.6) has a minimum over $\{x_0, \{f(k), d(k)\}_{k=0}^N\}$ if and only if $R_{\mathbf{y}_s}(k) = \langle \mathbf{y}_s(k), \mathbf{y}_s(k)\rangle$ has the same inertia with $Q_{vs}(k)$ and, if this is the case, we have*

$$\min J_N = \sum_{k=0}^{N} y_s^T(k) R_{\mathbf{y}_s}^{-1}(k) y_s(k). \tag{12.12}$$

Notably, it is difficult to determine the feasibility of the inertia condition and the positiveness of J_N in Lemma 12.1. To solve this problem, an alternative condition for the minimum is derived by using Krein space projection and innovation analysis in the following subsection.

12.3.2 The Minimum for J_N

We first turn to analyze the inertia condition in Lemma 12.1. Define the following innovations, for $i = 0, 1, \ldots, k+1$,

$$\begin{aligned}
\tilde{\mathbf{y}}_{f1}(i) &= \mathbf{y}_f(i) - \hat{\mathbf{y}}_f(i,1),\ R_{\tilde{\mathbf{y}}_1}(i) = \left\langle \tilde{\mathbf{y}}_{f1}(i), \tilde{\mathbf{y}}_{f1}(i) \right\rangle, \\
\tilde{\mathbf{y}}_2(i) &= \mathbf{y}(i) - C(i)\hat{\mathbf{x}}(i,2)\ R_{\tilde{\mathbf{y}}_2}(i) = \langle \tilde{\mathbf{y}}_2(i), \tilde{\mathbf{y}}_2(i) \rangle, \\
\mathbf{e}_1(i) &= \mathbf{x}(i) - \hat{\mathbf{x}}(i,1),\ P_1(i) = \langle \mathbf{e}_1(i), \mathbf{e}_1(i) \rangle, \\
\mathbf{e}_2(i) &= \mathbf{x}(i) - \hat{\mathbf{x}}(i,2),\ P_2(i) = \langle \mathbf{e}_2(i), \mathbf{e}_2(i) \rangle,
\end{aligned}$$

where

$$\begin{aligned}
\hat{\mathbf{y}}_f(i,1) &= \begin{bmatrix} C(i)\hat{\mathbf{x}}(i,1) \\ \hat{\mathbf{r}}(i,1) \end{bmatrix},\ \hat{\mathbf{r}}(i,1) = \text{Proj}\{\mathbf{r}(i)|\mathcal{L}\{\mathbf{y}_f(j)\}_{j=0}^{i-1}\}, \\
\hat{\mathbf{x}}(i,1) &= \text{Proj}\{\mathbf{x}(i)|\mathcal{L}\{\mathbf{y}_f(j)\}_{j=0}^{i-1}\}, \\
\hat{\mathbf{x}}(i,2) &= \text{Proj}\{\mathbf{x}(i)|\mathcal{L}\{\{\mathbf{y}_f(j)\}_{j=0}^{i-2}; \mathbf{y}(i-1)\}\}.
\end{aligned}$$

Note that

$$\langle \mathbf{r}(i), \mathbf{r}(j) \rangle = 0,\ \langle \mathbf{r}(i), \mathbf{y}(j) \rangle = 0,\ \forall\, i > j.$$

We have $\hat{\mathbf{r}}(i,1) = 0$ and

$$\begin{aligned}
\tilde{\mathbf{y}}_{f1}(i) &= \bar{C}(i)\mathbf{e}_1(i) + \mathbf{v}_s(i) + \bar{D}_f(i)\mathbf{f}(i), \\
R_{\tilde{\mathbf{y}}_1}(i) &= \bar{C}(i)P_1(i)\bar{C}^T(i) + \bar{D}_f(i)\bar{D}_f^T(i) + Q_{vs}(i), \\
\tilde{\mathbf{y}}_2(i) &= C(i)\mathbf{e}_2(i) + \mathbf{v}(i) + D_f(i)\mathbf{f}(i), \\
R_{\tilde{\mathbf{y}}_2}(i) &= C(i)P_2(i)C^T(i) + D_f(i)D_f^T(i) + I.
\end{aligned}$$

Moreover, $\mathcal{L}\{\{\tilde{\mathbf{y}}_{f1}(i)\}_{i=0}^{k-1}\}$ forms an orthogonal basis space of $\mathcal{L}\{\{\mathbf{y}_f(i)\}_{i=0}^{k-1}\}$ due to the construction of innovation $\tilde{\mathbf{y}}_{f1}(i)$. Hence, $\hat{\mathbf{x}}(i,1)$ $(i = 0, 1, \ldots, k-1)$, $\forall k \in \mathcal{S}$ can be calculated by

$$\begin{cases}
\hat{\mathbf{x}}(i+1,1) = \sum_{j=0}^{i} \left\langle \mathbf{x}(i+1), \tilde{\mathbf{y}}_{f1}(j) \right\rangle R_{\tilde{\mathbf{y}}_1}^{-1}(j)\tilde{\mathbf{y}}_{f1}(j) \\
\qquad\qquad\quad = A(i)\hat{\mathbf{x}}(i,1) + K_1(i)\tilde{\mathbf{y}}_{f1}(i), \\
\hat{\mathbf{x}}(0,1) = 0,
\end{cases} \tag{12.13}$$

where

$$K_1(i) = (A(i)P_1(i)\bar{C}^T(i) + B_f(i)\bar{D}_f^T(i))R_{\tilde{\mathbf{y}}_1}^{-1}(i), \tag{12.14}$$

$$\begin{aligned}
P_1(i+1) &= A(i)P_1(i)A^T(i) + B_f(i)B_f^T(i) \\
&\quad + B_d(i)B_d^T(i), -K_1(i)R_{\tilde{\mathbf{y}}_1}(i)K_1^T(i),
\end{aligned} \tag{12.15}$$

$$P_1(0) = P_0. \tag{12.16}$$

Using (12.13), we have

$$
\begin{aligned}
&\tilde{\mathbf{y}}_2(i) \in \mathcal{L}\{\{\mathbf{y}_f(j)\}_{j=0}^{i-2}; \mathbf{y}(i-1), \mathbf{y}(i)\} \subseteq \mathcal{L}\{\{\mathbf{y}_f(j)\}_{j=0}^{i-1}; \mathbf{y}(i)\} \\
\Rightarrow\ &\mathcal{L}\{\{\tilde{\mathbf{y}}_{f1}(j)\}_{j=0}^{i-1}; \tilde{\mathbf{y}}_2(i)\} \subseteq \mathcal{L}\{\{\mathbf{y}_f(j)\}_{j=0}^{i-1}; \mathbf{y}(i)\}, \\
&\mathbf{y}(i) \in \mathcal{L}\{\{\mathbf{y}_f(j)\}_{j=0}^{i-2}; \mathbf{y}(i-1), \tilde{\mathbf{y}}_2(i)\} \subseteq \mathcal{L}\{\{\tilde{\mathbf{y}}_{f1}(j)\}_{j=0}^{i-1}; \tilde{\mathbf{y}}_2(i)\} \\
\Rightarrow\ &\mathcal{L}\{\{\mathbf{y}_f(j)\}_{j=0}^{i-1}; \mathbf{y}(i)\} \subseteq \mathcal{L}\{\{\tilde{\mathbf{y}}_{f1}(j)\}_{j=0}^{i-1}; \tilde{\mathbf{y}}_2(i)\},
\end{aligned}
$$

which implies

$$
\mathcal{L}\{\{\tilde{\mathbf{y}}_{f1}(j)\}_{j=0}^{i-1}; \tilde{\mathbf{y}}_2(i)\} = \mathcal{L}\{\{\mathbf{y}_f(j)\}_{j=0}^{i-1}; \mathbf{y}(i)\}. \tag{12.17}
$$

Hence, $\hat{\mathbf{x}}(i, 2)$ $(i = k, k+1)$ can be calculated by

$$
\begin{cases}
\hat{\mathbf{x}}(i+1, 2) = \sum_{j=0}^{i-1} \langle \mathbf{x}(i+1), \tilde{\mathbf{y}}_1(j)\rangle R_{\tilde{\mathbf{y}}_1}^{-1}(j)\tilde{\mathbf{y}}_{f1}(j) \\
\qquad\qquad + \langle \mathbf{x}(i+1), \tilde{\mathbf{y}}_2(i)\rangle R_{\tilde{\mathbf{y}}_2}^{-1}(i)\tilde{\mathbf{y}}_2(i) \\
\qquad\qquad = A(i)\hat{\mathbf{x}}(i, 2) + K_2(i)\tilde{\mathbf{y}}_2(i), \\
\hat{\mathbf{x}}(i, 2) = \hat{\mathbf{x}}(i, 1),
\end{cases} \tag{12.18}
$$

where

$$
K_2(i) = (A(i)P_2(i)C^T(i) + B_f(i)D_f^T(i))R_{\tilde{\mathbf{y}}_2}^{-1}(i), \tag{12.19}
$$

$$
\begin{aligned}
P_2(i+1) &= A(i)P_2(i)A^T(i) + B_f(i)B_f^T(i) \\
&\quad + B_d(i)B_d^T(i) - K_2(i)R_{\tilde{\mathbf{y}}_2}(i)K_2^T(i),
\end{aligned} \tag{12.20}
$$

$$
P_2(k-1) = P_1(k-1). \tag{12.21}
$$

Furthermore, one can also have

$$
\mathcal{L}\{\{\tilde{\mathbf{y}}_{f1}(i)\}_{i=0}^{k-1}; \tilde{\mathbf{y}}_2(k), \tilde{\mathbf{y}}_2(k+1)\} = \mathcal{L}\{\{\mathbf{y}_f(i)\}_{i=0}^{k-1}; \mathbf{y}(k), \mathbf{y}(k+1)\}.
$$

The projection of $\mathbf{r}(k)$ onto $\mathcal{L}\{\{\mathbf{y}_f(i)\}_{i=0}^{k-1}; \mathbf{y}(k), \mathbf{y}(k+1)\}$ can be calculated by

$$
\begin{aligned}
\hat{\mathbf{r}}(k|k+1) &= \sum_{i=0}^{k-1} \langle \mathbf{r}(k), \tilde{\mathbf{y}}_{f1}(i)\rangle R_{\tilde{\mathbf{y}}_{f1}}^{-1}(i)\tilde{\mathbf{y}}_{f1}(i) + \langle \mathbf{r}(k), \tilde{\mathbf{y}}_2(k)\rangle R_{\tilde{\mathbf{y}}_2}^{-1}(k)\tilde{\mathbf{y}}_2(k) \\
&\quad + \langle \mathbf{r}(k), \tilde{\mathbf{y}}_2(k+1)\rangle R_{\tilde{\mathbf{y}}_2}^{-1}(k+1)\tilde{\mathbf{y}}_2(k+1) \\
&= (B_f(k) - K_2(k)D_f(k))^T C^T(k+1)R_{\tilde{\mathbf{y}}_2}^{-1}(k+1)\tilde{\mathbf{y}}_2(k+1) \\
&\quad + D_f^T(k)R_{\tilde{\mathbf{y}}_2}^{-1}(k)\tilde{\mathbf{y}}_2(k).
\end{aligned} \tag{12.22}
$$

Define innovations again, $\forall k \in \mathcal{S}$,

$$\tilde{\mathbf{r}}(k) = \mathbf{r}(k) - \hat{\mathbf{r}}(k|k+1),\ R_{\tilde{\mathbf{r}}}(k) = \langle \tilde{\mathbf{r}}(k), \tilde{\mathbf{r}}(k)\rangle,$$
$$\tilde{\mathbf{y}}_s(k) = \begin{cases} \tilde{\mathbf{y}}_r(k), & k \le N-1 \\ \tilde{\mathbf{y}}_2(N), & k = N \end{cases},\ \tilde{\mathbf{y}}_r(k) = \begin{bmatrix} \tilde{\mathbf{y}}_2(k) \\ \tilde{\mathbf{r}}(k) \end{bmatrix},$$
$$R_{\tilde{\mathbf{y}}_r}(k) = \langle \tilde{\mathbf{y}}_r(k), \tilde{\mathbf{y}}_r(k)\rangle,\ R_{\tilde{\mathbf{y}}_s}(k) = \langle \tilde{\mathbf{y}}_s(k), \tilde{\mathbf{y}}_s(k)\rangle.$$

Similar to the derivation of (12.17), one can further have

$$\mathcal{L}\{\{\tilde{\mathbf{y}}_r(i)\}_{i=0}^{k-2}; \tilde{\mathbf{y}}_2(k-1), \tilde{\mathbf{y}}_2(k)\} = \mathcal{L}\{\{\mathbf{y}_f(i)\}_{i=0}^{k-2}; \mathbf{y}(k-1), \mathbf{y}(k)\}. \tag{12.23}$$

Note also that $\langle \tilde{\mathbf{r}}(k), \hat{\mathbf{r}}(k|k+1)\rangle = 0$, which implies

$$\tilde{\mathbf{r}}(k) \perp \mathcal{L}\{\{\mathbf{y}_f(i)\}_{i=0}^{k-1}; \mathbf{y}(k), \mathbf{y}(k+1)\}. \tag{12.24}$$

In light of (12.23) and (12.24), one obtains

$$\tilde{\mathbf{r}}(k) \perp \mathcal{L}\{\{\tilde{\mathbf{y}}_r(i)\}_{i=0}^{k-1}; \tilde{\mathbf{y}}_2(k), \tilde{\mathbf{y}}_2(k+1)\}.$$

Therefore,

$$\begin{aligned} R_{\tilde{\mathbf{y}}_s}(k) &= \begin{bmatrix} \langle \tilde{\mathbf{y}}_2(k), \tilde{\mathbf{y}}_2(k)\rangle & \langle \tilde{\mathbf{y}}_2(k), \tilde{\mathbf{r}}(k)\rangle \\ \langle \tilde{\mathbf{r}}(k), \tilde{\mathbf{y}}_2(k)\rangle & \langle \tilde{\mathbf{r}}(k), \tilde{\mathbf{r}}(k)\rangle \end{bmatrix} \\ &= \mathrm{diag}(R_{\tilde{\mathbf{y}}_2}(k),\ R_{\tilde{\mathbf{r}}}(k)),\ k \le N-1 \end{aligned} \tag{12.25}$$

and $R_{\tilde{\mathbf{y}}_s}(k) = R_{\tilde{\mathbf{y}}_2}(N)$ for $k = N$, where

$$R_{\tilde{\mathbf{y}}_2}(k) = C(k)P_2(k)C^T(k) + D_f(k)D_f^T(k) + I, \tag{12.26}$$

$$\begin{aligned} R_{\tilde{\mathbf{r}}}(k) &= \langle \mathbf{r}(k), \mathbf{r}(k)\rangle - \langle \hat{\mathbf{r}}(k|k+1), \hat{\mathbf{r}}(k|k+1)\rangle \\ &= (1-\gamma^2)I - \{D_f^T(k)R_{\tilde{\mathbf{y}}_2}^{-1}(k)D_f(k) \\ &\quad +(B_f(k) - K_2(k)D_f(k))^T C^T(k+1) \\ &\quad \times R_{\tilde{\mathbf{y}}_2}^{-1}(k+1)C(k+1)(B_f(k) - K_2(k)D_f(k))\}. \end{aligned} \tag{12.27}$$

Based on the above results, the following theorem provides an alternative condition for the minimum of J_N in terms of innovation variance matrices.

Theorem 12.1 *For $k = 1, 2, \ldots, N$, suppose that $P_1(i)$ $(i = 0, 1, \ldots, k-1)$, P_2 (i) $(i = k, k+1)$, $R_{\tilde{\mathbf{y}}_2}(k)$ and $R_{\tilde{\mathbf{r}}}(k)$ are calculated by (12.15)–(12.16), (12.20)–(12.21), (12.26) and (12.27), respectively. Then J_N subject to (12.6) has a minimum with respective to $\{x_0, d_N, f_N\}$ if and only if $R_{\tilde{\mathbf{y}}_2}(k) > 0$ and $R_{\tilde{\mathbf{r}}}(k) < 0$. Moreover, the minimum of J_N is*

$$\min J_N = \sum_{k=0}^{N} \tilde{y}_2^T(k)R_{\tilde{\mathbf{y}}_2}^{-1}(k)\tilde{y}_2(k) + \sum_{k=0}^{N-1} \tilde{r}^T(k)R_{\tilde{\mathbf{r}}}^{-1}(k)\tilde{r}(k),$$

where

$$\tilde{y}_2(k) = y(k) - C(k)\hat{x}(k,2),\ \tilde{r}(k) = r(k) - \hat{r}(k|k+1),$$

$\hat{x}(k,2)$ *and* $\hat{r}(k|k+1)$ *are calculated by the formulas of* $\hat{\mathbf{x}}(k,2)$ *and* $\hat{\mathbf{r}}(k|k+1)$, *respectively.*

Proof Define

$$\begin{aligned}
\mathbf{y}_{sk} &= \left[\mathbf{y}_s^T(0)\ \mathbf{y}_s^T(1)\ \cdots\ \mathbf{y}_s^T(k)\right]^T,\\
R_{yk} &= \mathrm{diag}(R_{y_s}(0),\ R_{y_s}(1), \ldots,\ R_{y_s}(k)),\\
\tilde{\mathbf{y}}_{sk} &= \left[\tilde{\mathbf{y}}_s^T(0)\ \tilde{\mathbf{y}}_s^T(1)\ \cdots\ \tilde{\mathbf{y}}_s^T(k)\right]^T,\\
R_{\tilde{\mathbf{y}}_{sk}} &= \mathrm{diag}(R_{\tilde{\mathbf{y}}_s}(0),\ R_{\tilde{\mathbf{y}}_s}(1), \ldots,\ R_{\tilde{\mathbf{y}}_s}(k)),\\
y_{sk} &= \left[y_s^T(0)\ y_s^T(1)\ \cdots\ y_s^T(k)\right]^T,\\
\tilde{y}_{sk} &= \left[\tilde{y}_s^T(0)\ \tilde{y}_s^T(1)\ \cdots\ \tilde{y}_s^T(k)\right]^T,\\
\tilde{y}_s(i) &= \begin{cases} \left[\tilde{y}_2^T(i)\ \tilde{r}^T(i)\right]^T,\ i = 0 \le k-1 \\ \tilde{y}_2(k),\ i = k \end{cases}.
\end{aligned}$$

It is easy to see that

$$\mathbf{y}_{sk} = \Phi_k \tilde{\mathbf{y}}_{sk},\ y_{sk} = \Phi_k \tilde{y}_{sk},\ R_{yk} = \Phi_k R_{\tilde{\mathbf{y}}_{sk}} \Phi_k^T,$$

where

$$\begin{aligned}
\Phi_k &= \Psi_k \left[\phi_k(i,j)\right]_{k+1,k+1} \Psi_k^T,\\
\Psi_k &= \mathrm{diag}(I, \psi, \ldots, \psi),\ \psi = \begin{bmatrix} 0 & I \\ I & 0 \end{bmatrix},\\
\phi_k(i,i) &= \begin{bmatrix} I & 0 \\ \langle \mathbf{r}(i-2), \tilde{\mathbf{y}}_2(i-1)\rangle R_{\tilde{\mathbf{y}}_2}^{-1}(i-1) & I \end{bmatrix},\\
\phi_k(i,1) &= \begin{bmatrix} \langle \mathbf{y}(i-1), \tilde{\mathbf{y}}_2(0)\rangle R_{\tilde{\mathbf{y}}_2}^{-1}(0) \\ \langle \mathbf{r}(i-2), \tilde{\mathbf{y}}_2(0)\rangle R_{\tilde{\mathbf{y}}_2}^{-1}(0) \end{bmatrix},
\end{aligned}$$

for $i = 2, 3, \ldots, k+1$, and

$$\phi_k(i,j) = \begin{bmatrix} \langle \mathbf{y}(i-1), \tilde{\mathbf{y}}_2(j-1)\rangle R_{\tilde{\mathbf{y}}_2}^{-1}(j-1) & \langle \mathbf{y}(i-1), \tilde{\mathbf{r}}(j-2)\rangle R_{\mathbf{r}}^{-1}(j-2) \\ \langle \mathbf{r}(i-2), \tilde{\mathbf{y}}_2(j-1)\rangle R_{\tilde{\mathbf{y}}_2}^{-1}(j-1) & \langle \mathbf{r}(i-2), \tilde{\mathbf{r}}(j-2)\rangle R_{\tilde{\mathbf{r}}}^{-1}(j-2) \end{bmatrix}$$

for $i = 3, 4, \ldots, k+1,\ j = 2, 3, \ldots, i-1$. It yields that the eigenvalues of R_{yk} are same with the ones of $R_{\tilde{\mathbf{y}}_{sk}}$. Hence, $R_{ys}(k)$ has the same inertia with $R_{\tilde{\mathbf{y}}_s}(k)$. In view of (12.9) and (12.25), the inertias of $R_{\tilde{\mathbf{y}}_s}(k)$ and $Q_{vs}(k)$ coincide if and only if $R_{\tilde{\mathbf{y}}_2}(k) > 0$ and $R_{\tilde{\mathbf{r}}}(k) < 0$.

Applying Lemma 12.1, J_N has a minimum if and only if $R_{ys}(k)$ has the same inertia with $Q_{vs}(k)$, which is equivalent to $R_{\tilde{\mathbf{y}}_2}(k) > 0,\ R_{\tilde{\mathbf{r}}}(k) < 0$. Furthermore, we have

$$\begin{aligned}\min J_N &= y_{sN}^T R_{yN}^{-1} y_{sN} = \tilde{y}_{sN}^T R_{\tilde{\mathbf{y}}_{sN}}^{-1} \tilde{y}_{sN} \\ &= \sum_{k=0}^{N} \tilde{y}_2^T(k) R_{\tilde{\mathbf{y}}_2}^{-1}(k)\tilde{y}_2(k) + \sum_{k=0}^{N-1} \tilde{r}^T(k) R_{\tilde{\mathbf{r}}}^{-1}(k)\tilde{r}(k).\end{aligned}$$

This completes the proof. □

12.3.3 Design of an H_∞-FDF

Up to now, we have derived an equivalent sufficient and necessary condition for the minimum of J_N in Theorem 12.1. The remaining task of H_∞-FDF design is to find $H_i(k)$ $(i = 1, 2, 3)$, $V_1(k)$ and $V_2(k+1)$ such that the minimum of J_N is positive.

Choose the parameter matrices of FDF in (12.2) as

$$H_1(k) = K_2(k),\ V_1(k) = D_f^T(k) R_{\tilde{\mathbf{y}}_2}^{-1}(k), \tag{12.28}$$

$$H_2(k) = K_1(k)\left[I\ 0\right]^T - H_1(k), \tag{12.29}$$

$$H_3(k) = K_1(k)\left[0\ I\right]^T, \tag{12.30}$$

$$V_2(k+1) = (B_f(k) - H_1(k)D_f(k))^T C^T(k+1) R_{\tilde{\mathbf{y}}_2}^{-1}(k+1), \tag{12.31}$$

where $K_1(k)$, $K_2(k)$ and $R_{\tilde{\mathbf{y}}_2}(i)$ $(i = k, k+1)$ are calculated by (12.14), (12.19) and (12.26), respectively. It is easy to see from (12.2), (12.13), (12.18) and (12.22) that $r(k)$ obeys the same equations with those of $\hat{\mathbf{r}}(k|k+1)$. For $\hat{r}(k|k+1)$ being calculated by the projection formula of $\hat{\mathbf{r}}(k|k+1)$, we have $r(k) = \hat{r}(k|k+1)$, which implies that

$$\min J_N = \sum_{k=0}^{N} \tilde{y}_2^T(k) R_{\tilde{\mathbf{y}}_2}^{-1}(k)\tilde{y}_2(k) > 0.$$

From the above analysis, the following theorem is concluded.

Theorem 12.2 *For $k = 1, 2, \ldots, N$, if $R_{\tilde{\mathbf{y}}_2}(k) > 0$ and $R_{\tilde{\mathbf{r}}}(k) < 0$, then the observer-based FDF (12.2) with parameter matrices given in (12.28)–(12.31) is an H_∞-FDF satisfying (12.3).*

We now summarize the calculation of $H_i(k)$ $(i = 1, 2, 3)$, $V_1(k)$ and $V_2(k+1)$ into the Algorithm 12.3.5.

Algorithm 12.3.5 Calculation of matrices $H_i(k)$ $(i = 1, 2, 3)$, $V_1(k)$ and $V_2(k+1)$

1: Set $\gamma > 0$, $P_0 > 0$ and $k = 1$;
2: Calculate $K_1(i)$, $P_1(i)$ $(i \leq k-1)$ using (12.14)–(12.16);
3: Let $P_2(k-1) = P_1(k-1)$. Calculate $P_2(k)$, $K_2(k)$, $R_{\tilde{\mathbf{y}}_2}(k)$ $P_2(k+1)$, $R_{\tilde{\mathbf{y}}_2}(k+1)$ and $R_{\tilde{\mathbf{r}}}(k)$ by using (12.19)–(12.21) and (12.26)–(12.27), respectively;
4: If $R_{\tilde{\mathbf{y}}_2}(k) > 0$ and $R_{\tilde{\mathbf{r}}}(k) < 0$, go to Step 5; Otherwise, exit;
5: Calculate $H_i(k)$ $(i = 1, 2, 3)$, $V_1(k)$ and $V_2(k+1)$ using (12.28)–(12.31);
6: Let $k = k + 1$, go to Step 2 till $k = N$.

Remark 12.2 In Chap. 11, we have presented an H_∞ fault estimation method for LDTV system, in which the fault was treated as an extended state and the fault estimation was obtained by straightforward H_∞ filtering for the augmented high dimension system. This is different from the formulation in this chapter. It is interesting to note that the obtained residual $r(k)$ in (12.2) leads to the fault estimation in Chap. 11 by setting $H_2(k) = 0$, $H_3(k) = 0$ and $V_2(k+1) = 0$.

12.4 Numerical Examples

Below two examples are considered to show the effectiveness of the obtained results.

Example 12.1 Consider system (12.1) with

$$A(k) = \begin{bmatrix} 0.2e^{-k/100} & 0.6 & 0 \\ 0 & 0.5 & \sin(k) \\ 0 & 0 & 0.7 \end{bmatrix}, \quad B_f(k) = \begin{bmatrix} 0.2 \\ 1.8 \\ 0.3 \end{bmatrix}, \quad B_d(k) = \begin{bmatrix} 1.3 \\ 0.5 \\ 0.6 \end{bmatrix},$$
$$C(k) = \begin{bmatrix} -0.5 & 1.5 & 0 \end{bmatrix}, \quad D_f(k) = 0.5.$$

Set $P_0 = I, x_0 = \begin{bmatrix} 1 & -1 & 2 \end{bmatrix}^T, d(k) = 0.5\sin(2k)$ and $v(k) = 0.3\cos(k)$. By applying Algorithm 12.3.5, the achieved minimum γ is 0.6358. We design an H_∞-FDF for $\gamma = 0.6358$. The simulated $f(k)$ and the generated $r(k)$ are shown in Fig. 12.1. As a comparison, the fault estimation result by using the Algorithm 11.3.4 given in Chap. 11 is also considered and the achieved minimum γ is 1.0001. The fault estimation corresponding to $\gamma = 1.0001$ is shown in Fig. 12.2. It is seen that the fault estimation is fast to approach zero.

Example 12.2 Consider the numerical example in Chap. 11. For different $D_f(k)$, the achieved minimum γ using Algorithms 12.3.5 and 11.3.4 are shown in the Table 12.1. It is seen that the performance index of the H_∞-FDF method in this chapter is better than the one in Chap. 11.

For $D_f(k) = 8$ and $\gamma = 0.4$, Figs. 12.3 and 12.4 show the simulated $f(k)$, the generated $r(k)$ using Algorithms 12.3.5 and 11.3.4, respectively.

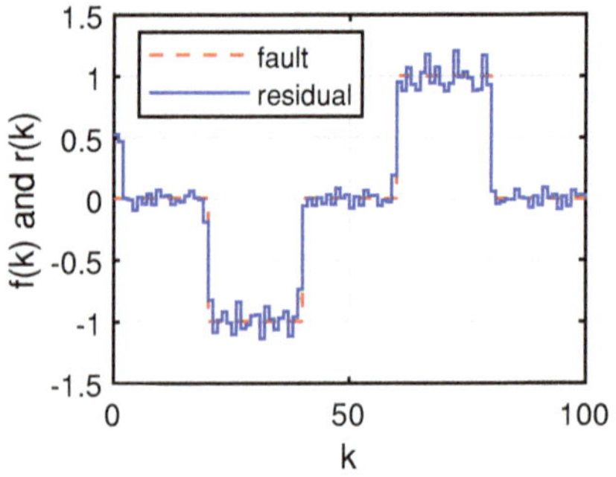

Fig. 12.1 The $f(k)$ and $r(k)$ in Example 12.1

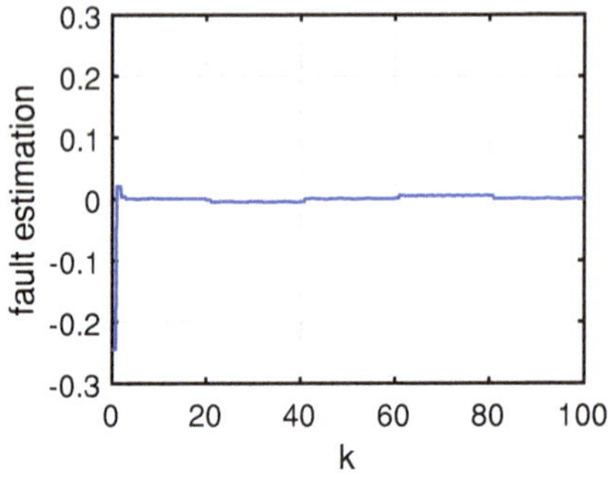

Fig. 12.2 The fault estimation in Example 12.1

Table 12.1 The achieved minimum γ for different $D_f(k)$

$D_f(k)$	8	4	2	1	0
H_∞-FDF method	0.3799	0.6465	0.7938	0.9059	0.9394
H_∞ fault estimation method	0.3960	0.7282	0.9693	1.0001	1.0001

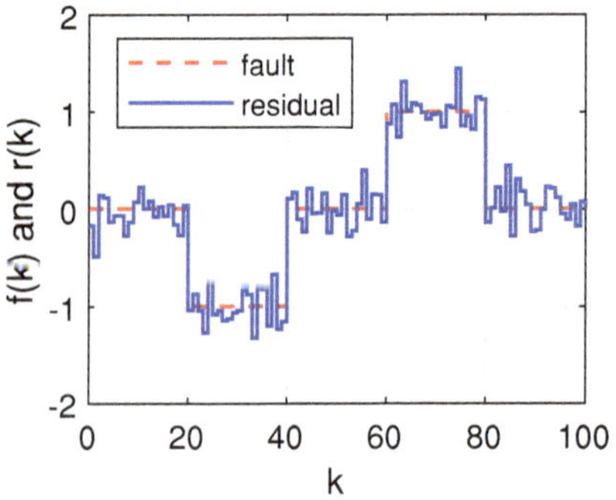

Fig. 12.3 The $f(k)$ and $r(k)$ in Example 12.2

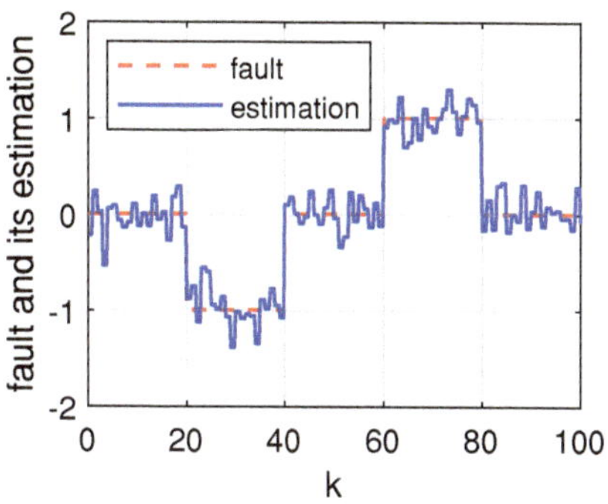

Fig. 12.4 The fault estimation in Example 12.2

12.5 Conclusion

This chapter has demonstrated an H_∞ filtering design scheme of observer-based FDF with the aid of Krein space techniques. The idea, similar to Chap. 11, is to convert the designing of H_∞-FDF into a minimum problem of an indefinite quadratic form in Krein space. By means of orthogonal projection and innovation analysis in Krein space, a sufficient and necessary condition for the minimum has been derived. On this basis, recursive realization algorithm of H_∞-FDF have been obtained by computing Riccati recursions. On the other hand, different from the H_∞ filtering based fault estimation method, the FDF concerned in this chapter is a closed-loop H_∞ smoothing formulation, in which the fault-free output is accommdated into the residual and the obtained residual is then feedback to estimate the system state. In such a way, better fault estimation results can be achieved in comparison with the scheme in Chap. 11. The numerical examples have demonstrated this fact.

References

1. Hassibi, B., Sayed, A. H., & Kailath, T. (1999). *Indefinite-quadratic estimation and control: A unified approach to* H_2 and H_∞ *theories*. SIAM.
2. Henry, D., & Zolghadri, A. (2006). Norm-based design of robust FDI schemes for uncertain systems under feedback control: Comparison of two approaches. *Control Engineering Practice, 14*(6), 1081–1097.
3. Tadmor, G., & Mirkin, L. (2005). H_∞ control and estimation with preview-Part II: fixed-size ARE solutions in discrete time. *IEEE Transactions on Automatic Control, 50*(1), 29–40.
4. Zhang, H., Xie, L., Soh, Y. C., et al. (2005). H_∞ fixed-lag smoothing for discrete linear time-varying systems. *Automatica, 41*(5), 839–846.
5. Zhang, H., & Xie, L. (2007). *Control and estimation of systems with input/output delays*. Berlin, Germany: Springer.
6. Chen, J., & Patton, R. J. (2000). Standard H_∞ filtering formulation of robust fault detection. *IFAC-PaperOnline, 33*(11), 261–266.
7. Zhong, M., Ding, S. X., Lam, J., et al. (2003). An LMI approach to design robust fault detection filter for uncertain LTI systems. *Automatica, 39*(3), 543–550.

8. Zhong, M., Liu, S., & Zhao, H. (2008). Krein space-based H_∞ fault estimation for linear discrete time-varying systems. *ACTA Automatica Sinica, 34*(12), 1529–1533.
9. Bolzern, P., Colaneri, P., & Nicolao, G. (2004). On discrete-time H_∞ fixed-lag smoothing. *IEEE Transactions on Signal Processing, 52*(1), 132–141.

Chapter 13
Krein Space Based H_∞ Fault Detection for LDTV Systems with Delayed State

In this chapter, we apply the Krein space based H_∞ filtering approach to FD for LDTV systems with delayed state. In the presence of delayed state, the design of an H_∞-FDF is converted into a minimum problem of an indefinite quadratic form. After introducing a corresponding Krein space stochastic system and building a relationship with projections, a sufficient and necessary condition on the existence of the minimum is derived by applying innovation analysis. A solution to the H_∞-FDF is then obtained in terms of matrix difference equations.

13.1 Introduction

Concerning the time-delay characteristic inherited in many engineering systems, e.g., chemical processes, long transmission lines in pneumatic, hydraulic and rolling mill systems, networked control systems, and traffic control systems, etc., research on FD for linear time-delay systems is a nontrivial topic. When it comes to LDTV systems with delayed state, we have demonstrated a robust FD scheme in Chap. 10, by directly designing an evaluation function in the framework of H_i/H_∞-optimization. Along the research line of H_∞ filtering, H_∞-optimal fault estimation and detection for LDTV systems have also been investigated respectively in Chaps. 11 and 12 by virtue of Krein space projection and innovation analysis.

As an application and extension of the results in Chap. 12, the objective of this chapter is to address FDF-based FD issues for LDTV systems with delayed state and l_2-norm bounded unknown inputs, by using Krein space based H_∞ filtering methd. To this end, we first show that the design of the targeting H_∞-FDF can be converted into an estimation problem of a corresponding stochastic system in Krein space, and then, by employing innovation analysis and orthogonal projection, both the existence condition and a solution to the H_∞-FDF are obtained in terms of Riccati recursions.

M. Zhong et al., *Fault Diagnosis for Linear Discrete Time-Varying Systems and Its Applications*, https://doi.org/10.1007/978-981-19-5438-2_13

13.2 Problem Formulation

Consider an LDTV system with delayed state in form of (10.1), i.e.,

$$\begin{cases} x(k+1) = A(k)x(k) + A_\tau(k)x(k-\tau) + B_d(k)d(k) + B_f(k)f(k), \\ y(k) = C(k)x(k) + v(k) + D_f(k)f(k), \end{cases} \tag{13.1}$$

where $x(k) \in \mathbb{R}^n$, $y(k) \in \mathbb{R}^q$, $d(k) \in \mathbb{R}^m$ and $v(k) \in \mathbb{R}^q$ are the state, measurement output, unknown input and measurement noise vectors, respectively, $f(k) \in \mathbb{R}^l$ is the fault vector to be detected, $x(k) = 0$ for $k \leq 0$, $\tau > 0$ is the state delay which is a constant integer, $A(k)$, $A_\tau(k)$, $B_d(k)$, $B_f(k)$, $C(k)$ and $D_f(k)$ are known matrices with appropriate dimensions. It is assumed that $d(k),\ v(k),\ f(k) \in l_{2,[0,N]}$.

Denote by $r(k)$ the residual generated at time step k. Let

$$w(k) = \begin{bmatrix} d^T(k) & f^T(k) & v^T(k) \end{bmatrix}^T.$$

A finite horizon H_∞ FD problem under consideration is stated as: for a given scalar $\gamma > 0$ and observation sequence $\{y(0), y(1), \ldots, y(N)\}$, design a residual generator such that

$$\sup_{\sum_{k=0}^{N} \|w(k)\|^2 \neq 0} \frac{\sum_{k=0}^{N-1} \|r(k) - f(k)\|^2}{\sum_{k=0}^{N} \|w(k)\|^2} < \gamma^2. \tag{13.2}$$

Note that the residual sequence $\{r(0), r(1), \ldots, r(k-1)\}$ is known when we construct residual $r(k)$. So we concern to generate the residual $r(k)$ by making use of $\{r(0), r(1), \ldots, r(k-1)\}$ and sequences $\{y(0), y(1), \ldots, y(k+1)\}$. For this purpose, $\{r(0), r(1), \ldots, r(k-1)\}$ is considered as a generalized observation sequence and a fiction unknown input $v_f(k)$ representing the error between $r(k)$ and $f(k)$ is introduced as

$$v_f(k) = r(k) - f(k),\ \forall k \in \mathcal{S}, \tag{13.3}$$

where $\mathcal{S}$ is a set with integers $0, 1, \ldots, N$. From (13.3) we have

$$r(k) = f(k) + v_f(k),\ \forall k \in \mathcal{S}. \tag{13.4}$$

For any $k \in \mathcal{S}$, combing (13.1) with (13.4) yields

$$\begin{cases} x(i+1) = A(i)x(i) + A_\tau(i)x(i-\tau) + B_f(i)f(i) + B_d(i)d(i), \\ y_s(i) = \widetilde{C}(i)x(i) + \widetilde{D}_f(i)f(i) + v_s(i),\ i = 0, 1, \ldots, k, \\ x(\theta) = 0,\ \text{ for } \theta \leq 0, \end{cases} \tag{13.5}$$

where $y_s,\ v_s$ are referred to (12.5), $\widetilde{C}(i) = C_s(i)$, $\widetilde{D}_f(i) = D_{sf}(i)$ with $C_s(i),\ D_{sf}(i)$ being given in (12.7).

We now turn to design a residual generator for system (13.5) and the following FDF is considered for this purpose

$$\begin{cases} \hat{x}(i+1) = A(i)\hat{x}(i) + A_\tau(i)\hat{x}(i-\tau) + K_0(i)(y_s(i) - \widetilde{C}(i)\hat{x}(i)), \\ \qquad + \sum_{j=1}^{\tau} K_{\tau j}(i)(y_s(i-j) - \widetilde{C}(i-j)\hat{x}(i-j)), \ i = 0, 1, \ldots, k \\ r(k) = V_0(k)(y(k) - C(k)\hat{x}(k)) + V_1(k+1)(y(k+1) - C(k+1)\hat{x}(k+1)), \\ \hat{x}(\theta) = 0, \ \text{for } \theta \le 0, \end{cases} \tag{13.6}$$

where $\hat{x}(i) \in \mathbb{R}^n$. Then the H_∞-FDF problem can be formulated as to find $K_{\tau j}(i)(j = 0, 1, \ldots, \tau)$, $V_0(k)$ and $V_1(k+1)$ such that (13.2) is satisfied.

Recall the definitions of f_N, d_N, v_{sN}, $Q_{v_s N}$ in (12.8). We then have

$$y_{sN} = H_{dN} d_N + H_{fN} f_N + v_{sN}, \tag{13.7}$$

$$\begin{bmatrix} d_N \\ f_N \\ y_{sN} \end{bmatrix} = \Gamma_N \begin{bmatrix} d_N \\ f_N \\ v_{sN} \end{bmatrix}, \tag{13.8}$$

where

$$\Gamma_N = \begin{bmatrix} I & 0 & 0 \\ 0 & I & 0 \\ H_{dN} & H_{fN} & I \end{bmatrix}, \ H_{fN} = \begin{bmatrix} \widetilde{D}_f(0) & 0 & \cdots & 0 \\ (2,1) & \widetilde{D}_f(1) & \cdots & 0 \\ \vdots & \ddots & \ddots & 0 \\ (N+1,1) & \cdots & (N+1,N) & \widetilde{D}_f(N) \end{bmatrix},$$

$$(i, j) = \widetilde{C}(i-1)\Phi(i-1, j)B_f(j-1); \ i = 2, 3, \ldots, N+1; \ j = 1, 2, \ldots, N,$$

the matrix H_{dN} is derived by replacing $\widetilde{D}_f$ and B_f in H_{fN} with 0 and B_d, respectively, and $\Phi(i, j)$ is given by

$$\begin{cases} \Phi(i+1, j) = A(i)\Phi(i, j) + A_\tau(i)\Phi(i-\tau, j) \\ \Phi(\theta, j) = 0(\theta < j), \ \Phi(j, j) = I, i = 0, 1, \ldots, N-1, \ j = 0, 1, \ldots, i. \end{cases} \tag{13.9}$$

Introduce

$$J_N = \sum_{k=0}^{N} \|w(k)\|^2 - \gamma^{-2} \sum_{k=0}^{N-1} \|v_f(k)\|^2, \tag{13.10}$$

which can be rewritten as

$$J_N = \begin{bmatrix} d_N \\ f_N \\ v_{sN} \end{bmatrix}^T \begin{bmatrix} I & 0 & 0 \\ 0 & I & 0 \\ 0 & 0 & Q_{v_s N} \end{bmatrix}^{-1} \begin{bmatrix} d_N \\ f_N \\ v_{sN} \end{bmatrix},$$

Algorithm 13.2.6 Calculation of residual sequences

1: When $y(k)$ is available one can calculate $r(k-1)$ by using the second equation of (13.6);
2: Update $y_s(k-1),\ y_s(k)$ with

$$y_s(k-1) = \begin{bmatrix} y(k-1) \\ r(k-1) \end{bmatrix},\ y_s(k) = y(k);$$

3: Calculate $\hat{x}(k)$ and $\hat{x}(k+1)$ by using the first equation of (13.6);
4: When $k \to k+1$ and $y(k+1)$ is available, calculate $r(k)$ by using the second equation of (13.6) and update $y_s(k)$ and $y_s(k+1)$ with

$$y_s(k) = \begin{bmatrix} y(k) \\ r(k) \end{bmatrix},\ y_s(k+1) = y(k+1);$$

5: Calculate $\hat{x}(k+1),\ \hat{x}(k+2)$ by using the first equation of (13.6), and then $r(k+1)$ can be obtained for the next sample $k+1 \to k+2$.

or, furthermore

$$J_N = \begin{bmatrix} d_N \\ f_N \\ y_{sN} \end{bmatrix}^T \left\{ \Gamma_N \begin{bmatrix} I & 0 & 0 \\ 0 & I & 0 \\ 0 & 0 & Q_{v_s N} \end{bmatrix} \Gamma_N^T \right\}^{-1} \begin{bmatrix} d_N \\ f_N \\ y_{sN} \end{bmatrix}. \tag{13.11}$$

It is obvious that (13.2) is satisfied if $J_N > 0$. We then formulate the H_∞-FDF problem as follow.

Problem 13.1 *Given system (13.1) and J_N in (13.11), construct an H_∞-FDF in form of (13.6) by finding $K_{\tau j}(i)(j = 0, 1, \ldots, \tau)$, $V_0(k)$ and $V_1(k+1)$ solving*

$$\min_{f_N, d_N} J_N, \ s.t. \ (13.5)$$

such that J_N at its minimum is positive.

Remark 13.1 For any given $k \in \mathcal{S}$, the dimension of $y_s(k)$ is different from that of $y_s(k-1)$ and the dimension of $y_s(i)$ is the same for all $i < k$. Observing that $y(k-1),\ \hat{x}(k-1),\ \hat{x}(k)$ are known at k, the calculation of residual sequence can be simply summarized as Algorithm 13.2.6.

Remark 13.2 The formulated H_∞ FD problem with design objective (13.2) is a finite horizon extension of H_∞ filtering formulation of FD in [7] to LDTV systems with delayed state. The philosophy to cast the H_∞-FDF into the minimum problem of indefinite quadratic form J_N is borrowed from the H_∞ estimation in Chap. 11 (or initially in [1]) for LDTV systems without time delay.

13.3 Design of H_∞ Fault Detection System in Krein Space

Applying the idea of H_∞-optimal FD in Chap. 12, in this section we first introduce an artificial Krein space stochastic system corresponding to (13.5) and convert the minimum problem of J_N into an orthogonal projection in Krein space. Then the existence condition on the minimum is derived and a solution to the H_∞-FDF parameter matrices achieving $J_N > 0$ is developed by using innovation analysis, in the existence of state delay.

13.3.1 The Minimum for J_N

We begin with introducing the following Krein space stochastic system corresponding to (13.5)

$$\begin{cases} \mathbf{x}(i+1) = A(i)\mathbf{x}(i) + A_\tau(i)\mathbf{x}(i-\tau) + B_f(i)\mathbf{f}(i) + B_d(i)\mathbf{d}(i), \\ \mathbf{y}_s(i) = \widetilde{C}(i)\mathbf{x}(i) + \widetilde{D}_f(i)\mathbf{f}(i) + \mathbf{v}_s(i), \\ \mathbf{x}(\theta) = 0, \text{ for } \theta \le 0, \end{cases} \tag{13.12}$$

where, for $i = 0, 1, \ldots, k$, $\forall k \in \mathcal{S}$, $A(i)$, $A_\tau(i)$, $B_f(i)$, $B_d(i)$, $\widetilde{C}(i)$ and $\widetilde{D}_f(i)$ are the same as in system (13.5), $\mathbf{d}(i)$, $\mathbf{f}(i)$ and $\mathbf{v}_s(i)$ are uncorrelated white noises with

$$\langle \begin{bmatrix} \mathbf{d}(i) \\ \mathbf{f}(i) \\ \mathbf{v}_s(i) \end{bmatrix}, \begin{bmatrix} \mathbf{d}(j) \\ \mathbf{f}(j) \\ \mathbf{v}_s(j) \end{bmatrix} \rangle = \begin{bmatrix} I\delta_{ij} & 0 & 0 \\ 0 & I\delta_{ij} & 0 \\ 0 & 0 & Q_{v_s}(i)\delta_{ij} \end{bmatrix}, \tag{13.13}$$

$$Q_{v_s}(i) = \begin{cases} \mathrm{diag}\{I \quad -\gamma^2 I\}, \ i = 0, 1, \ldots, k-1, \\ I, \ i = k, \end{cases} \tag{13.14}$$

and

$$\mathbf{y}_s(i) = \begin{cases} \mathbf{y}_f(i), \ i = 0, 1, \ldots, k-1, \\ \mathbf{y}(i), \ \ i = k, \end{cases}, \ \mathbf{y}_f(i) = \begin{bmatrix} \mathbf{y}(i) \\ \mathbf{r}(i) \end{bmatrix},$$
$$\mathbf{y}(i) = C(i)\mathbf{x}(i) + D_f(i)\mathbf{f}(i) + \mathbf{v}(i), \ \mathbf{r}(i) = \mathbf{f}(i) + \mathbf{v}_f(i),$$
$$\mathbf{v}_s(i) = \begin{cases} \begin{bmatrix} \mathbf{v}(i) \\ \mathbf{v}_f(i) \end{bmatrix}, \ i = 0, 1, \ldots, k-1 \\ \mathbf{v}(i), \ i = k \end{cases}.$$

Define

$$\mathbf{d}_N = \left[\mathbf{d}^T(0)\ \mathbf{d}^T(1) \cdots \mathbf{d}^T(N)\right]^T, \ \mathbf{f}_N = \left[\mathbf{f}^T(0)\ \mathbf{f}^T(1) \cdots \mathbf{f}^T(N)\right]^T,$$
$$\mathbf{v}_{sN} = \left[\mathbf{v}_s^T(0)\ \mathbf{v}_s^T(1) \cdots \mathbf{v}_s^T(N)\right]^T, \ \mathbf{y}_{sN} = \left[\mathbf{y}_s^T(0)\ \mathbf{y}_s^T(1) \cdots \mathbf{y}_s^T(N)\right]^T.$$

In view of (13.12) and (13.13), the J_N in (13.11) can be further rewritten as

$$J_N = \begin{bmatrix} d_N \\ f_N \\ y_{sN} \end{bmatrix}^T \langle \begin{bmatrix} \mathbf{d}_N \\ \mathbf{f}_N \\ \mathbf{y}_{sN} \end{bmatrix}, \begin{bmatrix} \mathbf{d}_N \\ \mathbf{f}_N \\ \mathbf{y}_{sN} \end{bmatrix} \rangle^{-1} \begin{bmatrix} d_N \\ f_N \\ y_{sN} \end{bmatrix}. \tag{13.15}$$

Similar to Lemma 12.1 (that is derived from the Lemma 3.2.4 in [1]), a solution to the minimum of J_N with respect to $\{f_N, d_N\}$ can be obtained in the following lemma.

Lemma 13.1 *For a given scalar $\gamma > 0$, J_N subject to (13.5) has a minimum over $\{f_N, d_N\}$ if and only if $R_{y_s N}$ and $Q_{v_s N}$ have the same inertia. If this is the case, then we have*

$$\min_{f_N, d_N} J_N = y_{sN}^T R_{y_s N}^{-1} y_{sN}, \tag{13.16}$$

where $R_{y_s N} =< \mathbf{y}_{sN}, \mathbf{y}_{sN} >$.

Proof Let

$$s_N = \begin{bmatrix} d_N \\ f_N \end{bmatrix}, \mathbf{s}_N = \begin{bmatrix} \mathbf{d}_N \\ \mathbf{f}_N \end{bmatrix}.$$

Then (13.15) can be rewritten as

$$J_N = \begin{bmatrix} s_N \\ y_{sN} \end{bmatrix}^T \begin{bmatrix} R_s & R_{sy} \\ R_{ys} & R_{y_s N} \end{bmatrix}^{-1} \begin{bmatrix} s_N \\ y_{sN} \end{bmatrix},$$

where $R_s =< \mathbf{s}_N, \mathbf{s}_N >$, $R_{sy} =< \mathbf{s}_N, \mathbf{y}_{sN} >= R_{ys}^T$. Note that

$$\begin{aligned} \begin{bmatrix} R_s & R_{sy} \\ R_{ys} & R_{y_s N} \end{bmatrix}^{-1} &= \begin{bmatrix} I & 0 \\ -R_{y_s N}^{-1} R_{ys} & I \end{bmatrix} \begin{bmatrix} R_s - R_{sy} R_{y_s N}^{-1} R_{ys} & 0 \\ 0 & R_{y_s N} \end{bmatrix}^{-1} \begin{bmatrix} I & -R_{sy} R_{y_s N}^{-1} \\ 0 & I \end{bmatrix} \\ &= \begin{bmatrix} I & -R_s^{-1} R_{sy} \\ 0 & I \end{bmatrix} \begin{bmatrix} R_s & 0 \\ 0 & R_{y_s N} - R_{ys} R_s^{-1} R_{sy} \end{bmatrix}^{-1} \begin{bmatrix} I & 0 \\ -R_{ys} R_s^{-1} & I \end{bmatrix}. \end{aligned}$$

Hence, J_N can be further rewritten as

$$\begin{aligned} J_N =&(s_N - R_{sy} R_{y_s N}^{-1} y_{sN})^T (R_s - R_{sy} R_{y_s N}^{-1} R_{ys})^{-1} (s_N - R_{sy} R_{y_s N}^{-1} y_{sN}) \\ &+ y_{sN}^T R_{y_s N}^{-1} y_{sN}. \end{aligned}$$

Applying Lemma 4.3 (i.e., Lemma 2.4.1 in [1]), it is known that the stationary point $s_N^* = R_{sy} R_{y_s N}^{-1} y_{sN}$ is a unique minimum if and only if $R_s - R_{sy} R_{y_s N}^{-1} R_{ys} > 0$. From the above triangular factorization, we know that $\text{diag}(R_s - R_{sy} R_{y_s N}^{-1} R_{ys}, R_{y_s N})$ and $\text{diag}(R_s, R_{y_s N} - R_{ys} R_s^{-1} R_{sy})$ are congruent. So

$$I_-[R_s - R_{sy} R_{y_s N}^{-1} R_{ys}] + I_-[R_{y_s N}] = I_-[R_s] + I_-[R_{y_s N} - R_{ys} R_s^{-1} R_{sy}],$$

where $\mathcal{I}_+(P)$ and $\mathcal{I}_-(P)$ respectively denote the numbers of positive negative eigenvalues of P. From (13.9) and (13.12), we have

$$\mathbf{y}_{sN} = H_{dN}\mathbf{d}_N + H_{fN}\mathbf{f}_N + \mathbf{v}_{sN},$$

where H_{dN} and H_{fN} are the same as in (13.7). Moreover, it follows from (13.13) that

$$R_s > 0,\ R_{y_sN} - R_{ys}R_s^{-1}R_{sy} = Q_{v_sN}.$$

Hence, $I_-[R_s - R_{sy}R_{y_sN}^{-1}R_{ys}] = 0$ holds if $I_-[R_{y_sN}] = I_-[Q_{v_sN}]$. As a result, we have $R_s - R_{sy}R_{y_sN}^{-1}R_{ys} > 0$.

Therefore, J_N has a minimizing solution over $\{f_N, d_N\}$ if and only if R_{y_sN} and Q_{v_sN} have the same inertia. Moreover, the minimum of J_N is given in (13.16). This completes the proof. □

Note that it is difficult to determine the feasibility of the inertia condition and the positiveness of J_N by directly applying Lemma 13.1. To solve this problem, an alternative condition for the minimum problem will be derived by using projection and innovation analysis in Krein space.

Denote by $\hat{\mathbf{x}}_1(i|j)$ the projection of $\mathbf{x}(i)$ $(i = 0, 1, \ldots, k)$ onto the linear space spanned by $\mathcal{L}\{\{\mathbf{y}_f(g)\}_{g=0}^{j}\}$, and $\hat{\mathbf{x}}_2(i)$ the projection of $\mathbf{x}(i)$ onto the linear space spanned by $\mathcal{L}\{\{\mathbf{y}_f(g)\}_{g=0}^{i-1}\}$. Let

$$\tilde{\mathbf{y}}_1(i) = \mathbf{y}_f(i) - \bar{C}(i)\hat{\mathbf{x}}_1(i|i-1),\ R_{\tilde{y}_1}(i) = \langle\tilde{\mathbf{y}}_1(i), \tilde{\mathbf{y}}_1(i)\rangle, \tag{13.17}$$

$$\tilde{\mathbf{y}}_2(i) = \mathbf{y}(i) - C(i)\hat{\mathbf{x}}_2(i), \qquad R_{\tilde{y}_2}(i) = \langle\tilde{\mathbf{y}}_2(i), \tilde{\mathbf{y}}_2(i)\rangle, \tag{13.18}$$

$$\mathbf{e}_1(i) = \mathbf{x}(i) - \hat{\mathbf{x}}_1(i|i-1), \qquad P_1(j,i) = \langle\mathbf{e}_1(j), \mathbf{e}_1(i)\rangle, \tag{13.19}$$

$$\mathbf{e}_2(i) = \mathbf{x}(i) - \hat{\mathbf{x}}_2(i), \qquad P_2(j,i) = \langle\mathbf{e}_2(j), \mathbf{e}_2(i)\rangle. \tag{13.20}$$

Then $\{\tilde{\mathbf{y}}_1(0), \tilde{\mathbf{y}}_1(1), \ldots, \tilde{\mathbf{y}}_1(k-1), \tilde{\mathbf{y}}_2(k), \tilde{\mathbf{y}}_2(k+1)\}$ is an innovation sequence which spans the same linear space as $\mathcal{L}\{\{\mathbf{y}_f(g)\}_{g=0}^{k+1}\}$ due to the construction of $\tilde{\mathbf{y}}_1(i)$ and $\tilde{\mathbf{y}}_2(i)$, please refer to Lemma 2.2.1 in [6]. Hence, projection $\hat{\mathbf{x}}_1(i+1|i)$ can be calculated by

$$\begin{aligned}
\hat{\mathbf{x}}_1(i+1|i) &= \sum_{j=0}^{i}\langle\mathbf{x}(i+1), \tilde{\mathbf{y}}_1(j)\rangle R_{\tilde{y}_1}^{-1}(j)\tilde{\mathbf{y}}_1(j) \\
&= \sum_{j=0}^{i}\langle A(i)\mathbf{x}(i) + A_\tau(i)\mathbf{x}(i-\tau), \tilde{\mathbf{y}}_1(j)\rangle R_{\tilde{y}_1}^{-1}(j)\tilde{\mathbf{y}}_1(j) \\
&\quad + \sum_{j=0}^{i}\langle B_f(i)\mathbf{f}(i) + B_d(i)\mathbf{d}(i), \tilde{\mathbf{y}}_1(j)\rangle R_{\tilde{y}_1}^{-1}(j)\tilde{\mathbf{y}}_1(j) \\
&= A(i)\hat{\mathbf{x}}_1(i|i) + A_\tau(i)\hat{\mathbf{x}}_1(i-\tau|i) + B_f(i)\bar{D}_f^T(i)R_{\tilde{y}_1}^{-1}(i)\tilde{\mathbf{y}}_1(i),
\end{aligned}$$

where

$$R_{\tilde{y}_1}(i) = \begin{bmatrix} C(i)P_1(i,i)C^T(i) + D_f(i)D_f^T(i) + I & D_f(i) \\ D_f^T(i) & I - \gamma^2 I \end{bmatrix}.$$

Note that

$$\begin{aligned} \hat{\mathbf{x}}_1(i-l|i) &= \sum_{j=0}^{i} \langle \mathbf{x}(i-l), \tilde{\mathbf{y}}_1(j)\rangle R_{\tilde{y}_1}^{-1}(j)\tilde{\mathbf{y}}_1(j) \\ &= \sum_{j=0}^{i-1} \langle \mathbf{x}(i-l), \tilde{\mathbf{y}}_1(j)\rangle R_{\tilde{y}_1}^{-1}(j)\tilde{\mathbf{y}}_1(j) + \langle \mathbf{x}(i-l), \tilde{\mathbf{y}}_1(i)\rangle R_{\tilde{y}_1}^{-1}(i)\tilde{\mathbf{y}}_1(i) \\ &= \hat{\mathbf{x}}_1(i-l|i-1) + R_1(i-l,i)\tilde{\mathbf{y}}_1(i), \end{aligned}$$

where

$$R_1(i-l,i) = \langle \mathbf{x}(i-l), \tilde{\mathbf{y}}_1(i)\rangle R_{\tilde{y}_1}^{-1}(i) = P_1(i-l,i)\bar{C}^T(i)R_{\tilde{y}_1}^{-1}(i). \quad (13.21)$$

We have

$$\begin{aligned} \hat{\mathbf{x}}_1(i|i) &= \hat{\mathbf{x}}_1(i|i-1) + R_1(i,i)\tilde{\mathbf{y}}_1(i), \\ \hat{\mathbf{x}}_1(i-\tau|i) &= \hat{\mathbf{x}}_1(i-\tau|i-\tau-1) + \sum_{j=0}^{\tau} R_1(i-\tau, i-j)\tilde{\mathbf{y}}_1(i-j). \end{aligned}$$

Thus, $\hat{\mathbf{x}}_1(i+1|i)(i = 0, 1, \ldots, k-1)$ can be further calculated by

$$\begin{cases} \hat{\mathbf{x}}_1(i+1|i) = A(i)\hat{\mathbf{x}}_1(i|i-1) + A_\tau(i)\hat{\mathbf{x}}_1(i-\tau|i-\tau-1) \\ \qquad + H_1(i)\tilde{\mathbf{y}}_1(i) + A_\tau(i)\sum_{j=1}^{\tau} R_1(i-\tau, i-j)\tilde{\mathbf{y}}_1(i-j), \\ \hat{\mathbf{x}}_1(-l|-l-1) = 0, \quad l = 0, 1, \ldots, \tau, \end{cases} \quad (13.22)$$

where

$$H_1(i) = A(i)R_1(i,i) + A_\tau(i)R_1(i-\tau,i) + B_f(i)\bar{D}_f^T(i)R_{\tilde{y}_1}^{-1}(i). \quad (13.23)$$

It follows from (13.12), (13.19) and (13.22) that

$$\begin{aligned} \mathbf{e}_1(i+1) = {} & A(i)\mathbf{e}_1(i) + A_\tau(i)\mathbf{e}_1(i-\tau) + B_f(i)\mathbf{f}(i) + B_d(i)\mathbf{d}(i) \\ & -H_1(i)\tilde{\mathbf{y}}_1(i) - A_\tau(i)\sum_{j=1}^{\tau} R_1(i-\tau, i-j)\tilde{\mathbf{y}}_1(i-j). \end{aligned}$$

Hence, for $i = 0, 1, \ldots, k-1$, $\forall k \in \mathcal{S}$, we have

$$
\begin{aligned}
P_1(i-\tau, i+1) =& \langle \mathbf{e}_1(i-\tau), \mathbf{e}_1(i+1)\rangle \\
=& P_1(i-\tau, i)A^T(i) + P_1^T(i-\tau, i-\tau)A_\tau^T(i) \\
& - P_1(i-\tau, i)\bar{C}^T(i)H_1^T(i) \\
& - \sum_{l=1}^{\tau} R_1(i-\tau, i-l)R_{\tilde{y}_1}(i-l)R_1^T(i-\tau, i-l)A_\tau^T(i),
\end{aligned} \tag{13.24}
$$

$$
\begin{aligned}
P_1(i, i+1) =& \langle \mathbf{e}_1(i), \mathbf{e}_1(i+1)\rangle \\
=& P_1(i, i)A^T(i) + P_1^T(i-\tau, i)A_\tau^T(i) \\
& - P_1(i, i)\bar{C}^T(i)H_1^T(i),
\end{aligned} \tag{13.25}
$$

$$
\begin{aligned}
P_1(i+1, i+1) =& \langle \mathbf{e}_1(i+1), \mathbf{e}_1(i+1)\rangle \\
=& A(i)P_1(i, i+1) + A_\tau(i)P_1(i-\tau, i+1) \\
& + B_f(i)B_f^T(i) + B_d(i)B_d^T(i) - B_f(i)\bar{D}_f^T(i)H_1^T(i).
\end{aligned} \tag{13.26}
$$

Similarly, projection $\hat{\mathbf{x}}_2(k+1)$ is calculated by

$$
\begin{cases}
\hat{\mathbf{x}}_2(k+1) = A(k)\hat{\mathbf{x}}_2(k) + A_\tau(k)\hat{\mathbf{x}}_1(k-\tau|k-\tau-1), \\
\qquad + H_2(k)\tilde{\mathbf{y}}_2(k) + A_\tau(k)\sum_{j=1}^{\tau} R_1(k-\tau, k-j)\tilde{\mathbf{y}}_1(k-j), \\
\hat{\mathbf{x}}_2(k) = \hat{\mathbf{x}}_1(k|k-1),
\end{cases} \tag{13.27}
$$

where

$$
H_2(k) = A(k)R_2(k, k) + A_\tau(k)R_2(k-\tau, k) + B_f(k)D_f^T(k)R_{\tilde{y}_2}^{-1}(k), \tag{13.28}
$$

$$
R_2(k-l, k) = P_1(k-l, k)C^T(k)R_{\tilde{y}_2}^{-1}(k), \tag{13.29}
$$

$$
R_{\tilde{y}_2}(k) = C(k)P_2(k, k)C^T(k) + D_f(k)D_f^T(k) + I. \tag{13.30}
$$

We then have

$$
\begin{aligned}
\mathbf{e}_2(k+1) =& A(k)\mathbf{e}_2(k) + A_\tau(k)\mathbf{e}_1(k-\tau) + B_f(k)\mathbf{f}(k) + B_d(k)\mathbf{d}(k) \\
& - H_2(k)\tilde{\mathbf{y}}_2(k) - A_\tau(k)\sum_{j=1}^{\tau} R_1(k-\tau, k-j)\tilde{\mathbf{y}}_1(k-j), \\
\mathbf{e}_2(k) =& \mathbf{e}_1(k).
\end{aligned}
$$

Therefore, $P_2(k+1, k+1)$ can be calculated by

$$\begin{aligned}
P_2(k-\tau,k+1) =&\langle \mathbf{e}_1(k-\tau), \mathbf{e}_2(k+1)\rangle \\
=&P_1(k-\tau,k)A^T(k) + P_1^T(k-\tau,k-\tau)A_\tau^T(k) \\
&-P_1(k-\tau,k)C^T(k)H_2^T(k) \\
&-\sum_{l=1}^{\tau} R_1(k-\tau,k-l)R_{\tilde{y}_1}(k-l)R_1^T(k-\tau,k-l)A_\tau^T(k), \quad (13.31)\\
P_2(k,k+1) =&\langle \mathbf{e}_2(k), \mathbf{e}_2(k+1)\rangle \\
=&P_2(k,k)A^T(k) + P_1^T(k-\tau,k)A_\tau^T(k) \\
&- P_2(k,k)C^T(k)H_2^T(k), \quad (13.32)\\
P_2(k+1,k+1) =&\langle \mathbf{e}_2(k+1), \mathbf{e}_2(k+1)\rangle \\
=&A(k)P_2(k,k+1) + A_\tau(k)P_2(k-\tau,k+1) \\
&+B_f(k)B_f^T(k)+B_d(k)B_d^T(k)-B_f(k)D_f^T(k)H_2^T(k), \quad (13.33)\\
P_2(k,k) =&P_1(k,k). \quad (13.34)
\end{aligned}$$

Now, we are ready to calculate the projection of $\mathbf{r}(k)$ onto $\mathcal{L}\{\{\mathbf{y}_f(g)\}_{g=0}^{k+1}\}$ by

$$\begin{aligned}
\hat{\mathbf{r}}(k) &= \sum_{i=0}^{k-1}\langle \mathbf{r}(k), \tilde{\mathbf{y}}_1(i)\rangle R_{\tilde{y}_1}^{-1}(i)\tilde{\mathbf{y}}_1(i) + \sum_{i=k}^{k+1}\langle \mathbf{r}(k), \tilde{\mathbf{y}}_2(i)\rangle R_{\tilde{y}_2}^{-1}(i)\tilde{\mathbf{y}}_2(i) \\
&= D_f^T(k)R_{\tilde{y}_2}^{-1}(k)\tilde{\mathbf{y}}_2(k) + \langle \mathbf{r}(k), \mathbf{e}_2(k+1)\rangle C^T(k+1)R_{\tilde{y}_2}^{-1}(k+1)\tilde{\mathbf{y}}_2(k+1) \\
&= D_f^T(k)R_{\tilde{y}_2}^{-1}(k)\tilde{\mathbf{y}}_2(k) \\
&\quad + \left(B_f^T(k) - D_f^T(k)H_2^T(k)\right)C^T(k+1)R_{\tilde{y}_2}^{-1}(k+1)\tilde{\mathbf{y}}_2(k+1). \quad (13.35)
\end{aligned}$$

For any $k \in \mathcal{S}$, define

$$\begin{aligned}
\tilde{\mathbf{y}}_s(i) &= \begin{cases} \tilde{\mathbf{y}}_f(i), & i=0,1,\dots,k-1 \\ \tilde{\mathbf{y}}_2(i), & i=k \end{cases}, \quad R_{\tilde{y}_s}(i) = \langle \tilde{\mathbf{y}}_s(i), \tilde{\mathbf{y}}_s(i)\rangle, \quad (13.36)\\
\tilde{\mathbf{r}}(i) &=\mathbf{r}(i) - \hat{\mathbf{r}}(i), \quad R_{\tilde{r}}(i) = \langle \tilde{\mathbf{r}}(i), \tilde{\mathbf{r}}(i)\rangle, \quad (13.37)\\
\tilde{\mathbf{y}}_f(i) &= \begin{bmatrix} \tilde{\mathbf{y}}_2(i) \\ \tilde{\mathbf{r}}(i) \end{bmatrix}, \quad R_{\tilde{y}_f}(i) = \langle \tilde{\mathbf{y}}_f(i), \tilde{\mathbf{y}}_f(i)\rangle. \quad (13.38)
\end{aligned}$$

Then the following lemma can be obtained.

Lemma 13.2 $\{\tilde{\mathbf{y}}_f(0), \tilde{\mathbf{y}}_f(1), \dots, \tilde{\mathbf{y}}_f(k-2), \tilde{\mathbf{y}}_2(k-1), \tilde{\mathbf{y}}_2(k)\}$ *is an innovation sequence which spans the same linear space as* $\mathcal{L}\{\{\mathbf{y}_f(g)\}_{g=0}^{k}\}$ *and*

$$\tilde{\mathbf{y}}_2(i) \perp \mathcal{L}\{\{\tilde{\mathbf{y}}_f(g)\}_{g=0}^{i-2},\ \tilde{\mathbf{y}}_2(i-1)\}, \quad (13.39)$$

$$\tilde{\mathbf{r}}(i-1) \perp \mathcal{L}\{\{\tilde{\mathbf{y}}_f(g)\}_{g=0}^{i-1},\ \tilde{\mathbf{y}}_2(i)\}. \quad (13.40)$$

Proof It is seen from (13.18) and (13.37) that $\tilde{\mathbf{y}}_2(i)$ and $\tilde{\mathbf{r}}(i)$ are the linear combination of sequence $\{\mathbf{y}_f(0), \mathbf{y}_f(1), \ldots, \mathbf{y}_f(i-2), \mathbf{y}(i-1), \mathbf{y}(i)\}$ and $\{\mathbf{y}_f(0), \mathbf{y}_f(1), \cdots, \mathbf{y}_f(i), \mathbf{y}(i+1)\}$, respectively. Therefore,

$$\mathcal{L}\{\{\tilde{\mathbf{y}}_f(g)\}_{g=0}^{k-2},\ \tilde{\mathbf{y}}_2(k-1),\ \tilde{\mathbf{y}}_2(k)\} \subseteq \mathcal{L}\{\{\tilde{\mathbf{y}}_f(g)\}_{g=0}^{k-2},\ \tilde{\mathbf{y}}(k-1),\ \tilde{\mathbf{y}}(k)\}.$$

Moreover, from (13.18) and (13.37), we have

$$\begin{aligned}
\mathbf{y}(0) &= \tilde{\mathbf{y}}_2(0) \in \mathcal{L}\{\tilde{\mathbf{y}}_2(0)\},\\
\mathbf{y}(1) &= \tilde{\mathbf{y}}_2(1) + \text{Proj}\{\mathbf{y}(1)|\mathbf{y}(0)\} \in \mathcal{L}\{\tilde{\mathbf{y}}_2(0), \tilde{\mathbf{y}}_2(1)\},\\
\mathbf{r}(0) &= \tilde{\mathbf{r}}(0) + \text{Proj}\{\mathbf{r}(0)|\mathbf{y}(0), \mathbf{y}(1)\} \in \mathcal{L}\{\tilde{\mathbf{y}}_f(0); \tilde{\mathbf{y}}_2(1)\},\\
\mathbf{y}(2) &= \tilde{\mathbf{y}}_2(2) + \text{Proj}\{\mathbf{y}(2)|\mathbf{y}_f(0); \mathbf{y}(1)\}\\
&= \tilde{\mathbf{y}}_2(2) + \langle \mathbf{y}(2), \tilde{\mathbf{y}}_f(0)\rangle R_{\tilde{y}_f}^{-1}(0)\tilde{\mathbf{y}}_f(0) + \langle \mathbf{y}(2), \tilde{\mathbf{y}}_2(1)\rangle R_{\tilde{y}_2}^{-1}(1)\tilde{\mathbf{y}}_2(1)\\
&\in \mathcal{L}\{\tilde{\mathbf{y}}_f(0); \tilde{\mathbf{y}}_2(1), \tilde{\mathbf{y}}_2(2)\},\\
\mathbf{r}(1) &= \tilde{\mathbf{r}}(1) + \text{Proj}\{\mathbf{r}(1)|\mathbf{y}_f(0); \mathbf{y}(1), \mathbf{y}(2)\}\\
&= \tilde{\mathbf{r}}(1) + \langle \mathbf{r}(1), \tilde{\mathbf{y}}_f(0)\rangle R_{\tilde{y}_f}^{-1}(0)\tilde{\mathbf{y}}_f(0) + \langle \mathbf{r}(1), \tilde{\mathbf{y}}_2(1)\rangle R_{\tilde{y}_2}^{-1}(1)\tilde{\mathbf{y}}_2(1)\\
&\quad + \langle \mathbf{r}(1), \tilde{\mathbf{y}}_2(2)\rangle R_{\tilde{y}_2}^{-1}(2)\tilde{\mathbf{y}}_2(2)\\
&\in \mathcal{L}\{\tilde{\mathbf{y}}_f(0), \tilde{\mathbf{y}}_f(1), \tilde{\mathbf{y}}_2(2)\},\\
&\ \vdots\\
\mathbf{y}(k-1) &= \tilde{\mathbf{y}}_2(k-1) + \text{Proj}\{\mathbf{y}(k-1)|\mathbf{y}_f(0), \ldots, \mathbf{y}_f(k-3); \mathbf{y}(k-2)\}\\
&= \tilde{\mathbf{y}}_2(k-1) + \sum_{j=0}^{k-3} \langle \mathbf{y}(k-1), \tilde{\mathbf{y}}_f(j)\rangle R_{\tilde{y}_f}^{-1}(j)\tilde{\mathbf{y}}_f(j)\\
&\quad + \langle \mathbf{y}(k-1), \tilde{\mathbf{y}}_2(k-2)\rangle R_{\tilde{y}_2}^{-1}(k-2)\tilde{\mathbf{y}}_2(k-2)\\
&\in \mathcal{L}\{\tilde{\mathbf{y}}_f(0), \ldots, \tilde{\mathbf{y}}_f(k-3); \tilde{\mathbf{y}}_2(k-2), \tilde{\mathbf{y}}_2(k-1)],\\
\mathbf{r}(k-2) &= \tilde{\mathbf{r}}(k-2) + \text{Proj}\{\mathbf{r}(k-2)|\mathbf{y}_f(0), \ldots, \mathbf{y}_f(k-3); \mathbf{y}(k-2), \mathbf{y}(k-1)\}\\
&= \tilde{\mathbf{r}}(k-2) + \sum_{j=0}^{k-3} \langle \mathbf{r}(k-2), \tilde{\mathbf{y}}_f(j)\rangle R_{\tilde{y}_f}^{-1}(j)\tilde{\mathbf{y}}_f(j) + < \mathbf{r}(k-2), \tilde{\mathbf{y}}_2(k-2) >\\
&\quad \times R_{\tilde{y}_2}^{-1}(k-2)\tilde{\mathbf{y}}_2(k-2) + \langle \mathbf{r}(k-2), \tilde{\mathbf{y}}_2(k-1)\rangle R_{\tilde{y}_2}^{-1}(k-1)\tilde{\mathbf{y}}_2(k-1)\\
&\in \mathcal{L}\{\tilde{\mathbf{y}}_f(0), \tilde{\mathbf{y}}_f(1), \ldots, \tilde{\mathbf{y}}_f(k-2); \tilde{\mathbf{y}}_2(k-1)\},\\
\mathbf{y}(k) &= \tilde{\mathbf{y}}_2(k) + \text{Proj}\{\mathbf{y}(k)|\mathbf{y}_f(0), \ldots, \mathbf{y}_f(k-2); \mathbf{y}(k-1)\}\\
&= \mathbf{y}_2(k) + \sum_{j=0}^{k-2} \langle \mathbf{y}(k), \tilde{\mathbf{y}}_f(j)\rangle R_{\tilde{y}_f}^{-1}(j)\tilde{\mathbf{y}}_f(j)\\
&\quad + \langle \mathbf{y}(k), \tilde{\mathbf{y}}_2(k-1)\rangle R_{\tilde{y}_2}^{-1}(k-1)\tilde{\mathbf{y}}_2(k-1)\\
&\in \mathcal{L}\{\tilde{\mathbf{y}}_f(0), \tilde{\mathbf{y}}_f(1), \ldots, \tilde{\mathbf{y}}_f(k-2); \tilde{\mathbf{y}}_2(k-1), \tilde{\mathbf{y}}_2(k)\}.
\end{aligned}$$

Hence,

$$\begin{aligned}&\mathcal{L}\{\mathbf{y}_f(0), \mathbf{y}_f(1), \cdots, \mathbf{y}_f(k-2); \mathbf{y}(k-1), \mathbf{y}(k)\}\\ &\subseteq \mathcal{L}\{\tilde{\mathbf{y}}_f(0), \tilde{\mathbf{y}}_f(1), \ldots, \tilde{\mathbf{y}}_f(k-2); \tilde{\mathbf{y}}_2(k-1), \tilde{\mathbf{y}}_2(k)\}.\end{aligned}$$

Thus, $\{\tilde{\mathbf{y}}_f(0), \tilde{\mathbf{y}}_f(1), \ldots, \tilde{\mathbf{y}}_f(k-2); \tilde{\mathbf{y}}_2(k-1), \tilde{\mathbf{y}}_2(k)\}$ spans the same linear space as $\mathcal{L}\{\mathbf{y}_f(0), \mathbf{y}_f(1), \ldots, \mathbf{y}_f(k-2); \mathbf{y}(k-1), \mathbf{y}(k)\}$.

On the other hand, $\tilde{\mathbf{y}}_2(i)$ and $\tilde{\mathbf{r}}(i-1)$ satisfy the following orthogonality conditions

$$\begin{aligned}\tilde{\mathbf{y}}_2(i) &\perp \mathcal{L}\{\mathbf{y}_f(0), \mathbf{y}_f(1), \ldots, \mathbf{y}_f(i-2); \mathbf{y}(i-1)\},\\ \tilde{\mathbf{r}}(i-1) &\perp \mathcal{L}\{\mathbf{y}_f(0), \mathbf{y}_f(1), \ldots, \mathbf{y}_f(i-2); \mathbf{y}(i-1), \mathbf{y}(i)\}.\end{aligned}$$

Therefore,

$$\begin{aligned}\tilde{\mathbf{y}}_2(i) &\perp \mathcal{L}\{\tilde{\mathbf{y}}_f(0), \tilde{\mathbf{y}}_f(1), \ldots, \tilde{\mathbf{y}}_f(i-2); \tilde{\mathbf{y}}_2(i-1)\},\\ \tilde{\mathbf{r}}(i-1) &\perp \mathcal{L}\{\tilde{\mathbf{y}}_f(0), \tilde{\mathbf{y}}_f(1), \ldots, \tilde{\mathbf{y}}_f(i-2); \tilde{\mathbf{y}}_2(i-1), \tilde{\mathbf{y}}_2(i)\}.\end{aligned}$$

This completes the proof. □

In light of (13.39)–(13.40), it is easy to have

$$\langle \tilde{\mathbf{y}}_2(i), \tilde{\mathbf{r}}(i)\rangle = 0.$$

Hence,

$$R_{\tilde{y}_s}(i) = \begin{cases} \mathrm{diag}\{R_{\tilde{y}_2}(i), R_{\tilde{r}}(i)\},\ i = 0, 1, \ldots, k-1,\\ R_{\tilde{y}_2}(i),\ i = k,\end{cases}$$

where $R_{\tilde{y}_2}(k)$ is given in (13.30), $R_{\tilde{r}}(k)$ is calculated by

$$\begin{aligned}R_{\tilde{r}}(k) &= \langle \tilde{\mathbf{r}}(k), \tilde{\mathbf{r}}(k)\rangle\\ &= \langle \mathbf{r}(k), \mathbf{r}(k)\rangle - \langle \hat{\mathbf{r}}(k), \hat{\mathbf{r}}(k)\rangle\\ &= (1-\gamma^2)I - D_f^T(k)R_{\tilde{y}_2}^{-1}(k)D_f(k) - \left(B_f^T(k) - D_f^T(k)H_2^T(k)\right)\\ &\times C^T(k+1)R_{\tilde{y}_2}^{-1}(k+1)C(k+1)\left(B_f(k) - H_2(k)D_f(k)\right).\end{aligned} \quad (13.41)$$

Let

$$\begin{aligned}\tilde{\mathbf{y}}_{sN} &= \left[\,\tilde{\mathbf{y}}_s^T(0)\ \tilde{\mathbf{y}}_s^T(1)\ \cdots\ \tilde{\mathbf{y}}_s^T(N)\,\right]^T,\\ R_{\tilde{y}_sN} &= \mathrm{diag}\{R_{\tilde{y}_s}(0), R_{\tilde{y}_s}(1), \ldots, R_{\tilde{y}_s}(N)\}.\end{aligned}$$

It is see from the proof of Theorem 13.1 that $R_{\tilde{y}_s N}$ and $R_{y_s N}$ have the same inertia. We are now in a position to provide an alternative inertia condition in terms of $R_{\tilde{y}_2}(k)$ and $R_{\tilde{r}}(k)$ for the minimum of J_N, as summarized in the following theorem.

Theorem 13.1 *Given $R_{\tilde{y}_2}(k)$ and $R_{\tilde{r}}(k)$ respectively by (13.30) and (13.41), for a scalar $\gamma > 0$, J_N subject to (13.5) has a minimum over $\{f_N, d_N\}$ if and only if $R_{\tilde{y}_2}(k) > 0$ $(k = 0, 1, \ldots, N)$ and $R_{\tilde{r}}(k) < 0$ $(k = 0, 1, \ldots, N-1)$. Moreover, the minimum for J_N is*

$$\min_{f_N, d_N} J_N = \sum_{k=0}^{N} \tilde{y}_2^T(k) R_{\tilde{y}_2}^{-1}(k) \tilde{y}_2(k) + \sum_{k=0}^{N-1} \tilde{r}^T(k) R_{\tilde{r}}^{-1}(k) \tilde{r}(k), \tag{13.42}$$

where

$$\tilde{y}_2(k) = y(k) - C(k)\hat{x}_2(k), \quad \tilde{r}(k) = r(k) - \hat{r}(k), \tag{13.43}$$

$H_2(k)$, $\hat{x}_2(k)$ and $\hat{r}(k)$ are respectively calculated by (13.28), (13.27) and (13.35).

Proof Let

$$\begin{aligned}
\mathbf{y}_{sk} &= \left[\mathbf{y}_s^T(0)\ \mathbf{y}_s^T(1) \cdots \mathbf{y}_s^T(k)\right]^T, \ \tilde{\mathbf{y}}_{sk} = \left[\tilde{\mathbf{y}}_s^T(0)\ \tilde{\mathbf{y}}_s^T(1) \cdots \tilde{\mathbf{y}}_s^T(k)\right]^T \\
y_{sk} &= \left[y_s^T(0)\ y_s^T(1) \cdots y_s^T(k)\right]^T, \ \tilde{y}_{sk} = \left[\tilde{y}_s^T(0)\ \tilde{y}_s^T(1) \cdots \tilde{y}_s^T(k)\right]^T, \\
R_{y_s k} &= \langle \mathbf{y}_{sk}, \mathbf{y}_{sk} \rangle, \ R_{\tilde{y}_s k} = \mathrm{diag}(R_{\tilde{y}_s}(0), R_{\tilde{y}_s}(1), \ldots, R_{\tilde{y}_s}(k)), \\
Q_{v_s k} &= \mathrm{diag}(Q_{v_s}(0),\ Q_{v_s}(1), \ldots, Q_{v_s}(k)), \\
\tilde{y}_s(i) &= \begin{cases} \left[\tilde{y}_2^T(i) \ \ \tilde{r}^T(i)\right]^T, \ i = 0, 1, \ldots, k-1, \\ \tilde{y}_2(i), i = k. \end{cases}
\end{aligned}$$

It is easy to have

$$\mathbf{y}_{sk} = \Phi_k \tilde{\mathbf{y}}_{sk}, \ R_{y_s k} = \Phi_k R_{\tilde{y}_s k} \Phi_k^T, \ y_{sk} = \Phi_k \tilde{y}_{sk}. \tag{13.44}$$

where

$$\Phi_k = \Psi_k \begin{bmatrix} I & 0 & 0 & \cdots & 0 \\ \phi_k(2,1) & \phi_k(2,2) & 0 & \cdots & 0 \\ \phi_k(3,1) & \phi_k(3,2) & \phi_k(3,3) & \ddots & \vdots \\ \vdots & \vdots & \vdots & \ddots & 0 \\ \phi_k(k+1,1) & \phi_k(k+1,2) & \phi_k(k+1,3) & \cdots & \phi_k(k+1,k+1) \end{bmatrix} \Psi_k^T,$$

$$\Psi_k = \mathrm{diag}(I, \psi, \ldots, \psi), \ \psi = \begin{bmatrix} 0 & I \\ I & 0 \end{bmatrix},$$

$$\phi_k(i,i) = \begin{bmatrix} I & 0 \\ \langle \mathbf{r}(i-2), \tilde{\mathbf{y}}_2(i-1) \rangle R_{\tilde{y}_2}^{-1}(i-1) & I \end{bmatrix},$$

$$\phi_k(i,1) = \begin{bmatrix} \langle \mathbf{y}(i-1), \tilde{\mathbf{y}}_2(0)\rangle R_{\tilde{y}_2}^{-1}(0) \\ \langle \mathbf{r}(i-2), \tilde{\mathbf{y}}_2(0)\rangle R_{\tilde{y}_2}^{-1}(0) \end{bmatrix}, \; i = 2, 3, \ldots, k+1,$$

$$\phi_k(i,j) = \begin{bmatrix} \langle \mathbf{y}(i-1), \tilde{\mathbf{y}}_2(j-1)\rangle R_{\tilde{y}_2}^{-1}(j-1) & \langle \mathbf{y}(i-1), \tilde{\mathbf{r}}(j-2)\rangle R_{\tilde{r}}^{-1}(j-2) \\ \langle \mathbf{r}(i-2), \tilde{\mathbf{y}}_2(j-1)\rangle R_{\tilde{y}_2}^{-1}(j-1) & \langle \mathbf{r}(i-2), \tilde{\mathbf{r}}(j-2)\rangle R_{\tilde{r}}^{-1}(j-2) \end{bmatrix},$$
$$i = 3, 4, \ldots, k+1, \; j = 2, 3, \ldots, i-1.$$

It is readily seen from (13.44) that $R_{y_s k}$ and $R_{\tilde{y}_s k}$ have the same inertia. Since both $R_{\tilde{y}_s k}$ and $Q_{v_s k}$ are block-diagonal matrices, it follows from (13.14) and (13.41) that $R_{\tilde{y}_s k}$ and $Q_{v_s k}$ have the same inertia for any $k = 0, 1, \ldots, N$ if and only if $R_{\tilde{y}_2}(i) > 0 (i = 0, 1, \ldots, k)$ and $R_{\tilde{r}}(i) < 0 (i = 0, 1, \ldots, k-1)$. Therefore, by applying Lemma 13.1, J_N subject to system (13.5) has a minimum over $\{f_N, d_N\}$ if and only if $R_{\tilde{y}_2}(k) > 0$ $(k = 0, 1, \ldots, N)$ and $R_{\tilde{r}}(k) < 0$ $(k = 0, 1, \ldots, N-1)$.

Moreover, the minimum of J_N is given by

$$\min_{f_N, d_N} J_N = y_{sN}^T R_{y_s N}^{-1} y_{sN} = \tilde{y}_{sN}^T R_{\tilde{y}_s N}^{-1} \tilde{y}_{sN}$$
$$= \sum_{k=0}^{N} \tilde{y}_2^T(k) R_{\tilde{y}_2}^{-1}(k) \tilde{y}_2(k) + \sum_{k=0}^{N-1} \tilde{r}^T(k) R_{\tilde{r}}^{-1}(k) \tilde{r}(k).$$

This completes the proof. □

13.3.2 Design of an H_∞-FDF

It is known from Theorem 13.1 that J_N has a minimum over $\{f_N, d_N\}$ if and only if $R_{\tilde{y}_2}(k) > 0$ $(k = 0, 1, \ldots, N)$ and $R_{\tilde{r}}(k) < 0$ $(k = 0, 1, \ldots, N-1)$ and, the minimum of J_N is given by (13.42). Now, the remaining task of H_∞-FDF design is to choose $K_{\tau j}(i)$ $(j = 0, 1, \ldots, \tau)$, $V_0(k)$ and $V_1(k+1)$ to ensure the positive minimum of J_N. Set

$$\begin{cases} K_0(i) = \begin{cases} H_1(i), & i = 0, 1, \ldots, k-1 \\ H_2(k), & i = k \end{cases}, \\ K_{\tau j}(i) = A_\tau(i) R_1(i-\tau, i-j), \quad j = 1, 2, \ldots, \tau, \\ V_0(k) = D_f^T(k) R_{\tilde{y}_2}^{-1}(k), \\ V_1(k+1) = [B_f^T(k) - D_f^T(k) H_2^T(k)] C^T(k+1) R_{\tilde{y}_2}^{-1}(k+1). \end{cases} \tag{13.45}$$

where $R_1(i-\tau, i-j)$, $H_1(i)$ $(i = 0, 1, \ldots, k-1;\; j = 1, 2, \ldots, \tau)$, $H_2(k)$ and $R_{\tilde{y}_2}(k)$ are computed by (13.21), (13.23), (13.28) and (13.30), respectively. In light of

Algorithm 13.3.7 Updating of matrix $K_0(i)$ $(i \leq k+1)$

1: Compute $H_1(k)$ using (13.23) and let $K_0(k) = H_1(k)$;
2: Compute $H_2(k+1)$ using (13.28) and let $K_0(k+1) = H_2(k+1)$;
3: $K_0(i)$ $(i < k)$ remains the value of last sample k, i.e. $K_0(i) = H_1(i)$ for $i < k$.

(13.6), (13.18), (13.27) and (13.35), it is easy to have $r(k) = \hat{r}(k)$ by using Theorem 13.1, where $\hat{r}(k)$ is calculated by the same formula of $\hat{\mathbf{r}}(k)$ in (13.35). In this case, (13.42) leads to

$$\min_{f_N, d_N} J_N = \sum_{k=0}^{N} \tilde{y}_2^T(k) R_{\tilde{y}_2}^{-1}(k) \tilde{y}_2(k) > 0.$$

which implies that the parameter matrices given in (13.45) deliver an H_∞-FDF satisfying (13.1). The Problem 13.1 is thus addressed.

Based on the above analysis, the following theorem can be concluded.

Theorem 13.2 *For a given scalar $\gamma > 0$, suppose that $P_1(i-j, i)$, $P_2(k-j, k)$, $R_1(i-\tau, i-j)$, $H_1(i)$ $(i = 0, 1, \ldots, k-1;\ j = 0, 1, \ldots, \tau)$, $H_2(k)$, $R_{\tilde{y}_2}(k)$ and $R_{\tilde{r}}(k)$ are calculated by (13.24)–(13.26), (13.31)–(13.34), (13.21), (13.23), (13.28), (13.30) and (13.41), respectively. Then an H_∞-FDF satisfying (13.2) exists if and only if $R_{\tilde{y}_2}(k) > 0$ $(k = 0, 1, \ldots, N)$ and $R_{\tilde{r}}(k) < 0$ $(k = 0, 1, \ldots, N-1)$. Moreover, the parameter matrices of the H_∞-FDF are given in (13.45).*

Remark 13.3 For any given $k \in \mathcal{S}$, the column of $K_0(k)$ is different from that of $K_0(k-1)$ and it is the same for all $i < k$. Similar to the case mentioned in Remark 13.1, we still use the same matrix for different k. When $k \to k+1$, the matrices $K_0(i)$ $(i \leq k+1)$ are updated by using Algorithm 13.3.7.

Remark 13.4 It is observed from the above results and proof procedure that, the design of H_∞-FDF concerned in this chapter is an extension of the method in Chap. 12 wherein the time delay is not taken into account, i.e., $\tau = 0$.

13.4 Numerical Examples

Two examples are given below to show the effectiveness of the presented method.

Example 13.1 Consider system (13.1) with parameters

$$A(k) = \begin{bmatrix} 0.1e^{-k} & 0.2\sin(k) \\ 0 & 0.6 \end{bmatrix}, \quad A_\tau(k) = \begin{bmatrix} 0.7 & 0.3\cos(k) \\ 0 & 0.9 \end{bmatrix},$$

$$B_f(k) = \begin{bmatrix} 0.9 \\ 1.8 \end{bmatrix}, \quad B_d(k) = \begin{bmatrix} 0.3 \\ 0.7 \end{bmatrix}, \quad C(k) = \begin{bmatrix} 1.4 & 2.8 \end{bmatrix}, \quad D_f(k) = 0.$$

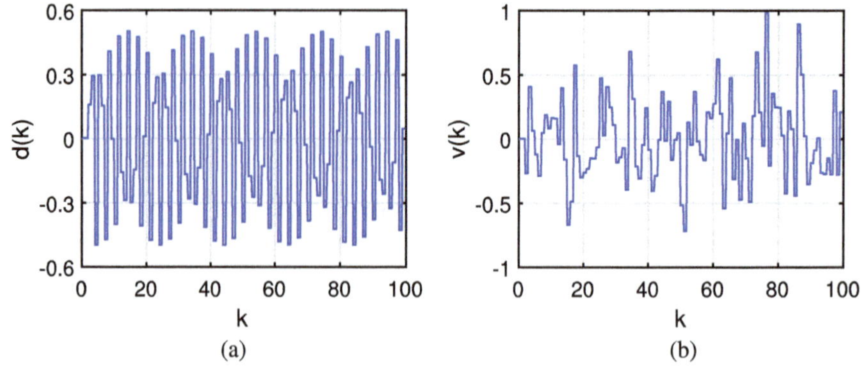

Fig. 13.1 **a** The unknown input $d(k)$; **b** The measurement noise $v(k)$

Set $N = 100, \tau = 2, \gamma = 0.6, P_1(j,l) = I\ (j < 0, l < 0)$ and the initial time step 3. For $k = 3, 4, \ldots, 100$, we calculate the parameter matrices of H_∞-FDF by applying Theorem 13.2, i.e., Eq. (13.45). It is verified that $R_{\tilde{y}_2}(k) > 0\ (k = 3, 4, \ldots, 100)$ and $R_{\tilde{r}}(k) < 0\ (k = 3, 4, \ldots, 99)$. For the sake of simplicity, only the following parameter matrices are presented and others are not listed here, i.e.,

$$R_{\tilde{y}_2}(3) = 48.4193,\ R_{\tilde{r}}(3) = -0.1797,\ V_0(3) = 0,\ V_1(4) = 0.1301,$$
$$H_1(3) = \begin{bmatrix} 0.0217\ 1.4063 \\ 0.3889\ 2.8125 \end{bmatrix},\ H_2(4) = \begin{bmatrix} -0.0007 \\ 0.1167 \end{bmatrix},$$
$$K_{\tau 1}(3) = \begin{bmatrix} 0.0399\ 0 \\ 0.2333\ 0 \end{bmatrix},\ K_{\tau 2}(3) = \begin{bmatrix} 0\ 0 \\ 0\ 0 \end{bmatrix}.$$

The unknown input $d(k)$ is simulated with $d(k) = 0.5\sin(2.2k)$ and the measurement noise $v(k)$ is a band-limited white noise with power 0.01 and sample time $0.1s$, as demonstrated in Fig. 13.1a, b, respectively. An impulse fault and a sine wave fault are respectively considered. The residuals $r(k)$ are obtained by using (13.6) with the aid of Remark 13.1. Figure 13.2 shows the two cases faults and residuals. It is seen from the simulation results that satisfactory residual signals can be obtained by using the designed H_∞-FDF.

Example 13.2 As a special case, we apply the obtained result to an LTI discrete system with delayed state. Suppose the system parameter matrices are the same with the numerical example in [13] and the time delay is $\tau = 2$, i.e.,

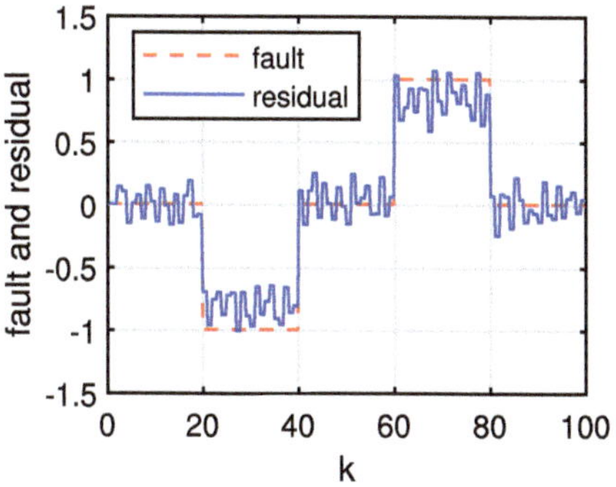

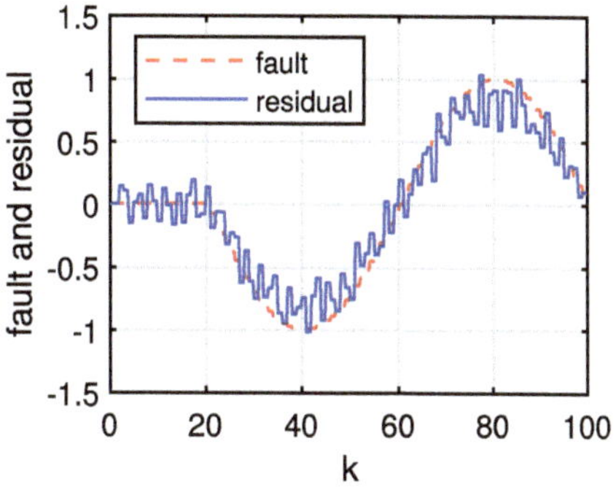

Fig. 13.2 **a** The impulse fault and residual; **b** The sine wave fault and residual

$$\begin{cases} x(k+1) = \begin{bmatrix} -0.4830 & -0.2478 \\ 0.4678 & 0.5242 \end{bmatrix} x(k) + \begin{bmatrix} -0.1204 & 0.0544 \\ -0.3946 & -0.3905 \end{bmatrix} x(k-2) \\ \qquad + \begin{bmatrix} -0.1268 \\ 0.4663 \end{bmatrix} f(k) + \begin{bmatrix} 0.4447 \\ -0.6628 \end{bmatrix} d(k), \\ y(k) = \begin{bmatrix} -0.4466 & -0.1644 \end{bmatrix} x(k) + 0.1724 d(k) - 0.7272 f(k), \\ x(k) = 0, \quad k \le 0. \end{cases}$$

We first design the fault detection filter using Theorem 13.2 for $\gamma = 0.9$. The following constant parameter matrices are obtained

$$R_{\tilde{y}_2} = 1.5416, \ R_{\tilde{r}} = -0.1528, \ V_0 = -0.4712, \ V_1 = -0.0103,$$
$$H_1 = \begin{bmatrix} 0.3041 & 0.4966 \\ -1.1593 & -1.9830 \end{bmatrix}, \ H_2 = \begin{bmatrix} 0.0701 \\ -0.2250 \end{bmatrix},$$
$$K_{\tau 1} = \begin{bmatrix} 0.0049 & 0.0189 \\ -0.0040 & -0.0152 \end{bmatrix}, \ K_{\tau 2} = \begin{bmatrix} -0.0038 & -0.0147 \\ -0.0086 & -0.0329 \end{bmatrix}.$$

Next, we use the finite frequency range fault estimator in [13] as a residual generator, i.e.,

$$\begin{cases} x_f(k+1) = \begin{bmatrix} 0.2587 & 0.3092 \\ 0.2080 & 0.2475 \end{bmatrix} x_f(k) + \begin{bmatrix} 10.1367 \\ 7.6504 \end{bmatrix} y(k), \\ r(k) = \begin{bmatrix} -0.6910 & 0.8369 \end{bmatrix} x_f(k) - 1.3020 y(k), \\ x_f(0) = 0. \end{cases}$$

Suppose the simulated fault is $f(k) = 5$ for $k \ge 20$ and $f(k) = 0$ elsewhere, $d(k) = \sin(k)$. Figure 13.3 show the absolute values of error $r(k) - f(k)$ for both the case of this chapter and [13]. The simulation results show that the achieved fault estimation

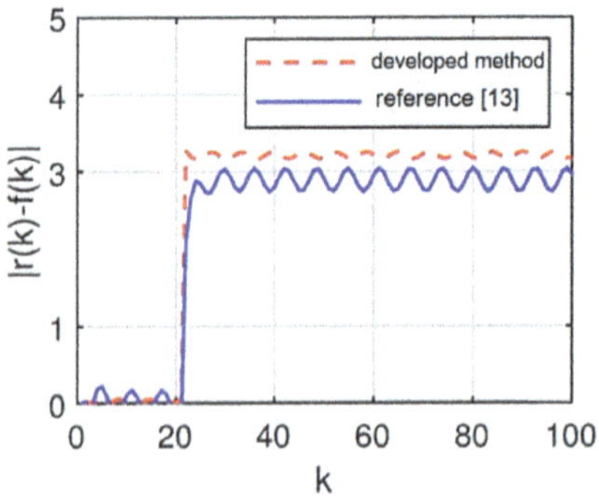

Fig. 13.3 The obtained $|r(k) - f(k)|$ of this chapter and [13]

level of this chapter is similar to the existing result in [13], but [13] is not applicable to time-varying systems. While, the method in this chapter can only be used to handle the case of known constant time-delay.

13.5 Conclusion

This chapter has confined to a Krein space based H_∞ FD method for LDTV systems with delayed state. A one-step smoother has been developed as a residual generator and the design of an observer-based H_∞-FDF has been converted into the minimum problem of an indefinite quadratic form. With the introduction of a Krein space stochastic system, such a minimum problem has been solved by using projection and innovation analysis in Krein space. A sufficient and necessary condition for the minimum and a solution to the H_∞-FDF have been derived in terms of Riccati equations. It is remarkable that the results in this chapter is, indeed, an application of the method in Chap. 12, concerning the LDTV systems with delayed state.

References

1. Hassibi, B., Sayed, A. H., & Kailath, T. (1999). *Indefinite-quadratic estimation and control: A unified approach to H_2 and H_∞ theories*. SIAM.
2. Jiang, B., Staroswiecki, M., & Cocquempot, V. 2002). Fault identification for a class of time-delay systems. In *Proceedings of the American Control Conference (IEEE Cat. No.CH37301)*, May 08–10, Anchorage, AK, UAS (Vol. 3, pp. 2239–3344).
3. Jiang, B., Staroswiecki, M., & Cocquempot, V. (2003). H_∞ fault detection filter design for linear discrete-time systems with multiple time delays. *International Journal of Systems Science, 34*(5), 365–373.
4. Tang, G., & Li, J. (2008). Optimal fault diagnosis for systems with delayed measurements. *IET Control Theory and Applications, 2*(11), 990–998.
5. Yang, H., & Saif, M. (1998). Observer design and fault diagnosis for state-retarded dynamical systems. *Automatica, 34*(2), 217–227.

6. Zhang, H., & Xie, L. (2007). *Control and estimation of systems with input/output delays*. Berlin, Germany: Springer.
7. Chen, J. (2008). Formulating and solving robust fault diagnosis problems based on a H_∞ setting. *IFAC Proceedings Volumes, 41*(2), 7259–7264.
8. Bai, L., Tian, Z., & Shi, S. (2007). RFDF design for linear time-delay systems with unknown inputs and parameter uncertainties. *International Journal of Systems Science, 38*(2), 139–149.
9. De la Sen, M. (2006). Stability of impulsive time-varying systems and compactness of the operators mapping the input space into the state and output spaces. *Journal of Mathematical Analysis and Applications, 321*(2), 621–650.
10. Zhong, M., Ding, S. X., Lam, J., et al. (2003). Fault detection filter design for LTI system with time delays. In *Proceedings of the 42th IEEE Conference on Decision and Control*, December 09–12, Hawaii, USA (Vol. 2, pp. 1467–1472).
11. Ding, S. X., Zhong, M., & Tang, B., et al. (2001). An LMI approach to the design of fault detection filter for time-delay LTI systems with unknown inputs. In *Proceedings of the American Control Conference. (Cat. No.01CH37148)*, June 25–27, Arlington, VA, USA (Vol. 3, pp. 2137–2142).
12. He, X., Wang, Z., Ji, Y., et al. (2008). Network-based fault detection for discrete-time state-delay systems: A new measurement model. *International Journal of Adaptive Control and Signal Processing, 22*(5), 510–528.
13. Li, X., & Yang, G. (2009). Fault estimation for discrete-time delay systems in finite frequency domain. In *Proceedings of the American Control Conference*, June 10–12, St. Louis, MO, USA (pp. 4328–4333).

Part IV
Parity Space-Based Fault Diagnosis for LDTV Systems

Chapter 14
Parity Space-Based Fault Detection for LDTV Systems with Unknown Input

Parity space-based method is one of the most important technique in model-based fault diagnosis. While for LDTV systems, one critical condrum encountered in parity space-based fault diagnosis is the high computation cost during online realization. In this chapter we confines to a parity space-based approach to FD for LDTV systems with unknown input, focusing on reducing the online computational burden. To this aim, the targeting parity space-based FD problem is first formulated as finding a minimum of quadratic form of unknown input. Then, by applying projection in Krein space, an optimal solution to the minimum problem is obtained via recursively computing innovations. It is shown that the reduction of computation cost is achievable, especially for high parity space order. A numerical example is given to demonstrate the applicability of the approach.

14.1 Introduction

Parity space-based FD method is characterized by its straightforward construction of a residual generator using the system input and output data and, the complete decoupling of residual to system initial state. For LTI systems, parity space-based FD issues have been well studied, see for example [3–7, 9, 11–14, 16]. In the early development, Wuennenberg and Patton proposed to formulate the parity space-based FD issue as minimizing the robustness/sensitivity ratio index and a solution was given by transforming the optimal selection of parity vector into a generalized eigenvalue-eigenvector problem [8, 10]. In the further investigations, the relationships among the design parameters of parity space, observer-based and factorization approaches have been established by Ding [11]. On this basis, a unified approach has been proposed under different types of performance indices.

M. Zhong et al., *Fault Diagnosis for Linear Discrete Time-Varying Systems and Its Applications*, https://doi.org/10.1007/978-981-19-5438-2_14

In parity space-based FD for LDTV systems, due to the time-varying characteristic of the system matrices, one serious difficulty, as mentioned in Sect. 3.3.4, is the high computation cost during online implementation, especially with the increase of the parity space order. Motivated by the following facts:

(i) a parity space-based residual generator is actually a deadbeat observer;

(ii) any design scheme using parity space relation can be applied to the observer-based residual generator, and vice verse, and

(iii) an optimal solution to observer-based H_i/H_∞-FDF can be derived by means of Krein space projection techniques and realized with higher computational efficiency, as demonstrated in Chaps. 6 and 8, in this chapter, by focusing on the real-time implementation problem, we attempt to give a new formulation of the parity space-based FD problem and derive a recursive realization algorithm of FD system by virtue of Krein space projection. The key tasks consist of

- establishing a relationship between parity space-based FD and a minimization problem of quadratic form of unknown input;
- the derivation of the recursive realization algorithm by means of Krein space projection techniques.

14.2 Problem Formulation

Recall in Sect. 3.3.4 that an LDTV system with state space model (3.3) can be described by the following input-output model

$$y_s(k) = H_{os}(k)x(k-s) + H_{us}(k)u_s(k) + H_{ds}(k)d_s(k) + H_{fs}(k)f_s(k), \quad (14.1)$$

where $x(k-s) \in \mathbb{R}^n$ is the state vector at time step $k-s$ for a given positive integer s, $y(k) \in \mathbb{R}^q$, $u(k) \in \mathbb{R}^p$, $d(k) \in \mathbb{R}^m$ and $f(k) \in \mathbb{R}^l$ are the measurement output, system input, unknown input and fault vectors, respectively, $y_s(k)$, $u_s(k)$, $d_s(k)$ and $f_s(k)$ are in form of $\theta_s = \left[\theta^T(k-s)\ \theta^T(k-s+1)\ \cdots\ \theta^T(k)\right]^T$ with $\theta = y,\ u,\ d,\ f$ in sequence, matrix $H_{os}(k)$ is given in (3.32), matrices $H_{us}(k)$, $H_{ds}(k)$ and $H_{fs}(k)$ are in form of (3.33). In the subsequent study in this chapter, we assume $d(k),\ f(k) \in l_{2,[0,N]}$ and $H_{ds}(k)$ is of full row rank.

Denote by $\mathcal{P}_s(k)$ the parity space of order $s(>n)$. Let $N_{bs}(k) \in \mathbb{R}^{\gamma\times(1+s)q}$ be the basis matrix of $\mathcal{P}_s(k)$, i.e., $N_{bs}(k)H_{os}(k) = 0$, and

$$\bar{H}_{us}(k) = N_{bs}(k)H_{us}(k),\ \bar{H}_{ds}(k) = N_{bs}(k)H_{ds}(k),\ \bar{H}_{fs}(k) = N_{bs}(k)H_{fs}(k).$$

We can construct a parity space-based residual generator as

$$\begin{aligned} r(k) &= P_s(k)(N_{bs}(k)y_s(k) - \bar{H}_{us}(k)u_s(k)) \\ &= P_s(k)(\bar{H}_{ds}(k)d_s(k) + \bar{H}_{fs}(k)f_s(k)), \end{aligned} \tag{14.2}$$

where $P_s(k) \in \mathbb{R}^{(s+1)q\times\gamma}$ with $P_s(k) \neq 0$ is a weighting matrix to be designed.

Without loss of generality, we can extend the well-established parity space-based residual generation scheme for LTI systems in [12] to LDTV systems. The design of $P_s(k)$ can be handled in the sense of maximizing the sensitivity/robustness ratio criterion. In this regard, we define

$$\|H_{rd}\|_{2,[k-s,k]} := \sup_{d_s(k)\neq 0} \frac{\|P_s(k)\bar{H}_{ds}(k)d_s(k)\|_2}{\|d_s(k)\|_2} = \bar{\sigma}(P_s(k)\bar{H}_{ds}(k)),$$

$$\|H_{rf}\|_{2,[k-s,k]} := \sup_{f_s(k)\neq 0} \frac{\|P_s(k)\bar{H}_{fs}(k)f_s(k)\|_2}{\|f_s(k)\|_2} = \bar{\sigma}(P_s(k)\bar{H}_{fs}(k)),$$

where $\bar{\sigma}(\cdot)$ denotes the maximal singular value of $(\cdot)$, $\|H_{rd}\|_{2,[k-s,k]}$ is used to evaluate the robustness of residual to unknown input in worst case and $\|H_{rf}\|_{2,[k-s,k]}$ measures the fault sensitivity in best case. The design of parity space-based residual generation is thus formulated as the following problem.

Problem 14.1 Given residual generator (14.2) for system (14.1), find matrix $P_s(k)$ such that the following robustness/sensitivity criterion is satisfied

$$J_{H_2/H_2} := \max_{P_s\neq 0} \frac{\|H_{rf}\|_{2,[k-s,k]}}{\|H_{rd}\|_{2,[k-s,k]}} = \max_{P_s\neq 0} \frac{\|P_s(k)\bar{H}_{fs}(k)\|_2}{\|P_s(k)\bar{H}_{ds}(k)\|_2}. \tag{14.3}$$

By directly extending the optimal solution of parity space-based residual generator for LTI systems (see Theorem 1 in [12]) to the LDTV case, we do an SVD on $\bar{H}_{ds}(k)$

$$\begin{cases} \bar{H}_{ds}(k) = U(k)\Sigma(k)V^T(k), \ U(k)U^T(k) = I, \ V^T(k)V(k) = I, \\ \Sigma(k) = \begin{bmatrix} S(k) & 0 \end{bmatrix}, \quad S(k) = \mathrm{diag}(\sigma_1(k), \sigma_2(k), \ldots, \sigma_\gamma(k)). \end{cases} \tag{14.4}$$

An optimal solution to Problem 14.1 can be obtained, as summarized in the following lemma and the proof is omitted here for simplicity.

Lemma 14.1 *Given residual generator (14.2) for system (14.1), then*

$$P_s(k) = S^{-1}(k)U^T(k) \tag{14.5}$$

solves the optimization problem (14.3) that results in, at the optimum,

$$J_{H_2/H_2} = \|S^{-1}(k)U^T(k)\bar{H}_{fs}(k)\|_2 = \bar{\sigma}(S^{-1}(k)U^T(k)\bar{H}_{fs}(k)).$$

Remark 14.1 In residual generator (14.2), $P_s(k)$ is supposed to be a matrix, so as to deliver good fault detectability for the faults lying in the kernel subspace of input and output. Meanwhile, we note that the optimal solution to problem (14.3) is not unique. A vector-valued solution of $P_s(k)$ to (14.3) can also be obtained, as given in the following lemma. Since the results is an intuitive extension of the Theorem 7.2 in [13], the proof is omitted here for the sake of simplicity.

Lemma 14.2 *Given residual generator (14.2) for system (14.1), then*

$$p_s(k) = v_s(k)S^{-1}(k)U^T(k) \tag{14.6}$$

solves the problem (14.3) and delivers

$$\max_{p_s \neq 0} \frac{\left\| p_s(k)\bar{H}_{fs}(k) \right\|_2}{\left\| p_s(k)\bar{H}_{ds}(k) \right\|_2} = \bar{\sigma}(S^{-1}(k)U^T(k)\bar{H}_{fs}(k)) = \lambda_m^{\frac{1}{2}}(k).$$

where $v_s(k) \in \mathbb{R}^\gamma$ solve the following generalized eigenvalue-eigenvector problem

$$v_s(k)(\lambda_m(k)I - S^{-1}(k)U^T(k)\bar{H}_{fs}(k)\bar{H}_{fs}^T(k)U(k)S^{-1}(k))v_s^T(k) = 0, \tag{14.7}$$

$\lambda_m(k)$ is the maximal eigenvalue of $S^{-1}(k)U^T(k)\bar{H}_{fs}(k)\bar{H}_{fs}^T(k)U(k)S^{-1}(k)$, $v_s(k)$ is the corresponding eigenvector satisfying $v_s(k)v_s^T(k) = 1$.

It is noticed that, the vector-valued solution specifies one "direction" with maximal gain λ_m, which, in comparison with the matrix-valued solution $P_s(k)$, might deliver poor fault detectability for the faults deviating such a direction. To overcome this deficiency, a wavelet transform aided parity space method will be introduced in Chap. 17. In this chapter, we mainly focus on the matrix-valued solution of $P_s(k)$.

Referring to $P_s(k)$ in (14.5), let

$$z_s(k) = \bar{H}_{ds}(k)d_s(k) + \bar{H}_{fs}(k)f_s(k), \tag{14.8}$$

which yields $r(k) = P_s(k)z_s(k) = S^{-1}(k)U^T(k)z_s(k)$. A standard evaluation function of the following form is generated

$$J_s(k) = r^T(k)r(k) \tag{14.9}$$

that in the fault-free case holds

$$\begin{aligned} J_s(k) &= z_s^T(k)U(k)S^{-2}(k)U^T(k)z_s(k) \\ &= z_s^T(k)(\bar{H}_{ds}(k)\bar{H}_{ds}^T(k))^{-1}z_s(k). \end{aligned} \tag{14.10}$$

Moreover, it is of interest to notice that $\forall d(k)$

$$z_s^T(k)(\bar{H}_{ds}(k)\bar{H}_{ds}^T(k))^{-1}z_s(k) \leq d_s^T(k)d_s(k).$$

The threshold setting with a zero FAR can be set as

$$J_{th} = \sup_{d_s(k)} d_s^T(k)d_s(k). \tag{14.11}$$

We can easily prove that the evaluation function $J_s(k)$ and threshold J_{th} solves the optimal FD problem in Definition 2.4, by noting $P_s(k)\bar{H}_{ds}(k)\bar{H}_{ds}^T(k)P_s^T(k) = I$ with $P_s(k)$ given in (14.5).

It is remarkable that the above FD scheme can be interpreted as an LDTV extension of the integrated design of parity space-based FD for LTI systems given in [13]. On the other hand, an online updating of $H_{os}(k)$, $H_{us}(k)$ and $H_{ds}(k)$ at every time step k is necessary for LDTV systems, and a real-time SVD should be carried out for an optimal parity space matrix $P_s(k)$. As a result, the online realization computation cost is high, especially with the increase of parity space order s. These observations motivate us to find a new way to handle parity space-based FD issues so as to implement the FD system easily with less computation load.

14.3 An Equivalent Optimization Formulation

We consider the following cost function

$$J(d_s(k)) = d_s^T(k)d_s(k) + \|z_s(k) - \bar{H}_{ds}(k)d_s(k)\|^2.$$

Given $z_s(k)$, it is known that (see [15])

$$\min_{d_s(k)} J(d_s(k)) \tag{14.12}$$

with the constraint (14.8) in the fault-free case is solved by

$$\hat{d}_s(k) = (I + \bar{H}_{ds}^T(k)\bar{H}_{ds}(k))^{-1}\bar{H}_{ds}^T(k)z_s(k) \tag{14.13}$$

and the minimum value of $J(d_s(k))$ is

$$J(\hat{d}_s(k)) = z_s^T(k)(I + \bar{H}_{ds}^T(k)\bar{H}_{ds}(k))^{-1}z_s(k).$$

It follows from

$$\begin{aligned}
&(I + \bar{H}_{ds}^T(k)\bar{H}_{ds}(k))^{-1}\bar{H}_{ds}^T(k) \\
&= [I - \bar{H}_{ds}^T(k)(I + \bar{H}_{ds}(k)\bar{H}_{ds}^T(k))^{-1}\bar{H}_{ds}(k)]\bar{H}_{ds}^T(k) \\
&= \bar{H}_{ds}^T(k)(\bar{H}_{ds}(k)\bar{H}_{ds}^T(k))^{-1}
\end{aligned}$$

that

$$\hat{d}_s(k) = \bar{H}_{ds}^T(k)(\bar{H}_{ds}(k)\bar{H}_{ds}^T(k))^{-1}z_s(k), \tag{14.14}$$

$$J(\hat{d}_s(k)) = z_s^T(k)(\bar{H}_{ds}(k)\bar{H}_{ds}^T(k))^{-1}z_s(k). \tag{14.15}$$

Note that $J(\hat{d}_s(k))$ given above is exactly $J_s(k)$ in (14.10). In other words, the Problem 14.1 can be addressed equally by solving the problem (14.12), i.e.,

$$J_s(k) = \min_{d_s(k)} J(d_s(k)) = J(\hat{d}_s(k)). \tag{14.16}$$

Our basic idea is to reformulate the integrated design of parity space-based FD systems as finding a minimum of $J(d_s(k))$. In order to solve the optimization problem (14.12), the well-established relationship between the minimum of quadratic form in Hilbert space and projection in Krein space is helpful. In the sequel, we give a parity space-based FD scheme in a Krein space framework.

14.4 A Krein Space Based Recursive Solution

In this section, we shall first introduce a Krein space model corresponding to (14.1) with $f(k) = 0$ and give an equivalent form of $J_s(k)$ in Krein space. Next, the minimum problem will be solved by applying the techniques of Krein space projection and innovation analysis, which allows us to compute $J_s(k)$ recursively.

14.4.1 An Equivalent Krein Space Problem

Introduce a Krein space system as follows

$$\begin{cases} \mathbf{x}(k+1) = A(k)\mathbf{x}(k) + B(k)\mathbf{u}(k) + B_d(k)\mathbf{d}(k), \\ \mathbf{y}(k) = C(k)\mathbf{x}(k) + D_d(k)\mathbf{d}(k), \\ \mathbf{x}(0) = \mathbf{x}_0, \end{cases}$$

where

$$\left\langle \begin{bmatrix} \mathbf{x}_0 \\ \mathbf{d}(i) \end{bmatrix}, \begin{bmatrix} \mathbf{x}_0 \\ \mathbf{d}(j) \end{bmatrix} \right\rangle - \begin{bmatrix} P_0 & 0 \\ 0 & I\delta_{ij} \end{bmatrix}.$$

It is easy to have

$$\mathbf{z}_s(k) = N_{bs}(k)(\mathbf{y}_s(k) - H_{os}(k)\mathbf{u}_s(k)) = \bar{H}_{ds}(k)\mathbf{d}_s(k), \tag{14.17}$$

$$R_{\mathbf{d}_s}(k) = I, \ \ R_{\mathbf{z}_s}(k) = \bar{H}_{ds}(k)\bar{H}_{ds}^T(k). \tag{14.18}$$

According to Theorem 4.1 and Corollary 4.1, we know that, if and only if $R_{\mathbf{d}_s}(k) > 0$, the projection of $\mathbf{d}_s(k)$ onto $\mathcal{L}\{\{\mathbf{z}(i)\}_{i=k-s}^{k}\}$ exists and is unique, and yields a unique minimum of the error Gramian

$$R_{\tilde{\mathbf{d}}_s}(k) = \left\langle \mathbf{d}_s(k) - \hat{\mathbf{d}}_s(k), \mathbf{d}_s(k) - \hat{\mathbf{d}}_s(k) \right\rangle,$$

where the projection $\hat{\mathbf{d}}_s(k)$ is given by

$$\hat{\mathbf{d}}_s(k) =< \mathbf{d}_s(k), \mathbf{z}_s(k) > R_{\mathbf{z}_s}^{-1}(k)\mathbf{z}_s(k) = \bar{H}_{ds}^T(k) R_{\mathbf{z}_s}^{-1}(k)\mathbf{z}_s(k). \quad (14.19)$$

Note that the calculation of $\hat{d}_s(k)$ in (14.13) follows the same formulae with $\hat{\mathbf{d}}_s(k)$ in (14.19). Moreover, $\bar{H}_{ds}(k)$ being full row rank implies $R_{\mathbf{z}_s}(k) > 0$. Similar to the relations of Krein space and deterministic problems mentioned in Sect. 4.3, we can conclude that, given $z_s(k)$ with $z_s(k) = \bar{H}_{ds}(k)d_s(k)$ and Krein space model (14.17) with Gramians (14.18), the minimum of $J(d_s(k))$ can be obtained by applying the projection of $\mathbf{d}_s(k)$ onto $\mathcal{L}\{\{\mathbf{z}(i)\}_{i=k-s}^{k}\}$. Thus, we can rewrite $J(\hat{d}_s(k))$ as

$$J(\hat{d}_s(k)) = z_s^T(k) R_{\mathbf{z}_s}^{-1}(k) z_s(k). \quad (14.20)$$

Substituting $\bar{H}_{ds}(k) = U(k)\Sigma(k)V^T(k)$ into (14.18) and (14.20) yields

$$\begin{aligned} R_{\mathbf{z}_s}(k) &= U(k)S^2(k)U^T(k), \\ J(\hat{d}_s(k)) &= z_s^T(k)U(k)S^{-2}(k)U^T(k)z_s(k) = J_s(k). \end{aligned}$$

which is identical to (14.16). Therefore, the following proposition is readily obtained.

Proposition 14.1 *Given $z_s(k)$ with $z_s(k) = \bar{H}_{ds}(k)d_s(k)$, one can construct a Krein space model (14.17) with Gramians (14.18). When $R_{\mathbf{z}_s}(k) > 0$, the $\hat{d}_s(k) = \bar{H}_{ds}^T(k)R_{\mathbf{z}_s}^{-1}(k)z_s(k)$ yields a minimum of $J(d_s(k))$ with the constraint of fault-free case (14.8) and the resulted minimum value $J(\hat{d}_s(k))$ can be defined as a standard evaluation function $J_s(k)$. With an SVD on $R_{\mathbf{z}_s}(k)$, i.e.,*

$$\begin{aligned} R_{\mathbf{z}_s}(k) &= U(k)S^2(k)U^T(k), \ \ U(k)U^T(k) = I, \\ S(k) &= \operatorname{diag}(\sigma_1(k), \sigma_2(k), \ldots, \sigma_\gamma(k)), \end{aligned}$$

$r(k) = P_s(k)z_s(k)$ with $P_s(k) = S^{-1}(k)U^T(k)$ delivers a solution of parity space-based FD satisfying (14.3).

Remark 14.2 Similar to the case of LTI systems in [12], Proposition 14.1 provides a unified design of FD systems under different types of performance indices related to parity space-based FD. So, although it seems that the fault sensitivity constraint is not considered when we construct $J_s(k)$ using $J(\hat{d}_s(k))$, the robustness/sensitivity criteria in (14.3) is also satisfied.

14.4.2 Recursive Computation of $J_s(k)$

So far, a relationship between parity space-based FD and projection in Krein space has been established. However, the computation cost of $J_s(k)$ remains expensive. To solve this problem, the techniques of projection and innovation analysis in Krein space are applied.

Denote by $\hat{\mathbf{x}}(k)$ and $\hat{\mathbf{y}}(k)$ the projection of $\mathbf{x}(k)$ and $\mathbf{y}(k)$ on space spanned by $\{\mathbf{y}(i)\}_{i=0}^{k}$, respectively. Introduce the following innovations

$$\begin{aligned}\mathbf{e}(k) &= \mathbf{x}(k) - \hat{\mathbf{x}}(k), \ \ P(k) = < \mathbf{e}(k), \ \mathbf{e}(k) >, \\ \tilde{\mathbf{y}}(k) &= \mathbf{y}(k) - \hat{\mathbf{y}}(k), \ \ R_{\tilde{\mathbf{y}}}(k) = < \tilde{\mathbf{y}}(k), \tilde{\mathbf{y}}(k) > .\end{aligned}$$

It is known from Chap. 5 that $\hat{\mathbf{x}}(k)$ and innovation $\tilde{\mathbf{y}}(k)$ can be obtained via

$$\begin{cases}\hat{\mathbf{x}}(k+1) = A(k)\hat{\mathbf{x}}(k) + B(k)\mathbf{u}(k) + L(k)\tilde{\mathbf{y}}(k), \\ \tilde{\mathbf{y}}(k) = \mathbf{y}(k) - C(k)\hat{\mathbf{x}}(k), \ \hat{\mathbf{x}}(0) = 0,\end{cases} \tag{14.21}$$

where

$$\begin{cases}L(k) = (A(k)P(k)C^T(k) + B_d(k)D_d^T(k))R_{\tilde{\mathbf{y}}}^{-1}(k), \\ R_{\tilde{\mathbf{y}}}(k) = C(k)P(k)C^T(k) + D_d(k)D_d^T(k), \\ P(k+1) = A(k)P(k)A^T(k) - L(k)R_{\tilde{\mathbf{y}}}(k)L^T(k) + B_d(k)B_d^T(k).\end{cases} \tag{14.22}$$

Thus we have

$$\begin{cases}\mathbf{e}(k+1) = A_l(k)\mathbf{e}(k) + B_{ld}(k)\mathbf{d}(k), \ \mathbf{e}(0) = \mathbf{x}_0, \\ \tilde{\mathbf{y}}(k) = C(k)\mathbf{e}(k) + D_d(k)\mathbf{d}(k),\end{cases}$$

where

$$A_l(k) = A(k) - L(k)C(k), \ B_{ld}(k) = B_d(k) - L(k)D_d(k).$$

Similar to (14.1), we can get

$$\tilde{\mathbf{y}}_s(k) = H_{l,os}(k)\mathbf{x}(k-s) + H_{l,ds}(k)\mathbf{d}_s(k),$$

where the $H_{l,os}(k)$ is constructed by replacing $A(k)$ with $A_l(k)$ in $H_{os}(k)$, the $H_{l,ds}(k)$ by replacing $\{A(k), B_d(k)\}$ with $\{A_l(k), B_{ld}(k)\}$ in $H_{ds}(k)$.

Denote the corresponding parity space of order s with $\mathcal{P}_{l,s}(k)$, its basis matrix with $N_{l,bs}(k)$. Thus we have

$$\begin{aligned}N_{l,bs}(k)H_{l,os}(k) = 0, \ \ \bar{H}_{l,ds}(k) &= N_{l,bs}(k)H_{l,ds}(k), \\ \tilde{\mathbf{z}}_s(k) = N_{l,bs}(k)\tilde{\mathbf{y}}_s(k) &= \bar{H}_{l,ds}(k)\mathbf{d}_s(k), \\ R_{\tilde{\mathbf{z}}_s}(k) = N_{l,bs}(k)R_{\tilde{\mathbf{y}}_s}(k)N_{l,bs}^T(k) &= \bar{H}_{l,ds}(k)\bar{H}_{l,ds}^T(k).\end{aligned}$$

Under the assumption of $\bar{H}_{l,ds}(k)$ being full row rank , the projection of $\mathbf{d}_s(k)$ onto $\mathcal{L}\{\{\tilde{\mathbf{z}}(i)\}_{i=k-s}^{k}\}$ is given by $\hat{\mathbf{d}}_s(k) = \bar{H}_{l,ds}^T(k)R_{\tilde{\mathbf{z}}_s}^{-1}(k)\tilde{\mathbf{z}}_s(k)$. We now calculate $\tilde{y}(k)$ by

$$\begin{cases} \hat{x}(k+1) = A(k)\hat{x}(k) + L(k)(y(k) - C(k)\hat{x}(k)), \\ \tilde{y}(k) = y(k) - C(k)\hat{x}(k), \ \hat{x}(0) = 0. \end{cases} \tag{14.23}$$

Let $\tilde{z}_s(k) = N_{l,bs}(k)\tilde{y}_s(k)$. Similar to (14.17), we can write

$$\tilde{z}_s(k) = \bar{H}_{l,ds}(k)d_s(k). \tag{14.24}$$

By applying Proposition 14.1, a solution of the minimum $J(d_s(k))$ with constraint (14.24) becomes

$$\hat{d}_s(k) = \bar{H}_{l,ds}^T(k)R_{\tilde{\mathbf{z}}_s}^{-1}(k)\tilde{z}_s(k) = \bar{H}_{l,ds}^T(k)(\bar{H}_{l,ds}(k)\bar{H}_{l,ds}^T(k))^{-1}\tilde{z}_s(k),$$

which follows the same formulae with $\hat{\mathbf{d}}_s(k)$. The $J_s(k)$ is then rewritten as

$$J_s(k) = \tilde{z}_s^T(k)R_{\tilde{\mathbf{z}}_s}^{-1}(k)\tilde{z}_s(k). \tag{14.25}$$

Note also that $\{\tilde{\mathbf{y}}(i)\}_{i=k-s}^{k}$ forms an orthogonal basis of linear space spanned by $\{\mathbf{y}(i)\}_{i=k-s}^{k}$. It is easy to have

$$R_{\tilde{\mathbf{y}}_s}(k) = \text{diag}(R_{\tilde{\mathbf{y}}}(k-s), R_{\tilde{\mathbf{y}}}(k-s+1), \ldots, R_{\tilde{\mathbf{y}}}(k)).$$

Rewrite $N_{l,bs}(k)$ as

$$N_{l,bs}(k) = \begin{bmatrix} n_l(k-s) \ n_l(k-s+1) \cdots n_l(k) \end{bmatrix}.$$

Then $R_{\tilde{\mathbf{z}}_s}(k)$ becomes

$$R_{\tilde{\mathbf{z}}_s}(k) = \sum_{i=k-s}^{k} n_l(i)R_{\tilde{\mathbf{y}}}(i)n_l^T(i). \tag{14.26}$$

Doing an SVD on $R_{\tilde{\mathbf{z}}_s}(k)$ yields

$$R_{\tilde{\mathbf{z}}_s}(k) = \tilde{U}(k)\tilde{S}^2(k)\tilde{U}^T(k), \ \tilde{U}(k)\tilde{U}^T(k) = I, \tag{14.27}$$

$$\tilde{S}(k) = \text{diag}(\varphi_1(k), \varphi_2(k), \ldots, \varphi_\gamma(k)). \tag{14.28}$$

We thus have

$$
\begin{aligned}
J_s(k) &= \tilde{z}_s^T(k)\left(\sum_{i=k-s}^{k} n_l(i)R_{\tilde{\mathbf{y}}}(i)n_l^T(i)\right)^{-1}\tilde{z}_s(k) \\
&= \tilde{z}_s^T(k)\tilde{U}(k)\tilde{S}^{-2}(k)\tilde{U}^T(k)\tilde{z}_s(k).
\end{aligned} \tag{14.29}
$$

Based on the above analysis, the recursive computation of $J_s(k)$ can be summarized into the following proposition.

Proposition 14.2 *Suppose that the innovations $\tilde{y}(k)$ are computed via the recursions (14.23), the matrix $R_{\tilde{\mathbf{z}}_s}(k)$ is nonsingular and has an SVD given by (14.27). Then $J_s(k)$ can be calculated by (14.29). Moreover, the $r(k) = \tilde{P}_s(k)\tilde{z}_s(k)$ with $\tilde{P}_s(k) = \tilde{S}^{-1}(k)\tilde{U}^T(k)$ delivers an optimal solution to parity space-based FD satisfying*

$$
\min_{\tilde{P}_s(k)\neq 0} \frac{||\tilde{P}_s(k)\bar{H}_{l,ds}(k)||_2}{||\tilde{P}_s(k)\bar{H}_{l,fs}(k)||_2}.
$$

In comparison with the traditional parity space-based FD method, the major difference of the above method lies in the computation of $R_{\tilde{\mathbf{z}}_s}(k)$ and $\bar{H}_{ds}(k)\bar{H}_{ds}^T(k)$. As additions are much faster than multiplications and divisions, here only the number of multiplications and divisions are used as the operation count. The computation cost of $R_{\tilde{\mathbf{z}}_s}(k)$ concerns with updating matrices $P(k)$, $L(k)$ $R_{\tilde{\mathbf{y}}}(k)$ by (14.22) and computing $R_{\tilde{\mathbf{z}}_s}(k)$ by (14.26). The number of major multiplications and divisions is

$$
MD_1 = 3n^3 + 3n^2q + 3nq^2 + n^2p + npq + q^2p + q^3 + (s+1)(\gamma q^2 + \gamma^2 q).
$$

Meanwhile, the computation cost of $\bar{H}_{ds}(k)\bar{H}_{ds}^T(k)$ concerns with updating $H_{ds}(k)$ and computing $\bar{H}_{ds}(k)\bar{H}_{ds}^T(k)$. The number of major multiplications is

$$
\begin{aligned}
MD_2 = &\frac{(s-2)(s-1)}{2}n^3 + \frac{(s-1)s}{2}n^2q + \frac{s(s+1)}{2}npq \\
&+(s+1)((s+1)\gamma pq + \gamma^2 p).
\end{aligned}
$$

In addition, the updating of $H_{us}(k)$ is not needed when we calculate $J_s(k)$ by using innovations $\tilde{z}_s(k)$. Because $H_{us}(k)$ can be obtained by replacing $\{B_d(k), D_d(k)\}$ with $\{B(k), 0\}$ in $H_{ds}(k)$, the additional number of major multiplications is $MD_3 = \frac{s(s+1)}{2}qnm$. So, with the presented parity space-based FD scheme, this part of major computation cost is changed to MD_1 from $MD_2 + MD_3$ at every time step. It is easy to see that $MD_1 < MD_2 + MD_3$ for $s > 3$.

Hence, with the developed method, not only the FD performance index in (14.3) is guaranteed, but also the reduction of computation cost is achievable. The computation of $J_s(k)$ and $r(k)$ is summarized in Algorithm 14.4.8.

Algorithm 14.4.8 Computation of $J_s(k)$ and $r(k)$

1: Compute $L(k)$, $R_{\tilde{\mathbf{y}}}(k)$, $P(k)$ by (14.22), generate innovations $\tilde{y}(k)$ by (14.23);
2: Update $H_{l,os}(k)$ and find $N_{l,s}(k)$;
3: Calculate $R_{\tilde{\mathbf{z}}_s}(k)$ by (14.26);
4: Do an SVD on $R_{\tilde{\mathbf{z}}_s}(k)$ and get $\tilde{U}(k)$, $\tilde{S}(k)$;
5: Calculate $J_s(k)$ by (14.29) and let $r(k) = \tilde{S}^{-1}(k)\tilde{U}^T(k)\tilde{z}_s(k)$.

Remark 14.3 Since $R_{\tilde{\mathbf{y}}_s}(k)$ is a diagonal matrix and $R_{\tilde{\mathbf{y}}}(i)$ $(i = 0, 1, \ldots, k)$ are obtained via recursions, the reduction of computation cost is achievable. In particular, only the computation of $R_{\tilde{\mathbf{z}}_s}(k)$ is required if a closed-loop system with observer (14.23) is available. In this case, the number of major multiplications and divisions of $R_{\tilde{\mathbf{z}}_s}(k)$ at every time step is $(s+1)(\gamma q^2 + \gamma^2 q)$.

14.5 A Numerical Example

To demonstrate the applicability pf the developed FD approach, we consider the LDTV system (3.3) with

$$A(k) = \begin{bmatrix} 0 & 1-0.4\cos(k) & 0 \\ 0 & 0.5-e^{-k/100} & 0.6 \\ -1 & 0 & 0 \end{bmatrix}, \ B_d(k) = \begin{bmatrix} 0.5 & 0 \\ 0.2 & 0 \\ 0 & 1 \end{bmatrix}, \ B(k) = \begin{bmatrix} 1 \\ 1 \\ 0 \end{bmatrix},$$

$$B_f(k) = \begin{bmatrix} 1 & 0 & 1 \end{bmatrix}^T, \ C(k) = \begin{bmatrix} 1 & 0.9-\frac{1}{k} & 0 \end{bmatrix}, \ D_d(k) = \begin{bmatrix} 0 & 0.5 \end{bmatrix}, \ D_f(k) = 0.5.$$

Set $s = 4$, $u(k) = 1$, and $d(k) = \begin{bmatrix} d_1(k) & 0.5\cos(k) \end{bmatrix}^T$, where $d_1(k) \sim \mathcal{N}(0, 0.5^2)$. The fault is simulated as

$$f(k) = \begin{cases} 1, & 20 \le k < 40, \\ -1, & 60 \le k < 80, \\ 0, & \text{others}. \end{cases}$$

Based on Proposition 14.2, $J_s(k)$ and $r(k)$ are obtained by using Algorithm 14.4.8.

Denote by $J_{sf}(k), r_f(k) = \begin{bmatrix} r_{1f}(k) & r_{2f}(k) \end{bmatrix}^T$ and $J_{s0}(k)$, $r(k) = \begin{bmatrix} r_1(k) & r_2(k) \end{bmatrix}^T$ the faulty case and fault-free case evaluation functions and residuals, respectively. A threshold is chosen as $J_{th,1} = \sup J_{s0}(k)$. The calculated $J_{sf}(k)$, $J_{s0}(k)$ and $J_{th,1} = 0.9355$ are shown Fig. 14.1. The $r_{1f}(k)$, $r_1(k)$ and $r_{2f}(k)$, $r_2(k)$ are shown in Fig. 14.2 and Fig. 14.3, respectively. It is notable that, by using the solution in Lemma 14.1, the obtained $J_s(k)$ and $r(k)$ are the same with the scheme in Krein space that are not listed here for simplicity.

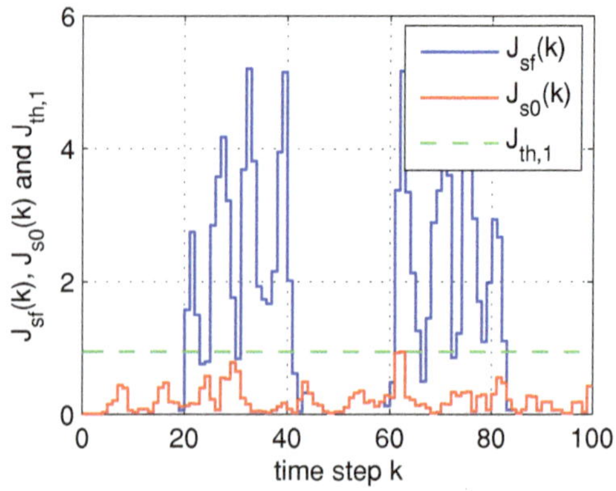

Fig. 14.1 The first case $J_s(k)$ and $J_{th,1}$

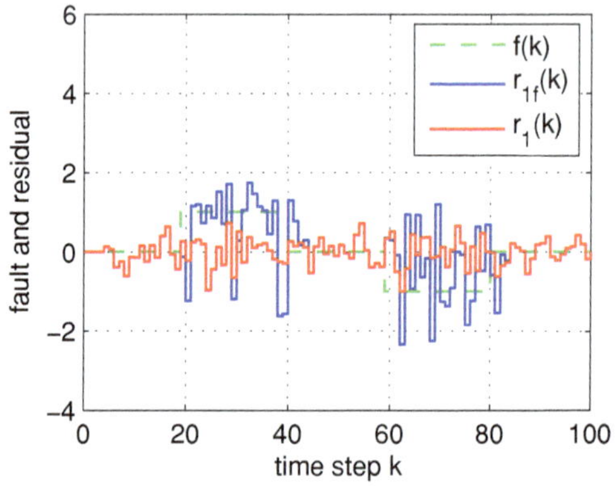

Fig. 14.2 The first case $r_{1f}(k)$, $r_1(k)$ and $f(k)$

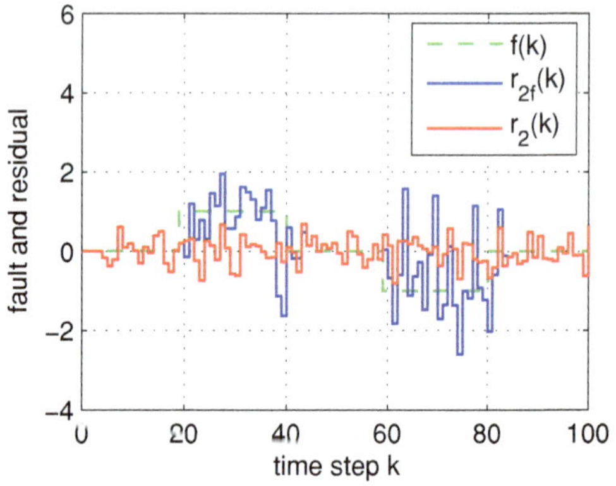

Fig. 14.3 The first case $r_{2f}(k)$, $r_2(k)$ and $f(k)$

Finally, we compare the major computation cost of the presented scheme with the original parity space-based FD. For $s = 4,\ n = 3,\ p = 2,\ q = 1$, we have

$$MD_1 = 174,\ MD_2 = 325,\ MD_3 = 30,$$
$$MD_1 = 174 < MD_2 + MD_3 = 355.$$

Obviously, the computation cost is reduced efficiently by using the recursive realization approach.

14.6 Conclusion

Although parity space-based FD for LDTV systems is straightforward, the online computation burden is usually heavy in the case of a high parity space order that is often necessary to guarantee FD performance. As a remedy, this chapter has presented a Krein space based design scheme of parity space-based FD for LDTV systems. We have demonstrated that the design of an optimal parity space-based residual is equivalent to finding a minimum of quadratic form of unknown input. By introducing a Krein space system and applying the techniques of projection and innovation analysis in Krein space, a solution to the problem has been obtained via recursions. As a result, the reduction of online computation cost can be achieved in the implementation of parity space-based FD.

References

1. Hassibi, B., Sayed, A. H., & Kailath, T. (1999). *Indefinite-quadratic estimation and control: A unified approach to H_2 and H_∞ theories*. SIAM.
2. Shen, B., Ding, S. X., & Wang, Z. (2013). Finite-horizon H_- fault estimation for linear discrete time-varying systems with delayed measurements. *Automatica, 49*(1), 293–296.
3. Chow, E., & Willsky, A. (1984). Analytical redundancy and the design of robust detection systems. *IEEE Transactions on Automatic Control, 29*(7), 603–614.
4. Blesa, J., Puig, V., Saludes, J., et al. (2014). Set-membership parity space approach for fault detection in linear uncertain dynamic systems. *International Journal of Adaptive Control and Signal Processing, 30*(2), 186–205.
5. Chen, J., & Patton, R. J. (1999). *Robust model-based fault diagnosis for dynamic systems*. Boston, MA: Kluwer Academic Publishers.
6. Gertler, J. (1998). *Fault detection and diagnosis in engineering systems*. New York: Marcel Dekker.
7. Gertler, J., Staroswiecki, M., & Shen, M. (2002). Direct design of structured residuals for fault diagnosis in linear systems. In *Proceedings of the American Control Conference (Cat. No.CH37301)*, May 08–10, Anchorage, USA (pp. 4519–4524).
8. Wuennenberg, J. (1990). *Observer-based fault detection in dynamic systems*, Ph.D. dissertation. University of Duisburg, Germany.
9. Wu, N., & Wang, Y. (1995). Robust failure detection with parity check on filtered measurements. *IEEE Transactions on Aerospace and Electronic Systems, 31*(1), 489–491.

10. Patton, R. J., & Chen, J. (1994). Parity space approach to model-based fault diagnosis. In *Proceedings of the SAFEPROCESS'1994*, Baden-Baden, Germany (pp. 256–261).
11. Ding, S. X., Guo, L., & Jeinsch, T. (1996). A characterization of parity space and its application to robust fault detection. *IEEE Transactions on Automatic Control, 44*(2), 337–343.
12. Ding, S. X., Ding, E. L., & Jeinsch, T. et al. (2000). An approach to a unified design of FDI systems. In *Proceedings of the 3rd Asian Control Conference*, Shanghai, China (pp. 2812–2817).
13. Ding, S. X. (2013). *Model-based fault diagnosis techniques: Design schemes, algorithms, and tools* (2nd ed.). London, UK: Springer.
14. Varrier, S., Koenig, D., & Martinez, J. (2012). A parity space-based fault detection on LPV systems: Approach for vehicle lateral dynamics control system. *IFAC Proceedings Volumes, 45*(20), 1191–1196.
15. Kailath, T., Sayed, A. H., & Hassibi, B. (1999). *Linear Estimation*. NJ: Prentice Hall.
16. Wang, Y., & Wu, N. (1993). An approach to configuration of robust control systems for robust failure detection. In *Proceedings of the 32nd IEEE Conference on Decision and Control*, December 15–17, San Antonio, TX, USA (pp. 1704–1709).

Chapter 15
Parity Space-Based Fault Estimation for LDTV Systems

This chapter gives a parity space-based approach to fault estimation for LDTV systems with unknown input. The core idea is to formulate the design of a parity space-based fault estimator as finding a minimum for a matrix quadratic form. It is proven that the minimum provides a unified solution to the fault estimation problem with different design criteria, such as the performance indices in terms of maximum singular value and Frobenius-norm. A necessary and sufficient condition for the minimum is derived. Also, an analytic solution to the parity space-based fault estimator is developed.

15.1 Introduction

In Chap. 14, we have contributed an optimal parity space-based approach to FD for LDTV systems, by converting the FD problem into a minimization problem of certain quadratic form and solved with Krein space projection technique. Aiming to provide more detailed information of a fault such as the location, size and duration, etc., fault estimation should also be carried out. In comparison with LTI systems, only a few studies of parity space-based fault estimation for LDTV systems have been reported [2–4, 6, 8]. For example, a recursive least squares algorithm aided parity space approach has been proposed by Han et al. [8] to actuator fault detection and estimation for a quadrotor unmanned aerial vehicle. In [6], the parity space approach has been generalized to fault estimation of Takagi-Sugeno fuzzy models, but only sufficient conditions on the existence of fault estimator have been obtained via LMI. In [5], under the condition of unknown input full decoupling being achievable, parity space-based FD technique for LTI systems has been extended to linear periodic systems and a periodically varying parity vector has been obtained.

Motivate by the idea in Chap. 14, in this chapter we cope with parity space-based fault estimation issues for LDTV systems. The relationship between parity

M. Zhong et al., *Fault Diagnosis for Linear Discrete Time-Varying Systems and Its Applications*, https://doi.org/10.1007/978-981-19-5438-2_15

space-based fault estimation and the minimum problem of a matrix quadratic builds the basis of this aim. Our work revolves around the following tasks:

- introduce some design criteria to evaluate the fault estimation performance and then formulate the design of a parity space-based fault estimator as solving a minimum problem of a matrix quadratic form;
- find a necessary and sufficient condition for the minimum and
- solve the minimum problem for a unified analytic solution to parity matrix.

15.2 Problem Formulation

Consider LDTV systems described by state space model (3.3) whose input-output model is given by (14.1), i.e.,

$$y_s(k) = H_{os}(k)x(k-s) + H_{us}(k)u_s(k) + H_{ds}(k)d_s(k) + H_{fs}(k)f_s(k), \quad (15.1)$$

where $x(k-s) \in \mathbb{R}^n$ is the state vector at time step $k-s$ for a given integer $s > 0$, $y(k) \in \mathbb{R}^q$, $u(k) \in \mathbb{R}^p$, $d(k) \in \mathbb{R}^m$ and $f(k) \in \mathbb{R}^l$ are the measurement output, system input, unknown input and fault vectors, respectively, $y_s(k), u_s(k), d_s(k)$ and $f_s(k)$ are in form of $\theta_s = \left[\theta^T(k-s)\ \theta^T(k-s+1) \cdots \theta^T(k)\right]^T$ with $\theta = y, u, d, f$ in sequence, matrix $H_{os}(k)$ is given in (3.32), matrices $H_{us}(k)$, $H_{ds}(k)$ and $H_{fs}(k)$ are in form of (3.33). In this chapter, we also assume $d(k), f(k) \in l_{2,[0,N]}$ and $H_{ds}(k)$ is of full row rank.

For the purpose of fault estimation, we suppose that process data in the time interval $[k-s, k]$, i.e. $y_s(k), u_s(k)$, are available. Our task is to use the following parity space-based fault estimator to reconstruct the fault vector,

$$\hat{f}(k) = W_s(k)(y_s(k) - H_{us}(k)u_s(k)), \quad (15.2)$$

$$= W_s(k)(H_{ds}(k)d_s(k) + H_{fs}(k)f_s(k)), \quad (15.3)$$

where $\hat{f}(k) \in \mathbb{R}^l$ is the fault estimate, $W_s(k) \in \mathbb{R}^{l\times(s+1)q}$ is referred to a parity matrix to be designed, s is the order of parity space satisfying $(s+1)q > n$ so as to ensures the existence of $W_s(k)$. Let

$$W_s(k) = P_s(k)N_{bs}(k), \quad (15.4)$$

where $N_{bs}(k) \in \mathbb{R}^{\beta_k\times(s+1)q}$ is a basis matrix of parity space $\mathcal{P}_s(k)$ of order s, i.e., $N_{bs}(k)H_{os}(k) = 0$, $N_{bs}(k)$ is full row rank and β_k equals the rank of the null matrix of $H_{os}(k)$, $P_s(k) \in \mathbb{R}^{l\times\beta_k}$. It turns out

$$\hat{f}(k) = P_s(k)N_{bs}(k)(H_{ds}(k)d_s(k) + H_{fs}(k)f_s(k)). \quad (15.5)$$

It is well-known that, for fault estimation in LTI systems, a weighting transfer function matrix is often added to specify the frequency range in which the fault will be estimated [1]. For LDTV systems, it is reasonable to introduce a weighting matrix $W_f \in \mathbb{R}^{l\times(s+1)l}$ to represent some time domain features of the fault which are of practical interests. For instance, setting $W_f = \begin{bmatrix} 0 \cdots 0 \; I \end{bmatrix}$ and $W_f = \begin{bmatrix} 0 \cdots 0 \; I \; 0 \end{bmatrix}$ deliver an optimal estimation of $f(k)$ and $f(k-1)$, respectively, or $W_f = \begin{bmatrix} 0 \cdots -I \; I \end{bmatrix}$ for an estimate of the change in $f(k)$ at each time instant, and $W_f = \begin{bmatrix} \frac{1}{s+1}I \cdots \frac{1}{s+1}I \; \frac{1}{s+1}I \end{bmatrix}$ for an estimate of the average value of the fault over the time interval $[k-s, k]$.

Denote by $r(k)$ the estimation error. We then have

$$r(k) = \hat{f}(k) - W_f f_s(k). \tag{15.6}$$

It follows from (15.3) and (15.6) that

$$\begin{aligned} r(k) &= P_s(k)N_{bs}(k)H_{ds}(k)d_s(k) + (P_s(k)N_{bs}(k)H_{fs}(k) - W_f)f_s(k) \\ &= \begin{bmatrix} P_s(k)N_{bs}(k)H_{ds}(k) & P_s(k)N_{bs}(k)H_{fs}(k) - W_f \end{bmatrix} \begin{bmatrix} d_s(k) \\ f_s(k) \end{bmatrix}, \end{aligned}$$

which can be rewritten as $r(k) = T_s(k)\omega_s(k)$ with

$$\begin{cases} T_s(k) = \begin{bmatrix} T_{ds}(k) & T_{fs}(k) \end{bmatrix}, \; \omega_s(k) = \begin{bmatrix} d_s(k) \\ f_s(k) \end{bmatrix}, \\ T_{ds}(k) = P_s(k)N_{bs}(k)H_{ds}(k), \\ T_{fs}(k) = P_s(k)N_{bs}(k)H_{fs}(k) - W_f. \end{cases} \tag{15.7}$$

Recall that the maximum singular value $\bar{\sigma}(T_s(k))$ represents the largest 2-norm induced gain from $\omega_s(k)$ to $r(k) = T_s(k)\omega_s(k)$. Denoting by $\sigma_i(T_s(k))$ the ith non-zero singular value of $T_s(k)$, the Frobenius-norm $||T_s(k)||_F = [\sum \sigma_i^2(T_s(k))]^{1/2}$ represents the norm induced gain in all directions of the subspace spanned by $T_s(k)$. Therefore, $\bar{\sigma}(T_s(k))$ and $||T_s(k)||_F$ are often used to evaluate the performance of the parity space-based fault estimator. Based on these evaluation criteria, one can find a solution for $P_s(k)$ by solving an optimization problem, such as

$$\min_{P_s(k)} \bar{\sigma}(T_s(k)) \text{ and } \min_{P_s(k)} \|T_s(k)\|_F.$$

Considering that the computation of $\sigma_i(T_s(k))$ is based on the matrix quadratic form

$$T_s(k)T_s^T(k) = T_{ds}(k)T_{ds}^T(k) + T_{fs}(k)T_{fs}(k),$$

this motivates us to introduce

$$T_{s\gamma}(k) = \begin{bmatrix} T_{ds}(k) & \gamma T_{fs}(k) \end{bmatrix}, \; J(P_s(k)) = T_{s\gamma}(k)T_{s\gamma}^T(k),$$

where $\gamma > 0$ is a given weighting coefficient and is used to provide us with additional design freedom. Observe that $J(P_s(k))$ is a matrix quadratic form in $P_s(k)$, i.e.

$$
\begin{aligned}
J(P_s(k)) = & \ P_s(k)N_{bs}(k)H_{ds}(k)H_{ds}^T(k)N_{bs}^T(k)P_s^T(k) \\
& +\gamma^2(P_s(k)N_{bs}(k)H_{fs}(k) - W_f)(P_s(k)N_{bs}(k)H_{fs}(k) - W_f)^T.
\end{aligned}
$$

Moreover, it is well known that, for any two different matrices $T_{s\gamma,1}(k)$ and $T_{s\gamma,2}(k)$ satisfying

$$T_{s\gamma,1}(k)T_{s\gamma,1}^T(k) > T_{s\gamma,2}(k)T_{s\gamma,2}^T(k),$$

it holds

$$\sigma_i(T_{s\gamma,1}(k)) > \sigma_i(T_{s\gamma,2}(k)), i = 1, ..., l \Rightarrow \sum_{i=1}^{l} \sigma_i^2(T_{s\gamma,1}(k)) > \sum_{i=1}^{l} \sigma_i^2(T_{s\gamma,2}(k)),$$

where it is supposed that

$$\sigma_i(T_{s\gamma,1}(k)) \geq \sigma_{i+j}(T_{s\gamma,1}(k)), \ \sigma_i(T_{s\gamma,2}(k)) \geq \sigma_{i+j}(T_{s\gamma,2}(k)),$$

for $i = 1, \cdots, l, \ j = 0, \cdots, l - i$. Therefore, a minimum of $J(P_s(k))$ provides a unified optimal solution to the following minimization problems

$$\min_{P_s(k)} \bar{\sigma}(T_{s\gamma}(k)) \ \text{ and } \ \min_{P_s(k)} \|T_{s\gamma}(k)\|_F.$$

Based on the above analysis, we now formulate the parity space-based fault estimator design problem as to find a minimum for $J(P_s(k))$.

15.3 Design of a Parity Space-Based Fault Estimator

In this section, we derive a sufficient and necessary condition for the minimum of matrix quadratic form $J(P_s(k))$ and present an analytic solution to $P_s(k)$.

By referring to the Definition 4.5 and Lemma 4.3, we present a sufficient and necessary condition for the minimum of matrix quadratic form $J(P_s(k))$ in the following theorem.

Theorem 15.1 *Given a scalar $\gamma > 0$ and matrices W_f, $N_{bs}(k)$, $H_{fs}(k)$ and $H_{ds}(k)$, the matrix*

$$P_s^*(k) = \gamma^2 W_f H_{fs}^T(k)N_{bs}^T(k)R_N^{-1}(k) \tag{15.8}$$

is a unique minimum of $J(P_s(k))$ if and only if

$$\operatorname{rank}\left(N_{bs}(k)\left[\,H_{ds}(k)\ H_{fs}(k)\,\right]\right) = \beta_k, \tag{15.9}$$

where

$$R_N(k) = N_{bs}(k)(H_{ds}(k)H_{ds}^T(k) + \gamma^2 H_{fs}(k)H_{fs}^T(k))N_{bs}^T(k). \tag{15.10}$$

Moreover, the fault estimate $\hat{f}(k)$ *is*

$$\hat{f}_o(k) = P_s^*(k)N_{bs}(k)(H_{ds}(k)d_s(k) + H_{fs}(k)f_s(k)). \tag{15.11}$$

Proof For all complex vectors $v(k) \in \mathbb{R}^{(s+1)l}$, we can write

$$v^T(k)J(P_s(k))v(k) = v^T(k)P_s(k)N_{bs}(k)H_{ds}(k)H_{ds}^T(k)N_{bs}^T(k)P_s^T(k)v(k)$$
$$+\gamma^2 v^T(k)(P_s(k)N_{bs}(k)H_{fs}(k) - W_f)(P_s(k)N_{bs}(k)H_{fs}(k) - W_f)^T v(k).$$

It is easy to see that $v^T(k)J(P_s(k))v(k)$ is a scalar quadratic form in $P_s^T(k)v(k)$. Thus we have

$$\frac{\partial(v^T(k)J(P_s(k))v(k))}{\partial(P_s^T(k)v(k))} = 2R_N(k)P_s^T(k)v(k) - 2\gamma^2 N_{bs}(k)H_{fs}(k)W_f^T v(k), \tag{15.12}$$

$$\frac{\partial^2(v^T(k)J(k)v(k))}{\partial(P_s^T(k)v(k))^2} = R_N(k), \tag{15.13}$$

where $R_N(k)$ is given in (15.10). According to Definition 4.5 and (15.12), a stationary point of $J(P_s(k))$ can be obtained from

$$(R_N(k)P_s^T(k) - \gamma^2 N_{bs}(k)H_{fs}(k)W_f^T)v(k) = 0.$$

Hence, for any given k, $P_s^*(k)$ given in (15.8) is a stationary point of $J(P_s(k))$ under the condition of $R_N(k)$ being nonsingular. Moreover, by using Lemma 4.3 and (15.13), a stationary point of $J(P_s(k))$ is a unique minimum if and only if $R_N(k) > 0$. On the other hand, $R_N(k) > 0$ holds if and only if (15.9) is satisfied. Therefore, $P_s^*(k)$ is a unique minimum of $J(P_s(k))$ if and only if (15.9) is satisfied. Substituting $P_s(k) = P_s^*(k)$ into (15.3) yields (15.11). This completes the proof. □

By using Theorem 15.1, a sufficient and necessary condition for the minimum of $J(P_s(k))$ and an optimal solution $P_s^*(k)$ is obtained. As a special case of Theorem 15.1, we now consider the case of unknown input completely decoupling.

If the basis matrix $N_{bs}(k)$ of $\mathcal{P}_s(k)$ satisfying

$$N_{bs}(k)\left[\,H_{os}(k)\ H_{ds}(k)\,\right] = 0, \tag{15.14}$$

then $T_{ds}(k) = 0$ and $\hat{f}(k)$ becomes

$$\hat{f}(k) = P_s(k)N_{bs}(k)H_{fs}(k)f_s(k). \tag{15.15}$$

Applying Theorem 15.1, we can conclude the following corollary.

Corollary 15.1 *Let $N_{bs}(k)$ be a basis matrix of $\mathcal{P}_s(k)$ satisfying (15.14). Then $P_s^*(k)$ given by*

$$P_s^*(k) = W_f H_{fs}^T(k)N_{bs}^T(k)\left(N_{bs}(k)H_{fs}(k)H_{fs}^T(k))N_{bs}^T(k)\right)^{-1}$$

is a unique minimum of $J(P_s(k))$ if and only if

$$\text{rank}(N_{bs}(k)H_{fs}(k)) = \beta_k. \tag{15.16}$$

Moreover, the fault estimate $\hat{f}(k)$ is

$$\hat{f}_o(k) = P_s^*(k)N_{bs}(k)H_{fs}(k)f_s(k). \tag{15.17}$$

Note that, under the condition of (15.14) being satisfied,

$$Q_s^*(k) = H_{fs}^T(k)N_{bs}^T(k)\left(N_{bs}(k)H_{fs}(k)H_{fs}^T(k))N_{bs}^T(k)\right)^{-1}$$

is the pseudo-inverse of $N_{bs}(k)H_{fs}(k)$ that solves optimization problem

$$\min_{Q_s(k)\in\mathbb{R}^{(s+1)l\times\beta_k}} \left\| \hat{f}_s(k) - f_s(k) \right\|_2$$
$$s.t. \;\; \hat{f}_s(k) = Q_s(k)N_{bs}(k)H_{fs}(k)f_s(k). \tag{15.18}$$

Hence, the optimal estimate for $f_s(k)$ can be written into

$$\begin{aligned}\hat{f}_s^*(k) &= Q_s^*(k)N_{bs}(k)H_{fs}(k)f_s(k)\\ &= Q_s^*(k)N_{bs}(k)(y_s(k) - H_{us}(k)u_s(k)).\end{aligned}$$

Thus we have

$$\hat{f}_o(k) = P_s^*(k)N_{bs}(k)(y_s(k) - H_{us}(k)u_s(k)) = W_f\hat{f}_s^*(k).$$

It implies that, under the condition of unknown input completely decoupling, the underlying meaning of the optimal solution (15.8) is the estimate of $f_s(k)$ in the sense of (15.18). It is worth mentioning that the parity space-based fault estimate $\hat{f}_o(k)$ solves the optimal fault estimation problem in Definition 2.6.

It should be pointed out that Theorem 15.1 and Corollary 15.1 deal with the problem of parity space-based fault estimation for given $N_{bs}(k)$ satisfying (15.4). Note that the decomposition of $W_s(k)$ given by (15.4) is not unique. The following analysis will show that the resulting estimation performance $J(P_s(k))$ is independent of the decomposition (15.4).

Let $N_{b1}(k)$ and $N_{b2}(k)$ be two different basis matrices of $\mathcal{P}_s(k)$ and $N_{b1}(k),\ N_{b2}(k)$ are full row rank. There must exist a non-singular matrix $\Phi(k)$ such that $N_{b2}(k) = \Phi(k)N_{b1}(k)$. Denote by $P_{s1}^*(k)$ and $P_{s2}^*(k)$ the corresponding optimal solutions of $P_s(k)$ obtained in Theorem 15.1. Then, for $i = 1, 2$, we have

$$\begin{aligned} P_{si}^*(k) &= \gamma^2 W_f H_{fs}^T(k) N_{bi}^T(k) R_{Ni}^{-1}(k), \\ R_{Ni}(k) &= N_{bi}(k)(H_{ds}(k)H_{ds}^T(k) + \gamma^2 H_{fs}(k)H_{fs}^T(k))N_{bi}^T(k). \end{aligned}$$

Note that

$$\begin{aligned} R_{N2}(k) &= \Phi(k)N_{b1}(k)(H_{ds}(k)H_{ds}^T(k) + \gamma^2 H_{fs}(k)H_{fs}^T(k))N_{b1}^T(k)\Phi^T(k) \\ &= \Phi(k)R_{N1}(k)\Phi^T(k), \\ P_{s2}^*(k) &= \gamma^2 W_f H_{fs}^T(k)N_{b1}^T(k)\Phi^T(k)\Phi^{-T}(k)R_{N1}^{-1}(k)\Phi^{-1}(k) \\ &= \gamma^2 W_f H_{fs}^T(k)N_{b1}^T(k)R_{N1}^{-1}(k)\Phi^{-1}(k) \\ &= P_{s1}^*(k)\Phi^{-1}(k). \end{aligned}$$

Hence,

$$\begin{aligned} J(P_{s2}^*(k)) &= P_{s2}^*(k)R_{N2}(k)(P_{s2}^*(k))^T \\ &= P_{s1}^*(k)\Phi^{-1}(k)\Phi(k)R_{N1}(k)\Phi^T(k)\Phi^{-T}(k)(P_{s1}^*(k))^T \\ &= P_{s1}^*(k)R_{N1}(k)(P_{s1}^*(k))^T = J(P_{s1}^*(k)), \end{aligned}$$

which implies that the value of $J(P_s(k))$ at its minimum is independent of the decomposition given by (15.4). We now conclude the following result.

Theorem 15.2 *Suppose that $N_{b1}(k)$ and $N_{b2}(k)$ are two different basis matrices of $\mathcal{P}_s(k)$ satisfying decomposition (15.4), and $P_{s1}^*(k),\ P_{s2}^*(k)$ are the minimums of $J(P_s(k))$ corresponding to $N_{b1}(k)$ and $N_{b2}(k)$, respectively. Then there exists a non-singular matrix $\Phi(k)$ such that*

$$N_{b2}(k) = \Phi(k)N_{b1}(k),\ \ P_{s2}^*(k) = P_{s1}^*(k)\Phi^{-1}(k)$$

and the value of $J(P_s(k))$ at its minimums $P_{s1}^*(k)$ and $P_{s2}^*(k)$ are the same.

15.4 Numerical Examples

The following three examples are illustrated to show the effectiveness of the designed fault estimator.

Example 15.1 For system (15.1) with the following parameter matrices

$$A(k) = \begin{bmatrix} -0.4\cos(k) & 0 \\ 0.5 - e^{-k} & 0.6 \end{bmatrix},\ B = \begin{bmatrix} 1 \\ 1 \end{bmatrix},\ B_d = \begin{bmatrix} 0.5 \\ 0.2 \end{bmatrix},\ B_f = \begin{bmatrix} 1 \\ 1 \end{bmatrix},$$
$$C(k) = \begin{bmatrix} 1 & -0.1 + \frac{k}{k+1} \end{bmatrix},\ D_d = 0.5,\ D_f = 1,$$

set $s = 3$ and $\gamma = 1$. Consider three different weighting matrices W_f, i.e.

$$\begin{aligned} \text{Case 1}: \ W_f &= \begin{bmatrix} 0 \ 0 \ 0 \ 1 \end{bmatrix}, \\ \text{Case 2}: \ W_f &= \begin{bmatrix} 0 \ 0 \ 1 \ 0 \end{bmatrix}, \\ \text{Case 3}: \ W_f &= \begin{bmatrix} 0 \ 0 \ 0.7 \ 0.3 \end{bmatrix}. \end{aligned}$$

We apply Theorem 15.1 to calculate $P_s^*(k)$ and $\hat{f}_o(k)$ at $k = 3, 4, \ldots, 100$. The simulated unknown input $d(k)$ is shown in Fig. 15.1. The optimal fault estimates $\hat{f}_o(k)$ in these three cases are shown in Figs. 15.2, 15.3 and 15.4, which denote the estimation of $f(k)$, $f(k-1)$ and $0.7f(k-1) + 0.3f(k)$, respectively. From these figures, one can clearly see that satisfactory fault estimation results can be obtained.

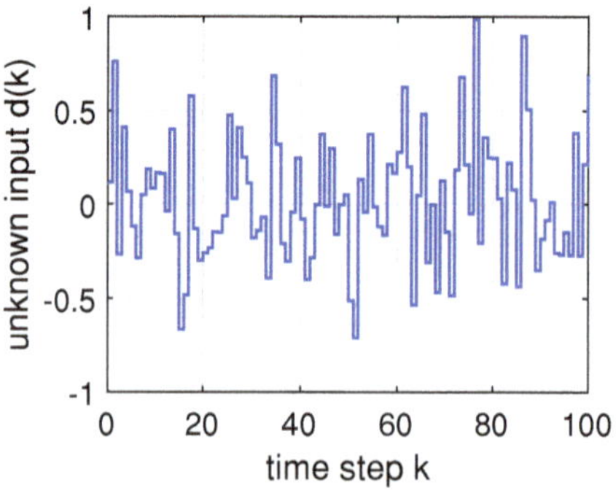

Fig. 15.1 The unknown input $d(k)$

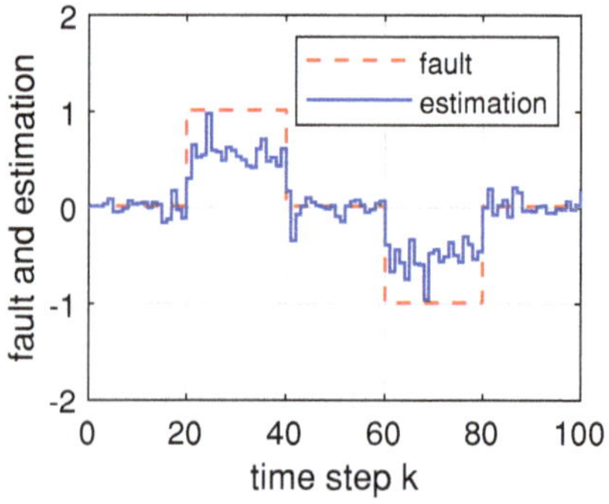

Fig. 15.2 The $f(k)$ and $\hat{f}_o(k)$ in Case 1

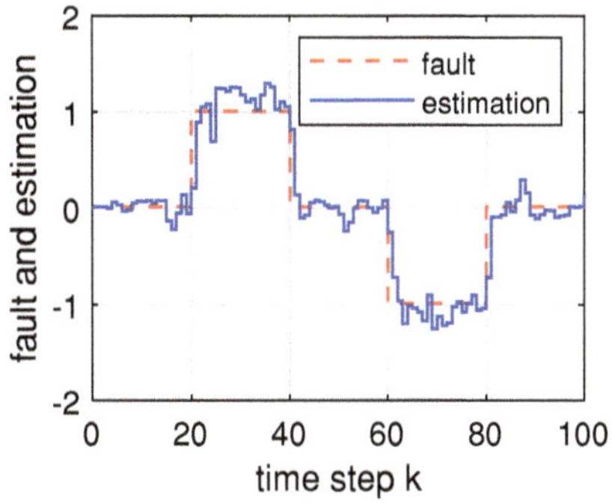

Fig. 15.3 The $f(k)$ and $\hat{f}_o(k)$ in Case 2

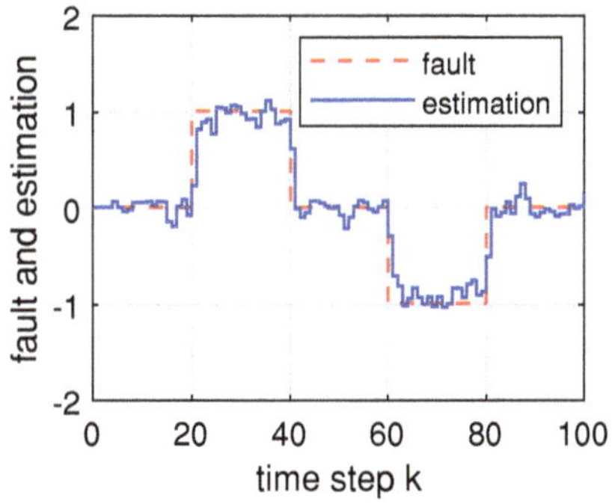

Fig. 15.4 The $f(k)$ and $\hat{f}_o(k)$ in Case 3

Example 15.2 Consider a 2-dimensional fault vector. The system parameter matrices are given as

$$A(k) = \begin{bmatrix} -0.4\cos(k) & 0 \\ 0.5 - e^{-k} & 0.6 \end{bmatrix}, \; B = \begin{bmatrix} 1 \\ 1 \end{bmatrix}, \; B_f = \begin{bmatrix} 1 & 0 \\ 0 & 1 \end{bmatrix}, \; B_d = \begin{bmatrix} 0.5 \\ 0.2 \end{bmatrix},$$
$$D_d = \begin{bmatrix} 0 \\ 0 \end{bmatrix}, \; C(k) = \begin{bmatrix} 1 & -0.1 + \frac{k}{k+1} \\ 0 & 1 \end{bmatrix}, \; D_f = \begin{bmatrix} 0 & 0 \\ 0 & 0 \end{bmatrix}.$$

Set $s = 3, \; \gamma = 3$ and

$$W_f = \begin{bmatrix} 0 & 0 & 0 & 0 & 1 & 0 & 0 & 0 \\ 0 & 0 & 0 & 0 & 0 & 1 & 0 & 0 \end{bmatrix}.$$

By applying Theorem 15.1, we calculate $P_s^*(k)$ and $\hat{f}_o(k) = \left[\, \hat{f}_{o,1}(k) \; \hat{f}_{o,2}(k) \,\right]^T$ at $k = 3, 4, \ldots, 100$. Suppose that the unknown input $d(k)$ is simulated as in Example 15.1 and $f(k) = \left[\, f_1(k) \; f_2(k) \,\right]^T$. We show $f_i(k)$ $(i = 1, 2)$ and the estimation $\hat{f}_{o,i}(k)$ $(i = 1, 2)$, respectively, in Figs. 15.5 and 15.6, from which it is again to see that satisfactory fault estimates can be obtained.

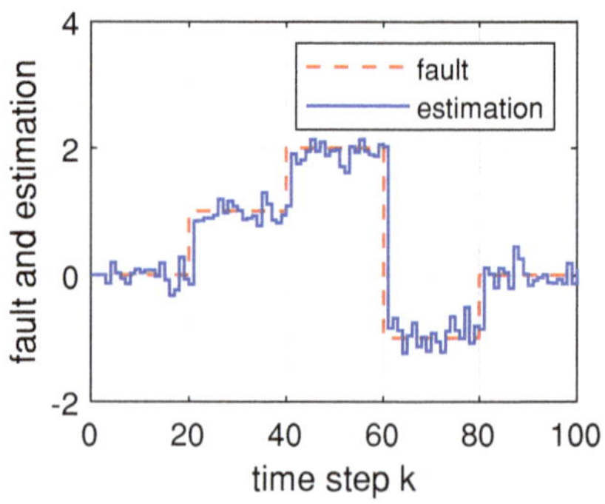

Fig. 15.5 The $f_1(k)$ and $\hat{f}_{o,1}(k)$

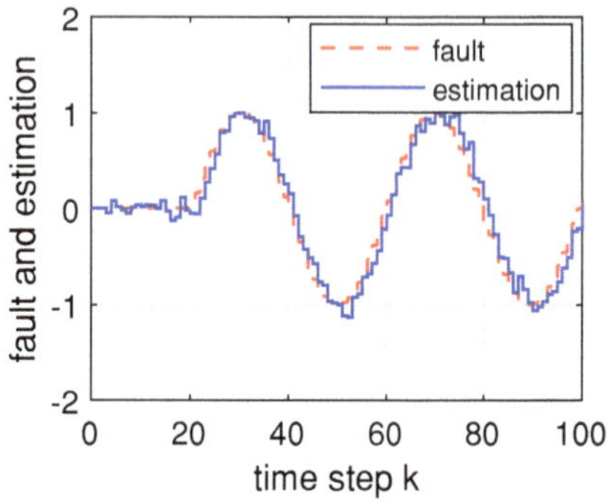

Fig. 15.6 The $f_2(k)$ and $\hat{f}_{o,2}(k)$

As mentioned in the introduction, in [5] the parity space-based fault detection technique for LTI systems have been extended to handle the fault detection problem for linear periodic systems, a special class of LTV systems. Under the condition of unknown input full decoupling being achievable, a periodically varying parity vector has been established for linear periodic systems. In order to apply the developed design method to linear periodic systems and compare with the FD results achieved in [5], we carry out the same example, i.e., Example 15.3.

Example 15.3 Consider a linear periodic system in [5] described by (3.3) with period 2 and

$$A(0) = \begin{bmatrix} 0.25 & 0.25 & 0.1 & -0.1 \\ 0.5 & 0.1 & 0.1 & 0.5 \\ 0.5 & -0.2 & 0.2 & 0.25 \\ 0.1 & 0 & 0.25 & 0.1 \end{bmatrix}, \ B(0) = \begin{bmatrix} 0.5 \\ 0.1 \\ 0.1 \\ 0.25 \end{bmatrix}, \ B_d(0) = \begin{bmatrix} 1.3 \\ 1.8 \\ 1.6 \\ 0.32 \end{bmatrix},$$

$$C(0) = \begin{bmatrix} 0.25 & 0.1 & 0.2 & 0.1 \\ -0.1 & 0.5 & 0.2 & 0.5 \\ 0.25 & 0.5 & -0.1 & 0.1 \end{bmatrix}, B_f(0) = \begin{bmatrix} 0.1 \\ -1 \\ 0.2 \\ 0.1 \end{bmatrix}, A(1) = \begin{bmatrix} 0.1 & 0.2 & 0.1 & -0.1 \\ -0.1 & 0.5 & 0 & 0.5 \\ 0.5 & 0.5 & 0.1 & 0.25 \\ 0 & 0.1 & 0.1 & 0.25 \end{bmatrix},$$

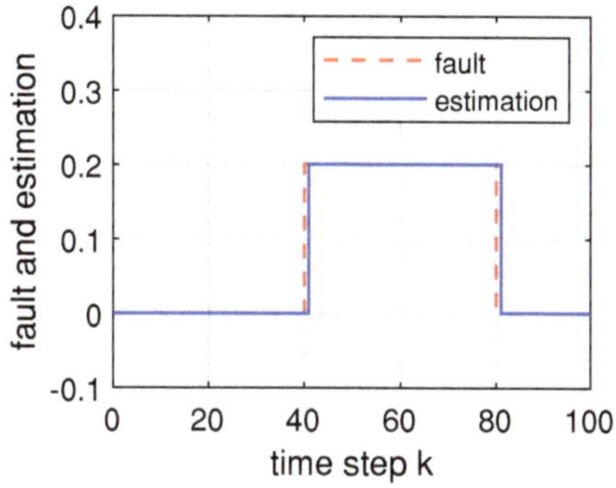

Fig. 15.7 The $f(k)$ and $\hat{f}_o(k)$ generated by using the presented method

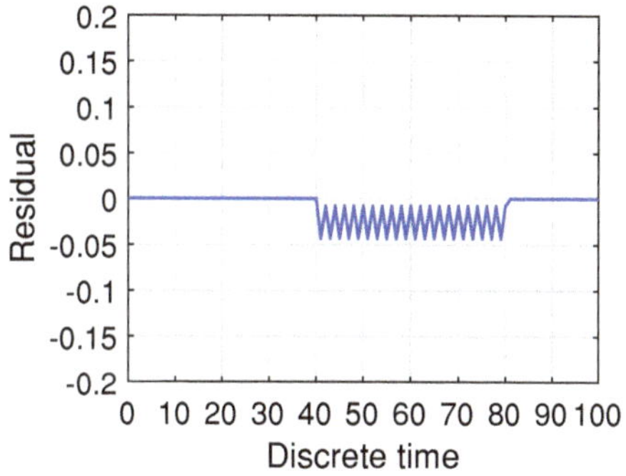

Fig. 15.8 The residual generated by using parity space approach in [5]

$$B(1) = \begin{bmatrix} 0.1 \\ 0.5 \\ 0.1 \\ 0.5 \end{bmatrix}, \ B_d(1) = \begin{bmatrix} 3.2 \\ 2 \\ -1 \\ -2 \end{bmatrix}, C(1) = \begin{bmatrix} 0.1 & 0.25 & 0.1 & -0.1 \\ 0.25 & 0.1 & 0.2 & 0.1 \\ 0.1 & 0.25 & -0.2 & 0.5 \end{bmatrix},$$

$$B_f(1) = \begin{bmatrix} 0.1 \\ -1 \\ 0.2 \\ 0.1 \end{bmatrix}, \ D_d(0) = D_d(1) = D_f(0) = D_f(1) = \begin{bmatrix} 0 \\ 0 \\ 0 \end{bmatrix}.$$

Set $s = 1$, $\gamma = 0.8$ and $W_f = [\,1\ 0\,]$. It is seen that the unknown input decoupling is achievable. Then we design the fault estimator by using Corollary 15.1. Assume that the control input is a step signal (step time at 0) of amplitude 1, the disturbance is $d(k) = \sin(0.01\pi k)$. The fault $f(k)$ and the obtained estimation $\hat{f}_o(k)$ are shown in Fig. 15.7. In order to comparing the proposed approach with the one in [5], the residual signal generated by parity spaced-based residual generator in [5] is also presented. When the fault in Fig. 15.7 occurs, the residual of [5] is shown in Fig. 15.8. It is evident that the results obtained by using our method is better than that of [5], no matter in the sense of fault estimation or FD.

15.5 Conclusion

A parity space-based fault estimation scheme has been presented in this chapter for LDTV systems. By formulating the design of a fault estimator as to find a minimum for the matrix quadratic form, we have derived a necessary and sufficient condition on the existence of the minimum. An analytic optimal solution to the parity matrix was derived that was proven to be not unique. Through numerical examples, we have shown that the unified solution works well both for the cases of unknown input full decoupling and a linear periodic system. On the other hand, we would like to mention that the parity relation based input-output model of the monitored dynamic system builds the basis of identifying a data-driven stable kernel representation of the monitored system [7, 9]. Based on this relation, the parity space-based fault detection and estimation schemes in this chapter and Chap. 14 have the potential to deal with data-driven fault detection and estimation issues.

References

1. Niemann, H., Saberi, A., Stoovogel, A., et al. (2000). Optimal fault estimation. *IFAC Proceedings Volumes, 33*(11), 267–272.
2. Blesa, J., Puig, V., Saludes, J., et al. (2016). Set-membership parity space approach for fault detection in linear uncertain dynamic systems. *International Journal of Adaptive Control and Signal Processing, 30*(2), 186–205.
3. Zhong, M., Ding, S. X., Han, Q.-L., et al. (2010). Parity space-based fault estimation for linear discrete time-varying systems. *IEEE Transactions on Automatic Control, 55*(7), 1726–1731.
4. Zhong, M., Song, Y., & Ding, S. X. (2015). Parity space-based fault detection for linear discrete time-varying systems with unknown input. *Automatica, 59*, 120–126.
5. Zhang, P., & Ding, S. X. (2007). Disturbance decoupling in fault detection of linear periodic systems. *Automatica, 43*(5), 1410–1417.
6. Nguang, S. K., Zhang, P., & Ding, S. X. (2007). Parity relation based fault estimation for nonlinear systems: An LMI approach. *International Journal of Automation and Computing, 4*(2), 164–168.
7. Ding, S. X. (2014). *Data-driven design of fault diagnosis and fault-tolerant control systems*. London, UK: Springer.
8. Han, W., Wang, Z., & Shen, Y. (2018). Fault estimation for a quadrotor unmanned aerial vehicle by integrating the parity space approach with recursive least squares. *Proceedings of the Institution of Mechanical Engineers, Part G: Journal of Aerospace Engineering, 232*(4), 783–796.
9. Jiang, Y., Yin, S., & Kaynak, O. (2021). Optimized design of parity relation-based residual generator for fault detection: Data-driven approaches. *IEEE Transactions on Industrial Informatics, 17*(2), 1449–1458.

Chapter 16
Event-Triggered Parity Space Approach to Fault Detection for LDTV Systems

This chapter demonstrates an application of parity space-based approach to event-triggered FD for linear discrete-time systems with initial motivation of decoupling the parity space-based residual generator from event-triggered transmission error. To this end, a linear discrete time system model with varying sampling periods is first presented for event-triggered FD and a new parity relation is established. Based on this, an event-triggered residual generator is constructed that is completely decoupled from event-triggered transmission error. The residual generator and event generator can thus be designed independently. The issue of residual evaluation is also considered in the event-triggering implementation.

16.1 Introduction

Thanks to the advantages of event-triggered communication in improving the utilization efficiency of network resources, investigation on event-triggered FD has become an active topic in recent years, see e.g., [2, 4–6, 9, 12, 13]. In Chap. 9, we have also demonstrated an observer-based H_i/H_∞-optimization approach to event-triggered FD for linear discrete-time systems, with main attention being paid to decoupling the event-triggered transmission error from residual and, at the same time, maximizing the sensitivity/robustness ratio criteria. Study on parity space-based FD for event-triggered systems is still at its initial stage.

In this chapter, we present a parity space-based approach to event-triggered FD. Our main attention will be focused on

- constructing an event-triggered residual generator based on parity relation;

M. Zhong et al., *Fault Diagnosis for Linear Discrete Time-Varying Systems and Its Applications*, https://doi.org/10.1007/978-981-19-5438-2_16

- designing an optimal parity space matrix to fully decouple the event-triggered transmission error from residual signal and, simultaneously, maximizing the fault sensitivity and the robustness of residual to unknown input and
- developing a residual evaluation method to FD aim.

16.2 Problem Formulation

Consider the following LTI system

$$\begin{cases} x(k+1) = Ax(k) + B_u u(k) + B_d d(k) + B_f f(k), \\ y(k) = Cx(k) + D_u u(k) + D_d d(k) + D_f f(k), \end{cases} \tag{16.1}$$

where $x(k) \in \mathbb{R}^n$, $u(k) \in \mathbb{R}^p$, $y(k) \in \mathbb{R}^q$, $d(k) \in \mathbb{R}^m$, $f(k) \in \mathbb{R}^l$ denote the system state, control input, measurement output, unknown input and fault vectors, respectively, A, B_u, B_d, B_f, C, D_u, D_d and D_f are known parameter matrices. Let $\theta_s(k) = [\theta^T(k-s)\ \theta^T(k-s+1)\ \cdots\ \theta^T(k)]^T$ with $\theta = u, y, d, f$ in sequence. A parity relation based input-out model of (16.1) can be expressed as

$$y_s(k) = H_{os}x(k-s) + H_{us}u_s(k) + H_{ds}d_s(k) + H_{fs}f_s(k) \tag{16.2}$$

where parity space order $s > n$ and

$$H_{os} = \begin{bmatrix} C \\ CA \\ \vdots \\ CA^s \end{bmatrix}, \quad H_{us} = \begin{bmatrix} D_u & 0 & 0 & \cdots & 0 \\ CB_u & D_u & 0 & \cdots & 0 \\ \vdots & \ddots & \ddots & \ddots & 0 \\ CA^{s-1}B_u & \cdots & CAB_u & CB_u & D_u \end{bmatrix},$$

H_{ds} and H_{fs} are constructed by replacing $\{B_u, D_u\}$ with $\{B_d, D_d\}$ and $\{B_f, D_f\}$ in H_{us}, respectively.

Introduce an event-triggered machine having the following event condition

$$||\Omega^{\frac{1}{2}}[y(k_i+j) - y(k_i)]|| < \epsilon^{\frac{1}{2}}||\Omega^{\frac{1}{2}}y(k_i)||, \tag{16.3}$$

where k_i is the ith release time instant of the event-triggered data transmission and $k_{i+1} > k_i$, $\epsilon > 0$ is a given scalar, Ω is a weighting positive definite matrix, $j = 1, 2, \ldots, \tau_M$ and τ_M denotes the maximum event interval. Then, when the measurement output $y(k_i)$ is released by the event detector, the next release instant time step k_{i+1} is determined by

$$k_{i+1} = k_i + \min_{j\geq 1}\{j | e_y^T(k_i)\Omega e_y(k_i) \geq \epsilon y^T(k_i)\Omega y(k_i)\}, \tag{16.4}$$

where $e_y(k_i) = y(k_i + j) - y(k_i)$. For $k \in [k_i, k_{i+1})$, the input data of the FD module are given as

$$\bar{y}(k) = y(k_i), \ \forall k \in [k_i, k_{i+1}). \tag{16.5}$$

Thus, we have

$$\bar{y}(k) = y(k_j) = y(k) + \delta_y(k), \ \forall k \in [k_i, k_{i+1}), \tag{16.6}$$

where $\delta_y(k)$ is the so-called event-triggered transmission error.

Denote by $N_{bs} \in \mathbb{R}^{\gamma \times (s+1)q}$ the base matrix of parity space $\mathcal{P}_s$, i.e., $N_{bs} H_{os} = 0$, where γ is the dimension of null space of H_{os}. Let P_s be an invertible matrix. Instead of using the output $y(k)$, the output of the event-triggered machine is applied for FD. An event-triggered residual generator is constructed as

$$r(k) = P_s N_{bs}(\bar{y}_s(k) - H_{us} u_s(k)), \tag{16.7}$$

where $\bar{y}_s(k) = \left[\, \bar{y}^T(k-s) \ \cdots \ \bar{y}^T(k-1) \ \bar{y}^T(k) \,\right]^T$. In view of (16.6), $\bar{y}_s(k)$ can thus be rewritten as

$$\bar{y}_s(k) = y_s(k) + \delta_{ys}(k), \tag{16.8}$$

where $\delta_{ys}(k) = \left[\, \delta_y^T(k-s) \ \cdots \ \delta_y^T(k-1) \ \delta_y^T(k) \,\right]^T$. Let

$$\bar{H}_{us} = N_{bs} H_{us}, \ \bar{H}_{ds} = N_{bs} H_{ds}, \ \bar{H}_{fs} = N_{bs} H_{fs}. \tag{16.9}$$

Substituting (16.2), (16.9) and (16.8) into (16.7) yields

$$r(k) = P_s(N_{bs}\delta_{ys}(k) + \bar{H}_{ds} d_s(k) + \bar{H}_{fs} f_s(k)). \tag{16.10}$$

It is obvious that the effect of $\delta_{ys}(k)$ on residual $r(k)$ cannot be eliminated by using such a parity space scheme, which may considerably degrade the performance of event-triggered FD. Aiming to decouple residual $r(k)$ from the transmission error $\delta_{ys}(k)$ and, at the same time, enhance the sensitivity of residual to fault and the robustness to unknown input, the main tasks of this chapter consist of the follow points:

- establish a linear discrete time system model with varying sampling periods to describe the system dynamics at every release instants;
- then we formulate the problem of event-triggered parity space-based FD as the one for non-uniformly sampled systems so that the generated residual can be completely decoupled from $\delta_{ys}(k)$;
- develop an event-triggered residual evaluation strategy for FD purpose.

16.3 Parity Space Based Design of an Event-Triggered FD System

As shown in (16.2), the first step of parity space-based FD is to establish a parity relation of the system under consideration and, based on this, to construct a residual generator. So, system modeling is the fundamental to event-triggered FD and plays an important rule in the design and online implementation stages. We start with establishing such an event-triggered system model concerning (16.1) and (16.6).

16.3.1 Event-Triggered System Model

Note in the framework of existing event-triggered FD that, the residual signal $r(k)$ is periodically updated at every time step k by utilizing $u(k)$ and $\bar{y}(k)$. Recalling (16.6) and combining with (16.1), we have

$$\begin{cases} x(k+1) = Ax(k) + B_u u(k) + B_d d(k) + B_f f(k), \\ \bar{y}(k) = Cx(k) + D_u u(k) + D_d d(k) + D_f f(k) + \delta_y(k). \end{cases} \tag{16.11}$$

The core of our idea is to handle an event-triggering mechanism as a sampling event and, based on it, to model an event-triggered system as a linear discrete time system with varying sampling periods, which is finally brought into the standard parity relation form that is widely adopted in parity space-based residual generation.

Denote by k_i the ith event-triggering instant and recall the notations $\underline{A}(k_i)$, $\underline{B}_\vartheta(k_i)$, $\underline{\vartheta}(k_i)$ in (9.5) and $\underline{D}_\vartheta(k_i)$ in (9.7) with ϑ standing for u, d and f in sequence. Under the consideration of event-triggering mechanism (16.4), the system (16.1) can be described by a linear discrete time system model with varying sampling periods (9.6), i.e.,

$$\begin{cases} x(k_{i+1}) = \underline{A}(k_i)x(k_i) + \underline{B}_u(k_i)\underline{u}(k_i) + \underline{B}_d(k_i)\underline{d}(k_i) + \underline{B}_f(k_i)\underline{f}(k_i), \\ y(k_i) = Cx(k_i) + \underline{D}_u(k_i)\underline{u}(k_i) + \underline{D}_d(k_i)\underline{d}(k_i) + \underline{D}_f(k_i)\underline{f}(k_i), \end{cases} \tag{16.12}$$

where $\underline{u}(k_i)$, $y(k_i)$ are the input and measurement output respectively, $\underline{d}(k_i)$ is the unknown input, $\underline{f}(k_i))$ is the fault to be detected.

Comparing with (16.11), it is clear that (16.12) describes exactly the event-triggered system dynamics without suffering from the influence of the event-triggered transmission error $\delta_y(k)$. We are now ready to solve the problem of event-triggered FD by applying traditional parity space approach to the FD-oriented event-triggered system model (16.12).

16.3.2 Event-Triggered Residual Generation

In the framework of event-triggering, we define the parity space order in terms of the triggering times and, for the sake of simplicity, it is still denoted by s. Similar to [2], by combining together the released measurement outputs from triggering instant k_{i-s} to instant k_i, the following event-triggered parity relation can be established

$$\begin{aligned} & \underline{y}_s(k_i) - \underline{H}_{us}(k_i)\underline{u}_s(k_i) \\ = & \underline{H}_{os}(k_i)x(k_{i-s}) + \underline{H}_{ds}(k_i)\underline{d}_s(k_i) + \underline{H}_{fs}(k_i)\underline{f}_s(k_i), \end{aligned} \tag{16.13}$$

where

$$\underline{y}_s(k_i) = \left[y^T(k_{i-s}) \ \ y^T(k_{i-s+1}) \ \ \cdots \ \ y^T(k_i) \right]^T, \tag{16.14}$$

and

$$\underline{H}_{os}(k_i) = \begin{bmatrix} C \\ C\underline{A}(k_{i-s}) \\ \vdots \\ C\underline{A}(k_{i-1})\underline{A}(k_{i-2})\cdots\underline{A}(k_{i-s}) \end{bmatrix},$$

$$\underline{H}_{\vartheta s}(k_i) = [h_{\vartheta s}(j_1, j_2)]_{(s+1)\times(s+1)},$$

$$h_{\vartheta s}(j_1, j_1) = \underline{D}_{\vartheta}(k_{i-s+j_1-1}), h_{\vartheta s}(j_1, j_1-1) = C\underline{B}_{\vartheta}(k_{i-s+j_1-2}),$$

$$h_{\vartheta s}(j_1, j_2) = \begin{cases} 0, \ \text{if } j_2 > j_1, \\ C\underline{A}(k_{i-s+j_1-2})\cdots\underline{A}(k_{i-s+j_2})\underline{B}_{\vartheta}(k_{i-s+j_2-1}), \ \text{if } j_2 \le j_1 - 2. \end{cases}$$

Based on this, the following event-triggered residual generator can be constructed

$$\begin{aligned} r_{k_i} &= V_{k_i}(\underline{y}_s(k_i) - \underline{H}_{us}(k_i)\underline{u}_s(k_i)) \\ &= V_{k_i}(\underline{H}_{os}(k_i)x(k_{i-s}) + \underline{H}_{ds}(k_i)\underline{d}_s(k_i) + \underline{H}_{fs}(k_i)\underline{f}_s(k_i)), \end{aligned} \tag{16.15}$$

where r_{k_i} is the event-triggered residual, V_{k_i} is the so-called event-triggered parity space matrix satisfying $V_{k_i}\underline{H}_{os}(k_i) = 0$.

Notice that

$$\begin{aligned} \underline{u}_s(k_i) &= \left[u^T(k_{i-s}) \cdots \underline{u}^T(k_{i-1}) \ \underline{u}^T(k_i) \right]^T \\ &= \left[u^T(k_{i-s}) \ u^T(k_{i-s}+1) \cdots u^T(k_i) \cdots u^T(k_{i+1}-1) \right]^T. \end{aligned}$$

It seems that the computation of residual r_{ki} is involved in the future control inputs $u(k_i+1), \ldots, u(k_{i+1}-1)$. To avoid this misleading phenomenon, one can see the following analysis.

In view of (9.5), we have

$$\begin{aligned} A^{k_j-k_{i-s}} &= \underline{A}(k_{j-1})\underline{A}(k_{j-2})\cdots\underline{A}(k_{i-s}), \\ \underline{B}_\vartheta(k_{j-1}) &= \left[\, A^{k_j-k_{j-1}-1}B_\vartheta \;\cdots\; AB_\vartheta \; B_\vartheta \,\right], \\ \underline{A}(k_{i-s+j_1-2})&\cdots\underline{A}(k_{i-s+j_2})\underline{B}_\vartheta(k_{i-s+j_2-1}) \\ &= \left[\, A^{k_{i-s+j_1-1}-k_{i-s+j_2-1}-1}B_\vartheta \;\cdots\; AB_\vartheta \; B_\vartheta \,\right]. \end{aligned}$$

Substituting into $\underline{H}_{os}(k_i)$ and $\underline{H}_{us}(k_i)$ yields

$$\underline{H}_{os}(k_i) = H^i_{os},\quad \underline{H}_{us}(k_i)\underline{u}_s(k_i) = H^i_{us}u_{is}(k_i),$$

where

$$\vartheta_{is}(k_i) = \left[\, \vartheta^T(k_{i-s}) \; \vartheta^T(k_{i-s}+1) \;\cdots\; \vartheta^T(k_i) \,\right]^T \tag{16.16}$$

for $\vartheta = u, d, f$ in sequence, H^i_{os} and H^i_{us} are given as

$$H^i_{os} = \left[\, C^T \; (CA^{\tau^1_{is}})^T \;\cdots\; (CA^{\tau^s_{is}})^T \,\right]^T, \tag{16.17}$$

$$H^i_{us} = \begin{bmatrix} D_u & 0 & \cdots & \cdots & 0 & \cdots & \cdots & 0 \\ CA^{\tau^1_{is}-1}B_u & \cdots CAB_u & CB_u & D_u & 0 & \cdots & 0 \\ CA^{\tau^2_{is}-1}B_u & \cdots & \cdots & CA^2B_u & CAB_u & CB_u & D_u & 0 \\ \cdots & \cdots & \cdots & \cdots & \cdots & \cdots & \cdots & \cdots \\ CA^{\tau^s_{is}-1}B_u & \cdots & \cdots & \cdots & CA^2B_u & CAB_u & CB_u & D_u \end{bmatrix}, \tag{16.18}$$

and $\tau^j_{is} = k_{i-s+j} - k_{i-s}$, $j = 1, 2, \ldots, s$. Furthermore, we have

$$H_{ds}(k_i)\underline{d}_s(k_i) = H^i_{ds}d_{is}(k_i),\quad H_{fs}(k_i)\underline{f}_s(k_i) = H^i_{fs}f_{is}(k_i),$$

where H^i_{ds} and H^i_{fs} are obtained by replacing $\{B_u, D_u\}$ with $\{B_d, D_d\}$ and $\{B_f, D_f\}$ in H^i_{us}, respectively. Thus, we can rewrite residual generator (16.15) as

$$\begin{aligned} r_{k_i} &= V_{k_i}(\underline{y}_s(k_i) - H^i_{us}u_{is}(k_i)) \\ &= V_{k_i}(H^i_{ds}d_{is}(k_i) + H^i_{fs}f_{is}(k_i)). \end{aligned} \tag{16.19}$$

Noticing (16.18) and (16.19), the future control inputs $u(k_i+1), \ldots, u(k_{i+1}-1)$ are not involved in the computation of residual r_{k_i}. Therefore, by using (16.19), the event-triggered residual signal r_{k_i} can be updated whenever the measurement output $y(k_i)$ is available to the FD module.

Denote by $\mathcal{P}_{si}$ the set of V_{k_i}. Let $N_{si} \in \mathbb{R}^{\gamma_i \times (1+s)q}$ be the basis matrix of parity space $\mathcal{P}_{si}$ satisfying $N_{si} H_{os}^i = 0$. Then V_{k_i} can be expressed as $V_{k_i} = P_{k_i} N_{si}$, $P_{k_i} \in \mathbb{R}^{\gamma_i \times \gamma_i}$. Let

$$\bar{H}_{ds}^i = N_{si} H_{ds}^i, \ \bar{H}_{fs}^i = N_{si} H_{fs}^i. \tag{16.20}$$

Rewrite (16.19) as

$$r_{k_i} = P_{k_i} (\bar{H}_{ds}^i d_{is}(k_i) + \bar{H}_{fs}^i f_{is}(k_i)). \tag{16.21}$$

Notice that r_{k_i} is completely decoupled from the event-triggered transmission error $\delta_{ys}(k)$. Similar to the existing time-triggered parity space-based FD, we can define the robustness of residual with respect to unknown input as

$$||P_{k_i} \bar{H}_{ds}^i||_2 := \sup_{d \neq 0} \frac{||P_{k_i} \bar{H}_{ds}^i d_{is}(k_i)||}{||d_{is}(k_i)||},$$

and the sensitivity of residual to fault as

$$||P_{k_i} \bar{H}_{fs}^i||_2 := \sup_{f \neq 0} \frac{||P_{k_i} \bar{H}_{fs}^i f_{is}(k_i)||}{||f_{is}(k_i)||},$$

or $\underline{\sigma}(P_{k_i} \bar{H}_{fs}^i)$. Thus, in the framework of event-triggering, the design of parity space-based residual generator can be formulated as to find P_{k_i} satisfying

$$\max_{P_{k_i}} \frac{||P_{k_i} \bar{H}_{fs}^i||_2}{||P_{k_i} \bar{H}_{ds}^i||_2}, \tag{16.22}$$

and/or

$$\max_{P_{k_i}} \frac{\underline{\sigma}(P_{k_i} \bar{H}_{fs}^i)}{||P_{k_i} \bar{H}_{ds}^i||_2}. \tag{16.23}$$

Do an SVD on $\bar{H}_{ds}^i$ as

$$\bar{H}_{ds}^i = U_i \Sigma_i V_i^T, \ U_i U_i^T = I, \ V_i^T V_i = I, \tag{16.24}$$

$$\Sigma_i = \begin{bmatrix} S_i & 0 \end{bmatrix}, \ S_i = \mathrm{diag}(\sigma_{i1}, \sigma_{i2}, \ldots, \sigma_{i\gamma_i}). \tag{16.25}$$

The following proposition can be straightforward to get according to the Theorem 7.1 in [7]. The proof is omitted here.

Proposition 16.1 *Doing an SVD on* $\bar{H}_{ds}^i$ *as in (16.24)–(16.25), the matrix* $P_{k_i} = S_i^{-1} U_i^T$ *provides an optimal solution to both the optimization problems (16.22) and (16.23), which results in* $||P_{k_i} \bar{H}_{ds}^i||_2 = 1$ *and*

$$\max_{P_{k_i}} \frac{||P_{k_i}\bar{H}^i_{fs}||_2}{||P_{k_i}\bar{H}^i_{ds}||_2} = ||S_i^{-1}U_i^T\bar{H}^i_{fs}||_2,$$

$$\max_{P_{k_i}} \frac{\underline{\sigma}(P_{k_i}\bar{H}^i_{fs})}{||P_{k_i}\bar{H}^i_{ds}||_2} = \underline{\sigma}(S_i^{-1}U_i^T\bar{H}^i_{fs}).$$

Furthermore, an event-triggered optimal parity matrix can be given by $V_{k_i} = S_i^{-1}U_i^TN_{si}$.

It should be pointed out that (16.19) provides a computation form of parity space-based residual generator for event-triggered FD. The relation in (16.21) shows that the event-triggered residual r_{k_i} is only involved with unknown inputs and faults over the time window $[k_{i-s}, k_i]$. Therefore, the generated residual r_{k_i} is completely decoupled from the event-triggered transmission error $\delta_{ys}(k)$. Moreover, by using Proposition 16.1, the calculation of parity space matrix V_{k_i} is independent of the event-triggering parameters ϵ and Ω. From this point of view, the design of residual generator and event detector can be carried out independently.

16.3.3 Residual Evaluation

After the design of an event-triggered residual generator, we now turn to the decision making unit for FD, which concerns with determining a residual evaluation function and a threshold. It is noted that the parity space order s of the traditional time-triggered schemes indicates the time window length of input/output data for FD. In the event-triggered implementation, however, the time interval from the $(i-s)$-th event to the ith event may not be a constant. In this chapter, an event-triggered residual evaluation function is defined as

$$J(r_{k_i}) = \frac{1}{k_i - k_{i-s} + 1}||r_{k_i}||^2. \tag{16.26}$$

Thus a threshold can be chosen as

$$J_{eth} = \sup_{f(k)=0} J(r_{k_i}). \tag{16.27}$$

Based on this, the detection of a fault can be performed by

$$\begin{cases} J(r_{k_i}) > J_{eth}, \text{ fault alarm,} \\ J(r_{k_i}) \le J_{eth}, \text{ no alarm.} \end{cases} \tag{16.28}$$

On the other hand, by applying Proposition 16.1, we have $||P_{k_i}\bar{H}^i_{ds}||_2 = 1$. Therefore, in the case of fault-free, we get

$$
\begin{aligned}
J(r_{k_i}) &= \frac{1}{k_i - k_{i-s} + 1} ||P_{k_i} \bar{H}^i_{ds} \underline{d}_s(k_i)||^2 \\
&\leq \frac{1}{k_i - k_{i-s} + 1} ||\underline{d}_s(k_i)||^2 \\
&= \frac{1}{k_i - k_{i-s} + 1} \sum_{j=k_{i-s}}^{k_i} ||d(j)||^2.
\end{aligned}
$$

Similar to the case of LDTV system in Chap. 14, such a fault-free case evaluation function $J(r_{k_i})$ is also a minimum estimation of the mean square of $d(j)$ over time window $[k_{i-s}, k_i]$. Under the assumption of $d(j)$ being norm bounded and

$$
\frac{1}{k_i - k_{i-s} + 1} \sum_{j=k_{i-s}}^{k_i} ||d(j)||^2 \leq \delta_d^2.
$$

A threshold with zero FAR can be further chosen as $J_{eth} = \delta_d^2$. In this regard, the solution $\{J(r_{k_i}), J_{eth}\}$ also delivers an optimal FD in the context of maximal fault detectability with a zero FAR, as defined in Definition 2.4. For a given threshold J_{eth}, the implementation of event-triggered FD is summarized in Algorithm 16.3.9.

Algorithm 16.3.9 The implementation of event-triggered FD

1: Set a parity space order s, an initial value $k_0 = i_0 \geq s$, and a threshold J_{eth};
2: At the transmission time instant k_i, update $\underline{y}_s(k_i)$, $u_{is}(k_i)$, H^i_{os}, H^i_{us}, H^i_{ds} and $\bar{H}^i_{ds}$ by using (16.14), (16.16)–(16.18) and (16.20), respectively;
3: Do an SVD on $\bar{H}^i_{ds}$ as in (16.24)–(16.25) to get U_i, V_i and S_i;
4: Compute parity space matrix V_{k_i} by applying Proposition 16.1;
5: Compute residual r_{k_i} and evaluate function $J(r_{k_i})$ by using (16.15) and (16.26), respectively;
6: Perform the detection of fault by using (16.28);
7: If the event-triggered condition (16.3) is violated, then $i = i + 1$ and a new event is generated based on (16.4). Update k_i, go to Step 2, till the end terminal of i.

It is notable that the matrices H^i_{os}, H^i_{ds}, H^i_{fs} and V_{k_i} are not constant. So the computational complexity is increased compared with the existing time-triggered parity space FD. Noting that the terms CA^j and CA^jB $(j = 0, 1, \ldots, k_i - k_{i-s})$ can be computed and stored in advance, the computational load of H^i_{os}, H^i_{ds} and H^i_{fs} can be ignored. Therefore, the additional calculation effort is mainly caused by the update of V_{k_i}, which includes the computation of N_{si} and an SVD on $\bar{H}^i_{ds}$. In other words, we would like to say that the complete decoupling of residual from the event-triggered transmission error $\delta_{ys}(k)$ can be achieved at the cost of moderate computational burden.

Moreover, we would like to emphasize that, once the system under consideration is modelled in the algebraic input-output model form (16.13) (or the residual generator form (16.15)), optimal residual generation issues can be addressed by selecting a parity vector or a matrix. There exist numerous design methods for this purpose.

For example, in [2] the optimal parity vector is designed by solving an eigenvector-eigenvalue problem. Differently, in this chapter an optimal parity matrix has been found by doing an SVD. The threshold setting issue is also taken into account in this regard. As a result, the overall FD system (residual generator, threshold setting and residual evaluation) delivers an optimal FD performance in the sense of maximizing fault detectability with a zero FAR.

16.4 Simulation Results

Consider the vehicle lateral dynamic system in [1] with the parameter matrices in model (16.1) given as

$$A=\begin{bmatrix}0.6333 & -0.0672\\ 2.0570 & 0.6082\end{bmatrix},\ B_u=\begin{bmatrix}-0.0653\\ 3.4462\end{bmatrix},\ B_d=\begin{bmatrix}0.1571 & 0.2395 & 0 & 0\\ 0.3977 & 0.5156 & 0 & 0\end{bmatrix},$$
$$C=\begin{bmatrix}-156.7658 & 1.2493\\ 0 & 1\end{bmatrix},\ D_u=D_f=\begin{bmatrix}56\\ 0\end{bmatrix},\ D_d=\begin{bmatrix}0 & 0 & 1 & 0\\ 0 & 0 & 0 & 1\end{bmatrix},\ B_f=B_u.$$

The control input $u(k)$ is assumed to be a reference input given by

$$u(k)=6\sin(0.2k),\ 0\le k\le 100.$$

The unknown inputs and measurement noises are simulated as

$$d(k)=[0.5\ \ 0.3\ \ 0.001\ \ 0.001]^T\varepsilon(k),$$

where $\varepsilon(k)$ denotes the signal distributed over the interval $[-1,1]$. The initial condition is set as $x(0)=[0.1\ 0.1]^T$. The fault $f(k)$ is simulated as the following pulse signal

$$f(k)=\begin{cases}0.12, & 40\le k<80,\\ 0, & \text{elsewhere.}\end{cases}$$

The first case we considered is to set the event-triggering parameters as $\Omega=I,\ \epsilon=0.1$ and $\tau_M=5$. The resulted inter-event intervals $k_{i+1}-k_i$ are depicted in Fig. 16.1. It is seen from Fig. 16.1 that the data transmission reduction is 20 times.

Next, we handle such an event-triggering mechanism as a sampling event and, based on this, to model the event-triggered system as a linear discrete time system with varying sampling periods, which is as shown in (16.12). Then an event-triggered residual generator can be constructed by (16.19), where the parity matrix can be computed recursively by applying Proposition 16.1. In order to perform online FD, we define a residual evaluation function as in (16.26), while a threshold J_{eth} is set as $J_{eth}=0.15$ with zero FAR guaranteed.

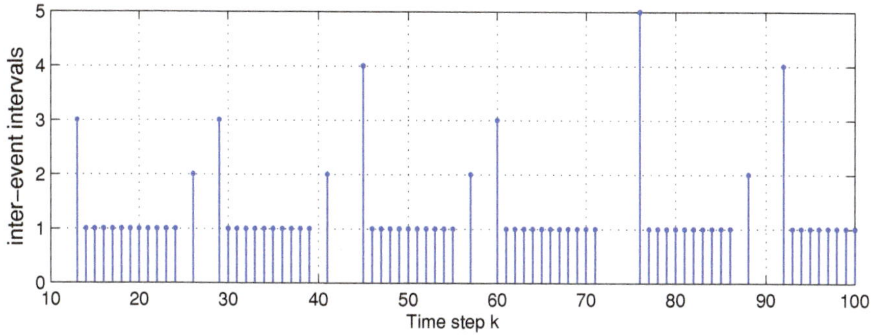

Fig. 16.1 The inter-event interval $k_{i+1} - k_i$ of case 1

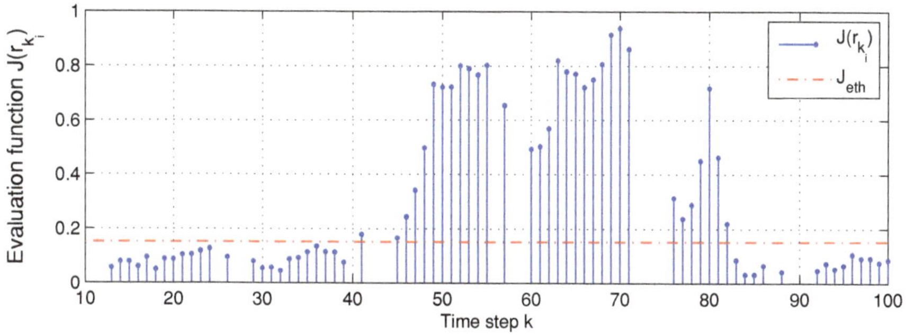

Fig. 16.2 The evaluation function $J(r_{k_i})$ of event-triggering case 1

Finally, the event triggered FD can be implemented by applying Algorithm 16.3.9. The evolution of evaluation function $J(r_{k_i})$ and the threshold are shown in Fig. 16.2. It can be seen that no measurement output is transmitted to the FD module at the fault setting initial time step $k = 40$, while the fault alarm signal appears at $k = 41$ and disappears at $k = 83$.

As a comparison, the aforementioned existing parity space approach is also considered, where the residual signals are generated by (16.7), the parity matrix W_s is designed by using Theorem 7.1 in [7]. The residual evaluation function is chosen as $J_t(r(k)) = 0.01r^T(k)r(k)$. A threshold is set as $J_{th} = 0.19$. The evolution of evaluation function $J_t(r(k))$ and threshold J_{th} are shown in Fig. 16.3. It can be seen that the fault alarm signals appear only at time steps $k = 45, 46, 47$ and 75, but false alarm signals also appear at $k = 92$ and 93. It is well known that FD performance is a comprehensive index which involves the missing fault alarm and the false fault alarm. Obviously, the influences of event-triggered transmission errors result in serious degradation of FD performance when residual generator (16.7) is applied.

On the other hand, with the increase of ϵ, substantial reduction of data transmission will be achieved. Although the design of residual generator and event detector can be carried out separately, the influence of event-triggering parameters on FD perfor-

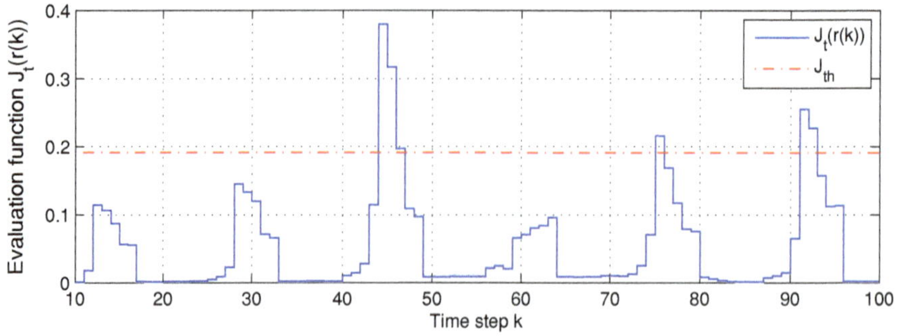

Fig. 16.3 The $J_t(r(k))$ of event-triggering case 1 with $J_{th} = 0.19$

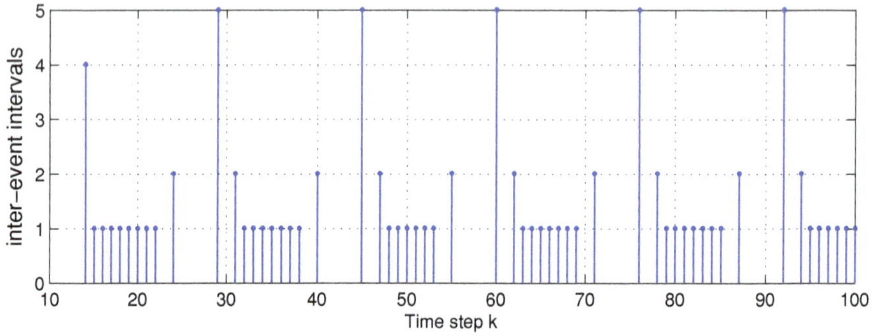

Fig. 16.4 The inter-event interval $k_{i+1} - k_i$ of case 2

mance are inevitable. To analyze the influence of event-triggering parameter ϵ on FD performance, the second case we considered is to set $\Omega = I$, $\epsilon = 0.2$ and $\tau_M = 5$. The inter-event intervals, the residual evaluation functions $J(r_{k_i})$ and $J_t(r(k))$ are shown in Figs. 16.4, 16.5, and 16.6, respectively. It can be seen from Fig. 16.4 that the reduction of data transmission is 33 times. With the given threshold $J_{eth} = 0.15$, missed fault alarm appears in Fig. 16.5. In addition, the simulation result in Fig. 16.6 shows that the maximum value of $J_t(r(k))$ in fault-free case is increased comparing with the case of $\epsilon = 0.1$. In this case, we set the threshold in Fig. 16.6 as $J_{th} = 0.4$. It is clear that, although such a threshold causes the result of no fault alarm in Fig. 16.6, the false fault alarm remains occurring. So, with the reduction of data transmission, some degradation of FD performance is inevitable.

To further analyze the influence of data transmission reduction, a case of intermittent fault is also considered as follows

$$f(k) = \begin{cases} 0.12, & \text{if } 40 \le k \le 42, \\ 0.12, & \text{if } 68 \le k \le 70, \\ 0, & \text{elsewhere.} \end{cases}$$

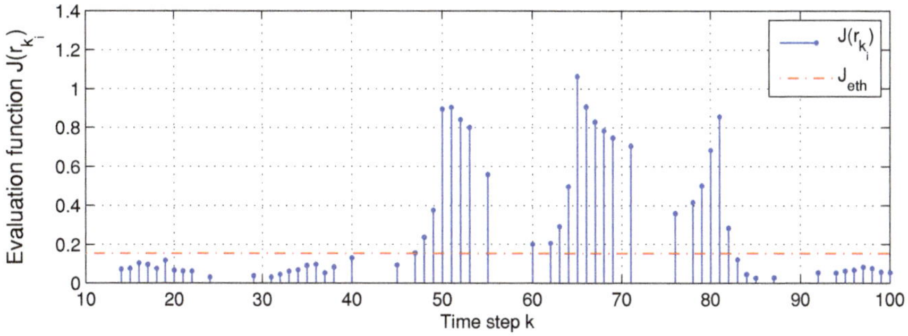

Fig. 16.5 The evaluation function $J(r_{k_i})$ of event-triggering case 2

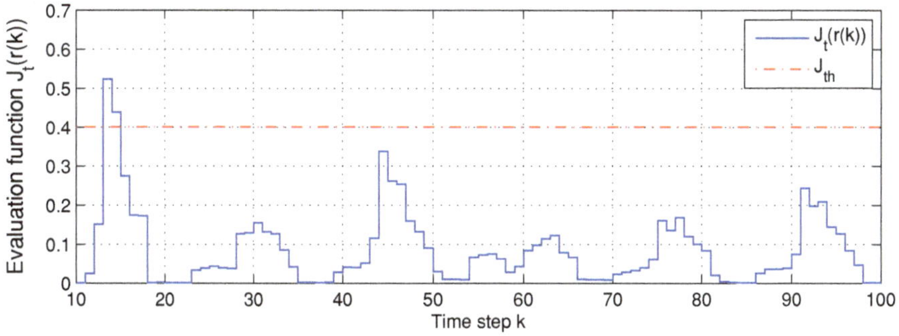

Fig. 16.6 The $J_t(r(k))$ of event-triggering case 2 with $J_{th} = 0.4$

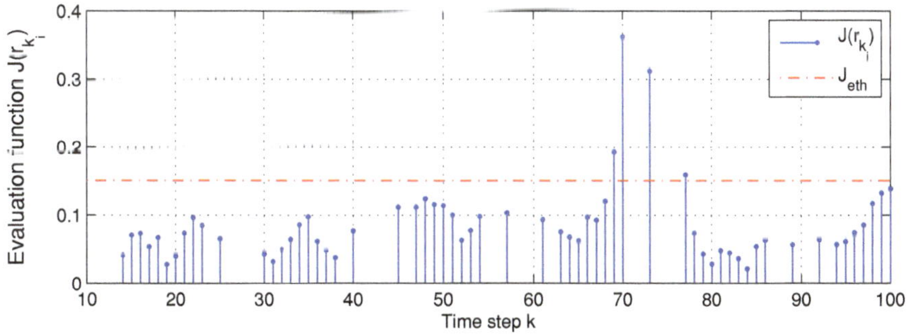

Fig. 16.7 The evaluation function $J(r_{k_i})$ of intermittent fault

The event-triggering parameters are set as the same with the second case, i.e., $\Omega = I$, $\epsilon = 0.2$ and $\tau_M = 5$. When the above intermittent fault is taken into account, the generated residual evaluation function $J(r_{k_i})$ is shown in Fig. 16.7.

Obviously, the measurement outputs at $k = 41, 42, 43$ and 44 are not transmitted to the FD module and, due to this, missing fault alarms are caused when the fault occurs at $k = 40, 41$ and 42. While, when a fault occurs at $k = 68, 69$ and 70, fault alarm starts at $k = 69$ and disappears at $k = 78$ (i.e., the third event-triggering instant after $k = 70$). Therefore, in practical applications, a trade-off between the reduction of data transmission and FD performance should be considered.

16.5 Conclusion

In this chapter, a parity space approach to event-triggered FD for linear discrete time systems has been investigated. For the purpose of efficient communication resource utilization, an event-triggering mechanism has been used to choose the necessary data of measurement outputs being transmitted to the FD module. The event-triggered system has been modeled as a linear discrete time system with varying sampling periods and, based on this, an event-triggered parity relation has been established. A parity space-based residual generator has been designed such that the generated residual is completely decoupled from the transmission errors. Furthermore, an event-triggered evaluation function has been introduced for event-triggered FD. Through the simulation examples, we can see that, compared with the existing parity space approach, the event-triggered FD performance improves, but the computation cost increases moderately. However, with the reduction of data transmission, the degradation of FD performance is inevitable. As possible future works corresponding to event-triggered FD, topics such as FD performance related event-triggering condition, the trade-off design between FD performance and data transmission reduction, the application of observer-based adaptive tracking control methods (see e.g., [11] and [3]), etc., remain to be investigated.

References

1. Qiu, A., Al-Dabbagh, A. W., & Chen, T. (2019). A tradeoff approach for optimal event-triggered fault detection. *IEEE Transactions on Industrial Electronics, 66*(3), 2111–2121.
2. Izadi, I., Shah, S. L., & Chen, T. (2008). Parity space fault detection based on irregulary sampled-data. In *Proceedings of the American Control Conference*, June 11–13, Washington, USA (pp. 2798–2803).
3. Ma, L., Zong, G., Zhao, X., et al. (2020). Observed-based adaptive finite-time tracking control for a class of nonstrict-feedback nonlinear systems with input saturation. *Journal of The Franklin Institute, 357*(16), 11518–11544.
4. Zhang, L., Liang, H., Sun, Y., et al. (2021). Adaptive event-triggered fault detection scheme for semi-markovian jump systems with output quantization. *IEEE Transactions on Systems, Man, and Cybernetics: Systems, 51*(4), 2370–2381.
5. Zhong, M., Ding, S. X., & He, X. (2020). An H_i/H_∞ optimization approach to event-triggered fault detection for linear discrete time systems. *IEEE Transactions on Automatic Control, 65*(10), 4464–4471.
6. Atitallah, M., Davoodi, M., & Meskin, N. (2018). Event-triggered fault detection for networked control systems subject to packet dropout. *Asian Journal of Control, 20*(6), 2195–2206.
7. Ding, S. X. (2013). *Model-based fault diagnosis techniques: Design schemes, algorithms, and tools* (2nd ed.). London, U.K.: Springer.

8. S. X. Ding, E. L. Ding, T. Jeinsch, et al. (2000). An approach to a unified design of FDI systems. In *Proceedings of the 3rd Asian Control Conference*, Shanghai China (pp. 2812–2817).
9. Wu, T., Li, F., Yang, C., et al. (2018). Event-based fault detection filtering for complexed networked jump systems. *IEEE/ASME Transactions on Mechatronics, 47*(8), 2299–2311.
10. Ren, W., Sun, S., Hou, N., et al. (2018). Event-triggered non-fragile H_∞ fault detection for discrete time-delayed nonlinear systems with channel fadings. *Journal of the Franklin Institute, 355*, 436–457.
11. Huo, X., Ma, L., Zhao, X., et al. (2019). Observer-based adaptive fuzzy tracking control of mimo switched nonlinear systems preceded by unknown backlash-like hysteresis. *Information Sciences, 490*, 369–386.
12. Wang, X., & Yang, G. (2019). Event-triggered H_∞ filtering for discrete-time T-S fuzzy systems via network delay optimization technique. *IEEE Transactions on Systems, Man, and Cybernetics: Systems, 49*(10), 2026–2035.
13. Wang, X., Fei, Z., Yan, H., et al. (2020). Dynamic event-triggered fault detection via zonotopic residual evaluation and its application to vehicle lateral dynamics. *IEEE Transactions on Industrial Informatics, 16*(11), 6952–6961.
14. Ju, Y., Wei, G., Ding, D., et al. (2019). Event-triggered distributed fault detection over sensor networks in finite-frequency domain. *IET Control Theory and Applications, 13*(4), 2261–2269.

Chapter 17
Stationary Wavelet Transform Aided Fault Detection for LDTV Systems

This chapter confines to a stationary wavelet transform (SWT) aided parity space approach to FD for LDTV systems, aiming at improving the FD performance of lower parity space order with an acceptable increase of computation load. In the sense of maximizing the sensitivity/robustness ratio criteria, a bank of parity relation based residual generators are first constructed. By virtue of the merits of wavelet transform in capturing the time-frequency features of faults, the selection of a bank of parity space vectors are then converted into solving a group of sensitivity/robustness ratio maximization problems. SVD based analytical solutions are derived by using the recursive algorithm of SWT. Compared with the traditional parity space-based method, this scheme, on the one hand, can improve the sensitivity/robustness ratio performance with a given parity space order and, on the other hand, can detect faults within wider frequency band at a zero FAR, while with acceptable increase of computation cost.

17.1 Introduction

In traditional parity space-based residual generation for dynamic systems, the so-called parity space vector or parity space matrix is usually designed in the context of minimizing the ratio of robustness index of residual to disturbance over the sensitivity index of residual to fault. Especially, when a parity space vector of higher order is adopted for decreasing the ratio criterion, it works in frequency domain as a narrow-band filter, which, unfortunately, will result in a poor fault detectability and higher computation cost. While a smaller parity space order means a lower computation load but a possible poor sensitivity/robustness criterion [3]. A remedy, as given in Chap. 14, is to convert the parity space-based FD issue into minimizing a quadratic form matrix so as to improve the computation efficiency. Despite of this result, how

M. Zhong et al., *Fault Diagnosis for Linear Discrete Time-Varying Systems and Its Applications*, https://doi.org/10.1007/978-981-19-5438-2_17

to compromise the FD performance and the computation load with respect to parity space order remains a nontrivial topic and practically meaningful.

With the rising surge of applying the techniques of signal processing, statistical learning and data mining for fault diagnosis, various hybrid FD schemes have been developed aiming at immunize the defects of conventional model-based methods [4, 13, 16]. Among them, wavelet transform, as a powerful signal processing tool, is characterized by its advantages of capturing time-frequency features of a signal and fast algorithms for online implementation. A series of wavelet transform aided FD methods have been developed in the past two decades, see e.g., [12, 13, 15, 16], while most of the results are given for LTI systems. For example, in [16] Ye et al. proposed to combine SWT with parity space based approach to FD for LTI systems, wherein a bank of parity space vectors of smaller parity space order were designed achieving a lower missed detection rate with a lower computational load. Such a success motivates us to address parity space based FD issues for LDTV systems by means of SWT.

In this chapter, we combine SWT with parity space-based method to deal with FD issues for LDTV systems. The main tasks to this aim consist of

- constructing a bank of parity space-based residual generator by using SWT;
- establishing criteria to measure the robustness of the SWT aided residuals against unknown input and the sensitivity to fault, and then formulating the selection of the bank of parity space vectors as solving a bank of sensitivity/robustness ratio index maximization problems and
- Developing recursive algorithm for online FD purpose.

17.2 Preliminaries of Wavelet Transform

Wavelet transform is a powerful tool for signal analysis and has been widely used in various fields such as signal processing [5, 6, 9, 11], fault diagnosis [2, 10, 12, 13, 15, 16] and pattern recognition [8], etc. In comparison with the Fourier transform and short-time Fourier transform, wavelet transform can depict the local information of a signal in time-frequency domain and has the multiscale resolution analysis (MRA) ability. Meanwhile, various fast implementation algorithms for wavelet transform have been proposed successively [6]. To declare the basic idea of wavelet transform more clearly, below we introduce the definition of wavelet at first.

Given a mother wavelet $\phi(t)$ of zero average satisfying

$$\int_{-\infty}^{\infty} \phi(t)dt = 0,$$

by dilating $\phi(t)$ with a scale parameter $a > 0$ and translating it by $\tau \in \mathbb{R}$, a family of wavelet basis functions based on the mother wavelet are defined as follows

$$\phi_{a,\tau}(t) = \frac{1}{\sqrt{a}}\phi\left(\frac{t-\tau}{a}\right).$$

By scaling the parameter a with a factor 2^j, a discrete wavelet scaled by 2^j can be obtained as

$$\psi_{j,k}(n) = \sqrt{2^{-j}}\psi\left(2^{-j}n - k\right),$$

where $j = 1, 2, \ldots, m,, k = 1, 2, \ldots, N$. Correspondingly, the discrete wavelet transform (DWT) of a signal $z(k)$ at scale j is defined by

$$T_z(j,k) = \langle z(n), \psi_{j,k}(n)\rangle = \sum_n z(n)\psi^*_{j,k}(n-k), \tag{17.1}$$

where $\psi^*_{j,k}$ is the conjugation of $\psi_{j,k}$.

Regarding the computation of DWT, Mallat fast algorithm [6] is one of the most popular choices. Denote by $c_{j,k}$ and $d_{j,k}$ the approximate and detailed coefficients of DWT at scale j. We have

$$c_{j+1,k} = \sum_n l_0(n-2k)c_{j,n},\ d_{j+1,k} = \sum_n h_0(n-2k)c_{j,n}, \tag{17.2}$$

where l_0 and h_0 are the primary low-pass and high-pass filters, respectively, that are determined by the selected mother wavelet.

In Mallat fast algorithm for DWT, the down sampling is performed to guarantee the equivalence of the lengths of original signal before and after DWT. Such a handling, unfortunately, would cause the missing of low frequency information of signal as the increase of scale j. To overcome this drawback, another fast algorithm named SWT [6, 7] is prone to be adopted for its shift-invariant property, i.e., the SWT coefficients of a delayed signal are just a time-shifted version of the original one [7]. In SWT, instead of doing down-sampling to the coefficients of DWT at each scale as done in Mallat fast algorithm, a bank of filters at different scale are utilized by performing interpolation to l_0 and h_0, i.e.,

$$l_{j+1}(n) = l_0(2^j n),\ h_{j+1}(n) = h_0(2^j n). \tag{17.3}$$

Then the coefficients of SWT can be computed by

$$c_{j,k} = c_{0,k} * g_{l,j},\ d_{j,k} = c_{0,k} * g_{b,j}, \tag{17.4}$$

where $c_{0,k}$ is the original signal to be analyzed, $*$ is the convolution operator and

$$g_{l,j} = l_1 * l_2 \ldots * l_{j-1} * l_j,\ g_{b,j} = l_1 * l_2 \ldots * l_{j-1} * h_j. \tag{17.5}$$

In this context, filters $g_{l,j}$ and $g_{b,j}$ can be regarded as the low-pass and high-pass filters at scale j, respectively. From the computation viewpoint, we can achieve $c_{j,k}$, $d_{j,k}$ in a recursive manner, i.e.,

$$c_{j,k} = \sum_n l_0(n) c_{j-1,k+2^j n}, \ d_{j,k} = \sum_n h_0(n) c_{j-1,k+2^j n}. \tag{17.6}$$

17.3 Problem Formulation

Given LDTV systems represented by state space model (3.3), the system dynamics within time interval $[k-s, k]$ can be described by the following input-output model

$$y_s(k) = H_{os}(k)x(k-s) + H_{us}(k)u_s(k) + H_{ds}(k)d_s(k) + H_{fs}(k)f_s(k), \tag{17.7}$$

where $x(k-s) \in \mathbb{R}^n$ is the state vector at time step $k-s$ for a given positive integer s, $y(k) \in \mathbb{R}^q$, $u(k) \in \mathbb{R}^p$, $d(k) \in \mathbb{R}^m$ and $f(k) \in \mathbb{R}^l$ are the measurement output, system input, unknown input and fault vectors, respectively, $y_s(k)$, $u_s(k)$, $d_s(k)$ and $f_s(k)$ are in form of $\theta_s = \left[\theta^T(k-s)\ \theta^T(k-s+1) \cdots \theta^T(k)\right]^T$ with $\theta = y,\ u,\ d,\ f$ in sequence, matrix $H_{os}(k)$ is given in (3.32), matrices $H_{us}(k)$, $H_{ds}(k)$ and $H_{fs}(k)$ are in form of (3.33). It is assumed that $d(k),\ f(k) \in l_{2,[0,N]}$.

Let $N_{bs}(k) \in \mathbb{R}^{\gamma \times q(s+1)}$ is a matrix in null space of $H_{os}(k)$, i.e., $N_{bs}(k)H_{os}(k) = 0$, and

$$\bar{H}_{us}(k) = N_{bs}(k)H_{us}(k),\ \bar{H}_{ds}(k) = N_{bs}(k)H_{ds}(k),\ \bar{H}_{fs}(k) = N_{bs}(k)H_{fs}(k).$$

A parity space-based residual generator can be constructed as follows

$$\begin{aligned} r_s(k) &= p_s(k)\left(N_{bs}(k)y_s(k) - \bar{H}_{us}(k)u_s(k)\right) \\ &= p_s(k)\left(\bar{H}_{ds}(k)d_s(k) + \bar{H}_{fs}(k)f_s(k)\right), \end{aligned} \tag{17.8}$$

where $r_s(k)$ is the residual signal, $p_s(k) \in R^\gamma$ is the parity space vector to be designed. The design of residual generator (17.8) hence lies in the design of parity space vector $p_s(k)$.

As demonstrated in Lemma 14.2 in Chap. 14, an optimal solution of $p_s(k)$ can be obtained by solving a generalized eigenvalue-eigenvector problem via SVD, i.e.,

$$p_s(k) = v_s(k)S^{-1}(k)U^T(k), \tag{17.9}$$

where $S(k)$ and $U(k)$ are given in (14.4), $v_s(k) \in \mathbb{R}^\gamma$ satisfies $v_s(k)v_s^T(k) = 1$ and solves (14.7). In this regard, $p_s(k)$ delivers the maximal sensitivity/robustness ratio criterion

$$I(p_s(k)) := \max_{p_s(k)} \frac{\left\| p_s(k)\bar{H}_{fs}(k)\right\|_2}{\left\| p_s(k)\bar{H}_{ds}(k)\right\|_2} = \bar{\sigma}(S^{-1}(k)U^T(k)\bar{H}_{fs}(k)). \tag{17.10}$$

Note that the solution (17.9) is an intuitive extension of the vector-valued optimal parity space vector for LTI systems (see Theorem 7.2 in [3]). Regarding of this, the following issues should be considered:

- the sensitivity/robustness ratio criterion $I(p_s(k))$ gets larger (better) with the increase of the parity space order s, i.e.,

$$I(p_{s+1}(k)) \geq I(p_s(k)).$$

  Meanwhile, the computation cost becomes larger, which is not desired;
- the parity space vector $p_s(k)$ functions as a bandpass filter with a certain central frequency ω_0. Only the faults lie in this band can be detected well. Otherwise, a high missed detection rate would be caused;
- Moreover, the larger of the parity space order s, the narrower of the band of the bandpass filter corresponding to $p_s(k)$.

Hence, how to design a parity space-based FD system for LDTV systems so as to achieve satisfactory FD performance with lower computation load remains an open and challenging task.

17.4 SWT Aided Parity Space Approach to Fault Detection

Thanks to the MRA and fast realization algorithm of SWT, in this section we address the parity space vector design issues by means of SWT. The method is motivated by the idea in [16] for LTI systems.

17.4.1 SWT Aided Residual Generators

At first, we perform wavelet transform to the parity space-based residual signal $r_s(k)$ and construct a bank of residual generators by

$$\begin{cases} r_s(k) = p_s(k)\left[N_{bs}(k)y_s(k) - \bar{H}_{us}(k)u_s(k)\right], \\ r_{s,j}(k) = T^d_{r_s}(j,k),\ j = 1,\ 2,\ \ldots,\ j_m, \\ r_{s,j_m+1}(k) = T^a_{r_s}(j_m,k), \end{cases} \tag{17.11}$$

where j_m is the maximal scale of SWT, $r_{s,j}(k)$ denotes the detailed coefficients of SWT for $r_s(k)$ at scale j, $r_{s,j_m+1}(k)$ is the approximate coefficient of the SWT of $r_s(k)$ at scale j_m.

By using the shift-invariant property of SWT, the residuals in (17.11) can be implemented by using a bank of filters, i.e.,

$$\begin{cases} r_{s,j}(k) = r_s(k) * g_{b,j}, \ j = 1, 2, \ldots, j_m, \\ r_{s,j_m+1}(k) = r_s(k) * g_{l,j_m}, \end{cases} \tag{17.12}$$

where $g_{b,j}, \ j = 1, 2, \ldots, j_m$ and g_{l,j_m} are given in (17.5).

According to the definition of convolution, it follows from (17.8) and (17.12) that

$$\begin{aligned} r_{s,j}(k) &= \sum_{i=0}^{\alpha_j} g_{b,j}(i) p_s(k) \left[\bar{H}_{d,s}(k) d_s(k-i) + \bar{H}_{f,s}(k) f_s(k-i)\right] \\ &= \sum_{i=0}^{\alpha_j} g_{b,j}(i) p_s(k) \bar{H}_{d,s}(k) d_s(k-i) + \sum_{i=0}^{\alpha_j} g_{b,j}(i) p_s(k) \bar{H}_{f,s}(k) f_s(k-i), \end{aligned} \tag{17.13}$$

$$\begin{aligned} r_{s,j_m+1}(k) &= \sum_{i=0}^{\alpha_{jm}} \left[\bar{H}_{d,s}(k) d_s(k-i) + \bar{H}_{f,s}(k) f_s(k-i)\right] \\ &= \sum_{i=0}^{\alpha_{jm}} g_{l,j_m}(i) p_s(k) \bar{H}_{d,s}(k) d_s(k-i) + \sum_{i=0}^{\alpha_{jm}} g_{l,j_m}(i) p_s(k) \bar{H}_{f,s}(k) f_s(k-i), \end{aligned} \tag{17.14}$$

where $\alpha_j, \ j = 1, 2, \ldots, j_m$ are the maximal settling time of filters $g_{b,j}$ and g_{l,j_m}, i.e. $g_{b,j}(i) \to 0$, when $i \geq \alpha_j$, $g_{l,j_m}(i) \to 0$, when $i \geq \alpha_{j_m}$.

To gain a deeper insight into the formulations (17.13) and (17.14), we introduce the following notations

$$\begin{aligned} M_{b,j}^f(i) &= \begin{bmatrix} 0_f & \ldots & 0_f & g_{b,j}(i) I_f & 0_f & \ldots & 0_f \end{bmatrix}, \ j = 1, 2, \ldots, j_m, \\ M_{b,j}^d(i) &= \begin{bmatrix} 0_d & \ldots & 0_d & g_{b,j}(i) I_d & 0_d & \ldots & 0_d \end{bmatrix}, \ j = 1, 2, \ldots, j_m, \\ M_{l,j_m}^f(i) &= \begin{bmatrix} 0_f & \ldots & 0_f & g_{l,j_m}(i) I_f & 0_f & \ldots & 0_f \end{bmatrix}, \\ M_{l,j_m}^d(i) &= \begin{bmatrix} 0_d & \ldots & 0_d & g_{l,j_m}(i) I_d & 0_d & \ldots & 0_d, \end{bmatrix}, \end{aligned}$$

where $0_d \in \mathbb{R}^{q(s+1) \times q}$, $0_f \in \mathbb{R}^{l(s+1) \times l}$ are zero matrices, $I_d \in \mathbb{R}^{q(s+1) \times q(s+1)}$, $I_f \in \mathbb{R}^{l(s+1) \times l(s+1)}$ ate identical matrices. For the shift-invariant property of SWT and convolution, we have, for the first item of (17.13), that

$$
\begin{aligned}
& g_{b,j}(i) p_s(k-i) \bar{H}_{ds}(k-i) d_s(k-i) \\
&= p_s(k-i) \bar{H}_{ds}(k-i) \left[\underbrace{0_d \ \ldots \ 0_d}_{(\alpha_j - i)} \ \ g_{b,j}(i) I_d \ \ \underbrace{0_d \ \ldots \ 0_d}_{i} \right] \begin{bmatrix} d(k-s-\alpha_j) \\ \vdots \\ d_s(k-i) \\ \vdots \\ d(k) \end{bmatrix}, \\
&= p_s(k-i) \bar{H}_{ds}(k-i) M^d_{b,j}(i) d_{s+\alpha_j}(k).
\end{aligned}
$$

Then we have

$$
\begin{aligned}
& \sum_{i=0}^{\alpha_j} g_{b,j}(i) p_s(k-i) \bar{H}_{ds}(k-i) d_s(k-i) \\
&= \sum_{i=0}^{\alpha_j} p_s(k-i) \bar{H}_{ds}(k-i) M^d_{b,j}(i) d_{s+\alpha_j}(k) \\
&= \left[p_s(k) \bar{H}_{ds}(k) M^d_{b,j}(0) + \sum_{i=1}^{\alpha_j} p_s(k-i) \bar{H}_{ds}(k-i) M^d_{b,j}(i) \right] d_{s+\alpha_j}(k) \\
&= [p_s(k) \quad 1] \begin{bmatrix} \bar{H}_{ds}(k) M^d_{b,j}(0) \\ \sum_{i=1}^{\alpha_j} p_s(k-i) \bar{H}_{ds}(k-i) M^d_{b,j}(i) \end{bmatrix} d_{s+\alpha_j}(k) \\
&= \tilde{p}_s(k) \bar{H}^{b,j}_{ds}(k) d_{s+\alpha_j}(k),
\end{aligned}
$$

where

$$
\begin{aligned}
\tilde{p}_s(k) &= [p_s(k) \quad 1], \\
\bar{H}^{b,j}_{ds}(k) &= \begin{bmatrix} \bar{H}_{ds}(k) M^d_{b,j}(0) \\ \sum_{i=1}^{\alpha_j} p_s(k-i) \bar{H}_{ds}(k-i) M^d_{b,j}(i) \end{bmatrix}.
\end{aligned} \tag{17.15}
$$

Similarly, denote

$$
\begin{aligned}
\bar{H}^{b,j}_{fs}(k) &= \begin{bmatrix} \bar{H}_{fs}(k) M^f_{b,j}(0) \\ \sum_{i=1}^{\alpha_j} p_s(k-i) \bar{H}_{fs}(k-i) M^f_{b,j}(i) \end{bmatrix}, \ j = 1, 2, \ldots, j_m, \\
\bar{H}^{l,j_m}_{fs}(k) &= \begin{bmatrix} \bar{H}_{ds}(k) M^d_{l,j_m}(0) \\ \sum_{i=1}^{\alpha_j} p_s(k-i) \bar{H}_{ds}(k-i) M^d_{l,j_m}(i) \end{bmatrix},
\end{aligned}
$$

$$\bar{H}_{fs}^{l,j_m}(k) = \begin{bmatrix} \bar{H}_{fs}(k)M_{l,j_m}^{f}(0) \\ \sum_{i=1}^{\alpha_j} p_s(k-i)\bar{H}_{fs}(k-i)M_{l,j_m}^{f}(i) \end{bmatrix}.$$

It follows

$$\sum_{i=0}^{\alpha_j} g_{b,j}(i)p_s(k-i)\bar{H}_{fs}(k-i)f_s(k-i) = \tilde{p}_s(k)\bar{H}_{fs}^{b,j}(k)f_{s+\alpha_j}(k),$$
$$\sum_{i=0}^{\alpha_{jm}} g_{l,j_m}(i)p_s(k-i)\bar{H}_{ds}(k-i)d_s(k-i) = \tilde{p}_s(k)\bar{H}_{ds}^{l,j_m}(k)d_{s+\alpha_{jm}}(k),$$
$$\sum_{i=0}^{\alpha_{jm}} g_{l,j_m}(i)p_s(k-i)\bar{H}_{fs}(k-i)f_s(k-i) = \tilde{p}_s(k)\bar{H}_{fs}^{l,j_m}(k)f_{s+\alpha_{jm}}(k).$$

The residual generator (17.12) can thus be rewritten as follows

$$\begin{cases} r_{s,j}(k)=\tilde{p}_s(k)[\bar{H}_{ds}^{b,j}(k)d_{s+\alpha_j}(k)+\bar{H}_{fs}^{b,j}(k)f_{s+\alpha_j}(k)],\ j=1,2,\ldots,j_m, \\ r_{s,j_m+1}(k) = \tilde{p}_s(k)[\bar{H}_{ds}^{l,j_m}(k)d_{s+\alpha_{jm}}(k) + \bar{H}_{fs}^{l,j_m}(k)f_{s+\alpha_{jm}}(k)]. \end{cases} \tag{17.16}$$

Up to now, we have formulated the design of $p_s(k)$ as $\tilde{p}_s(k)$. Below we give an optimal design of $\tilde{p}_s(k)$ in the sense of maximizing a group of ratio-type sensitivity/robustness criteria.

17.4.2 Design of the SWT Aided Parity Space Vectors

Towards enhancing the robustness of the bank of residuals $r_{s,j}(k)$, $j = 1, 2, \ldots j_m + 1$ to unknown input and the sensitivity to fault, vectors $\tilde{p}_s(k)$ can be designed by solving the following optimization problems

$$\begin{cases} I(\tilde{p}_s(k)) := \max\limits_{\tilde{p}_s(k)} \dfrac{\tilde{p}_s(k)\bar{H}_{fs}^{b,j}(k)\left(\bar{H}_{fs}^{b,j}(k)\right)^T \tilde{p}_s^T(k)}{\tilde{p}_s(k)\bar{H}_{ds}^{b,j}(k)\left(\bar{H}_{ds}^{b,j}(k)\right)^T \tilde{p}_s^T(k)},\ j = 1, 2, \ldots, j_m, \\ I(\tilde{p}_s(k)) := \max\limits_{\tilde{p}_s(k)} \dfrac{\tilde{p}_s(k)\bar{H}_{fs}^{l,j_m}(k)\left(\bar{H}_{fs}^{l,j_m}(k)\right)^T \tilde{p}_s^T(k)}{\tilde{p}_s(k)\bar{H}_{ds}^{l,j_m}(k)\left(\bar{H}_{ds}^{l,j_m}(k)\right)^T \tilde{p}_s^T(k)}. \end{cases} \tag{17.17}$$

For ease of declaration, we denote by $\tilde{p}_{s,j}(k)$, $j = 1, 2, \ldots, j_m + 1$ the optimal solutions of $\tilde{p}_s(k)$ by solving the corresponding jth problems in (17.17), respectively. The optimization problems in (17.17) become

$$\begin{cases} I(\tilde{p}_{s,j}(k)) := \max\limits_{\tilde{p}_{s,j}(k)} \dfrac{\tilde{p}_{s,j}(k)\bar{H}_{fs}^{b,j}(k)\left(\bar{H}_{fs}^{b,j}(k)\right)^T \tilde{p}_{s,j}^T(k)}{\tilde{p}_{s,j}(k)\bar{H}_{ds}^{b,j}(k)\left(\bar{H}_{ds}^{b,j}(k)\right)^T \tilde{p}_{s,j}^T(k)}, \; j = 1, 2, \ldots, j_m, \\ I(\tilde{p}_{s,j_m+1}(k)) := \max\limits_{\tilde{p}_{s,j_m+1}(k)} \dfrac{\tilde{p}_{s,j_m+1}(k)\bar{H}_{fs}^{l,j_m}(k)\left(\bar{H}_{fs}^{l,j_m}(k)\right)^T \tilde{p}_{s,j_m+1}^T(k)}{\tilde{p}_{s,j_m+1}(k)\bar{H}_{ds}^{l,j_m}(k)\left(\bar{H}_{ds}^{l,j_m}(k)\right)^T \tilde{p}_{s,j_m+1}^T(k)}. \end{cases} \tag{17.18}$$

It is clear that individual optimization problem in (17.18) can be addressed by directly using Lemma 14.2. To this end, do SVDs on $\bar{H}_{ds}^{b,j}(k)$, $j = 1, 2, \ldots, j_m$ and $\bar{H}_{ds}^{l,j_m}(k)$ at instance time k by

$$\begin{aligned} &\bar{H}_{ds}^{b,j}(k) = U_j(k)\left[S_j(k) \;\; 0\right]V_j^T(k), \; j = 1, 2, \ldots, j_m, \\ &\bar{H}_{ds}^{l,j_m}(k) = U_{j_m+1}(k)\left[S_{j_m+1}(k) \;\; 0\right]V_{j_m+1}^T(k), \\ &U_j(k)U_j^T(k) = I, \; V_j^T(k)V_j(k) = I, \; j = 1, 2, \ldots, j_m + 1. \end{aligned}$$

The optimal solution of $\tilde{p}_{s,j}(k)$ can thus be obtained as

$$\tilde{p}_{s,j}(k) = \bar{\tilde{p}}_{s,j}(k)S_j^{-1}(k)U_j^T(k) = \left[\tilde{p}_{s,j,1}(k) \;\; \tilde{p}_{s,j,2}(k)\right], \tag{17.19}$$

where $\tilde{p}_{s,j,1}(k) \in \mathbb{R}^{\gamma-1}$, $\tilde{p}_{s,j,2}(k) \in \mathbb{R}$, $\bar{\tilde{p}}_{s,j}(k)$ solves the following generalized eigenvalue-eigenvector problem

$$\begin{cases} \bar{\tilde{p}}_{s,j}(k)\left(G_j^{fs}(k) - \lambda_{m,j}(k)I\right)\bar{\tilde{p}}_{s,j}(k) = 0, \\ \bar{\tilde{p}}_{s,j}(k)\bar{\tilde{p}}_{s,j}(k)^T = 1, \;\; j = 1, 2, \ldots, j_m, \\ \bar{\tilde{p}}_{s,j_m+1}(k)\left(G_{j_m+1}^{fs}(k) - \lambda_{m,j_m+1}(k)I\right)\bar{\tilde{p}}_{s,j_m+1}(k) = 0, \\ \bar{\tilde{p}}_{s,j_m+1}(k)\bar{\tilde{p}}_{s,j_m+1}(k)^T = 1, \end{cases}$$

with

$$G_j^{fs}(k) = S_j^{-1}(k)U_j^T(k)\bar{H}_{fs}^{b,j}(\bar{H}_{fs}^{b,j})^T U_j(k)S_j^{-1}(k),$$

$$G_{j_m+1}^{fs}(k) = S_{j_m+1}^{-1}(k)U_{j_m+1}^T(k)\bar{H}_{fs}^{l,j_m+1}(\bar{H}_{fs}^{l,j_m+1})^T U_{j_m+1}(k)S_{j_m+1}^{-1}(k)$$

and $\lambda_{m,j}(k)$ being the maximal eigenvalue of matrix $G_j^{fs}(k)$ for $j = 1, 2, \ldots, j_m$, $\lambda_{m,j_m+1}(k)$ the one of $G_{j_m+1}^{fs}(k)$. At the optimum, we have

$$I(\tilde{p}_{s,j}(k)) = \lambda_{m,j}(k), \; j = 1, 2, \ldots, j_m + 1.$$

Moreover, noting from (17.15) that $\tilde{p}_{s,j}(k)$ should be in the form of

$$\tilde{p}_{s,j}(k) = [p_{s,j}(k) \;\; 1]$$

with $p_{s,j}(k)$ being the solution of $p_s(k)$ in the jth residual generator $r_{s,j}(k)$ in (17.16), we can get the optimal solution of $p_{s,j}(k)$ from (17.19) by

$$p_{s,j}(k) = \frac{\tilde{p}_{s,j,1}(k)}{\tilde{p}_{s,j,2}(k)},\ j = 1,\ 2,\ \ldots,\ j_m + 1. \tag{17.20}$$

In this regard, the residual generator (17.11) can be constructed by

$$\begin{cases} \tilde{r}_{s,j}(k) = p_{s,j}(k)\left[N_{bs}(k)y_s(k) - \bar{H}_{us}(k)u_s(k)\right], \\ r_{s,j}(k) = T^d_{\tilde{r}_{s,j}}(j,k),\ j = 1,\ 2,\ \ldots,\ j_m, \\ \tilde{r}_{s,j_m+1}(k) = p_{s,j_m+1}(k)\left[N_{bs}(k)y_s(k) - \bar{H}_{us}(k)u_s(k)\right], \\ r_{s,j_m+1}(k) = T^a_{\tilde{r}_{s,jm+1}}(j_m, k). \end{cases} \tag{17.21}$$

Refer to the recursive algorithm (17.6), the bank of residuals can be realization online recursively by

$$\begin{cases} T^d_{\tilde{r}_{s,j}}(j,k)_{j,k} = \sum\limits_n h_0(n) T^d_{\tilde{r}_{s,j}}(j-1,k)_{j,k+2^j n}, \\ T^a_{\tilde{r}_{s,jm+1}}(j_m,k)_{j,k} = \sum\limits_n l_0(n) T^a_{\tilde{r}_{s,jm+1}}(j_m-1,k)_{j,k+2^j n}. \end{cases} \tag{17.22}$$

In the stage of residual evaluation, we adopt the following bank of evaluation functions and thresholds with zero FAR

$$J(r_{s,j}(k)) = \sum_{i=k-N+1}^{k} r^T_{s,j}(i) r_{s,j}(i),\ J_{th,j} = \sup_{f(k)=0} J(r_{s,j}(k)), \tag{17.23}$$

where $j = 1,\ 2,\ \ldots,\ j_m + 1$. The occurrence of a fault can then be detected by using the following decision logic

$$\begin{cases} \exists j,\ J(r_{s,j}(k)) > J_{th,j}, & \text{fault alarm}, \\ \forall j,\ J(r_{s,j}(k)) \le J_{th,j}, & \text{no alarm}. \end{cases} \tag{17.24}$$

As a summary, the algorithm of SWT aided parity space FD for LDTV systems is summarized in Algorithm 17.4.10.

Remark 17.1 By doing SWT on the original residual signal, a bank of parity space vectors $p_{s,j}(k)$, $j = 1, 2, \ldots, j_m + 1$ are introduced to generate a group of residuals. These parity space vectors, similar to $p_s(k)$, function as a bank of bandpass filters with different center frequencies. Hence, for an identical s, faults within a wider range of frequency band can be detected by using the bank of residuals $r_{s,j}(k)$, $j = 1, 2, \ldots, j_m + 1$, which means the improvement of fault detectability.

Algorithm 17.4.10 SWT aided parity space FD for LDTV systems

1: Select appropriate wavelet basis and determine l_0 and h_0. Set parity space order $s > n$, the maximal wavelet transform scale j_m;
2: Construct matrices $H_{ds}(k)$, $H_{fs}(k)$, $H_{us}(k)$, $H_{os}(k)$ and compute $N_{bs}(k)$, $\bar{H}_{ds}(k)$, $\bar{H}_{fs}(k)$ and $\bar{H}_{us}(k)$;
3: Compute $g_{b,j}$, g_{l,j_m} and construct matrices $M^d_{b,j}$, $M^f_{b,j}$, M^d_{l,j_m}, M^f_{l,j_m}, $\bar{H}^{b,j}_{ds}(k)$, $\bar{H}^{b,j}_{fs}(k)$, $\bar{H}^{l,j_m}_{ds}(k)$, $\bar{H}^{l,j_m}_{fs}(k)$, $j = 1, 2, \cdots, j_m + 1$;
4: Compute the optimal solutions $\tilde{p}_{s,j}(k)$, $j = 1, 2, \cdots, j_m + 1$ with (17.19) by solving the generalized eigenvalue-eigenvector problems;
5: Compute the optimal solutions $p_{s,j}(k)$ with (17.20) and then compute $r_{s,j}(k)$ in (17.21) with (17.22) for $j = 1, 2, \cdots, j_m + 1$;
6: Perform residual evaluation by using (17.23) and (17.24);
7: At time step $k + 1$, assign $k + 1 \rightarrow k$, repeat the above steps. During this procedure, Algorithm 3.3.1 can be applied to update the matrices $H_{ds}(k)$, $H_{fs}(k)$, $H_{us}(k)$ and $H_{os}(k)$.

17.4.3 Selection of Wavelet Basis and the Maximal Scale

It is worth mentioning that the crux of using SWT to improve the performance of parity space-based FD with lower parity space order s lies in the application of multiscale narrow-band filters to the parity relation based residual and the recursive algorithm of SWT. To achieve a satisfactory performance criteria in (17.17) and an acceptable online implementation cost simultaneously, the selection of wavelet basis and the determination of the maximum scale j_m for SWT, undoubtedly, are significant. For this consideration, in this subsection we investigate the selection of wavelet basis and the maximum scale j_m for SWT.

Firstly, the wavelet basis should be selected according to the mathematics properties of wavelet basis [1, 6] remembering the requirements for both satisfactory FD performance and an easy online realization. The following two points should be emphasized:

- The selected wavelet basis should be self-similar with the fault signal as much as possible such that the fault information could be well extracted for FD purpose. This is required for the fact that wavelet transform coefficients measure the similarity between the wavelet basis and the underlying residual signal $r_s(k)$ subject to fault. In case of unknown type of fault in prior, a trail and error procedure can be carried out to determine an appropriate wavelet basis.
- The size of compact support of the wavelet basis should be as small as possible to achieve a lower computational expenses and smaller delay of FD time. Because a smaller size of compact support implies a smaller number of nonzero coefficients of the filters h_0 and l_0, which obviously delivers a lower computation cost regarding updating the parameter matrices in (17.17) and online SWT. Simultaneously, a smaller time delay of FD can be expected.

In practice, the family of Daubechies (dbN) discrete wavelet basis are usually adopted in FD field thanks to its properties of orthogonality and compact supports, e.g. $db1$, $db2$, $db5$, etc., as shown in Fig. 17.1. Especially, $db1$ is a popular option

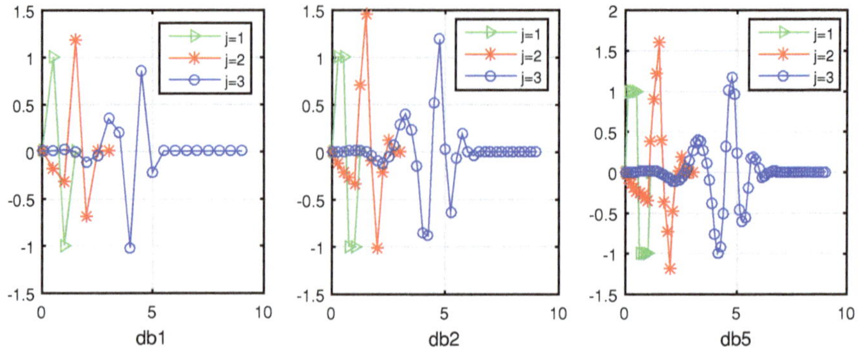

Fig. 17.1 Daubechies (dbN) wavelet basis for $N = 1, 2, 5$ at scale $j = 1, 2, 3$

for singularity signal detection due to the shortest size of compact support with $l_0 = \left\{1/\sqrt{2}, 1/\sqrt{2}\right\}, h_0 = \left\{1/\sqrt{2}, -1/\sqrt{2}\right\}$.

Secondly, as stated before, one problem encountered in parity space-based FD for LDTV systems is the heavy online computation cost concerning updating matrices $\bar{H}_{os}(k)$, $\bar{H}_{us}(k)$, $\bar{H}_{ds}(k)$, $\bar{H}_{fs}(k)$ and solving multiple optimization problems (17.17) at each time instance. Regarding updating matrices $\bar{H}_{os}(k)$, $\bar{H}_{us}(k)$, $\bar{H}_{ds}(k)$, $\bar{H}_{fs}(k)$, Algorithm 3.3.1 can be applied. When the wavelet basis is selected, the sizes of filters l_0 and h_0 are determined. It is seen from (17.17) that the computation load depends on the maximal scale j_m. More specifically, a smaller criterion $I(p_{s,j}(k))$ can be achieved with a larger j_m, while a higher computation cost and a slower response speed to FD. For these observations, we can generally compute the performance criterion (17.10) by using the traditional parity space-based method for a given parity space order s. The obtained result can be regarded as an acceptable reference value. Then determining a j_m such that the minimum criterion (17.17) is no larger than the reference value while the online computation load is acceptable.

17.5 A Numerical Example

To illustrate the effectiveness of the presented FD approach, we consider an LDTV system (3.3) with

$$A(k) = \begin{bmatrix} 0.25 & 0.75 - e^{-k/100} & 0 \\ 0 & -0.6 & 0 \\ 1 - 0.1\cos(0.1k) & 0 & 0.5 \end{bmatrix}, B(k) = \begin{bmatrix} 1 \\ 1 \\ 0 \end{bmatrix}$$

$$C(k) = \begin{bmatrix} 1 \\ 0 \\ 0.5 - \frac{k}{k+1} \end{bmatrix}^T, B_d(k) = \begin{bmatrix} 0.8 & 0.2 \\ 0.3 & 0.5 \\ 0 & 0.6 \end{bmatrix}, B_f(k) = \begin{bmatrix} 1 \\ 0 \\ 1 \end{bmatrix},$$

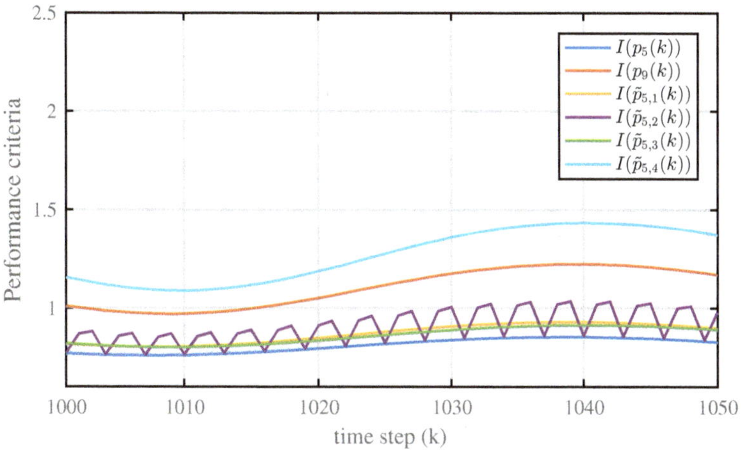

Fig. 17.2 Performance indices $I(\tilde{p}_{s,j}(k)),\ j = 1, 2, \ldots, j_m + 1\ (s = 5,\ j_m = 3)$ using the SWT aided approach and $I(p_s(k))$ ($s = 5$ and $s = 9$) using the traditional parity space method

$$D_d(k) = [0 \quad 0.5],\ \ D_f(k) = 0.25,\ \ D(k) = 0.$$

Let $u(k) = 1, d(k) = [d_1(k) \ \ d_2(k)]^T$, where $d_1(k) \sim \mathcal{U}[-0.05, 0.05]$ is a uniformly distributed random signal, $d_2(k) = 0.1\cos(k)$. For verification purpose, we consider the following faulty case

$$f(k) = \begin{cases} 0.1, & k \in [500, 1000], \\ 0.2sin(0.2\pi k), & k \in [1500, 2000], \\ 0.2sin(0.4\pi k), & k \in [2500, 3000], \\ 0, & \text{others}. \end{cases} \tag{17.25}$$

To achieve satisfactory FD performance with acceptable computation cost, we adopt the $db1$ wavelet as the wavelet basis. Set $s = 5,\ j_m = 3$. The performance indices $I(\tilde{p}_{s,j}(k)),\ j = 1, 2, \ldots j_m + 1$ in (17.18) can be obtained. The results are demonstrated in Fig. 17.2, in which the values of index $I(p_s(k))$ in (17.10) with $s = 5$ and $s = 9$ by using the traditional parity space-based method are also given. It is seen that the criterion $I(\tilde{p}_{5,4}(k))$ is larger than $I(p_5(k))$ and $I(p_9(k))$, which means that, the utilization of SWT improves the fault sensitivity and the robustness performance of traditional parity space based FD system with small parity space order.

For online FD, we set $N = 10$ and perform FD by using Algorithm 17.4.10. As a comparison, the traditional parity space-based method is also simulated with $J(r_s(k)) = ||r_s(k)||_{2,N},\ J_{th,tra} = \sup_{f=0} ||r_s(k)||_{2,N}$ for $s = 5$ and $s = 9$, respectively. The FD results are given in Figs. 17.3 and 17.4. We can see that the SWT aided method can detect the faults well by virtue of a lower missed detection rate in comparison with the traditional parity space-based method.

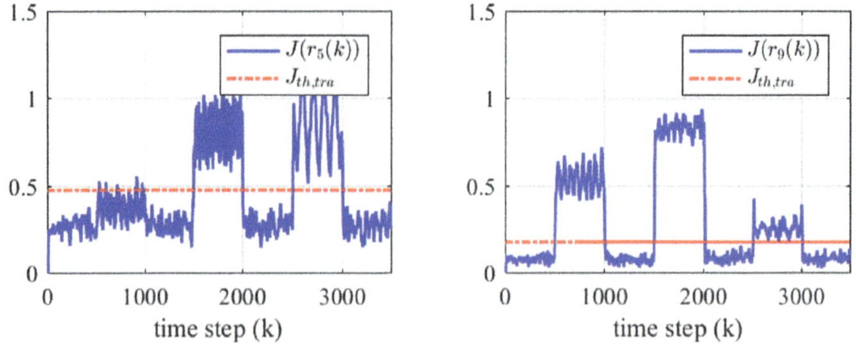

Fig. 17.3 FD results using traditional parity space-based method with $s = 5$ and $s = 9$

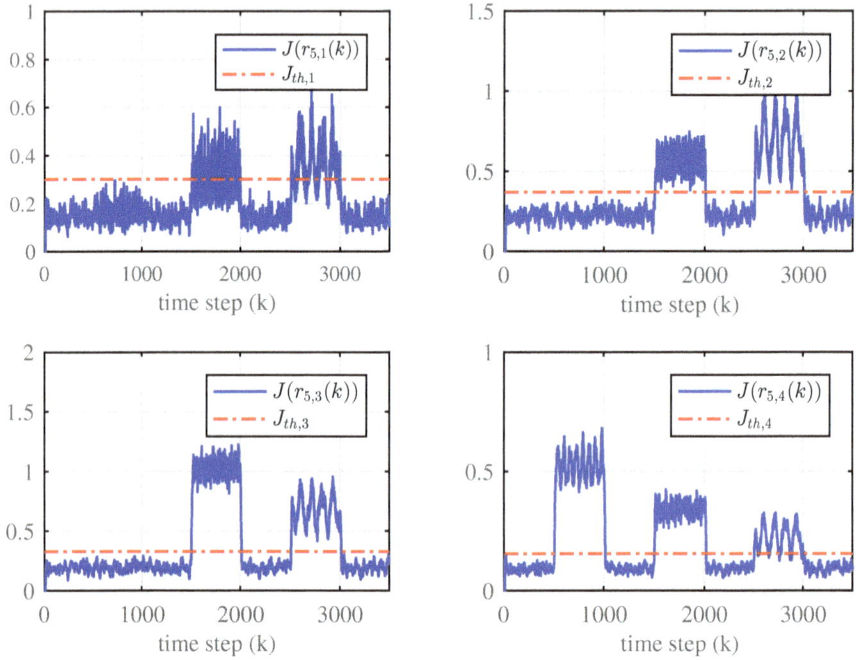

Fig. 17.4 FD results using SWT aided method with $s = 5$, $j_m = 3$

To show the outperformed robustness of SWT aided solution over the traditional parity space-based method, we consider the case of $d_1(k) \sim \mathcal{U}[-0.15\ 0.15]$. The FD results for faulty situation (17.25) are shown in Figs. 17.5 and 17.6. We can see that the performance of traditional parity space-based method becomes poor. In comparison, the SWT aided method can detect faults well with an acceptable rates of missed detection.

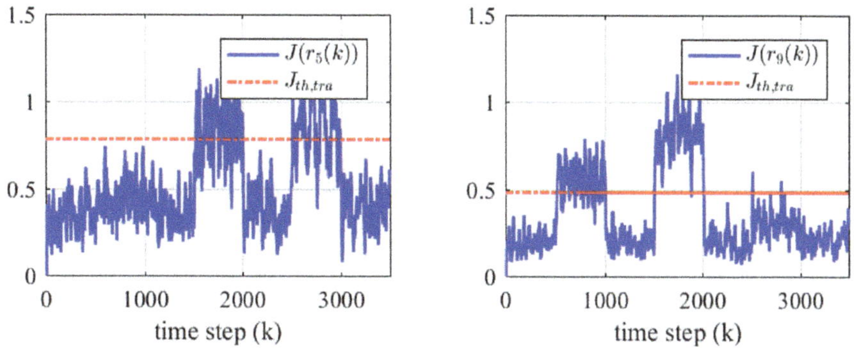

Fig. 17.5 FD results using traditional parity space-based method with $s = 5$ and $s = 9$ for the case with $d_2 \sim \mathcal{U}[-0.15,\ 0.15]$

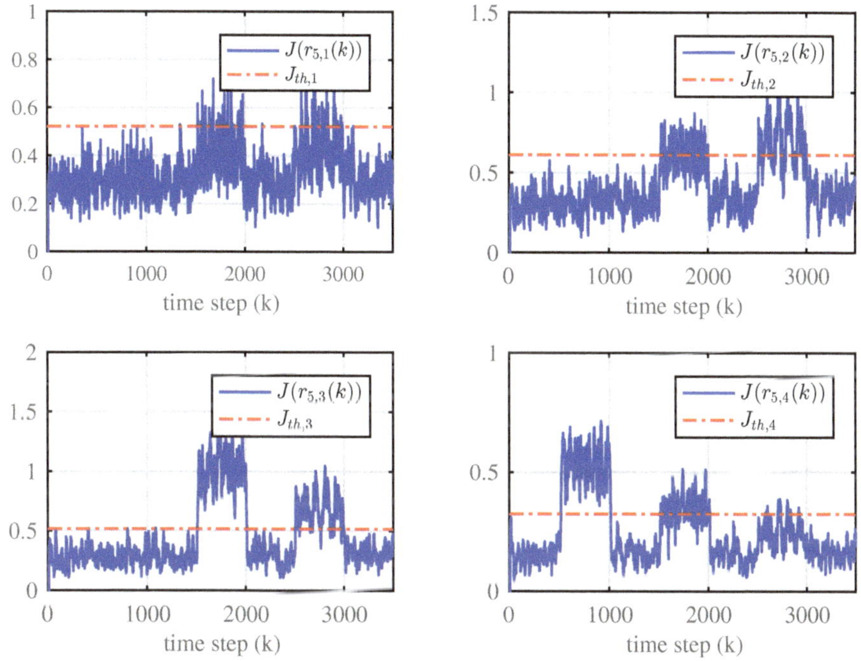

Fig. 17.6 FD results using SWT aided method with $s = 5$, $j_m = 3$ for the case with $d_2 \sim \mathcal{U}[-0.15,\ 0.15]$

17.6 Conclusion

This chapter has focused on an SWT aided parity space approach to FD for LDTV systems, aiming at achieving good FD performance with smaller online computation cost. To this end, SWT on the parity relation based residual signal has been first performed. Then the residual generation issue has been formulated as the selection of wavelet basis for SWT and the design of multiscale parity space vectors in the sense of maximizing the sensitivity/robustness criteria. By solving the group of maximization problems via SVD, a bank of residual generators have been constructed and a recursive online realization algorithm has been developed. In such a way, the developed scheme can achieve satisfactory sensitivity/robustness criteria for a smaller parity space order with acceptable computation cost. Simultaneously, with a bank of residuals being applied, faults in a wider frequency band can be detected, which implies that the SWT aided parity space approach outperforms the traditional method due to better fault detectability with an identical FAR.

References

1. Altay, Ö., & Kalenderli, Ö. (2014). Wavelet base selection for de-noising and extraction of partial discharge pulses in noisy environment. *IET Science, Measurement & Technology, 9*(3), 276–284.
2. Botre, C., Mansouri, M., Karim, M. N., et al. (2017). Multiscale PLS-based GLRT for fault detection of chemical processes. *Journal of Loss Prevention in the Process Industries, 46,* 143–153.
3. Ding, S. X. (2013). *Model-based fault diagnosis techniques: Design schemes, algorithms, and tools* (2nd ed.). London, U.K.: Springer.
4. Ding, S. X., Li, L., & Krüger, M. (2019). Application of randomized algorithms to assessment and design of observer-based fault detection systems. *Automatica, 107,* 175–182.
5. Gilles, J. (2013). Empirical wavelet transform. *IEEE transactions on signal processing, 61*(16), 3999–4010.
6. Mallat S. (2009). *A wavelet tour of signal processing*. 3rd edn., Elsevier.
7. Nason, G. P., & Silverman, B. W. (1995). *The stationary wavelet transform and some statistical applications* (pp. 281–299). New York, USA: Springer.
8. Parikh, U. B., Das, B., & Maheshwari, R. P. (2008). Combined wavelet-SVM technique for fault zone detection in a series compensated transmission line. *IEEE Transactions on Power Delivery, 23*(4), 1789–1794.
9. Renaud, O., Starck, J.-L., & Murtagh, F. (2005). Wavelet-based combined signal filtering and prediction. *IEEE Transactions on Systems, Man, and Cybernetics, Part B (Cybernetics), 35*(6), 1241–1251.
10. Sundaravaradan, N., Meyur, R., Rajaraman, P. et al. (2016). A wavelet based novel technique for detection and classification of parallel transmission line faults. In *Proceedings of the 2016 International Conference on Signal Processing, Communication, Power and Embedded System (SCOPES)*, October 03–05, Paralakhemundi, India (pp. 1951–1955).
11. Tahani, B., Boumedyen, B., Naceur, A. M., et al. (2017). Multiple fault detection based on wavelet denoising: Application on wind turbine system. In *Proceedings of the 25th Mediterranean Conference on Control and Automation*, July 03–06, Valletta, Malta (pp. 419–423).

12. Ye, H., & Wang, Y. (2006). Application of parity relation and stationary wavelet transform to fault detection of networked control systems. In *Proceedings of the 1st IEEE Conference on Industrial Electronics and Applications*, May 24–26, Singapore (pp. 1–6).
13. Ye, H., Zhang, P., Ding, S. X., et al. (2000). A time-frequency domain fault detection approach based on parity relation and wavelet transform. In *Proceedings of the 39th IEEE Conference on Decision and Control*, December 12–15, Sydney, NSW, Australia (Vol. 4, pp. 4156–4161).
14. Ye, H., Ding, S. X., & Wang, G. (2002). Integrated design of fault detection systems in time-frequency domain. *IEEE Transactions on Automatic Control, 47*(2), 384–390.
15. Ye, H., Wang, G., Ding, S. X., et al. (2002). An IIR filter based parity space approach for fault detection. *IFAC Proceedings Volumes, 35*(1), 155–160.
16. Ye, H., Wang, G., & Ding, S. X. (2004). A new parity space approach for fault detection based on stationary wavelet transform. *IEEE Transactions on Automatic Control, 49*(2), 281–287.

Part V
Applications in Discrete-Time Nonlinear Systems

Chapter 18
Extended H_-/H_∞-Optimal Fault Detection for a Class of Nonlinear Systems

The objective of this chapter is to extend H_-/H_∞-optimal FD approach to a class of discrete-time nonlinear systems subject to unknown inputs. To this end, an extended H_-/H_∞ optimization problem is formulated to the design of an FDF-based residual generator and a solution is derived by recursive computation that is feasible both for the case with stochastic disturbances and $l_{2,[0,N]}$-norm bounded unknown input. It is further shown that the local optimal solution is not unique and, for an arbitrary feasible filter gain matrix, a dynamic post-filter is also presented. In the stage of residual evaluation, an evaluation function for robust FD in the framework of l_2-norm estimation of unknown input is developed. Such an estimation leads to a unified design of evaluation function for FD using optimal H_-/H_∞-FDF or extended Kalman filter based FDF. Moreover, a randomized algorithms aided probabilistic approach to threshold determination and FD performance verification is suggested towards a trade-off between FAR and FDR. In the end, we demonstrate the applicability of the presented results through a nonlinear UAV control system.

18.1 Introduction

Nonlinear plants exist widely in industrial systems such as chemical processes, manufacturing systems, power systems, aerocrafts, economic systems, and communication systems. Study on fault diagnosis for nonlinear systems has also attracted increasing attention over the past years, see e.g., [1–3, 6–8, 10, 12, 15, 18, 20]. As well known, for perfect residual generation purpose, unknown input decoupling is a standard technique and can be achieved by using eigenstructure assignment technique or, alternatively, unknown input observer schemes. However, exactly decoupling of unknown input is usually not achievable for a practical system due to the existence of modeling errors and nonlinearity, such as a UAV system [9].

M. Zhong et al., *Fault Diagnosis for Linear Discrete Time-Varying Systems and Its Applications*, https://doi.org/10.1007/978-981-19-5438-2_18

Under the assumption of state perturbations and measurement noise having known Guassian probability distributions, an extended Kalman filter (EKF) for nonlinear system can be usually used as a residual generator. The innovation generated by EKF can thus be monitored by statistical tests. For the case of unknown input being l_2-norm bounded, the problem of robust FD can be considered in the framework of H_∞ norm and addressed by using the techniques of H_∞ filtering and H_∞-optimization. In [2, 7, 14, 18], FD observers were designed for nonlinear systems described by fuzzy Takagi-Sugeno models. References [3, 15] dealt with the problem of robust FD for nonlinear systems satisfying Lipschitz conditions by extension of linear observer. References [12, 20] dealt with the problem of FD for LPV systems by using techniques of LMI and, therefore, the solutions were conservative. More recently, in [1] the problem of FD for discrete-time nonlinear systems was investigated by using mixed H_-/H_∞ optimization and sufficient conditions for the solvability were provided in the form of two coupled Hamilton-Jacobi inequalities. It should be pointed out that obtaining an explicit relation for suitable filter gain matrix for nonlinear systems is a challenging task. Moreover, similar to the case of linear system, once an FDF is designed and l_2-norm based evaluation function is applied for FD purpose, a threshold is usually determined by taking into account the worst-case l_2-norm boundedness of the unknown inputs and, hence, it leads to a zero FAR but poorer FDR on the other hand. In order to achieve FD with acceptable FAR and FDR, randomized algorithms for analysis and control of uncertain systems in [17] provides us with a power tool.

As an extended application of the H_-/H_∞-optimal fault diagnosis methods presented in the former chapters, our main attention in this chapter is paid to an extended H_-/H_∞-optimal approach to FD for a class of nonlinear discrete-time systems, that involves the following issues:

- an extended H_-/H_∞-FDF is constructed and a recursive algorithm is correspondingly developed for online realization;
- a dynamic post-filter is presented concerning the nonuniqueness of the local optimal H_-/H_∞-FDF;
- an evaluation function is constructed by directly using a minimal estimation of unknown input and a threshold is determined for FD based on a randomized algorithms aided probabilistic analysis with an achievable FAR level.

The applicability of the obtained results is also demonstrated based on a UAV control system.

18.2 Problem Formulation

Consider the following nonlinear discrete-time system

$$\begin{cases} x(k+1) = a(x(k), u(k)) + g_d(x(k))d(k) + g_f(x(k))f(k), \\ y(k) = h(x(k)) + h_f(x(k))f(k) + v(k), x(0) = x_0, \end{cases} \tag{18.1}$$

where $x(k) \in \mathbb{R}^n$, $u(k) \in \mathbb{R}^p$, $y(k) \in \mathbb{R}^q$, $d(k) \in \mathbb{R}^m$, $v(k) \in \mathbb{R}^q$ and $f(k) \in \mathbb{R}^l$ denote the state, control input, measurement output, unknown input, measurement noise and the fault vectors, respectively. It is assumed that a, h, g_d, g_f, and h_f are smooth functions and their required partial derivatives exist.

For residual generation purpose, the following observer-based FDF is considered

$$\begin{cases} \hat{x}(k+1) = a(\hat{x}(k), u(k)) + L(k)\tilde{y}(k), \\ \tilde{y}(k) = y(k) - h(\hat{x}(k)), \ \hat{x}(0) = 0, \\ r(k) = W(k)\tilde{y}(k), \end{cases} \tag{18.2}$$

where $r(k)$ is the residual, $L(k)$ and $W(k)$ are the observer gain matrix and post-filter to be determined.

Let $e(k) = x(k) - \hat{x}(k)$. Expanding the nonlinear functions in Taylor series about the state estimate $\hat{x}(k)$, respectively, it yields

$$\begin{aligned} a(x(k), u(k)) &= a(\hat{x}(k), u(k)) + \frac{\partial a}{\partial x}|_{x(k)=\hat{x}(k)} e(k) + ..., \\ h(x(k)) &= h(\hat{x}(k)) + \frac{\partial h}{\partial x}|_{x(k)=\hat{x}(k)} e(k) + ..., \\ g_d(x(k)) &= g_d(\hat{x}(k)) + ..., \ g_f(x(k)) = g_f(\hat{x}(k)) + ..., \\ h_f(x(k)) &= h_f(\hat{x}(k)) + ..., \end{aligned}$$

where, for any function $a(x, u)$ with known u, $\frac{\partial a}{\partial x} = \left[\frac{\partial a_i}{\partial x_j}\right] \in R^{n \times n}$ denotes the Jacobian matrix of $a(x, u)$. Dropping the high order terms in the above Taylor series expansion for the nonlinearities and letting

$$F(k) = \frac{\partial a}{\partial x}|_{x(k)=\hat{x}(k)}, \ C(k) = \frac{\partial h}{\partial x}|_{x(k)=\hat{x}(k)}, \tag{18.3}$$

$$G_f(k) = g_f(\hat{x}(k)), \ G_d(k) = g_d(\hat{x}(k)), \ H_f(k) = h_f(\hat{x}(k)), \tag{18.4}$$

we then have

$$\begin{aligned} a(x(k), u(k)) &\simeq a(\hat{x}(k), u(k)) + F(k)e(k), \\ h(x(k)) &\simeq h(\hat{x}(k)) + C(k)e(k), \\ g_d(x(k)) &\simeq G_d(k), \ g_f(x(k)) \simeq G_f(k). \end{aligned}$$

Subtracting (18.2) from (18.1) yields the following approximation of estimation error dynamics

$$\begin{cases} e(k+1) = F_L(k)e(k) + G_d(k)d(k) - L(k)v(k) + G_{Lf}(k)f(k), \\ \tilde{y}(k) = C(k)e(k) + v(k) + H_f(k)f(k), \\ r(k) = W(k)\tilde{y}(k), \\ e(0) = x_0, \end{cases} \tag{18.5}$$

where

$$F_L(k) = F(k) - L(k)C(k),\ \ G_{Lf}(k) = G_f(k) - L(k)H_f(k).$$

In the stage of residual evaluation, an evaluation function J_N of residual over a finite time range (e.g., $[0, N]$ or $[k-N, k]$), and an appropriate threshold J_{th} should be determined towards satisfactory FD performance, such that the occurrence of a fault can be detected by performing

$$\begin{cases} J_N \geq J_{th}, & \text{fault alarm,} \\ J_N < J_{th}, & \text{no alarm.} \end{cases} \tag{18.6}$$

It is noteworthy that the design of an extended H_-/H_∞-optimal FD system for nonlinear plant (18.1) means the design of $W(k),\ L(k)$ in (18.2) for an FDF-based residual generation, the residual evaluation function J_N and the threshold J_{th}.

18.3 Extended H_-/H_∞-Optimal Fault Detection

In this section, we demonstrate two design schemes respectively concerning the unknown input $d(k)$ being stochastic noises and $l_{2,[0,N]}$-norm bounded disturbances. Then these two cases are further unified by converting the FD problem as into a direct construction issue of residual evaluation function.

18.3.1 An EKF Scheme for Stochastic Disturbances

Firstly, we consider the case that $d(k)$ and $v(k)$ are white noises with

$$\mathbb{E}\left[\begin{bmatrix} d(i) \\ v(i) \\ x_0 \end{bmatrix}\begin{bmatrix} d(i) \\ v(i) \\ x_0 \end{bmatrix}^T\right] = \begin{bmatrix} Q(i)\delta_{ij} & & \\ & R(i)\delta_{ij} & \\ & & I \end{bmatrix},$$

where δ_{ij} is the Kronecker delta function (identically zero except when $i = j$) and $Q(i) > 0,\ R(i) > 0$. By applying the well-known EKF, matrices $L(k)$ and $W(k)$ in (18.2) can be designed as follows

$$L(k) = F(k)P(k)C^T(k)R_{\tilde{y}}^{-1}(k), \tag{18.7}$$

$$R_{\tilde{y}}(k) = C(k)P(k)C^T(k) + R(k), \tag{18.8}$$

$$\begin{aligned} P(k+1) = &F(k)P(k)F(k) - L(k)R_{\tilde{y}}(k)L^T(k) \\ &+ G_d(k)Q(k)G_d^T(k),\ \ P(0) = I, \end{aligned} \tag{18.9}$$

$$W(k) = R_{\tilde{y}}^{-1/2}(k). \tag{18.10}$$

In fault-free case, the residual $r(k)$ is also a white noise with $\mathbb{E}[r(i)r^T(j)] = I\delta_{ij}$. One of the simplest possible FD strategy is to choose an evaluation function as

$$J_N = \sum_{k=0}^{N} r^T(k)r(k) \tag{18.11}$$

and a threshold with an acceptable FAR $\alpha > 0$ as

$$J_{th,1} = \chi^2_\alpha((N+1)q).$$

Thus, the occurrence of a fault can be detected by applying the following χ^2-test

$$\begin{cases} J_N \geq J_{th,1}, & \text{fault alarm,} \\ J_N < J_{th,1}, & \text{no alarm.} \end{cases}$$

The solution $\{J_N,\ J_{th,1}\}$ achieves an optimal FD defined in Definition 2.2.

18.3.2 An Extended H_-/H_∞ Scheme for Norm-Bounded Disturbances

Consider the case that disturbances are $l_{2,[0,N]}$-norm bounded, i.e., $d(k),\ v(k) \in l_{2,[0,N]}$ and

$$x_0^T x_0 + \sum_{k=0}^{N} (d^T(k)d(k) + v^T(k)v(k)) \leq \delta^2.$$

Let $w(k) = \left[\, d^T(k)\ v^T(k) \,\right]^T$. Define

$$\|G_{rw}\|_{\infty,[0,N]} = \sup_{f(k)=0} \frac{\sum_{k=0}^{N} r(k)^T r(k)}{x_0^T x_0 + \sum_{k=0}^{N} w(k)^T w(k)},$$

$$\|G_{rf}\|_{-,[0,N]} = \inf_{x_0=0,\,w(k)=0} \frac{\sum_{k=0}^{N} r(k)^T r(k)}{\sum_{k=0}^{N} f(k)^T f(k)},$$

where $\|G_{rw}\|_{\infty,[0,N]}$ is used as the robustness index of residual against unknown input in worst case and $\|G_{rf}\|_{-,[0,N]}$ is an index of worst-case fault sensitivity. An H_-/H_∞-optimal FDF for nonlinear system (18.1), i.e., the so-called extended H_-/H_∞-FDF for FD, can then be designed by solving

$$\max_{L(k),W(k)} \frac{\|G_{rf}\|_{-,[0,N]}}{\|G_{rw}\|_{\infty,[0,N]}}. \tag{18.12}$$

Noticing that the error dynamics (18.5) is an LDTV system, we can design $L(k)$ and $W(k)$ by applying the results in Chap. 6. A local optimal H_-/H_∞-FDF satisfying (18.12) is derived next, which is the so-called extended H_-/H_∞-FDF. Let

$$E_d(k) = [G_d(k)\ \ 0],\ \ F_d(k) = [0\ \ I],$$
$$G_{Ld} = E_d(k) - L(k)F_d(k) = [G_d(k)\ \ -L(k)].$$

The dynamics of residual generator in (18.5) can be further rewritten as follows

$$\begin{cases} e(k+1) = F_L(k)e(k) + G_{Ld}(k)w(k) + G_{Lf}(k)f(k), \\ \tilde{y}(k) = C(k)e(k) + F_d(k)w(k) + H_f(k)f(k), \\ r(k) = W(k)\tilde{y}(k), \\ e(0) = x_0. \end{cases} \tag{18.13}$$

First, we consider the case of $L(k)$ being given by

$$L(k) = (F(k)P(k)C^T(k) + E_d(k)F_d^T(k))R_{\tilde{y}}^{-1}(k), \tag{18.14}$$

where

$$R_{\tilde{y}}(k) = C(k)P(k)C^T(k) + F_d(k)F_d^T(k) \tag{18.15}$$
$$P(k+1) = F(k)P(k)F^T(k) + E_d(k)E_d^T(k), \\ - L(k)R_{\tilde{y}}(k)L^T(k),\ \ P(0) = I. \tag{18.16}$$

Let

$$\Phi_L(j,k) = F_L(j-1)F_L(j-2)\cdots F_L(k),$$
$$\Phi_L(j,j) = I,\ \ j = 1, 2, \ldots, N;\ \ k = 0, 1, \ldots, j-1,$$
$$f_N = \left[f^T(0)\ f^T(1)\ \cdots\ f^T(N)\right]^T,$$
$$w_N = \left[w^T(0)\ w^T(1)\ \cdots\ w^T(N)\right]^T,$$
$$r_N = \left[r^T(0)\ r^T(1)\ \cdots\ r^T(N)\right]^T.$$

We can rewrite (18.13) as

$$r_N = W_N(\Omega x_0 + \Gamma_w w_N + \Gamma_f f_N), \tag{18.17}$$

where

$$W_N = \mathrm{diag}(W(0), W(1), \ldots, W(N)),\ \ \Omega = \begin{bmatrix} C(0) \\ C(1)\Phi_L(1,0) \\ \vdots \\ C(N)\Phi_L(N,0) \end{bmatrix},$$

$$\Gamma_w = [\Gamma_w(j,k)]_{(N+1)\times(N+1)}, \ \Gamma_w(j,j) = F_d(j-1),$$
$$\Gamma_w(j,k) = C(j-1)\Phi_L(j-1,k)G_{Ld}(k-1) \text{ for } k < j,$$
$$\Gamma_w(j,k) = 0 \text{ for } k > j,$$

and Γ_f is constructed by replacing $\{G_{Ld}(k), F_d(j)\}$ in Γ_w with $\{G_{Lf}(k), G_f(j)\}$. It is clear that

$$\sup_{f(k)=0} \frac{\sum_{k=0}^{N} r(k)^T r(k)}{\sum_{k=0}^{N}(x_0^T x_0 + w(k)^T w(k))} = \|W_N \begin{bmatrix} \Omega \ \Gamma_w \end{bmatrix}\|_2^2,$$

$$\inf_{x_0=0,d(k)=0} \frac{\sum_{k=0}^{N} r(k)^T r(k)}{\sum_{k=0}^{N} f(k)^T f(k)} = \|W_N \Gamma_f\|_-^2.$$

By straightly application of Theorem 6.1 in Chap. 6, the filter gain matrix $L(k)$ given by (18.14)–(18.16) with post filter $W(k) = R_{\tilde{y}}^{-1/2}(k)$ provides a solution to the following optimization problem

$$\max_{L(k),W(k)} \frac{\|W_N \Gamma_f\|_-}{\|W_N \begin{bmatrix} \Omega \ \Gamma_w \end{bmatrix}\|_2}. \tag{18.18}$$

Moreover, with approximations of local linearization, it is clear that

$$\|G_{rw}\|_{\infty,[0,N]} = \sup_{f(k)=0} \frac{\sum_{k=0}^{N} r(k)^T r(k)}{\sum_{k=0}^{N}(x_0^T x_0 + w(k)^T w(k))} = \|W_N \begin{bmatrix} \Omega \ \Gamma_w \end{bmatrix}\|_2^2,$$

$$\|G_{rf}\|_{-,[0,N]} = \inf_{x_0=0,d(k)=0} \frac{\sum_{k=0}^{N} r(k)^T r(k)}{\sum_{k=0}^{N} f(k)^T f(k)} = |W_N \Gamma_f\|_-^2,$$

which implies that $L(k)$ and $W(k)$ deliver a local optimal H_-/H_∞-FDF satisfying (18.12), i.e., an extended H_-/H_∞-FDF. Now the following proposition is readily obtained.

Proposition 18.1 *With approximations of local linearization, the observer gain matrix $L(k)$ given by (18.14)–(18.16) and post-filter $W(k) = R_{\tilde{y}}^{-1/2}(k)$ provide a local solution for the H_-/H_∞ optimization problem (18.12). Moreover, $\|G_{rw}\|_{\infty,[0,N]} = 1$.*

As shown in Chap. 6 that the solution of optimization problem (18.32) subject to (18.13) is not unique. Consider an arbitrary gain matrix $L_d(k)$ such that $F(k) - L_d(k)C(k)$ is exponentially stable. The corresponding innovation $\tilde{y}_a(k)$ is given by

$$\begin{cases} \hat{x}_a(k+1) = a(\hat{x}_a(k), u(k)) + L_a(k)\tilde{y}_a(k), \\ \tilde{y}_a(k) = y(k) - h(\hat{x}_a(k)). \end{cases} \tag{18.19}$$

Let $e_a(k) = x(k) - \hat{x}_a(k)$. Applying linearization approximations and subtracting (18.19) from (18.1) yields

$$\begin{cases} e_a(k+1) = F_{L_a}(k)e_a(k) + G_{L_ad}(k)d(k) + G_{L_af}(k)f(k), \\ \tilde{y}_a(k) = C(k)e_a(k) + F_d(k)d(k) + H_f(k)f(k), \end{cases} \tag{18.20}$$

where

$$\begin{aligned} F_{L_a}(k) &= F(k) - L_a(k)C(k), \\ G_{L_ad}(k) &= G_d(k) - L_a(k)F_d(k), \\ G_{L_af}(k) &= G_f(k) - L_a(k)H_f(k). \end{aligned}$$

It is known from Theorem 6.2 in Chap. 6 that, there exists $\mathcal{Q}_L(k) : \tilde{y}_a(k) \to \tilde{y}(k)$ satisfying

$$\begin{cases} \eta(k+1) = (F(k) - L(k)C(k))\eta(k) + (L_a(k) - L(k))\tilde{y}_a(k), \\ \tilde{y}(k) = C(k)\eta(k) + \tilde{y}_a(k), \\ \eta(0) = 0, \end{cases} \tag{18.21}$$

where $L(k)$ is given by (18.14)–(18.16), $\tilde{y}(k)$ is the same with (18.13). By applying Proposition 18.1, innovation $\tilde{y}(k)$ with $r(k) = R_{\tilde{y}}^{-1/2}(k)\tilde{y}(k)$ provides a local optimal H_-/H_∞-FDF. Therefore, the innovation $\tilde{y}_a(k)$ with (18.21) and $r(k) = R_{\tilde{y}}^{-1/2}(k)\tilde{y}(k)$ is also a solution of local optimal H_-/H_∞-FDF, while (18.21) with $r(k) = R_{\tilde{y}}^{-1/2}(k)\tilde{y}(k)$ is the so-called dynamic post-filter. We now summarize the design of dynamic post-filter as the following proposition.

Proposition 18.2 *For an arbitrary gain matrix $L_d(k)$ such that $F(k) - L_d(k)C(k)$ is exponentially stable, the dynamic optimal post-filter $\mathcal{W}_d(k) : \tilde{y}(k) \to r(k)$ described by*

$$\begin{cases} \eta(k+1) = F_L(k)\eta(k) + (L_d(k) - L(k))\tilde{y}_a(k) \\ r(k) = R_{\tilde{y}}^{-1/2}(k)(C(k)\eta(k) + \tilde{y}_a(k)) \end{cases}$$

delivers a solution for the H_-/H_∞ optimization problem (18.12), where $L(k)$ is recursively calculated by (18.14)–(18.16).

Remark 18.1 The residual generator (18.2) is nonlinear, although Proposition 18.1 and Proposition 18.2 are obtained on the basis of Taylor approximations. When $d(k)$ is white noise, $L(k)$ given by (18.14)–(18.16) is in fact an EKF of (18.1) as discussed in Sect. 18.3.1. Moreover, if in this case the second-order term of estimation error remains large, we can regard it as a generalized unknown input which is $l_{2,[0,N]}$-norm bounded. Then, the results in Propositions 18.1 and 18.2 can be further applied also.

Regarding residual evaluation, we consider using the evaluation function given in (18.11). It holds in fault-free case that

$$J_N \leq \|G_{rw}\|_{\infty,[0,N]}(x_0^T x_0 + \sum_{k=0}^{N} w^T(k)w(k)) = x_0^T x_0 + \sum_{k=0}^{N} w^T(k)w(k) \leq \delta^2.$$

Therefore, a threshold can be chosen as $J_{th,2} = \delta^2$ towards a zero FAR by using

$$\begin{cases} J_N < J_{th,2}, & \text{no fault,} \\ J_N \geq J_{th,2}, & \text{fault alarm,} \end{cases}$$

Also, we can proven that the solution $\{J_N,\ J_{th,2}\}$ delivers an optimal FD defined in Definition 2.4.

18.4 A Unified Design Framework for Residual Evaluation

It is remarkable that $J_N = \sum_{k=0}^{N} r^T(k)r(k)$ provides us with an evaluation function for both the EKF based FD and extended H_-/H_∞-optimal FD, while a threshold should be determined according to a prior knowledge of unknown input $d(k)$ and $v(k)$. However, since the probability distribution of $d(k)$ and $v(k)$ may not be exactly known in practice, it is necessary to check the whiteness of the residuals $r(k)$ for seeing whether an EKF is working satisfactorily. In addition, for the case of l_2-norm bounded $d(k)$ and $v(k)$, the boundedness δ^2 can only be approached in the worst case. $J_{th,2}$ is no doubt a conservative choice and such an H_-/H_∞ optimization approach to FD often leads to poorer fault detectability. To address threshold determination issues towards a trade-off between FAR and fault detectability, we will present an FD scheme by directly estimating the residual evaluation function and then setting the threshold in the framework of probabilistic analysis.

Let $d_N = \left[d^T(0)\ d^T(1)\ \cdots\ d^T(N)\right]^T,\ v_N = \left[v^T(0)\ v^T(1)\ \cdots\ v^T(N)\right]^T$ and

$$J(x_0, d_N, v_N) = x_0^T x_0 + d_N^T d_N + v_N^T v_N.$$

By applying projection in Krein space, it will be shown next that $J(x_0, d_N, v_N)$ subject to (18.5) has a minimum and its value at the minimum equals to J_N. From this point of view, the generation of evaluation function can be formulated as to find a minimum of $J(x_0, d_N, v_N)$ and, based on this, a relationship between the minimum problem of $J(x_0, d_N, v_N)$ and the extended H_-/H_∞- and/or H_∞/H_∞-optimal FD can be established for nonlinear system (18.1). Furthermore, to get a less conservative determination of threshold, the problem of probabilistic analysis on evaluation function J_N is addressed for the case of $d(k),\ v(k) \in l_{2,[0,N]}$.

In the case of fault free, an approximation of estimation error dynamics in (18.5) becomes

$$\begin{cases} e(k+1) = F_L(k)e(k) + G_d(k)d(k) - L(k)v(k), \\ \tilde{y}(k) = C(k)e(k) + v(k), \\ e(0) = x_0. \end{cases} \tag{18.22}$$

It yields

$$\tilde{y}_N = \Omega x_0 + \Gamma_d d_N + \Gamma_v v_N, \tag{18.23}$$

where

$$\Gamma_v = [\Gamma_v(j,k)]_{(N+1)\times(N+1)}, \ \Gamma_v(j,j) = I, \Gamma_v(j,k) = 0 \text{ for } k > j,$$
$$\Gamma_v(j,k) = -C(j-1)\Phi_L(j-1,k)L(k-1) \text{ for } k < j$$

and Γ_d is constructed by replacing $\Gamma_v(j,k)$ and $\Gamma_v(j,j)$ in Γ_v with $C(j-1)\Phi(j-1,k)G_d(k-1)$ and 0, respectively. Thus, $J(x_0, d_N, v_N)$ subject to (18.23) can be rewritten as

$$J(x_0, d_N, v_N) = \begin{bmatrix} x_0 \\ d_N \\ \tilde{y}_N \end{bmatrix}^T \Xi_N^{-1} \begin{bmatrix} x_0 \\ d_N \\ \tilde{y}_N \end{bmatrix} = \tilde{J}(x_0, d_N, \tilde{y}_N),$$

where

$$\Xi_N = \begin{bmatrix} I & 0 & 0 \\ 0 & I & 0 \\ \Omega & \Gamma_d & \Gamma_v \end{bmatrix} \begin{bmatrix} I & 0 & 0 \\ 0 & I & 0 \\ \Omega & \Gamma_d & \Gamma_v \end{bmatrix}^T.$$

With given $\tilde{y}_N$, the minimum of quadratic form $\tilde{J}(x_0, d_N, \tilde{y}_N)$ can be obtained by applying the techniques of Krein space projection. For this purpose, introduce the following Krein space system

$$\begin{cases} \mathbf{z}(k+1) = F(k)\mathbf{z}(k) + G_d(k)\mathbf{d}(k), \\ \mathbf{y}_z(k) = C(k)\mathbf{z}(k) + \mathbf{v}(k), \\ \mathbf{z}(0) = \mathbf{z}_0, \end{cases} \tag{18.24}$$

where $\mathbf{x}_0$ and $\mathbf{d}(k)$ are vectors in Krein space with Gramian

$$\left\langle \begin{bmatrix} \mathbf{x}_0 \\ \mathbf{d}(i) \\ \mathbf{v}(i) \end{bmatrix}, \begin{bmatrix} \mathbf{x}_0 \\ \mathbf{d}(j) \\ \mathbf{v}(j) \end{bmatrix} \right\rangle = \begin{bmatrix} I & 0 & 0 \\ 0 & I\delta_{ij} & 0 \\ 0 & 0 & I\delta_{ij} \end{bmatrix}.$$

Denote by $\hat{\mathbf{z}}(k)$ the projection of $\mathbf{z}(k)$ onto subspace $\mathcal{L}\{\{\mathbf{y}_z(k)\}_{k=0}^N\}$. We then have

$$\begin{cases} \hat{\mathbf{z}}(k+1) = F(k)\hat{\mathbf{z}}(k) + K_p(k)(\mathbf{y}(k) - C(k)\hat{\mathbf{z}}(k)), \\ \tilde{\mathbf{y}}_z(k) = \mathbf{y}_z(k) - C(k)\hat{\mathbf{z}}(k), \\ \hat{\mathbf{z}}(0) = 0, \end{cases} \tag{18.25}$$

where

$$\begin{aligned} K_p(k) &= <\mathbf{z}(k+1), \tilde{\mathbf{y}}_z(k)> R_{\tilde{\mathbf{y}}_z}^{-1}(k) \\ &= F(k)P_z(k)C^T(k)R_{\tilde{\mathbf{y}}_z}^{-1}(k), \end{aligned} \tag{18.26}$$

$$R_{\tilde{\mathbf{y}}_z}(k) = C(k)P_z(k)C^T(k) + I, \tag{18.27}$$

$$\begin{aligned} P_z(k+1) &= F(k)P_z(k)F^T(k) + G_d(k)G_d^T(k) \\ &\quad - K_p(k)R_{\tilde{\mathbf{y}}_z}(k)K_p^T(k), \;\; P_z(0) = I. \end{aligned} \tag{18.28}$$

Let $\tilde{\mathbf{z}}(k) = \mathbf{z}(k) - \hat{\mathbf{z}}(k)$. Subtracting (18.25) from (18.24) yields

$$\begin{cases} \tilde{\mathbf{z}}(k+1) = F(k)\tilde{\mathbf{z}}(k) + K_p(k)\tilde{\mathbf{y}}_z(k) + G_d(k)\mathbf{d}(k), \\ \tilde{\mathbf{y}}_z(k) = C(k)\tilde{\mathbf{z}}(k) + \mathbf{v}(k), \\ \tilde{\mathbf{z}}(0) = \mathbf{z}_0. \end{cases} \tag{18.29}$$

Similar to (18.23), we can get

$$\tilde{\mathbf{y}}_N = \Omega\mathbf{x}_0 + \Gamma_d\mathbf{d}_N + \Gamma_v\mathbf{v}_N. \tag{18.30}$$

Since $\{\tilde{\mathbf{y}}_z(i)\}_{i=0}^N$ forms an orthogonal basis of subspace $\mathcal{L}\{\{\mathbf{y}_z(i)\}_{i=0}^N\}$, the Gramian of $\tilde{\mathbf{y}}_{zN}$ and cross Gramians of $\tilde{\mathbf{y}}_{zN}$, $\mathbf{x}_0$, and $\mathbf{d}_N$ can be given as

$$\begin{aligned} R_{\tilde{\mathbf{y}}_{zN}} &= \mathrm{diag}(R_{\tilde{\mathbf{y}}_z}(0), R_{\tilde{\mathbf{y}}_z}(1), \ldots, R_{\tilde{\mathbf{y}}_z}(N)) \\ &= \Omega\Omega^T + \Gamma_d\Gamma_d^T + \Gamma_v\Gamma_v^T, \\ R_{\mathbf{d}_N\tilde{\mathbf{y}}_{zN}} &= R_{\tilde{\mathbf{y}}_{zN}\mathbf{d}_N}^T = \Gamma_d, \\ R_{\mathbf{x}_0\tilde{\mathbf{y}}_{zN}} &= R_{\tilde{\mathbf{y}}_{zN}\mathbf{x}_0}^T = \Omega. \end{aligned}$$

Observing that Γ_v is full row rank, we then have $\Gamma_v\Gamma_v^T > 0$. Therefore, the minimum of $\tilde{J}(x_0, d_N, \tilde{y}_N)$ exists and is unique. Moreover, the minimum can be obtained by using the formulae of the projections of $\mathbf{x}_0$, $\mathbf{d}_N$, $\mathbf{v}_N$ onto subspace spanned by $\{\tilde{\mathbf{y}}_z(i)\}_{i=0}^N$, i.e.,

$$\hat{x}_0 = \Omega^T R_{\tilde{\mathbf{y}}_{zN}}^{-1}\tilde{y}_N, \;\; \hat{d}_N = \Gamma_d^T R_{\tilde{\mathbf{y}}_{zN}}^{-1}\tilde{y}_N, \;\; \hat{v}_N = \Gamma_v^T R_{\tilde{\mathbf{y}}_{zN}}^{-1}\tilde{y}_N,$$

$$\tilde{J}(\hat{x}_0, \hat{d}_N, y_N) = \tilde{y}_N^T R_{\tilde{\mathbf{y}}_{zN}}^{-1}\tilde{y}_N = \sum_{k=0}^{N} \tilde{y}^T(k)R_{\tilde{\mathbf{y}}_z}^{-1}(k)\tilde{y}(k).$$

Therefore, the minimal estimation of $J(x_0, d_N, v_N)$ subject to (18.5) with $f(k) = 0$ can be recursively computed by

Algorithm 18.4.11 A unified design of evaluation function

1: Calculate $F(k),\ C(k),\ G_d(k)$ by using (18.4)–(18.3);
2: Update $K_p(k),\ R_{\tilde{\mathbf{y}}_z}(k)$ and $P_z(k)$ by using (18.27)–(18.29) and let $L(k) = K_p(k)$;
3: If $R_{\tilde{\mathbf{y}}_z}(k) > 0$ is satisfied, then we calculate $\hat{x}(k+1)$ and innovation $\tilde{y}(k)$ by using (18.2);
4: Calculate J_N recursively by using (18.32).

$$J(\hat{x}_0, \hat{d}_N, \hat{v}_N) = \sum_{k=0}^{N} \tilde{y}^T(k) R_{\tilde{\mathbf{y}}_z}^{-1}(k) \tilde{y}(k).$$

In view of (18.7)–(18.10) and (18.26)–(18.28), we have

$$L(k) = K_p(k),\ W(k) = R_{\tilde{\mathbf{y}}_z}^{-1/2}(k),\ \ J(\hat{x}_0, \hat{d}_N, \hat{v}_N) = J_N$$

for $Q(k) = I$ and $R(k) = I$. It implies that $(L(k), W(k))$ provides a minimal estimate of $J(x_0, d_N, v_N)$ subject to (18.5) with $f(k) = 0$. Hence, the evaluation function J_N can be obtained as

$$J_N = \sum_{k=0}^{N} \tilde{y}^T(k) R_{\tilde{\mathbf{y}}_z}^{-1}(k) \tilde{y}(k). \tag{18.31}$$

On this basis, the following proposition is now readily to be concluded.

Proposition 18.3 *The minimum of $J(x_0, d_N, v_N)$ subject to (18.5) with $f(k) = 0$ provides us with a unified design of the evaluation function of EKF and extended H_i/H_∞-optimal FD. Moreover, such an evaluation function J_N can be computed recursively by Algorithm 18.4.11.*

We have demonstrated in the above that the design of evaluation function can be formulated as to find a minimal estimation of $J(x_0, d_N, v_N)$ subject to (18.1) and, for the case of $Q(k) = I,\ \ R(k) = I$, the estimation $J(\hat{x}_0, \hat{d}_N, \hat{v}_N)$ equals to the evaluation function of EKF based FD. We now consider the case of $d(k),\ v(k) \in l_{2,[0,N]}$. Define the best-case fault sensitivity index by

$$\|G_{rf}\|_{\infty,[0,N]} = \sup_{x_0=0, w(k)=0} \frac{\sum_{k=0}^{N} r^T(k) r(k)}{\sum_{k=0}^{N} f^T(k) f(k)}.$$

Applying the optimal FD results for LDTV systems in Chap. 6, the residual $r(k)$ generated by (18.2) with $L(k) = K_p(k),\ W(k) = R_{\tilde{\mathbf{y}}_z}^{-1/2}(k)$ satisfies both the following worst-case and best-case sensitivity/robustness ratio criteria

$$\max_{L(k), W(k)} \frac{\|G_{rf}\|_{-,[0,N]}}{\|G_{rw}\|_{\infty,[0,N]}} \ \text{and} \ \max_{L(k), W(k)} \frac{\|G_{rf}\|_{\infty,[0,N]}}{\|G_{rw}\|_{\infty,[0,N]}}, \tag{18.32}$$

while the l_2-norm type evaluation function J_N equals to $J(\hat{x}_0, \hat{d}_N, \hat{v}_N)$. Similar to the case of LDTV systems, residual generator (18.2) satisfying (18.32) is also called an extended H_i/H_∞-FDF for nonlinear system (18.1). Since $J(\hat{x}_0, \hat{d}_N, \hat{v}_N)$ can be used as an evaluation function of the EKF based FD and the extended H_i/H_∞-optimal FD, we say that Algorithm 18.4.11 gives a unified design of evaluation function.

Remark 18.2 It is interesting that $J(\hat{x}_0, \hat{d}_N, \hat{v}_N)$ can be a unified form of evaluation function both for the cases of EKF based FD and extended H_i/H_∞-optimal FD. In such a framework, the evaluation function is directly computed, whilst the stage of residual generation is not necessary again. From the viewpoint of practical application, the unified design of evaluation function is undoubtedly simpler and easier to carry out. Besides, Algorithm 18.4.11 is an extension of the results in Chap. 8 for LDTV systems to the case of nonlinear systems. Moreover, since J_N is obtained via recursive computation, Algorithm 18.4.11 is easy to implement.

18.5 Threshold Setting and Probabilistic Performance Analyzing

For a given evaluation function, threshold determination becomes crucial for FD. Recall that, under the assumption of unknown inputs being white noises, J_N generated by EKF follows approximately a central χ^2 distribution. With an accepted FAR α, the threshold can be chosen as $J_{th,1} = \chi^2_\alpha((N+1)q)$. When a fault f presents, the FDR is

$$\text{FDR} = \Pr\{J_N > J_{th,1} | f \neq 0\}.$$

When it comes to the case of $d(k), v(k) \in l_{2,[0,N]}$, under the assumption of $J(x_0, d_N, v_N) \leq \delta^2$, one widely used choice is $J_{th,2} = \delta^2$. However, such a zero FAR strategy may lead to poorer FDR. In order to improve the FDR, one way is to set a threshold as $J_{th,I} < \delta^2$, yet a zero FAR cannot be guaranteed. To achieve acceptable levels of FAR and FDR, a randomized algorithms aided threshold setting and performance evaluation of FD are given in this section.

Suppose that a threshold less than δ^2 is chosen as $J_{th,3}$. An FAR is unavailable if the statistic properties of $d(k)$ and $v(k)$ are not exactly known. From the viewpoint of practical application, we can evaluate the FAR considering the following three typical cases: the probability density function (PDF) of $J(x_0, d_N, v_N)$ is known, the PDF of $J(x_0, d_N, v_N)$ is unknown, and both of the upper bound and PDF of $J(x_0, d_N, v_N)$ are unknown.

Algorithm 18.5.12 FAR for the case of knowing PDF of $J(x_0, w_N, v_N)$

1: Set a threshold $J_{th,3} < \delta^2$ and the PDF of $J(x_0, w_N, v_N)$ over $(0,\ \delta^2]$, i.e., $f_J(\eta)$;
2: Calculate the probability P_M using (18.35)
3: Return an upper bound of the achievable $FAR = P_M$.

18.5.1 Case I: The PDF of $J(x_0, d_N, v_N)$ is Known

Consider $J(x_0, d_N, v_N)$ is a random variable with known density function. The key to evaluate the FAR for a given threshold $J_{th,3}$ is the probability of $J(x_0, d_N, v_N)$ satisfying

$$\Pr\{J(x_0, d_N, v_N) > J_{th,3}\}.$$

In the case of fault free, it is known from Proposition 18.3 that

$$J(x_0, d_N, v_N) \geq J_N.$$

We then have

$$\Pr\{J_N > J_{th,3} | f = 0\} \leq \Pr\{J(x_0, d_N, v_N) > J_{th,3}\},$$

which implies that FAR $\leq \Pr\{J(x_0, d_N, v_N) > J_{th,3}\}$. Especially, under the assumption of the $l_{2,[0,N]}$-norm of unknown input being uniformly distributed over $[\delta_l^2, \delta^2]$, the FAR satisfies

$$\text{FAR} \leq \Pr\{J(x_0, d_N, v_N) > J_{th,3}\} = \frac{\delta^2 - J_{th,3}}{\delta^2 - \delta_l^2}. \tag{18.33}$$

In summary, the FAR with threshold $J_{th,3}$ can be evaluated by the following proposition. The corresponding algorithm is given in Algorithm 18.5.12.

Proposition 18.4 *For a given threshold $J_{th,3}$, under the assumption of $J(x_0, d_N, v_N)$ being distributed over $(0, \delta^2]$ with PDF $f_J(\eta)$, the FAR satisfies*

$$FAR \leq P_M$$

with

$$P_M = \Pr\{J(x_0, d_N, v_N) > J_{th,3}\} = \int_{J_{th,3}}^{\delta^2} f_J(\eta) d\eta. \tag{18.34}$$

Remark 18.3 As a special case, if the $J(x_0, w_N, v_N)$ is uniformly distributed over $[\delta_l^2, \delta^2]$, then, similar to 18.33, the upper bound of FAR is obtained as

$$P_{FAR} = \Pr\{J(x_0, w_N, v_N) > J_{th}\} = \frac{\delta^2 - J_{th}}{\delta^2 - \delta_l^2}.$$

18.5.2 Case II: The PDF of $J(x_0, d_N, v_N)$ is Unknown

Consider a more general case of probability distribution of $J(x_0, d_N, v_N)$ being unknown. Inspired by the randomized algorithms for analysis and control of uncertain systems in [17], we evaluate the FAR for a given $J_{th,I}$ by using an empirical method. For this purpose, M independent identically distributed (i.i.d.) samples of J_N are generated firstly by using

$$J_N^{(i)} = \sup_{f(k)=0} \sum_{k=k_i}^{N+k_i} \tilde{y}^T(k) R_{\bar{\mathbf{y}}_z}^{-1}(k) \tilde{y}(k), \tag{18.35}$$

where k_i is the ith sampling initial time step. Define

$$\mu(i) = \begin{cases} 1, \text{ if } J_N^{(i)} > J_{th,I} \\ 0, \text{ if } J_N^{(i)} \leq J_{th,I} \end{cases}, \quad i = 1, 2, \ldots, M. \tag{18.36}$$

Then the empirical FAR can be evaluated by

$$\hat{P}_{FAR} = \frac{1}{M} \sum_{i=1}^{M} \mu(i) \text{ for } f(k) = 0. \tag{18.37}$$

With given probability levels ϵ_I, $\beta \in (0, 1)$, it is known by applying the probabilistic performance verification algorithm in [17] that, if the well known Chernoff bound is satisfied, i.e.,

$$M \geq \frac{1}{2\epsilon_I^2} \log \frac{2}{\beta}, \tag{18.38}$$

then $\hat{P}_{FAR}$ returns with probability at least $1 - \beta$ a level the estimate of FAR such that

$$|P_{FAR} - \hat{P}_{FAR}| < \epsilon_I, \tag{18.39}$$

where

$$P_{FAR} = \Pr\{J_N > J_{th,I} | f(k) = 0\}.$$

Algorithm 18.5.13 FAR for the case of unknown PDF of $J(x_0, w_N, v_N)$

1: Given ϵ_I, $\beta \in (0, 1)$, set the number M of samples according to Chernoff bound (18.39);
2: Generate M i.i.d samples of fault-free case J_N by applying Algorithm 18.4.11 and compute

$$J_N^{(i)} = \sup_{f(k)=0} \sum_{k=k_i}^{N+k_i} r^T(k)r(k);$$

3: Estimate the empirical FAR by using (18.37)–(18.38) and then return $\hat{P}_{FAR}$.

As a result, we have the following proposition. Correspondingly, an estimate of FAR is given by Algorithm 18.5.13.

Proposition 18.5 *Given $\epsilon_I \in (0, 1)$, $\beta \in (0, 1)$ and M i.i.d samples, the $\hat{P}_{FAR}$ returns with probability at least $1 - \beta$ an estimate of P_{FAR} such that (18.39) is satisfied if (18.38) holds true.*

18.5.3 Case III: The Norm Bound δ Is Unknown

Consider the probabilistic determination of threshold in the case of δ being unknown. For M i.i.d samples of J_N, the empirical maximum of J_N can be evaluated respectively as

$$\hat{\delta}_N^2 = \max_i J_N^{(i)}. \tag{18.40}$$

For a given ϵ_{II} and $\nu \in (0, 1)$, by applying the algorithm for probabilistic worst case performance analysis in [17], the estimation $\hat{\delta}_N$ returns with probability at least $1 - \nu$ a level such that

$$\Pr\{J_N \leq \hat{\delta}_N^2 | f(k) = 0\} \geq 1 - \epsilon_{II}. \tag{18.41}$$

if the sample size M satisfies

$$M \geq \frac{\log \nu}{\log(1 - \epsilon_{II})}. \tag{18.42}$$

It means that, with confidence greater than $1 - \nu$, the J_N is below $\hat{\delta}_N^2$ with probability at least $1 - \epsilon_{II}$. Therefore, if a threshold is set as $J_{th,II} = \hat{\delta}_N^2$ and the detection of a fault is performed as

$$\begin{cases} J_N < J_{th,II}, & \text{no alarm,} \\ J_N \geq J_{th,II}, & \text{fault alarm,} \end{cases} \tag{18.43}$$

then $\Pr\{J_N > J_{th,II} | f = 0\} < \epsilon_{II}$, which implies that the FAR is less than ϵ_{II} with confidence level at least $1 - \nu$. On this basis, we have the following proposition. A probabilistic estimate of the worst-case threshold without knowing exact upper bound of unknown input is summarize in Algorithm 18.5.14.

Proposition 18.6 *Given ϵ_{II}, $\nu \in (0, 1)$ and M i.i.d samples satisfying (18.42), if a threshold is chosen as $J_{th,II}$, then $\hat{\delta}_N^2$ returns an empirical estimation of the maximum of J_N with confidence level at least $1 - \nu$, while the corresponding FAR is less than ϵ_{II}.*

Moreover, noticing that $J_{th,II}$ is an empirical maximum of J_N, the threshold $J_{th,II}$ may lead to poorer FDR. Similarly, the empirical mean and variance of J_N can be evaluated respectively as

$$\hat{J}_N = \frac{1}{M}\sum_{i=1}^{M} J_N^{(i)}, \ \mathrm{Var}(J_N) = \frac{1}{M}\sum_{i=1}^{M}(J_N^{(i)} - \hat{J}_N)^2.$$

To improve the FDR with an acceptable FAR, we can choose a threshold less than $J_{th,II}$ in the framework of empirical probability analysis, such as

$$J_{th,III} = \hat{J}_N + 3(\mathrm{Var}(\hat{J}_N))^{1/2}.$$

Further probability analysis on the possible FAR can be performed similarly to $J_{th,I}$, which is not discussed for simplicity.

As a short summary, if the stochastic properties of x_0, $d(k)$ and $\upsilon(k)$ are known, a deterministic threshold corresponding the acceptable FAR can be obtained with the aid of probability distribution of $J(x_0, w_N, \upsilon_N)$. When the PDF of J_N is not exactly known, Proposition 18.5 provides us with a way to answer if the achieved FAR is acceptable, while Proposition 18.6 gives a probabilistic worst case estimate of J_N and probabilistic estimates of threshold and FAR. Especially, if the sample data are sufficient, the empirical mean and variance of J_N can also be used as a support of threshold determination, such as the alternative choice of $J_{th,III}$.

Algorithm 18.5.14 FAR for the case of δ being unknown

1: Given ϵ_{II}, $\nu \in (0, 1)$, set the number M of samples according to (18.42);
2: Generate M i.i.d samples of fault-free case J_N by applying Algorithm 18.4.11;
3: Compute $\hat{\delta}_N^2 = \max_i J_N^{(i)}$;
4: Set $J_{th,II} = \hat{\delta}_N^2$ and perform FD by using (18.43);
5: Return a probabilistic FAR with $P_{FAR} < \epsilon_{II}$ at least confidence level $1 - \nu$.

18.6 Application to a UAV Control System

In order to show the effectiveness of the above results, simulation study on a longitudinal control system of a UAV is demonstrated in this section.

Let $V,\ \alpha,\ q,\ \theta,\ H$ be the airspeed, angle of attack, body-axis pitch rate, pitch angle and height. $\delta_z,\ \delta_{th}$ are the elevator deflection and throttle setting. $V_{dx},\ a_{dx},\ V_{dy}$ and a_{dy} stand for the wind velocity and wind acceleration of horizontal and vertical directions, respectively. Define

$$\begin{aligned} X(t) &= \begin{bmatrix} V\ \alpha\ q\ \theta\ H \end{bmatrix}^T,\ u(t) = \begin{bmatrix} \delta_z\ \delta_{th} \end{bmatrix}^T, \\ d(t) &= \begin{bmatrix} a_{dx}\ a_{dy}\ v_{dy} \end{bmatrix}^T,\ y(t) = \begin{bmatrix} V\ q\ \theta\ H \end{bmatrix}^T. \end{aligned}$$

The control system block diagram is shown in Fig. 18.1. When an elevator fault is taken into account, the longitudinal control system nonlinear model of the UAV can be given by

$$\begin{cases} \dot{X}(t) = F(X(t)) + g(X(t))u(t) + g_d(X(t))d(t) + g_f(X(t))f(t), \\ y(t) = C(t)X(t) + v(t), \\ X(0) = X_0, \end{cases}$$

where

$$F(X(t)) = \begin{bmatrix} -0.0162V^2(1.5551\alpha^2 + 0.0748\alpha - 0.0868)- \\ g\sin(\theta - \alpha) \\ -V(0.1070\alpha + 0.0141) + \frac{g\cos(\theta-\alpha)}{V} + 1.02q \\ 0.0159V^2(-0.6390\alpha + 0.0604 - \frac{2.3569}{V}) \\ q \\ V\sin(\theta - \alpha) \end{bmatrix},$$

$$g(X(t)) = \begin{bmatrix} 0 & 0.0930 \\ 0.0162V(0.0029\alpha - 0.0052) & 0 \\ 0.0159V^2(-0.0722\alpha^2 + 0.0188\alpha - 0.016) & 0 \\ 0 & 0 \\ 0 & 0 \end{bmatrix},$$

$$g_d(X(t)) = \begin{bmatrix} -\cos(\theta-\alpha) & -\sin(\theta-\alpha) & 0 \\ -\frac{\sin(\theta-\alpha)}{V} & \frac{\cos(\theta-\alpha)}{V} & 0 \\ 0 & 0 & 0 \\ 0 & 0 & 0 \\ 0 & 0 & 1 \end{bmatrix},\ C(t) = \begin{bmatrix} 1\,0\,0\,0\,0 \\ 0\,0\,1\,0\,0 \\ 0\,0\,0\,1\,0 \\ 0\,0\,0\,0\,1 \end{bmatrix},$$

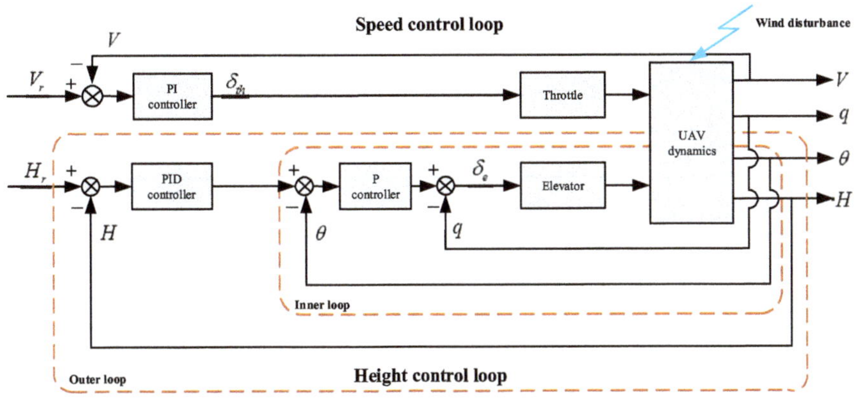

Fig. 18.1 The control block diagram of a UAV

$$g_f(X(t)) = \begin{bmatrix} 0 \\ 0.0162V(0.0029\alpha - 0.0052) \\ 0.0159V^2(-0.0722\alpha^2 + 0.0188\alpha - 0.016) \\ 0 \\ 0 \end{bmatrix}.$$

We first consider the case of $\delta^2 = 0.2$. Let the sampling period be $T = 0.02$s and $N = 100$. When a threshold is chosen as $J_{th,2} = 0.2$, the FAR is zero, but the FDR is poorer. In order to improve the FDR, we set a threshold as $J_{th,3} = 0.18$. Under the assumption of $J(x_0, d_N, v_N)$ being uniformly distributed over $[0.05, 0.2]$, it is known from Proposition 18.4 that the FAR is less than 0.13, while a threshold greater than 0.185 should be chosen for an acceptable FAR 0.1.

Next, we consider the more general case of δ being unknown exactly. The V_{dx}, V_{dy}, a_{dx}, and a_{dy} are simulated as in Fig. 18.2. The measurement noise is simulated as Gaussian signal with mean value 0 and variance 0.0001. At every time step k, the evaluation function $J_N(k)$ is computed by using Algorithm 18.4.11, i.e.,

$$J_N(k) = \sup_{f(i)=0} \sum_{i=k-N}^{k} \tilde{y}^T(i) R_{\tilde{\mathbf{y}}_z}^{-1}(i) \tilde{y}(i).$$

The method is implemented into a Simulink model built up by using Matlab Tool Box. Set $M = 10000$ and generate M samples of fault-free case J_N, i.e., $J_N^{(i)}$. It is obtained that

$$\hat{\delta}_N^2 = 0.1, \;\; \hat{J}_N = 0.0544, \;\; \text{Var}(J_N) = 1.1793 \times 10^{-4}.$$

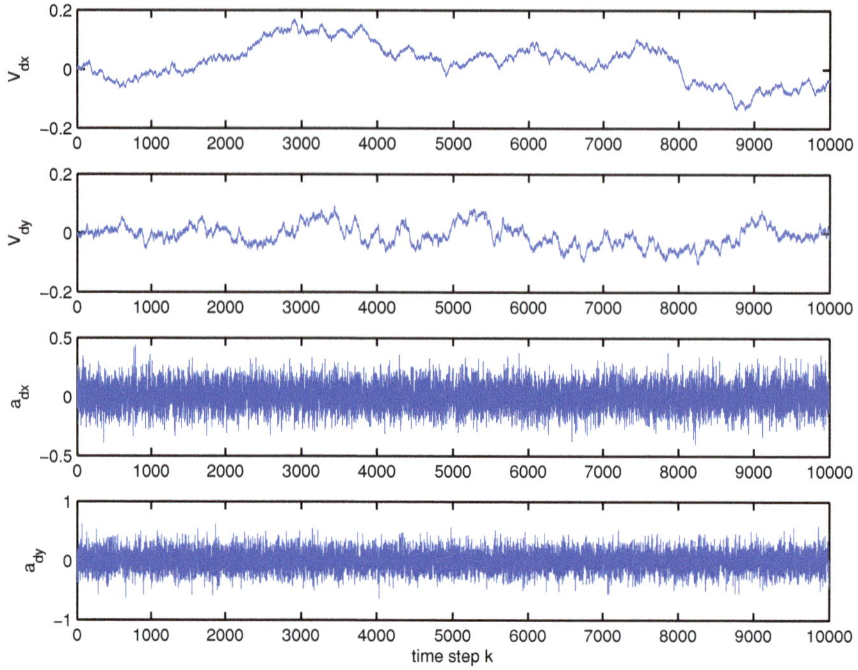

Fig. 18.2 The simulated wind velocities and accelerations

Table 18.1 The estimated values of P_{FAR}

Thresholds	0.0653	0.0870	0.1
$\hat{P}_{FAR}$	0.15	0.03	0

For the following three different thresholds

$$J_{th,I} = \hat{J}_N + (\mathrm{Var}(J_N))^{1/2} = 0.0653,$$
$$J_{th,II} = \hat{\delta}_N^2 = 0.1, \ J_{th,III} = \hat{J}_N + 3(\mathrm{Var}(J_N))^{1/2} = 0.0870,$$

we can calculate the empirical FAR, i.e., $\hat{P}_{FAR}$, which are listed in Table 18.1. By applying Proposition 18.5, the simulation returns with probability at least 0.95 an estimate of FAR such that $|P_{FAR} - \hat{P}_{FAR}| < 0.009$. Moreover, we can conclude from Proposition 18.6 that, with confidence at least 0.9995 a level, the fault-free case J_N satisfies $\Pr\{J_N \leq \hat{\delta}_N^2\} \geq 0.9992$.

Furthermore, to show the effectiveness of the selected thresholds, three different elevator faults resulted by the loss-of-effectiveness are simulated as follows.

- Case 1: 20% of the loss-of-effectiveness over $(20, 60)s$;
- Case 2: 15% of the loss-of-effectiveness over $(80, 120)s$;
- Case 3: 10% of the loss-of-effectiveness over $(140, 180)s$.

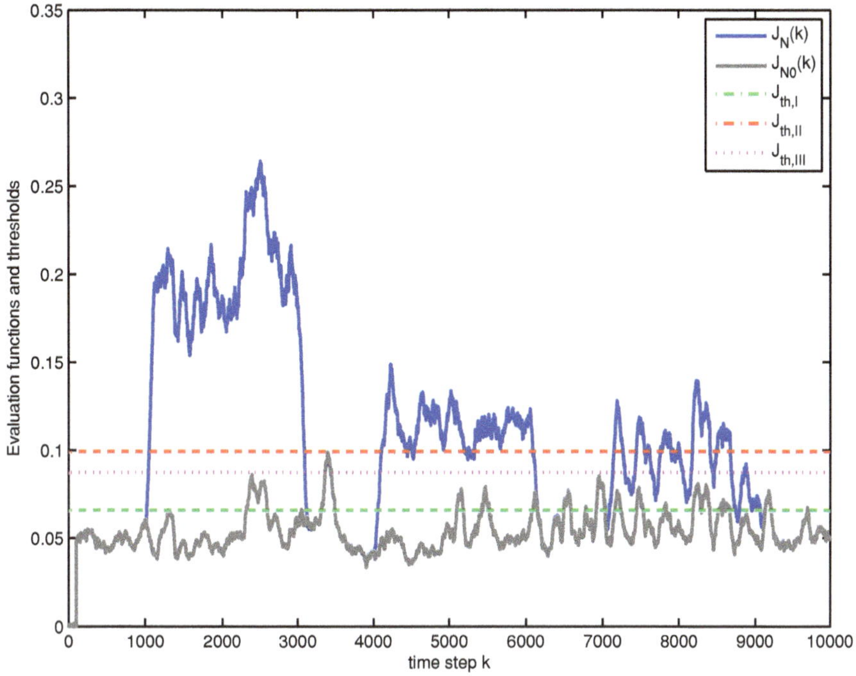

Fig. 18.3 The evolution of $J_N(k)$ and the thresholds

The evolutions of both the faulty case and fault free case $J_N(k)$ are shown in Fig. 18.3, i.e., the $J_N(k)$ and $J_{N0}(k)$. It is seen from Fig. 18.3 and Table 18.1 that the FDR of 10 percent of the loss-of-effectiveness is poorer if the threshold is chosen as $J_{th,II}$, while the FAR may not be acceptable for threshold $J_{th,I}$. So, $J_{th,III}$ is a choice of achieving some tradeoff between FAR and FDR in the simulation.

On the other hand, it is known from Proposition 18.3 that the FDF (18.2) with $L(k) = K_p(k)$, $W(k) = R_{\tilde{\mathbf{y}}_z}^{-1/2}(k)$ delivers an extended H_i/H_∞ optimization based residual generator. As a comparison, the generated residuals $r(k) = [r_1(k)\ r_2(k)\ r_3(k)\ r_4(k)]^T$ are also presented in Fig. 18.4. Although the influence of the fault in Case 3 on the residuals is not clear, the fault can be detected if threshold $J_{th,II}$ is chosen.

18.7 Conclusion

This chapter has demonstrated an extended H_-/H_∞ optimization approach to robust FD for a class of nonlinear systems. An observer-based FDF has been constructed as a residual generator, the design of which has been formulated as an H_-/H_∞ optimization problem towards maximizing the sensitivity/robustness ratio criterion.

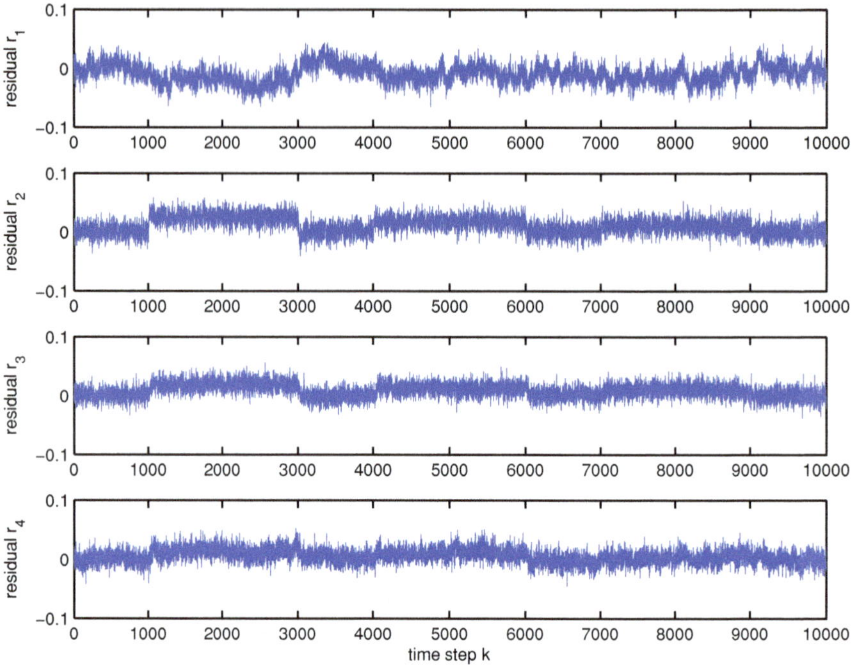

Fig. 18.4 The faulty case residuals

Concerning the unknown input being either white noises or $l_{2,[0,N]}$-norm bounded disturbances, an EKF-based scheme and an H_-/H_∞-optimal FD method have been correspondingly given by recursive computation. Especially, in the case with $l_{2,[0,N]}$-norm bounded disturbances, a local optimal solution to extended H_-/H_∞-FDF has been investigated and a dynamic post-filter has been developed for an arbitrary feasible filter gain matrix. The results can be regarded as an extension of the method in Chap. 6 for LDTV systems to the nonlinear case. Moreover, inspired by the idea of Chap. 8, a unified design of evaluation function have been given, the key of which lies in finding a minimum of a certain quadratic form of residual evaluation function. An optimal solution has be obtained by means of Krein space techniques. It has been shown that such a framework works both for the EKF based FD and extended H_i/H_∞ optimal FD. Regarding threshold determination issues, a probabilistic method using randomized algorithms has been presented towards a trade-off between FAR and FDR. In particular, under the condition of the l_2-norm boundedness of unknown inputs is unavailable, the probabilistic determination of a threshold has been addressed. A simulation study on a longitudinal control system of a UAV has been demonstrate to show the effectiveness of the methods.

References

1. Khan, A. Q., Abid, M., & Ding, S. X. (2014). Fault detection filter design for discrete-time nonlinear systems: A mixed H_-/H_∞ optimization. *Systems Control Letter, 67*(1), 46–54.
2. Jiang, B., Zhang, K., & Shi, P. (2011). Integrated fault estimation and accommodation design for discrete-time Takagi-Sugeno fuzzy systems with actuator faults. *IEEE Transactions on Fuzzy Systems, 19*(2), 291–304.
3. Jiang, B., Staroswiecki, M., & Cocquempot, V. (2006). Fault accommodation for nonlinear dynamic systems. *IEEE Transactions on Automatic Control, 51*(9), 1578–1583.
4. Jiang, B., & Chowdhury, F. N. (2005). Fault estimation and accommodation for linear MIMO discrete-time systems. *IEEE Transactions on Control Systems Technology, 13*(3), 493–499.
5. Shen, B., Ding, S. X., & Wang, Z. (2013). Finite-horizon H_∞ fault estimation for linear discrete time-varying systems with delayed measurements. *Automatica, 49*(1), 293–296.
6. Persis, C. D., & Isidori, A. (2001). A geometric approach to nonlinear fault detection and isolation. *IEEE Transactions on Automatic Control, 46*(6), 853–864.
7. Yang, G., & Wang, H. (2015). Fault detection and isolation for a class of uncertain state-feedback fuzzy control systems. *IEEE Transactions on Fuzzy Systems, 23*(1), 139–151.
8. Hwang, I., Kim, S., Kim, Y., et al. (2010). A survey of fault detection, isolation, and reconfiguration methods. *IEEE Transactions on Control Systems Technology, 18*(3), 636–653.
9. Marzat, J., Piet-Lahanier, H., Damongeot, F., et al. (2012). Model-based fault diagnosis for aerospace systems: A survey. *Journal of Aerospace Engineering, 226*(10), 1329–1360.
10. Bokor, J. (2009). Fault detection and isolation in nonlinear systems. *IFAC Proceedings Volumes, 42*(8), 1–11.
11. Chen, J., Cao, Y., & Zhang, W. (2015). A fault detection observer design for LPV systems in finite frequency domain. *International Journal of Control, 88*(3), 571–584.
12. Chen, L., & Patton, R. J. (2012). A mixed H_-/H_∞ LPV approach to adaptive fault compensation for a nonlinear UAV. *IFAC Proceedings Volumes, 45*(20), 830–835.
13. Abid, M., Chen, W., Ding, S. X., et al. (2011). Optimal residual evaluation for nonlinear systems using post-filter and threshold. *International Journal of Control, 84*(3), 526–539.
14. Chadli, M., Abdo, A., & Ding, S. X. (2013). H_-/H_∞ fault detection filter design for discrete-time Takagi-Sugeno fuzzy system. *Automatica, 49*(7), 1996–2005.
15. Demetriou, M. A., & Armaou, A. (2012). Dynamic online robust detection and accommodation of incipient component faults for nonlinear dissipative distributed processes. *International Journal of Robust Nonlinear Control, 22*(1), 3–23.
16. Patton, R. J., Frank, P. M., & Clark, R. N. (2000). *Issues of fault diagnosis for dynamic systems*. London, U.K.: Springer.
17. Tempo, R., Calafiore, G., & Dabbene, F. (2012). *Randomized algorithms for analysis and control of uncertain systems with applications* (2nd ed.). London, U.K.: Springer.
18. Aouaouda, S., Chadli, M., Shi, P., et al. (2015). Discrete-time H_-/H_∞ sensor fault detection observer design for nonlinear systems with parameter uncertainty. *International Journal of Robust Nonlinear Control, 25*(3), 339–361.
19. Ding, S. X. (2013). *Model-based fault diagnosis techniques: Design schemes, algorithms, and tools* (2nd ed.). London, U.K.: Springer.
20. Wang, Z., Rodrigues, M., Theilliol, D., et al. (2014). Fault estimation for a class of discrete-time nonlinear systems. *IFAC Proceedings Volumes, 47*(3), 8018–8023.

Chapter 19
An H_i/H_∞-Optimal Fault Diagnosis Scheme for Satellite Attitude Control Systems

Fault diagnosis for satellite attitude control systems is practically meaningful for improving the safety and reliability of satellites. In observer-based robust FD, one major concern is to answer if the choice of a threshold satisfies an acceptable trade-off between FAR and FDR. Besides, knowledge of fault isolability is useful for answering how difficult it is to isolate a fault from another one. With these in mind, the design and performance analysis of an H_i/H_∞-optimal fault diagnosis system are performed in this chapter for satellite attitude control systems, which concerns with fault detectability and fault isolability. An extended H_i/H_∞-FDF is first constructed as a residual generator for the nonlinear satellite attitude control system, then a recursive algorithm is developed for online realization. Regarding the uncertain statistical characteristics of the unknown inputs, randomized algorithms are applied to verify the achievable FAR for a prescribed threshold. Moreover, a contribution analysis based method is used for fault isolation of satellite attitude control systems. A simulation study is also performed to verify the obtained results.

19.1 Introduction

Satellites have been increasingly used for various purposes such as remote sensing, television, telephony, data communication, etc. The satellite attitude control system, as an essential part of a satellite, is of great importance to support these services. However, this system inevitably manifests various types of faults such as actuator faults, sensor faults in gyros, magnetometers and star trackers, which undoubtedly can degrade the system behavior and even cause a catastrophe. In response to the requirements for improving the safety and reliability of satellite attitude control systems, model-based fault diagnosis for such systems have received more and more attention, see, for example [12, 20, 25, 26, 29, 32] and references therein. Most

M. Zhong et al., *Fault Diagnosis for Linear Discrete Time-Varying Systems and Its Applications*, https://doi.org/10.1007/978-981-19-5438-2_19

of the results are devoted to handling FD issues. Concerning the importance of FD performance, increasing efforts have been made on the analysis of fault detectability and isolability in recent years. In [24], fault detectability and isolability are expressed as structural properties of the system under consideration, while FAR and FDR are used as quantitative evaluation of fault detectability. In recent years, different types of measures have been proposed to evaluate fault detectability and isolability without designing a diagnosis algorithm, see [6, 23, 29]. As mentioned in [6], the main limition of fault diagnosis performance is the model uncertainties, including modeling error, process and measurement noises. Recent works on observer design for system uncertainty, external disturbance, and actuator fault have promised the progress of observer-based FD and fault tolerant control. For example, [3] proposed a novel approach to handle system uncertainty and external disturbances by using observer technique. Reference [4] dealt with the problem of simultaneously fault tolerant control and uncertainty attenuation for a class of nonlinear systems. When taking into account the model uncertainties and fault time profiles, knowledge of achievable fault detectability and isolability is useful for evaluating how much fault diagnosis performance can be gained [9]. In the design process of fault diagnosis strategy, however, it is hard to quantitatively analyze the achievable FD performance for satellite attitude control systems with unknown uncertainties.

On the other hand, a satellite attitude control system is a typical nonlinear system. Regarding model-based nonlinear fault diagnosis, the methodology of EKF has been extensively used under the assumption of process noise and measurement noise having known Guassian probability distributions. Based on this, the well-established χ^2 test can be used to detect the occurrence of a fault. In [27], an intelligent particle filter was developed for FD of nonlinear non-Gaussian system. Concerning LDTV systems subject to l_2-norm bounded unknown inputs, as demonstrated in Part II, the indices H_-- and/or H_∞- norm are usually applied to measure the sensitivity of residual to fault and the robustness of residual to unknown input, such that the design of an observer-based FD system can be formulated as achieving a trade-off between sensitivity and robustness. As an extension of this idea, an extended H_-/H_∞-optimal FD method has been developed in Chapter 18 and successfully applied to a UVA control system, which provides us with an alternative way to handle FD issues of nonlinear satellite attitude control systems.

Moreover, it should be pointed out that, the main real challenge in practical aerospace engineering application of the existing theoretical methodologies of FD is the threshold setting, which caused easily a higher FAR at the on-orbit stage. The actual characteristics of the process noises in deep space may be cyclic, constant or random, considering the influences of aerodynamic effects, gravity gradient, magnetic moment, solar effects, etc. Also, the measurement noises of the star sensors and gyroscopes in deep space environment are complex and uncertain. In H_i/H_∞-optimal FD, a threshold is usually determined by taking into account the worst-case l_2-norm boundedness of unknown inputs and, hence, leads to zero FAR but poorer FDR. In this situation, randomized algorithms can be utilized towards a trade-off between FAR and FDR, as seen in Sect. 18.5.

Also, fault isolation is crucial for removing the detected fault and eliminating its damage to a practical system. Especially, regarding the system nonlinear characteristics, the existence of unknown uncertainties as well as fault propagation in a closed-loop system, perfect fault isolation is much difficult. To solve this problem, the combination of model-based FD and statistic learning methodologies offers possible ways [7, 8].

In this chapter, we focus on the H_i/H_∞-optimization based design and probabilistic performance analysis of observer-based FD system for satellite attitude control systems. On the basis of constructing an extended H_i/H_∞-FDF for residual generation, randomized algorithms presented in Chap. 18 are applied to achieve a probabilistic estimation of the worst-case threshold that, at the same time, can guarantee an acceptable level of FAR without knowing the l_2-norm boundedness of unknown input. For the purpose of fault isolation for the satellite attitude control systems, a contribution analysis based scheme is further developed. A simulation study is finally demonstrated for effectiveness verification purpose.

19.2 System Description

The schematic diagram of the considered satellite attitude control system is shown in Fig. 19.1, which mainly includes the attitude sensors, attitude controller and actuator. The attitude sensors consist of infrared earth sensor, sun sensor, star sensor and gyroscope. The attitude controller is composed of the original controller, fault diagnosis subsystem and fault tolerant controller, while the actuators include momentum wheel, thruster, magnetorque and control moment gyroscope. The three axes of satellite body-fixed frame O_b, X_b, Y_b, Z_b are defined as satellite's principal axes of inertia, which is called as inertia principal axes system. The reference coordinate system is chosen as orbital reference frame O, X_g, Y_g, Z_g, which is centered at the mass center of the satellite. The frame axes are aligned with the nadir vector (pointing directly to the center of the Earth) and along with the satellite orbital velocity, forming a right-handed system. The conversion relation between body-fixed reference frame and orbital reference frame is defined as satellite attitude.

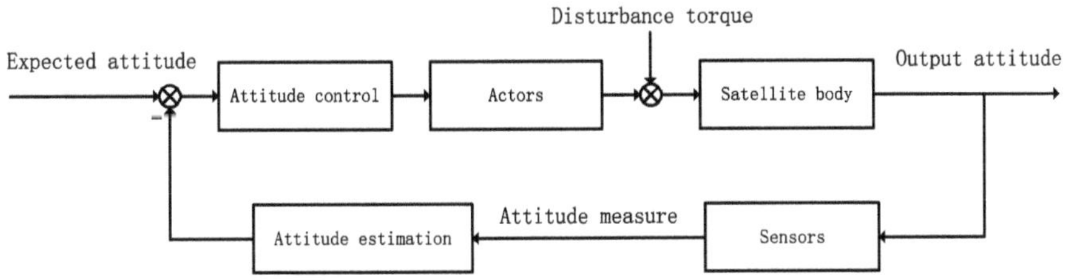

Fig. 19.1 The schematic of satellite attitude control system

Denote by φ, θ and ψ the rolling angle, pitching angle and yaw angle, respectively. Let

$$\omega(t) = \left[\omega_x(t)\ \omega_y(t)\ \omega_z(t)\right]^T$$

be the rotation velocity of satellite with respect to orbital reference frame, where $\omega_x(t)$, $\omega_y(t)$, $\omega_z(t)$ stand for the projections of attitude rate vector onto the three principal axes of inertia, respectively. Define the state vector $x(t)$ and control input vector $u(t)$ as

$$x(t) = \left[\varphi(t)\ \theta(t)\ \psi(t)\ \omega_x(t)\ \omega_y(t)\ \omega_z(t)\right]^T,$$
$$u(t) = \left[T_x(t)\ T_y(t)\ T_z(t)\right]^T,$$

respectively. Under the assumption of no fault being taken into account, the following nonlinear model was used in [29] to describe the dynamics of a kind of rigid satellite attitude control systems

$$\dot{x}(t) = q(x(t)) + \alpha_u u(t) + d(t), \tag{19.1}$$

where $q(x(t)) = \left[q_1\ q_2\ q_3\ q_4\ q_5\right]^T$ with

$$q_1 = \omega_0\psi(t) + \omega_x(t),\ q_2 = \omega_0 + \omega_y(t),\ q_3 = -\omega_0\varphi + \omega_z(t),$$
$$q_4 = (I_y - I_z)\omega_y(t)\omega_z(t)/I_x,\ q_5 = (I_z - I_x)\omega_x(t)\omega_z(t)/I_y,$$
$$q_6 = (I_x - I_y)\omega_y(t)\omega_x(t)/I_z,\ \alpha_u = \begin{bmatrix} 0_{3\times 3} \\ \mathrm{diag}\{1/I_x,\ 1/I_y,\ 1/I_z\} \end{bmatrix},$$

$d(t)$ is the disturbance vector, I_i $(i = x, y, z)$ stands for the moment of inertia of satellite in three principal axes of inertia, $\omega_i(t)$ is the projection of attitude rate vector onto the three principal axes of inertia, $T_i(t)$ denotes the control moments' components along principal axes of inertia. The satellite's attitude angle and rate are measured respectively by star sensors and gyroscopes, which can be described by

$$y(t) = x(t) + v(t),$$

where $v(t)$ denotes the measurement noise vector.

Without loss of generality, in this chapter only the cases of actuators' faults and sensors' faults are considered, which include the momentum wheels' faults, star sensors' faults and gyroscopes' faults. Suppose that both the actuators' faults and sensors' faults are represented by additive signals. Then the satellite attitude control system with additive faults can be further modeled as

$$\begin{cases} \dot{x}(t) = q(x(t)) + \alpha_u u(t) + \alpha_f f(t) + d(t), \\ y(t) = Cx(t) + D_f f(t) + v(t), \end{cases} \tag{19.2}$$

where $f(t) = \left[\, f_1(t)\ f_2(t) \cdots f_9(t) \,\right]^T$, $f_i(t)$ $(i = 1, 2, \ldots, 9)$ stands for the momentum wheels' faults, star sensors' faults and gyroscopes' faults, respectively, and

$$\alpha_f = \left[\alpha_u \ 0_{6\times 6}\right],\ D_f = \left[\, 0_{6\times 3}\ I \,\right],\ C = I.$$

Taking into account the unavoidable influences of system disturbance, measurement noise as well as model error for FD performance, the main attention of this chapter will be paid to the design and performance analysis of observer-based FD for satellite attitude control systems. For this purpose, in the following section, an extended H_i/H_∞-FDF is firstly designed for residual generation.

19.3 Extended H_i/H_∞-Optimal Fault Detection

Suppose that the sampling period is T_s and a zero-order-hold is chosen for system model (19.2). The following discrete time system model can be obtained

$$\begin{cases} x(k+1) = \phi(x(k)) + B_u u(k)) + B_f f(k) + w(k), \\ y(k) = Cx(k) + D_f f(k) + v(k), \end{cases} \tag{19.3}$$

where $\phi(x(k)) = x(k) + T_s q(x(k))$, $B_u = T_s \alpha_u$, $B_f = T_s \alpha_f$, $w(k) \in l_{2,[0,N]}$ stands for the l_2-norm bounded external disturbance, $v(k) \in l_{2,[0,N]}$ is the measurement noise.

In the stage of residual generation, the following FDF is constructed

$$\begin{cases} \hat{x}(k+1) = \phi(\hat{x}(k)) + B_u u(k)) + L(k)\tilde{y}(k), \\ \tilde{y}(k) = y(k) - C\hat{x}(k),\ \hat{x}(0) = 0, \\ r(k) = W(k)\tilde{y}(k), \end{cases} \tag{19.4}$$

where $L(k)$ and $W(k)$ are the observer gain matrix and post-filter, respectively.

Expanding the nonlinear function $\phi(x(k))$ in Taylor series about the state estimate $\hat{x}(k)$ and dropping the high order terms, we get

$$\begin{aligned} \phi(x(k)) &= \phi(\hat{x}(k)) + \frac{\partial \phi(x)}{\partial x}|x = \hat{x}(k) + \cdots \\ &\simeq \phi(\hat{x}(k)) + A(k)(x(k) - \hat{x}(k)), \end{aligned} \tag{19.5}$$

where

$$A(k)=I+T_s\frac{\partial q(x)}{\partial x}|_{x=\hat{x}(k)}=\begin{bmatrix} 1 & 0 & \omega_0 T_s & T_s & 0 & 0 \\ 0 & 1 & 0 & 0 & T_s & 0 \\ -\omega_0 T_s & 0 & 1 & 0 & 0 & T_s \\ 0 & 0 & 0 & 1 & a\hat{\omega}_z(k) & a\hat{\omega}_y(k) \\ 0 & 0 & 0 & b\hat{\omega}_z(k) & 1 & b\hat{\omega}_x(k) \\ 0 & 0 & 0 & c\hat{\omega}_y(k) & c\hat{\omega}_x(k) & 1 \end{bmatrix}, \tag{19.6}$$

$$a=\frac{I_y-I_z}{I_x}T_s,\ b=\frac{I_z-I_x}{I_y}T_s,\ c=\frac{I_x-I_y}{I_z}T_s,$$

and $\frac{\partial q}{\partial x}=\left[\frac{\partial q_i}{\partial x_j}\right]\in R^{n\times n}$ denotes the Jacobian matrix of $q(x(k))$. Let $e(k)=x(k)-\hat{x}(k)$. It follows from (19.3)–(19.5) that

$$\begin{cases} e(k+1)=(A(k)-L(k)C)e(k)+w(k)-L(k)v(k)+(B_f-L(k)D_f)f(k), \\ r(k)=W(k)(Ce(k)+D_f f(k)+v(k)),\ e(0)=x_0. \end{cases} \tag{19.7}$$

Since it is usually impossible to completely decouple the residual from unknown input and initial state, the matrices $L(k)$, $W(k)$ can be designed in the sense of H_i/H_∞-optimization.

Define the H_∞ robustness index as

$$\gamma_d^2=\sup\frac{\sum_{k=0}^{N}r_d^T(k)r_d(k)}{J(x_0,w_N,v_N)},$$

where $r_d(k)=r(k)|_{f=0}$ and

$$J(x_0,w_N,v_N)=x_0^Tx_0+\sum_{k=0}^{N}(w^T(k)R^{-1}(k)w(k)+v^T(k)Q^{-1}(k)v(k)),$$

$R(k)$ and $Q(k)$ are weighting matrices. While, the H_∞ and H_- sensitivity indices are, respectively, defined as

$$\gamma_f^2=\sup_{f(k)}\frac{\sum_{k=0}^{N}r_f^T(k)r_f(k)}{\sum_{k=0}^{N}f^T(k)f(k)},\ \gamma_{f-}^2=\inf_{f(k)}\frac{\sum_{k=0}^{N}r_f^T(k)r_f(k)}{\sum_{k=0}^{N}f^T(k)f(k)},$$

where $r_f(k)=r(k)|_{x_0,w,v=0}$, γ_f and γ_{f-} denote the best-case and worst-case sensitivity indices to fault, respectively. Thus, the problem of H_i/H_∞-optimization based FD can be formulated as to find $L(k)$ and $W(k)$ such that

$$\max_{L(k),W(k)}\frac{\gamma_f}{\gamma_d}\ \text{and}\ \max_{L(k),W(k)}\frac{\gamma_{f-}}{\gamma_d}. \tag{19.8}$$

Algorithm 19.3.15 Extended H_i/H_∞ optimization based FD

1: Set a finite horizon level $N > 0$, weighting matrices $R(k)$, $Q(k)$, and threshold J_{th};
2: Compute $L(k)$ and $W(k)$ by applying Proposition 19.1;
3: Generate residual $r(k)$ by using (19.4) and compute the evaluation function J_N by (19.13);
4: Detect the occurrence of a fault based on (19.14).

Noticing that (19.3) is nonlinear, the FDF given by (19.4) satisfying criteria (19.8) is an extended H_i/H_∞-FDF.

By applying Proposition 18.1, the optimal solution of $L(k)$ and post-filter $W(k)$ can be obtained by the following proposition.

Proposition 19.1 *The observer gain matrix $L(k)$ and post-filter $W(k)$ given by*

$$L(k) = A(k)P(k)C^T R_{\tilde{y}}^{-1}(k), \tag{19.9}$$

$$P(k+1) = A(k)P(k)A^T(k) - L(k)R_{\tilde{y}}(k)L^T(k) + R(k), \tag{19.10}$$

$$P(k_0) = I, \tag{19.11}$$

$$W(k) = R_{\tilde{y}}^{-\frac{1}{2}}(k), \tag{19.12}$$

with $R_{\tilde{y}}(k) = CP(k)C^T + Q(k) > 0$ provide an optimal solution to the optimization problem (19.8) subject to (19.7). Moreover, the achieved robustness index is $\gamma_d = 1$.

Under the assumption of $w(k)$, $v(k) \in l_{2,[0,N]}$, the following residual evaluation function is considered

$$J_N = \sum_{k=0}^{N} r^T(k)r(k). \tag{19.13}$$

For a prescribed given threshold J_{th}, the occurrence of a fault can be detected by comparing J_N with J_{th}, i.e.,

$$\begin{cases} J_N > J_{th}, & \text{fault alarm,} \\ J_N \leq J_{th}, & \text{no alarm.} \end{cases} \tag{19.14}$$

Hence, with the obtained $L(k)$ and $W(k)$ in (19.9)–(19.12), we generate residual signals by using (19.4) and calculate the evaluation function J_N by using (19.13). For a given threshold J_{th}, the implementation of extended H_i/H_∞-optimal FD is summarized in Algorithm 19.3.15.

Remark 19.1 Given $J(x_0, w_N, v_N) \leq \delta^2$, the choice of $J_{th} = \delta^2$ and J_N in (19.13) lead to an optimal FD regarding maximal fault detectability at a zero level of FAR, as defined in Definition 2.4.

On the basis of constructing an extended H_i/H_∞-FDF for satellite attitude control system (19.2), below we focus on probabilistic analysis of fault diagnosis performance, including randomized algorithms based analysis of achievable FAR for a given threshold (or the determination of a threshold for an acceptable FAR) and a contribution analysis based fault isolation.

19.4 Randomized Analysis of FAR

In Sect. 18.5, a successful application of randomized algorithms for threshold setting and FAR performance evaluation have been demonstrated. Along such a line, we consider the following three typical scenarios.

Case I: The PDF of $J(x_0, w_N, v_N)$ is known

Assuming that $J(x_0, w_N, v_N)$ is randomly distributed with known PDF, it is known from Proposition 19.1 that in the fault-free case,

$$J_N \leq \gamma_d^2 J(x_0, w_N, v_N) \leq \delta^2.$$

Therefore, $\Pr\{J_N > J_{th} | f = 0\} \leq \Pr\{J(x_0, w_N, v_N) > J_{th}\}$ holds. Suppose that $J(x_0, w_N, v_N)$ is distributed over $(0, \ \delta^2]$ with PDF $f_J(\eta)$. For a given threshold J_{th}, the FAR can be evaluated by using Algorithm 18.5.12.

Case II: The PDF of $J(x_0, w_N, v_N)$ is unknown

For the case of unknown PDF of $J(x_0, w_N, v_N)$, an effective way of improving the fault detectability is to reduce the threshold. To verify if the FAR is acceptable, an empirical method using the randomized algorithms can be applied to this end, as demonstrated in Algorithm 18.5.13.

Case III: The Case of δ Being Unknown

Consider the case of δ being unknown. It is known that, in the fault-free case, J_N provides a minimum estimation of $J(x_0, w_N, v_N)$. Based on this, a probabilistic estimation of the worst-case threshold, according to Proposition 18.6, can be derived guaranteeing an acceptable level of FAR, as summarized in Algorithm 18.5.14.

In these three typical scenarios, Case II concerns to verify the achievable FAR for a given threshold, while Algorithm 18.5.14 for Case III provides a way to estimate the worst-case threshold for guaranteeing an acceptable level of FAR. Algorithms 18.5.13 and 18.5.14 are respectively the extended applications of randomized Algorithm 10.1 and 10.2 in [22] for probabilistic performance verification.

19.5 Contribution Analysis Based Fault Isolation

We now turn our attention to the problem of fault isolation for the satellite attitude control system. Let

$$r(k) = \begin{bmatrix} r_1(k) \; r_2(k) \; \cdots \; r_6(k) \end{bmatrix}^T.$$

The evaluation function J_N can be rewritten as

$$J_N = \sum_{j=1}^{6} J_{Nj}, \; J_{Nj} = \sum_{k=0}^{N} r_j^2(k). \tag{19.15}$$

For the purpose of fault isolation, analysis on contribution of evaluation function is first considered. Introduce

$$\gamma_j = \frac{J_{Nj}}{J_N}, \; j = 1, 2, \ldots, 6, \tag{19.16}$$

where $0 \le \gamma_j \le 1$ is used to represent the contribution of r_j on J_N. Let

$$\gamma_j^{(0)} = \sup_{f=0} \gamma_j, \; j = 1, 2, \ldots, 6.$$

Choose the jth contribution threshold as $\gamma_{j,th} \ge \gamma_j^{(0)}$. When a fault f_i with a given magnitude $\bar{f}_i$ is taken into account, the contribution index of f_i on J_{Nj} is represented by

$$\gamma_j^{(i)} = \begin{cases} \gamma_j, & \text{if } \gamma_j > \gamma_{j,th}, \\ 0, & \text{if } \gamma_j \le \gamma_{j,th}. \end{cases}$$

Define the following contribution vector as an indicator of fault f_i

$$\begin{bmatrix} \gamma_1^{(i)} \; \gamma_2^{(i)} \; \cdots \; \gamma_6^{(i)} \end{bmatrix},$$

where $\gamma_j^{(i)}$ $(j = 1, 2, \ldots, 6)$ also represents the contribution of residual $r_j(k)$ on J_N. Based on this, we can perform the isolation of fault f_i according to the matching degree of contribution vector $\begin{bmatrix} \gamma_1 \; \gamma_2 \; \cdots \; \gamma_6 \end{bmatrix}$ with $\begin{bmatrix} \gamma_1^{(i)} \; \gamma_2^{(i)} \; \cdots \; \gamma_6^{(i)} \end{bmatrix}$.

Remark 19.2 The presented fault isolation for observer-based FD is similar to the data-driven FD using contribution plot in principal component analysis. Instead by directly analyzing the contribution of the ith fault variable, we consider to construct the contribution vector $\begin{bmatrix} \gamma_1^{(i)} \; \gamma_2^{(i)} \; \cdots \; \gamma_6^{(i)} \end{bmatrix}$ as an indicator of the ith fault. When a fault alarm is triggered, we can calculate $\begin{bmatrix} \gamma_1 & \gamma_2 & \cdots & \gamma_6 \end{bmatrix}$ and identify the fault variable f_i by comparing with $\begin{bmatrix} \gamma_1^{(i)} & \gamma_2^{(i)} & \cdots & \gamma_6^{(i)} \end{bmatrix}$, while the matching degree represents the isolability of fault f_i.

19.6 Simulation Results

To demonstrate the effectiveness of the developed algorithms, the satellite attitude control system given in [29] is considered. The parameters are set as

$$I_x = 12.50\text{kg} \cdot \text{m}^2,\ I_y = 13.70\text{kg} \cdot \text{m}^2, I_z = 15.90\text{kg} \cdot \text{m}^2,\ \omega_0 = 0.001\text{rad/s}.$$

The sampling period is $T_s = 0.01$s. The control input is simulated by a proportional controller with gain matrix $-[35I \quad 30I]$. Set $N = 10,\ R(k) = 2.25 \times 10^{-10} I,\ Q(k) = \sigma_v^2,\ \sigma_v^2 = \text{diag}\{8.70 \times 10^{-6} I, 4.20 \times 10^{-5} I\}$.

Firstly, we design an extended H_i/H_∞-FDF for the satellite attitude control systems by using Proposition 19.1. The detection of a fault is carried out by using Algorithm 19.3.15. Set $\delta_M^2 = 400$. A threshold is chosen as $J_{th,1} = 400$. The distur-

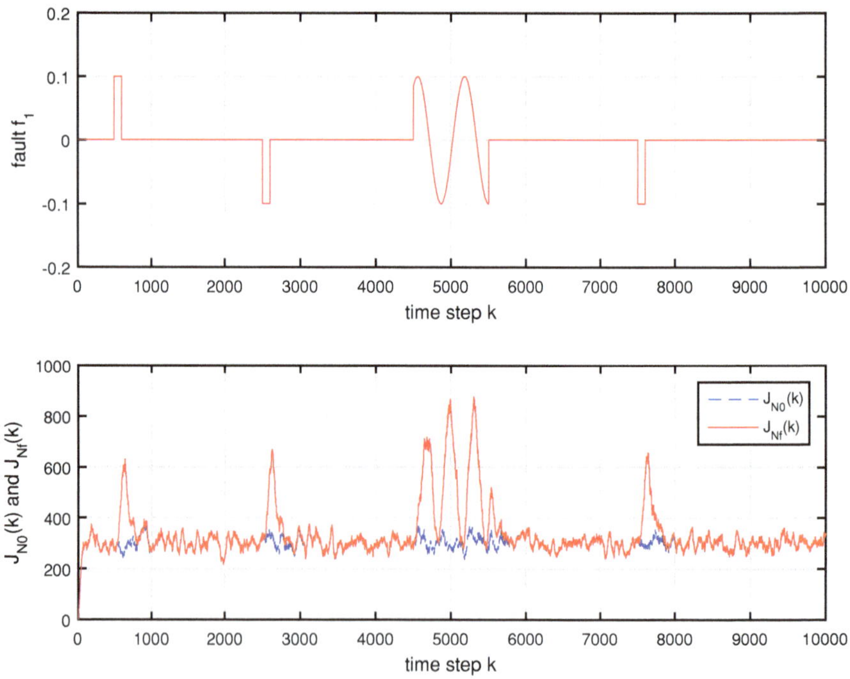

Fig. 19.2 The fault $f_1(k)$, $J_{NO}(k)$ and the corresponding $J_{Nf}(k)$

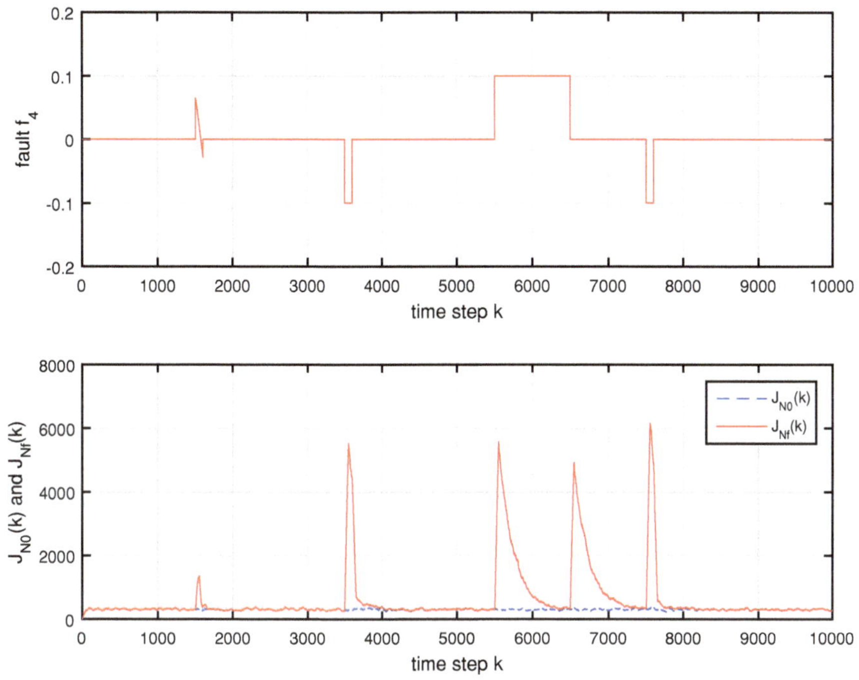

Fig. 19.3 The fault $f_4(k)$, $J_{N0}(k)$ and the corresponding $J_{Nf}(k)$

bances and measurement noises are simulated as $w(k) \in \mathcal{N}(0_{6\times1}, 2.25 \times 10^{-10} I)$, $v(k) \in \mathcal{N}(0_{6\times1}, \sigma_v^2)$. Here we only given the results of actuator fault $f_1(k)$ and sensor faults $f_4(k)$, $f_7(k)$ simplicity. The simulated faults and generated evaluation functions are shown in Figs. 19.2, 19.3 and 19.4, of which the $J_{N0}(k)$ and $J_{Nf}(k)$ are given by

$$\begin{cases} J_{N0}(k) = \sum\limits_{j=k-N}^{k} r^T(j) r(j), \text{ no fault}, \\ J_{Nf}(k) = \sum\limits_{j=k-N}^{k} r^T(j) r(j), \ f(j) \neq 0. \end{cases}$$

It is seen from the simulation results that zero FAR is achieved in the fault free case, while FDR is "1" when the simulated faults are considered. For $J_{th,2} = 350$. Set $M = 10000$. By applying Algorithm 18.5.13, an empirical estimation of the upper bound FAR is obtained as $\hat{P}_{FAR} = 0.016$, which guarantees that $|P_{FAR} - \hat{P}_{FAR}| < 0.009$ with a confidence level at least 0.95.

Next, we consider the case of both the PDF and the upper bound of $J(x_0, w_N, v_N)$ being unavailable. Set $M = 10000$, $\epsilon_{II} = 0.001$, $\nu = 0.0005$. By utilizing

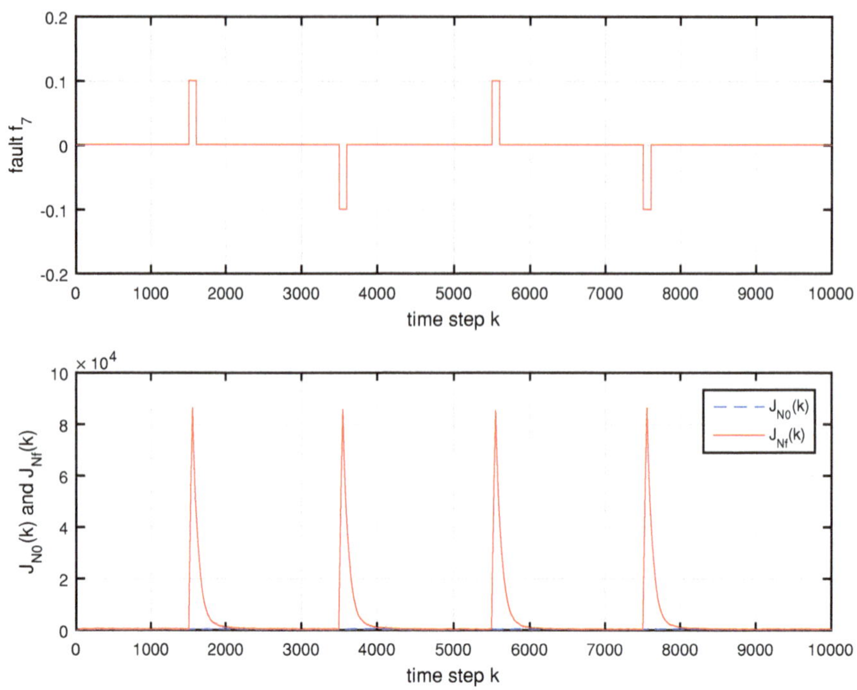

Fig. 19.4 The fault $f_7(k)$, $J_{N0}(k)$ and the corresponding $J_{Nf}(k)$

Table 19.1 The nonzero terms $\gamma_j^{(i)}$ of the contribution vectors

f_1	0	0	0	$\gamma_4^{(1)}$	0	0
f_2	0	0	0	0	$\gamma_5^{(2)}$	0
f_3	0	0	0	0	0	$\gamma_6^{(3)}$
f_4	$\gamma_1^{(4)}$	0	0	0	0	0
f_5	0	$\gamma_2^{(5)}$	0	0	0	0
f_6	0	0	$\gamma_3^{(6)}$	0	0	0
f_7	$\gamma_1^{(7)}$	0	0	$\gamma_4^{(7)}$	0	0
f_8	0	$\gamma_2^{(8)}$	0	0	$\gamma_5^{(8)}$	0
f_9	0	0	$\gamma_3^{(9)}$	0	0	$\gamma_6^{(9)}$

Algorithm 18.5.14, we get an empirical estimation $\hat{\delta}_N^2 = 405.2$ and, for $J_{th,3} = \hat{\delta}_N^2$, an estimation of FAR is $P_{FAR} < 0.001$ at least confidence level 0.9995.

Finally, we turn to analyze the fault isolability of the extended H_i/H_∞ optimization based FD for the considered satellite attitude control system. For $\gamma_{j,th} = 0.5$, the obtained contribution vectors $\left[\gamma_1^{(i)}\ \gamma_2^{(i)}\ \cdots\ \gamma_6^{(i)}\right]$ corresponding to the simulated f_i $(i = 1, 2, \ldots, 9)$ are listed in Table 19.1.

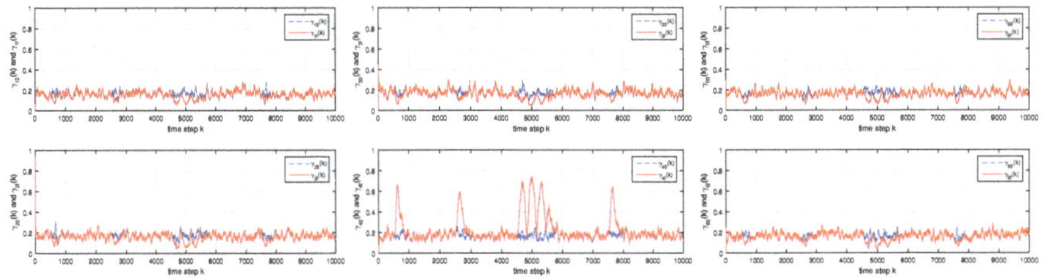

Fig. 19.5 The evolutions of $\gamma_{j0}(k), \gamma_{jf}(k)$ $(j = 1, 2, \ldots, 6)$ for $f_1(k)$

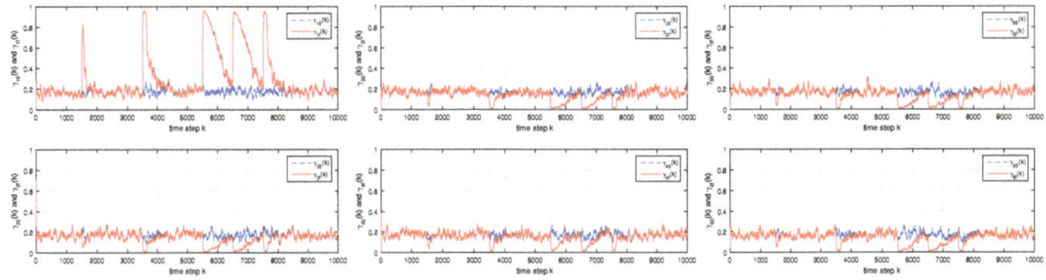

Fig. 19.6 The evolutions of $\gamma_{j0}(k), \gamma_{jf}(k)$ $(j = 1, 2, \ldots, 6)$ for $f_4(k)$

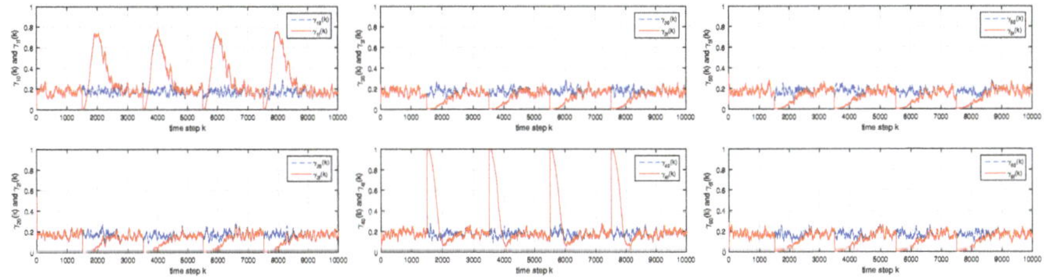

Fig. 19.7 The evolutions of $\gamma_{j0}(k), \gamma_{jf}(k)$ $(j = 1, 2, \ldots, 6)$ for $f_7(k)$

The results in Table 19.1 show that fault isolability can be guaranteed by using the extended H_i/H_∞ optimization based FD scheme. For example, based on analyzing the contribution of evaluation function, if only γ_4 is the major contribution, then fault f_1 is isolated. If both γ_1 and γ_4 are the major contributions, then the isolated fault is f_7. To show the results more clearly, the evolutions of γ_j $(j = 1, 2, \ldots, 6)$ for faults f_1, f_4 and f_7 are also shown in Figs. 19.5, 19.6 and 19.7, of which $\gamma_{j0}(k)$ and $\gamma_{jf}(k)$ stand for the fault free case and faulty case of γ_j, respectively. The simulation results also show that f_1, f_4 and f_7 are fault isolable.

It is shown from the simulation results that the actuator faults and sensor faults are detectable and, for a given threshold, the achievable FAR can be estimated in a probabilistic framework. About the fault isolability analysis, however, we can only say that the fault is isolable if only one single fault is taken into account at the same time. Moreover, both the choice of J_{th} and $\gamma_{j,th}$ $(j = 1, 2, \ldots, 6)$ are relevant to $J(x_0, w_N, v_N)$ in practice and, for simplicity, only the normal distributed $w(k)$ and $v(k)$ are considered in the simulation.

Remark 19.3 It is notable that only the case of one single fault is considered in the presented contribution analysis based fault isolation method, which cannot meet the requirements for practical applications in aerospace engineering, regarding the case of simultaneously occurring faults as well as fault propagation in a closed-loop system. Study on integrating of model-based FD with stochastic learning techniques hence can be considered in future research.

19.7 Conclusion

In this chapter, we have focused on the design and probabilistic performance analysis of extended H_i/H_∞-optimization based fault detection and isolation system for a satellite attitude control system. An extended H_i/H_∞-FDF for the targeting nonlinear system has been designed as a residual generator and a recursive algorithm has been given for online realization. Regarding FD performance analysis, the estimation of FAR for a given threshold and the determination of a threshold with an acceptable FAR level have been addressed in a probabilistic framework. Randomized analysis algorithms presented in Chap. 18 have been applied according to the characteristics of external disturbances and measurement noises. A contribution analysis based method has also been given for fault isolability analysis, wherein the isolation of fault has been performed by analyzing the matching degree with the corresponding contribution vector. Finally, the applicability of the presented method has been verified.

References

1. Marino, A., Pierri, F., & Arrichiello, F. (2017). Distributed fault detection isolation and accommodation for homogeneous networked discrete-time linear systems. *IEEE Transactions on Automatic Control, 62*(9), 4840–4847.
2. Khan, A. Q., Abid, M., & Ding, S. X. (2014). Fault detection filter design for discrete-time nonlinear systems-A mixed H_-/H_∞ optimization. *Systems and Control Letters, 67*(1), 46–54.
3. Xiao, B., & Yin, S. (2016). Velocity-free fault tolerant and uncertainty attenuation control for a class of nonlinear systems. *IEEE Transactions on Industrial Electronics, 63*(7), 4400–4411.
4. Xiao, B., Yin, S., & Kaynak, O. (2016). Tracking control of robotic manipulators with uncertain kinematics and dynamics. *IEEE Transactions on Industrial Electronics, 63*(10), 6439–6449.
5. Sun, C., Ma, M., Zhao, Z., et al. (2018). Sparse deep stacking network for fault diagnosis of motor. *IEEE Transactions on Industrial Informatics, 14*(7), 3261–3270.
6. Eriksson, D., Frisk, E., & Krysander, M. (2013). A method for quantitative fault diagnosability analysis of stochastic linear descriptor models. *Automatica, 49*(6), 1591–1600.
7. Jung, D., & Sundstrom, C. (2019). A combined data-driven and model-based residual selection algorithm for fault detection and isolation. *IEEE Transactions on Control Systems Technology, 27*(2), 616–630.
8. Guo, D., Zhong, M., Ji, H., et al. (2018). A hybrid feature model and deep learning based fault diagnosis for unmanned aerial vehicle sensors. *Neurocomputing, 319*, 155–163.

9. Fu, F., Wang, D., Li, W., et al. (2018). Evaluation of fault diagnosability for dynamic systems with unknown uncertainties. *IEEE Access, 6*, 16737–16745.
10. Zhu, G., Li, Z., & Wu, N. (2018). Model-based fault identification of discrete event systems using partially observed Petri nets. *Automatica, 96*, 201–212.
11. Ma, H., Liu, Y., Li, T., et al. (2019). Nonlinear high-gain observer-based diagnosis and compensation for actuators and sensors faults in a quadrotor unmanned aerial vehicle. *IEEE Transactions on Industrial Informatics, 15*(1), 550–562.
12. Song, H., Han, P., Zhang, J. et al. (2018). Fault diagnosis method for closed-loop satellite attitude control systems based on a fuzzy parity equation. *International Journal of Distributed Sensor Networks, 14*(10) Doc.1550147718805938.
13. Chen, J., & Patton, R. J. (1999). *Robust model-based fault diagnosis for dynamic systems*. Boston, MA: Kluwer Academic Publishers.
14. Zhang, K., Jiang, B., Yan, X., et al. (2017). Incipient voltage sensor fault isolation for rectifier in railway electrical traction systems. *IEEE Transactions on Industrial Electronics, 64*(8), 6763–6774.
15. Chadli, M., Abdo, A., & Ding, S. X. (2013). H_-/H_∞ fault detection filter design for discrete-time Takagi-Sugeno fuzzy system. *Automatica, 49*(7), 1996–2005.
16. Zhong, M., Liu, H., & Song, N. (2015). On designing an extended H_-/H_∞-FDF for a class of nonlinear systems. *IFAC-PapersOnLine, 48*(21), 707–712.
17. Zhong, M., Ding, S. X., & Zhou, D. (2016). A new scheme of fault detection of linear discrete time-varying systems. *IEEE Transactions on Automatic Control, 61*(9), 2597–2602.
18. Zhong, M., Zhang, L., Ding, S. X., et al. (2017). A probabilistic approach to robust fault detection for a class of nonlinear systems. *IEEE Transactions on Industrial Electronics, 64*(5), 3930–3939.
19. Davoodi, M. R., Meskin, N., & Khorasani, K. (2017). Event-triggered multiobjective control and fault diagnosis: A unified framework. *IEEE Transactions on Industrial Informatics, 13*(1), 298–311.
20. Shen, Q., Yue, C., & Goh, C. (2019). Active fault-tolerant control system design for spacecraft attitude maneuvers with actuator saturation and faults. *IEEE Transactions on Industrial Electronics, 66*(5), 3763–3772.
21. Zhang, Q. (2018). Adaptive Kalman filter for actuator fault diagnosis. *Automatica, 93*, 333–342.
22. Tempo, R., Calafiore, G., & Dabbene, F. (2012). *Randomized algorithms for analysis and control of uncertain systems with applications* (2nd ed.). London, U.K.: Springer.
23. Sharifi, R., & Langari, R. (2011). Isolability of faults in sensor fault diagnosis. *Mechanical Systems and Signal Processing, 25*(7), 2733–2744.
24. Ding, S. X. (2013). *Model-based fault diagnosis techniques: Design schemes, algorithms, and tools* (2nd ed.). London, U.K.: Springer.
25. Yin, S., Xiao, B., Ding, S. X., et al. (2016). A review on recent development of spacecraft attitude fault tolerant control system. *IEEE Transactions on Industrial Electronics, 63*(5), 3311–3320.
26. Nasrolahi, S., & Abdollahi, F. (2018). Sensor fault detection and recovery in satellite attitude control. *Acta Astronautica, 145*, 275–283.
27. Yin, S., & Zhu, X. (2015). Intelligent particle filter and its application to fault detection of nonlinear system. *IEEE Transactions on Industrial Electronics, 62*(6), 3852–3861.
28. Ding, S. X., Jeinsch, T., Frank, P. M., et al. (2000). A unified approach to the optimization of fault detection systems. *International Journal of Adaptive Control Signal Processing, 14*(7), 725–745.
29. Li, W., Wang, D., & Liu, C. (2015). Quantitative fault diagnosis ability evaluation for control systems with disturbances. *Control Theory & Applications (in Chinese), 32*(6), 744–752.
30. Chen, Y., Fink, O., & Sansavini, G. (2018). Combined fault location and classification for power transmission lines fault diagnosis with integrated feature extraction. *IEEE Transactions on Industrial Electronics, 65*(1), 561–569.
31. Wang, Z., Shi, P., & Lim, C. (2017). H_-/H_∞ fault detection observer in finite frequency domain for linear parameter-varying descriptor systems. *Automatica, 86*, 38–45.

32. Gao, Z., Zhou, Z., & Jiang, G. (2018). Active fault tolerant control scheme for satellite attitude systems: multiple actuator faults case. *International Journal of Control Automation and Systems, 16*(4), 1794–1804.
33. Gao, Z., Cecati, C., & Ding, S. X. (2015). A survey of fault diagnosis and fault-tolerant techniques-Part I: Fault diagnosis with model-based and signal-based approaches. *IEEE Transactions on Industrial Electronics, 62*(6), 3757–3767.

Chapter 20
Extended H_i/H_∞-Optimal Fault Detection for INS/GPS Integrated Systems

This chapter aims to deal with FD issues for an integrated system of inertial navigation system and global position system (INS/GPS). To this end, state space models of the INS/GPS integrated system are first established. An observer-based FDF for residual generation is then constructed with the observer gain matrix and the post-filter being designed in the context of H_i/H_∞-optimization. By utilizing techniques of innovation analysis and orthogonal projection, a solution to the extended H_i/H_∞-FDF is given in terms of Riccati recursions, followed by a residual evaluation strategy. Finally, a flight experiment is implemented to show the practical applicability of the method.

20.1 Introduction

Inertial navigation system (INS) provides a self-contained independent means for geographical position, velocity, and attitude by using inertial sensors, i.e. gyroscopes and accelerometers. It has been used as one of the complete autonomy navigation systems with short-term high precision, but the accuracy of the system degrades with time [14]. The global positioning system (GPS) is used to provide a three-dimensional position and velocity data with high long-term accuracy, but it may suffer from outliers, outages, jamming and multipath effects in some cases and has a low update frequency [8]. Thus, an integration of INS and GPS, i.e. INS/GPS-integrated navigation system, can exploit advantages of the two systems to achieve long-term high-accuracy measurements of the system motion, see [5, 7, 10, 15, 16] and references therein.

In order to improve the system safety and reliability, more and more efforts have also been devoted to fault diagnosis for INS/GPS integrated systems, see for example [2, 9, 11, 17, 19, 22, 23] and references therein. In [2, 23], Kalman filter was

M. Zhong et al., *Fault Diagnosis for Linear Discrete Time-Varying Systems and Its Applications*, https://doi.org/10.1007/978-981-19-5438-2_20

adopted for fault estimation of INS/GPS integrated system subject to GPS outliers. An adaptive two-stage extended Kalman filter (ATEKF) was proposed in [9] for nonlinear INS/GPS integrated system with inertial sensor faults. In [17], the well known unscented Kalman filter (UKF) was used to generate residual and, based on this, the presence of a sensor fault was detected by using χ^2-test. Reference [19] dealt with the problem of detecting and identifying interference/jamming and spoofing in the differential GPS signals. In practice, however, statistic properties of both process disturbances and measurement noises are usually not exactly known for a real INS/GPS system, especially the INS/GPS-integrated system for aerial mapping, i.e., the so-called position and orientation measurement system (POS), see for example [2, 5]. As known from [3, 14], POS is usually exposed outside of the aircraft and, therefore, the influences of environment temperature, variation and electromagnetic interference are serious. In the case of statistic properties of system noises being unavailable, the methods of Kalman filter (similarly either EKF or UKF) is not applicable to FD for the INS/GPS-integrated system.

The objective of this chapter is to handle FD issues for INS/GPS-integrated system in the framework of H_i/H_∞-optimization, which is an another engineering application of the extended H_i/H_∞-optimal FD method given in Chap. 18. The tasks to this aim consist of

- establishing state space models of the nonlinear INS/GPS integrated system;
- constructing an extended H_i/H_∞-FDF for residual generation and developing a recursive algorithm for online realization;
- designing appropriate residual evaluation function and threshold towards an optimal FD.

Finally, a flight experimental study is carried out for verification purpose.

20.2 System Modeling

We start with a brief introduction to a strap-down INS/GPS integrated system of POS. The inertial measurement unit (IMU) of INS is composed of three laser gyroscopes and three quartz accelerometers. The local level east-north-up (ENU) frame is selected as navigation frame (n-frame). The body frame (b-frame) has its origin at the centroid of the vehicle. An earth-centered Earth-fixed frame (e-frame) has its origin at the center of the Earth and is rotating with respect to the earth. The geocentric inertial frame (i-frame) is the same with e-frame but not rotating with the fixed stars. For ease of subsequent description, below we will demonstrate three models of INS/GPS integrated systems.

20.2.1 Model I

Without loss of generality, the biases of gyroscopes and accelerometers are classified into unknown constant signals and noises. According to [6] and [24], the nonlinear strap-down INS error model can be obtained as

$$\begin{cases} \dot{\phi} = (I - C_n^{n,ins})\omega_{in}^n + \delta\omega_{in}^n + C_b^n\varepsilon + C_b^n\varepsilon_w, \\ \delta\dot{V} = (I - C_n^{n,ins})C_b^n\hat{f}_b - \delta[(2\omega_{ie}^n + \omega_{en}^n) \times V^n] + C_b^n\nabla + C_b^n\nabla_w, \\ \delta\dot{p} = \delta V + \rho \times \delta p, \end{cases} \tag{20.1}$$

where the description for the involved notations is given in Table 20.1, and

$$\begin{aligned}
\phi &= [\,\phi_E\ \phi_N\ \phi_U\,]^T,\ \delta V = [\,\delta V_E\ \delta V_N\ \delta V_U\,]^T \\
\delta p &= [\,\delta L\ \delta\lambda\ \delta h\,]^T,\ \hat{f}_b = [\,f_E\ f_N\ f_U\,]^T \\
\varepsilon &= [\varepsilon_x\ \varepsilon_y\ \varepsilon_z]^T,\ \nabla = [\nabla_x\ \nabla_y\ \nabla_z]^T \\
\omega_{ie} &= [\,0\ \Omega_N\ \Omega_U\,]^T,\ \Omega_N = \omega_e \cos L,\ \Omega_U = \omega_e \sin L \\
\omega_{in} &= [-\dot{L}\ \dot{\lambda}\cos L + \Omega_N\ \dot{\lambda}\sin L + \Omega_U]^T,\ \dot{L} = V_N/(R_M + h) \\
\dot{\lambda} &= V_E \sec L/(R_N + h),\ R_M = R_e(1 - 2e + 3e\sin L^2) \\
R_N &= R_e(1 + e\sin L^2),\ \rho = [-\dot{L}\ \dot{\lambda}\cos L\ \ -\dot{\lambda}\sin L]^T, \\
C_b^n &= \begin{bmatrix} \cos\gamma\cos\psi - \sin\gamma\sin\theta\sin\psi & -\cos\theta\sin\psi & \sin\gamma\cos\psi + \cos\gamma\sin\theta\sin\gamma \\ \cos\gamma\sin\psi + \sin\gamma\sin\theta\cos\psi & \cos\theta\cos\psi & \sin\gamma\sin\psi - \cos\gamma\sin\theta\cos\psi \\ -\sin\gamma\cos\theta & \sin\theta & \cos\gamma\cos\theta \end{bmatrix},
\end{aligned}$$

$$C_n^{n,ins} = \begin{bmatrix} \cos\phi_N\cos\phi_U - \sin\phi_N\sin\phi_E\sin\phi_U & \cos\phi_N\sin\phi_U + \sin\phi_N\sin\phi_E\cos\phi_U & -\sin\phi_N\cos\phi_E \\ -\cos\phi_E\sin\phi_U & \cos\phi_E\cos\phi_U & \sin\phi_E \\ \sin\phi_N\cos\phi_U + \cos\phi_N\sin\phi_E\sin\phi_U & \sin\phi_N\sin\phi_U - \cos\phi_N\sin\phi_E\cos\phi_U & \cos\phi_N\cos\phi_E \end{bmatrix}. \tag{20.2}$$

Let

$$x(t) = [\,\phi^T\ (\delta V)^T\ (\delta p)^T\ \varepsilon^T\ \nabla^T\,]^T, \tag{20.3}$$

$$\vartheta(t) = [\,\varepsilon_w^T\ \nabla_w^T\,]^T. \tag{20.4}$$

Table 20.1 Parameters of the INS/GPS system

Variables	Description
$\phi_E,\ \phi_N,\ \phi_U$	Attitude errors
$\delta V_E,\ \delta V_N,\ \delta V_U$	Velocity errors
$\delta L,\ \delta\lambda,\ \delta h$	Position errors in latitude, longitude and height directions, respectively
$f_E,\ f_N,\ f_U$	Specific forces
$\varepsilon_x,\ \varepsilon_y,\ \varepsilon_z$	Constant gyroscope bias in the body frame b
$\nabla_x,\ \nabla_y,\ \nabla_z$	Accelerometer bias in the body frame b
$\varepsilon_w,\ \nabla_w$	Gyroscopes and accelerometers noises, respectively
ω_{ie}	Local Earth rotation rate vector in the inertial frame i
ω_e	Local Earth rotation rate
ω_{in}	Rotation velocity of the navigation frame n relative to inertial frame i expressed in n-frame
R_e	Semi-major axis of the ellipsoid with e denoting the eccentricity
C_b^n	Attitude transformation matrix between b-frame and n-frame
ψ, θ, γ	Heading, pitch and roll angles in the n-frame, respectively
$C_n^{n,ins}$	A small rotation of frame n computed by INS relative to the true frame

The subscripts E, N, U denote the east, north and up components in the n-frame

The continuous-time nonlinear state equation of INS/GPS integrated system is then given by

$$\dot{x}(t) = f_c(x(t)) + B_c(t)\vartheta(t), \tag{20.5}$$

where $f_c(x)$ and $B_c(t)$ are given by

$$f_c(x) = \begin{bmatrix} (I - C_n^{n,ins})\omega_{in}^n + F_{11}\cdot\delta V + F_{12}\cdot\delta p \\ (I - C_n^{n,ins})C_b^n f^b + F_{21}\cdot\delta V + F_{22}\cdot\delta p \\ F_{31}\cdot\delta V + F_{32}\cdot\delta p \\ 0_{6\times1} \end{bmatrix},$$

$$F_{11} = \begin{bmatrix} 0 & -1/(R_M+h) & 0 \\ 1/(R_N+h) & 0 & 0 \\ \tan L/(R_N+h) & 0 & 0 \end{bmatrix},$$

$$F_{12} = \begin{bmatrix} 0 & 0\ 0 \\ -\Omega_U & 0\ 0 \\ \Omega_N + \dot{\lambda}\sec^2 L & 0\ 0 \end{bmatrix},$$

$$F_{21} = \begin{bmatrix} \dot{L}\tan L - \dot{h} & 2\Omega_U + \dot{\lambda}\sin L & -\omega_{inN} - \Omega_N \\ -2\omega_{inU} & -\dot{h} & -\dot{L} \\ 2\omega_{inN} & 2\dot{L} & 0 \end{bmatrix},$$

$$F_{22} = \begin{bmatrix} (\dot{\lambda}\sec^2 L + 2\Omega_N)V_N + 2\Omega_U V_U & 0 & 0 \\ -(\dot{\lambda}\sec^2 L + 2\Omega_N)V_E & 0 & 0 \\ -2\Omega_U V_E & 0 & 0 \end{bmatrix},$$

$$F_{31} = \begin{bmatrix} 0 & 1/(R_M + h) & 0 \\ \sec L/(R_N + h) & 0 & 0 \\ 0 & 0 & 1 \end{bmatrix},$$

$$F_{32} = \begin{bmatrix} 0 & 0 & 0 \\ \dot{\lambda}\sin L/\cos^2 L & 0 & 0 \\ 0 & 0 & 0 \end{bmatrix}, \quad B_c(t) = \begin{bmatrix} C_b^n & 0_{3\times 3} \\ 0_{3\times 3} & C_b^n \\ 0_{9\times 3} & 0_{9\times 3} \end{bmatrix}. \tag{20.6}$$

As usual, one can treat the differences between the output of INS and output of GPS as the measurement output of the INS/GPS integrated system, i.e.,

$$y(t) = [\,(\delta V)^T \ (\delta p)^T\,]^T, \tag{20.7}$$

with $\delta V = V_{INS}(t) - V_{GPS}(t),\ \delta p = p_{INS}(t) - p_{GPS}(t)$. We then have the following measurement output equation

$$y(t) = C(t)x(t) + v(t), \tag{20.8}$$

where

$$C(t) = \begin{bmatrix} 0_{3\times 3} & I_{3\times 3} & 0_{3\times 3} & 0_{3\times 6} \\ 0_{3\times 3} & 0_{3\times 3} & I_{3\times 3} & 0_{3\times 6} \end{bmatrix} \tag{20.9}$$

and $v(t)$ is the measurement noise. So, the INS/GPS integrated system can be described by the nonlinear state space model (20.5) and (20.8), i.e.,

$$\begin{cases} \dot{x}(t) = f_c(x(t)) + B_c(t)\vartheta(t), \\ y(t) = C(t)x(t) + v(t). \end{cases} \tag{20.10}$$

20.2.2 Model II

Consider to model the INS/GPS integrated system with the local ENU coordinate frame being used as the navigation frame n. According to [1], the INS dynamic model can also be expressed by the following equations

$$\begin{cases} \dot{\phi} = \delta\omega_{in} - \omega_{in} \times \phi + C_b^n \varepsilon, \\ \delta\dot{V} = f_s \times \phi - \delta\,[(\omega_{ie} + \omega_{in}) \times V] + C_b^n \nabla, \\ \delta\dot{p} = \delta v + \rho \times \delta p, \end{cases} \tag{20.11}$$

where the notations are referred to Table 20.1 and (20.2).

Let

$$x(t) = [\phi^T \quad \delta V^T \quad \delta p^T]^T, \; f(t) = [\varepsilon^T \quad \nabla^T]^T.$$

Then the stochastic linear time-varying state space equation of INS/GPS integrated system can be given by

$$\dot{x}(t) = A(t)x(t) + B_f(t)f(t) + B_w(t)w(t), \tag{20.12}$$

where $w(t)$ is the white noise process of the inertial sensors with zero mean and covariance matrix $Q(t)$, the matrices $A(t)$, $B_f(t)$ and $B_\omega(t)$ are

$$A(t) = \begin{bmatrix} A_1 & A_2 & A_3 \\ A_4 & A_5 & A_6 \\ 0_{3\times3} & I_{3\times3} & A_7 \end{bmatrix},$$

$$A_1 = \begin{bmatrix} 0 & \Omega_U + \dot{\lambda}\sin L & -(\Omega_N + \dot{\lambda}\cos L) \\ -(\Omega_U + \dot{\lambda}\sin L) & 0 & -\dot{L} \\ \Omega_N + \dot{\lambda}\cos L & \dot{L} & 0 \end{bmatrix},$$

$$A_2 = \begin{bmatrix} 0 & -1/(R_M + h) & 0 \\ 1/(R_N + h) & 0 & 0 \\ \tan L/(R_N + h) & 0 & 0 \end{bmatrix},$$

$$A_3 = \begin{bmatrix} 0 & 0 & 0 \\ -\Omega_U & 0 & 0 \\ \Omega_N + \dot{\lambda}\sec L & 0 & 0 \end{bmatrix}, \; A_4 = \begin{bmatrix} 0 & -f_U & f_N \\ f_U & 0 & -f_E \\ -f_N & f_E & 0 \end{bmatrix},$$

$$A_5 = \begin{bmatrix} \dot{L}\tan L - \dot{h} & 2\Omega_U + \dot{\lambda}\sin L & -2\Omega_N - \dot{\lambda}\cos L \\ -2(\Omega_U + \dot{\lambda}\sin L) & -\dot{h} & -\dot{L} \\ 2(\Omega_N + \dot{\lambda}\cos L) & 2\dot{L} & 0 \end{bmatrix},$$

$$A_6 = \begin{bmatrix} 2(\Omega_N v_N + \Omega_U v_U) + \dot{\lambda} v_N \sec L & 0 & 0 \\ -(2\Omega_N + \dot{\lambda}\sec L)v_E & 0 & 0 \\ -2\Omega_U v_E & 0 & 0 \end{bmatrix},$$

$$A_7 = \begin{bmatrix} 0 & \dot{\lambda}\sin L & \dot{\lambda}\cos L \\ -\dot{\lambda}\sin L & 0 & \dot{L} \\ -\dot{\lambda}\cos L & -\dot{L} & 0 \end{bmatrix}, \; B_f(t) = B_w(t) = \begin{bmatrix} C_b^n & 0_{3\times3} \\ 0_{3\times3} & C_b^n \\ 0_{3\times3} & 0_{3\times3} \end{bmatrix}.$$

The measurement of the INS/GPS integrated system is obtained from the velocity and position differences between the INS and the GPS solutions. Then, by defining the system output as

$$y(t) = [\,(\delta V)^T \; (\delta p)^T\,]^T,$$

the following 15-dimension state space model of INS/GPS integrated system is often used

$$\begin{cases} \dot{x}(t) = A(t)x(t) + B_f(t)f(t) + B_w(t)w(t), \\ y(t) = C(t)x(t) + v(t), \end{cases} \tag{20.13}$$

where $v(t)$ denotes the measurement noise vector and

$$C(t) = \begin{bmatrix} 0_{3\times3} & I_{3\times3} & 0_{3\times3} \\ 0_{3\times3} & 0_{3\times3} & I_{3\times3} \end{bmatrix}.$$

20.2.3 Model III

Using the ENU coordinate frame as the navigation frame n, according to [5], the system dynamic model of the INS/GPS integrated system can be simply described by the following nonlinear INS equations

$$\begin{cases} \dot{\phi} = (I - C_n^p)\omega_{in} + \Delta\omega_{in} + C_b^n\varepsilon, \\ \delta\dot{V} = (I - C_p^n)f_s - \Delta[(\omega_{ie} + \omega_{in}) \times V] + C_b^n\nabla, \\ \delta\dot{p} = \delta v, \end{cases} \tag{20.14}$$

where the notations are same with the corresponding ones in Table 20.1 and (20.2) with $C_n^p = C_n^{n,ins}$.

By defining the system state and the output variables respectively as (20.3) and (20.7), the state space model of system (20.14) can be obtained as

$$\begin{cases} \dot{x}(t) = f(t, x(t)) + G(t)\vartheta(t) \\ \qquad = f_N(t, x(t)) + f_L(t)x(t) + G(t)\vartheta(t), \\ y(t) = C(t)x(t) + v(t). \end{cases} \tag{20.15}$$

where $C(t)$ is given in (20.9) and

$$f_N(t, x(t)) = \begin{bmatrix} (I - C_n^p)\omega_{in} \\ (I - C_p^n)f_s \\ 0_{9\times1} \end{bmatrix}, \; f_L(t) = \begin{bmatrix} 0_{3\times3} & A_1 & A_2 & C_b^n \\ 0_{3\times3} & A_3 & A_4 & C_b^n \\ 0_{3\times3} & A_5 & A_6 & 0_{3\times3} \end{bmatrix}, \; G(t) = \begin{bmatrix} C_b^n & 0_{3\times3} \\ 0_{3\times3} & C_b^n \\ 0_{9\times3} & 0_{9\times3} \end{bmatrix},$$

$$A_1 = \begin{bmatrix} 0 & -1/R_M & 0 \\ 1/R_N & 0 & 0 \\ \tan L/R_N & 0 & 0 \end{bmatrix}, \; A_2 = \begin{bmatrix} 0 & 0 & 0 \\ -\omega_N & 0 & 0 \\ \dot{\lambda}\sec L + \omega_N & 0 & 0 \end{bmatrix},$$

$$A_3 = \begin{bmatrix} \dot{L}\tan L - v_U/R_M & 2\omega_U + \dot{\lambda}\cos L & -2\omega_N - \dot{\lambda}\cos L \\ -2\omega_U - \dot{\lambda}\sin L & -v_U/R_M & -\dot{L} \\ 2(\omega_N + \dot{\lambda}\cos L) & 2\dot{L} & 0 \end{bmatrix},$$

$$A_4 = \begin{bmatrix} 2\omega_N v_N + \dot{L} v_E \sec^2 L + 2\omega_U v_U & 0 & 0 \\ -(2\omega_N v_E + \dot{\lambda} v_E \sec L) & 0 & 0 \\ -2\omega_U v_E & 0 & 0 \end{bmatrix},$$

$$A_5 = \begin{bmatrix} 0 & 1/R_M & 0 \\ \sec L/R_N & 0 & 0 \\ 0 & 0 & 1 \end{bmatrix}, \quad A_6 = \begin{bmatrix} 0 & 0 & -\dot{L}/R_M \\ \dot{\lambda}\tan L & 0 & -\dot{\lambda}/R_N \\ 0 & 0 & 0 \end{bmatrix}.$$

20.3 Problem Statement

In this chapter the INS/GPS integrated system modeled in form of (20.10) is concerned. It is well known that, if the stochastic properties of $w(t)$ and $v(t)$ are available, the INS/GPS integrated system can achieve a long-term high accuracy measurement by applying advanced filter techniques, such as the well known EKF or UKF. However, the properties of random error may be uncertain and variable in practical missions due to the influences of gyroscope and accelerometer degradations, thermal changes and electromagnetic interference [4, 12, 21], etc. Especially, the random error often leads to unexpected "bias" on the measurement of angular rate/acceleration and, if the "bias" is too serious to be compensated efficiently, it may even cause the IMU to malfunction. In addition, the GPS outlier is another possibly encountered error, resulting in evident GPS performance degradation [2, 23]. There is no doubt that the occurrence of faults in INS/GPS integrated system is inevitable. In particular, the INS/GPS integrated system used for POS often works in certain extreme conditions. With the increasing demands on high accuracy and reliability of POS, it is crucial to detect fault of INS/GPS integrated system [18].

Here we mainly consider the faults being described by unexpected "bias" of the gyroscopes and accelerometers, and the additive faults of GPS. When the potential faults are taken into account, a faulty case nonlinear state space model of the INS/GPS integrated system can be obtained as

$$\begin{cases} \dot{x}(t) = f_c(x(t)) + B_{cf}(t) f_1(t) + B_c(t)\vartheta(t), \\ y(t) = C(t)x(t) + D_{cf}(t) f_2(t) + v(t), \end{cases} \tag{20.16}$$

where $f_1(t)$ and $f_2(t)$ are the inertial sensor faults and GPS faults, respectively, $B_{cf}(t)$ and $D_{cf}(t)$ are known matrices with $B_{cf}(t) = B_c(t)$ and $D_{cf}(t) = I$.

Suppose that the sampling period is T_{GPS}, i.e., the sampling period of GPS, and a zero-order-hold is chosen for the model (20.16). Denote by $\Psi((k+1)T_{GPS}, kT_{GPS})$

the state transition matrix of (20.16). The faulty case discrete-time model of the INS/GPS integrated system is now obtained as

$$\begin{cases} x(k+1) = \varphi_F(x(k)) + B_f(k)f(k) + B(k)\vartheta(k), \\ y(k) = C(k)x(k) + D_f(k)f(k) + v(k), \end{cases} \tag{20.17}$$

where

$$f(k) = [f_1^T(k)\ f_2^T(k)]^T, \quad \varphi_F(x(k)) = \Psi((k+1)T_{GPS}, kT_{GPS}),$$

$$\begin{aligned} B(k) &= \int_{kT_{GPS}}^{(k+1)T_{GPS}} \Psi((k+1)T_{GPS}, \tau)B_c(\tau)d\tau, \\ B_f(k) &= \left[\int_{kT_{GPS}}^{(k+1)T_{GPS}} \Psi((k+1)T_{GPS}, \tau)B_{cf}(\tau)d\tau \ \ 0\right], \\ C(k) &= C(kT_{GPS}), \ D_f(k) = \begin{bmatrix} 0 & D_{cf}(kT_{GPS}) \end{bmatrix}. \end{aligned}$$

Hence, given nonlinear system (20.17), the main problem to be addressed in this chapter is to design an FD system to achieve satisfactory performance regarding sensitivity to faults and robustness against disturbances.

20.4 Design of an Extended H_i/H_∞-Optimal FD System

Given nonlinear system (20.17), the following observer-based FDF is considered as a residual generator

$$\begin{cases} \hat{x}(k+1) = \varphi_F(\hat{x}(k)) + L(k)\tilde{y}(k), \\ \tilde{y}(k) = y(k) - C(k)\hat{x}(k), \\ r(k) = W(k)\tilde{y}(k), \end{cases} \tag{20.18}$$

where $r(k)$ is the residual, $L(k)$ the gain matrix and $W(k)$ the post-filter. The modified INS/GPS integration architecture with FD is shown in Fig. 20.1.

Without loss of generality, it is assumed that $\vartheta(k)$, $v(k)$ subject by INS/GPS integrated system are energy bounded , i.e., $\vartheta(k),\ v(k) \in l_{2,[0,N]}$. Define

$$||G_{rd}||_{\infty,[0,N]} = \sup_{f=0} \frac{\sum_{k=0}^{N} ||r(k)||^2}{x_0^T \Sigma_0^{-1} x_0 + \sum_{k=0}^{N}(||\vartheta(k)||^2_{\Theta^{-1}} + ||v(k)||^2_{\Xi^{-1}})},$$

$$||G_{rf}||_{\infty,[0,N]} = \sup_{x_0,w,v=0} \frac{||r(k)||^2}{||f(k)||^2}, \ ||G_{rf}||_{-,[0,N]} = \inf_{x_0,w,v=0} \frac{\sum_{k=0}^{N} \|r(k)\|^2}{\sum_{k=0}^{N} \|f(k)\|^2},$$

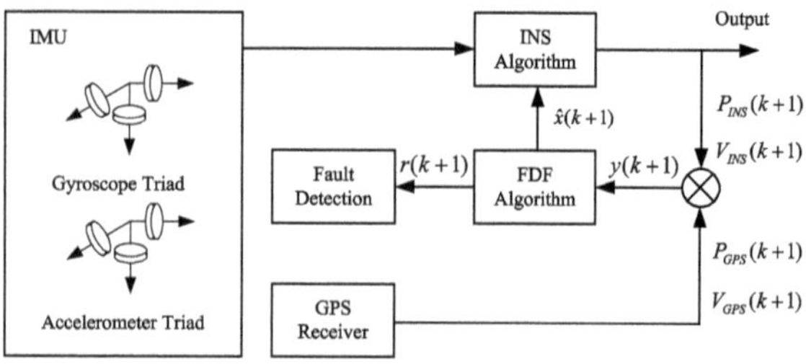

Fig. 20.1 The INS/GPS integration architecture with FD

where

$$\sum_{k=0}^{N} ||\vartheta(k)||^2_{\Theta^{-1}} = \sum_{k=0}^{N} \vartheta^T(k)\Theta^{-1}\vartheta(k) < +\infty,$$
$$\sum_{k=0}^{N} ||v(k)||^2_{\Xi^{-1}} = \sum_{k=0}^{N} v^T(k)\Xi^{-1}v(k) < +\infty$$

with Σ_0, Θ, Ξ being the weighting matrices. In the H_i/H_∞-optimization framework, we formulate the design of the FDF for system (20.17) as to find $L(k)$ and $W(k)$ satisfying

$$\max_{L(k),W(k)} \frac{\|G_{rf}\|_{\infty,[0,N]}}{\|G_{rd}\|_{\infty,[0,N]}}, \quad \text{and/or} \quad \max_{L(k),W(k)} \frac{\|G_{rf}\|_{-,[0,N]}}{\|G_{rd}\|_{\infty,[0,N]}}. \tag{20.19}$$

In the stage of residual evaluation, we suppose

$$x_0^T \Sigma_0^{-1} x_0 + \sum_{k=0}^{N} (\vartheta^T(k)\Theta^{-1}\vartheta(k) + v^T(k)\Xi^{-1}v(k)) \le \delta_d^2. \tag{20.20}$$

The $l_{2,[0,N]}$-norm type evaluation function J_N and threshold $J_{th} = J_{th,0}$ with a zero FAR can be selected as

$$J_N = \frac{1}{N+1}\sum_{k=0}^{N} r^T(k)r(k), \tag{20.21}$$

$$J_{th,0} = \frac{1}{N+1}||G_{rd}||_{\infty,[0,N]}\delta_d^2, \tag{20.22}$$

respectively. The presence of a fault can thus be detected by using

$$\begin{cases} J_N \le J_{th}, & \text{no alarm,} \\ J_N > J_{th}, & \text{fault alarm.} \end{cases} \tag{20.23}$$

Remark 20.1 It should be pointed out that the statistical properties of IMU random errors may be uncertain for a POS used in aerial remote imaging, due to the influences of the flight environment. Here, the weighting matrices $\Sigma_0,\ \Theta,\ \Xi$ in (20.20) are chosen as the normal case covariance matrices of $x_0,\ \vartheta(k)$ and $v(k)$, respectively. Note that, if $\Sigma_0 = I,\ \Theta = I,\ \Xi = I$, the FDF satisfying (20.19) delivers the extended H_-/H_∞-FDF of the method in Chap. 18.

20.4.1 EKF Based FDF

Firstly, we consider an EKF based solution to the FD for INS/GPS integrated system. It is assumed that $w(k),\ v(k)$ are zero mean Gaussian random variables satisfying

$$\mathbb{E}\left\langle \begin{bmatrix} x_0 \\ \vartheta(i) \\ v(i) \end{bmatrix}, \begin{bmatrix} x_0 \\ \vartheta(j) \\ v(j) \\ 1 \end{bmatrix} \right\rangle = \begin{bmatrix} P_0 & 0 & 0 & 0 \\ 0 & Q(i)\delta_{ij} & 0 & 0 \\ 0 & 0 & R(i)\delta_{ij} & 0 \end{bmatrix}, \tag{20.24}$$

where $Q(k)$ and $R(k)$ denote covariance matrices of inertial sensor random errors and GPS measurement noises, respectively. Thus the observer gain matrix $L(k)$ can be calculated recursively by applying EKF [20]

$$L(k) = F(k)P(k)C^T(k)R_{\tilde{y}}^{-1}(k), \tag{20.25}$$

$$R_{\tilde{y}}(k) = C(k)P(k)C^T(k) + R(k), \tag{20.26}$$

$$\begin{aligned} P(k+1) &= F(k)P(k)F^T(k) + B(k)Q(k)B^T(k), \\ &\quad -L(k)R_{\tilde{y}}(k)L^T(k),\ P(0) = P_0, \end{aligned} \tag{20.27}$$

where $F(k)$ is the Jacobian matrix of $\varphi_F(x(k))$, i.e.,

$$F(k) = \frac{\partial \varphi_F}{\partial x}|_{x(k)=\hat{x}(k)}.$$

Let $W(k) = R_{\tilde{y}}^{-1/2}(k)$. In the case of fault free, $r(k)$ is also Gaussian white. By using J_N in (20.21) as an evaluation function, J_N in fault-free case follows central χ^2 distribution with freedom degrees of 6. By setting a threshold as $J_{th,1} = \chi^2_{6,\epsilon}$ for a preset FAR $\epsilon > 0$, the presence of a fault can be detected by the using (20.23) with $J_{th} = J_{th,1}$ with FAR $= \epsilon$.

20.4.2 An Extended H_i/H_∞-FDF

It should be pointed out that the IMU is fixed to the imaging sensor and usually exposed outside of the aircraft for INS/GPS integrated system applied in POS. So, the statistical characteristics of process noise $\vartheta(k)$ and measurement noise $\upsilon(k)$ are often uncertain and not exactly known. In this subsection, we turn to the design of an extended H_i/H_∞-FDF for the INS/GPS integrated system with $\vartheta(k), \upsilon(k) \in l_{2,[0,N]}$.

Without loss of generality, it is assumed that $\vartheta(k)$ and $\upsilon(k)$ satisfy (20.20) with $\Sigma_0 = P_0,\ \Theta = Q(k),\ \Xi = R(k)$. Recall that EKF is widely used in INS/GPS integrated system and its stability could be guaranteed in general [13]. Our basic idea is to choose the $L(k)$ given by (20.25)–(20.27) as the gain matrix of (20.18). Then the remaining task is to find a post-filter $W(k)$ such that the $(L(k), W(k))$ is a solution of the optimization problem (20.19). We thus can choose the evaluation function and threshold as (20.21)–(20.22) and detect the occurrence of a fault by using (20.23).

For ease of notation, we define

$$
\begin{aligned}
&\bar{w}(k) = Q^{-1/2}(k)\vartheta(k),\ \ \bar{\upsilon}(k) = R^{-1/2}(k)\upsilon(k),\\
&w(k) = [\bar{w}^T(k)\ \ \bar{\upsilon}^T(k)]^T,\ e(k) = x(k) - \hat{x}(k),\\
&\bar{B}(k) = B(k)Q^{1/2}(k),\ \ \bar{D}_\upsilon(k) = R^{1/2}(k),\\
&B_L(k) = [\bar{B}(k) \quad -L(k)\bar{D}_\upsilon(k)],\ \ D_L(k) = [I \quad \bar{D}_\upsilon(k)],\\
&\varphi_F(x(k)) \simeq \varphi_F(\hat{x}(k)) + \frac{\partial \varphi_F}{\partial x}|_{x(k)=\hat{x}(k)} e(k).
\end{aligned}
$$

Subtracting (20.18) from (20.17) yields

$$
\begin{cases}
e(k+1) = (F(k) - L(k)C(k))e(k) + B_L(k)w(k)\\
\qquad\qquad +(B_f(k) - L(k)D_f(k))f(k),\\
\tilde{y}(k) = C(k)e(k) + D_L(k)w(k) + D_f(k)f(k),
\end{cases}
\tag{20.28}
$$

which is an LDTV system. Recalling the formulations in Sect. 18.3.2, the system (20.28) can be further represented by

$$
r_N = W_N(\Omega x_0 + \Gamma_w w_N + \Gamma_f f_N), \tag{20.29}
$$

where the involved parameters $r_N,\ w_N,\ f_N,\ W_N,\ \Omega,\ \Gamma_w$ and Γ_f are referred to Sect. 18.3.2.

The robustness and fault sensitivity criteria can be rewritten as

$$
\begin{aligned}
\|G_{rd}\|_{\infty,[0,N]} &= \sup_{w \in l_2[0,N]} \frac{\|r_{dN}\|^2}{x_0^T P_0^{-1} x_0 + \|w_N\|^2},\\
\|G_{rf}\|_{\infty,[0,N]} &= \sup_{f \in l_2[0,N]} \frac{\|r_{fN}\|^2}{\|f_N\|^2},\quad \|G_{rf}\|_{-,[0,N]} = \inf_{f \in l_2[0,N]} \frac{\|r_{fN}\|^2}{\|f_N\|^2},
\end{aligned}
$$

respectively, where $r_{dN} = W_N(\Omega x_0 + \Gamma_w w_N)$, $r_{fN} = W_N \Gamma_f f_N$. Let

$$G_{rd,N} = W_N \left[\Omega P_0^{1/2} \; \Gamma_w \right], \quad G_{rf,N} = W_N \Gamma_f, \quad R_{rdN} = G_{rd,N} G_{rd,N}^T.$$

We have

$$\begin{aligned} \|G_{rd}\|_{\infty,[0,N]} &= \bar{\sigma}(G_{rd,N}), \\ \left\|G_{rf}\right\|_{\infty,[0,N]} &= \bar{\sigma}(G_{rf,N}), \\ \left\|G_{rf}\right\|_{-,[0,N]} &= \underline{\sigma}(G_{rf,N}). \end{aligned}$$

Thus the optimization problem (20.19) are equivalent respectively to

$$\max_{L(k),W(k)} \frac{\bar{\sigma}(G_{rf,N})}{\bar{\sigma}(G_{rd,N})} \quad \text{and/or} \quad \max_{L(k),W(k)} \frac{\underline{\sigma}(G_{rf,N})}{\bar{\sigma}(G_{rd,N})}. \tag{20.30}$$

It is known from the statistical characteristics of the EKF innovations that

$$R_{\tilde{y}}(k) = C(k)P(k)C^T(k) + R(k), \;\; R_{\tilde{y}_N} = \mathrm{diag}(R_{\tilde{y}}(0), R_{\tilde{y}}(1), \cdots, R_{\tilde{y}}(N)).$$

Set the post-filter as $W(k) = R_{\tilde{y}}^{-1/2}(k)$. We then have

$$\begin{aligned} & R_{rdN} = W_N(\Omega P_0 \Omega^T + \Gamma_w \Gamma_w^T) W_N^T = W_N R_{\tilde{y}_N} W_N^T = I. \\ & \|G_{rd}\|_{\infty,[0,N]} = \bar{\sigma}(G_{rd,N}) = \lambda_{max}(R_{rdN}) = 1. \end{aligned}$$

Hence,

$$\frac{\bar{\sigma}(G_{rf,N})}{\bar{\sigma}(G_{rd,N})} = \bar{\sigma}(G_{rf,N}), \quad \frac{\underline{\sigma}(G_{rf,N})}{\bar{\sigma}(G_{rd,N})} = \underline{\sigma}(G_{rf,N}). \tag{20.31}$$

To show that the $L(k)$ given by (20.25)–(20.27) and the $W(k) = R_{\tilde{y}}^{-1/2}(k)$ satisfy (20.19), we now consider any other given matrices $L_a(k)$ $(k = 0, 1, \cdots, N)$ which guarantee the stability of system $e(k+1) = (F(k) - L_a(k)C(k))e(k)$. The error dynamic model (20.28) with $L(k) = L_a(k)$ becomes

$$\begin{cases} e_a(k+1) = A_{L_a}(k)e_a(k) + \bar{B}_a(k)\bar{w}(k) \\ \qquad\qquad - L_a(k)\bar{D}_v\bar{v}(k) + B_{fa}(k)f(k) \\ \tilde{y}_a(k) = C(k)e_a(k) + \bar{D}_v\bar{v}(k) + D_f(k)f(k) \end{cases} \tag{20.32}$$

where

$$A_{L_a}(k) = F(k) - L_a(k)C(k), \quad B_{fa}(k) = B_f(k) - L_a(k)D_f(k).$$

It is known from Proposition 18.2 that there exists a mapping $\mathcal{M}_a(k): \tilde{y}_a(k) \rightarrow \tilde{y}(k)$, i.e.,

$$\begin{cases} \eta(k+1) = (F(k) - L(k)C(k))\eta(k) + (L_a(k) - L(k))\tilde{y}_a(k), \\ \tilde{y}(k) = C(k)\eta(k) + \tilde{y}_a(k), \quad \eta(0) = 0. \end{cases}$$

If the post-filter is set as $W(k) = W_a(k)$, we can further get

$$r_{aN} = W_{aN}\tilde{y}_{aN} = W_{aN}M_{aN}\tilde{y}_N = W_{aN}M_{aN}W_N^{-1}r_N,$$

where $W_{aN} = \text{diag}\{W_a(0), W_a(1)..., W_a(N)\}$, M_{aN} is obtained by replacing $B_f(k), D_f(k), L(k)$ in Γ_f with $L(k), I$ and $L_a(k))$, respectively. Let $M_{WN} = W_{aN}M_{aN}W_N^{-1}$, we then have

$$\frac{\bar{\sigma}(G_{raf,N})}{\bar{\sigma}(G_{rad,N})} = \frac{\bar{\sigma}(M_{WN}G_{rf,N})}{\bar{\sigma}(M_{WN}G_{rd,N})} = \frac{\bar{\sigma}(M_{WN}G_{rf,N})}{\bar{\sigma}(M_{WN})} \leq \bar{\sigma}(G_{rf,N}),$$
$$\frac{\underline{\sigma}(G_{raf,N})}{\bar{\sigma}(G_{rad,N})} = \frac{\underline{\sigma}(M_{WN}G_{rf,N})}{\bar{\sigma}(M_{WN}G_{rd,N})} = \frac{\underline{\sigma}(M_{WN}G_{rf,N})}{\bar{\sigma}(M_{WN})} \leq \underline{\sigma}(G_{rf,N}).$$

Observing (20.31), it is easy to see that the $L(k)$ given by (20.25)–(20.27) and $W(k) = R_{\tilde{y}}^{-1/2}(k)$ provide us with a solution to (20.30). That means that the EKF with $W(k) = R_{\tilde{y}}^{-1/2}(k)$ is an extended H_i/H_∞-FDF. Moreover, since in the fault free case

$$J_N \leq \frac{1}{N+1}||G_{rd}||_{\infty,[0,N]}\delta_d^2 = \frac{1}{N+1}\delta_d^2$$

and $\|G_{rd}\|_{\infty,[0,N]} = 1$, we can choose a threshold as $J_{th} = J_{th,2} = \frac{1}{N+1}\delta_d^2$ and, based on this, the detection of a fault can be carried out by using (20.23).

For an INS/GPS integrated system subject to energy bounded random errors and measurement noises, the extended H_i/H_∞-optimization based FD is summarized in Algorithm 20.4.16.

Remark 20.2 We would like to say that it is difficult to find a stable residual generator satisfying (20.19). After building a relationship between EKF and the extended H_i/H_∞-FDF, a feasible solution of H_i/H_∞-optimization based FD is obtained via recursive computation. It is known that EKF is widely used for INS/GPS integrated system subject to (20.24) and it is easily implemented in practice. For the case of $\vartheta(k), \upsilon(k) \in l_{2,[0,N]}$, residual generator (20.18) with (20.25)–(20.27) delivers an extended H_i/H_∞-FDF and an optimal trade-off between the sensitivity of residual to fault and the robustness to unknown input can be achieved. For the purpose of real application, we can directly calculate the evaluation function J_N by using Algorithm 20.4.16, while the threshold can be determined based on analyzing the characteristics of $\vartheta(k)$ and $\upsilon(k)$. Randomized algorithms given in Chap. 18 can be well applied to this end.

Algorithm 20.4.16 Extended H_i/H_∞-optimization based FD for INS/GPS integrated systems

1: Let $k=0$, and set $\hat{x}(0)=0$;
2: Calculate $L(k)$ and $W(k)$ using (20.25)–(20.27) with $W(k)=R_{\tilde{y}}^{-1/2}(k)$;
3: Update $y(k+1)$ using (20.7);
4: Calculate $\hat{x}(k+1)$ and $r(k+1)$ using (20.18);
5: Correct the solution by using $\hat{x}(k+1)$, i.e., $\bar{C}_b^n=(I-\hat{\phi})\times C_b^n$, $\bar{V}=V-\delta V$ and $\bar{P}=p-\delta p$, respectively;
6: Reset $\hat{x}(k+1)=[0_{1\times 9}, \varepsilon^T(k+1), \nabla^T(k+1)]$ and calculate J_N by using (20.21);
7: Choose a threshold $J_{th}=\frac{1}{N+1}\delta_d^2$ and detect the occurrence of a fault using (20.23);
8: Let $k=k+1$, go to Step 2 until $k=5\times 10^5$.

20.5 Experimental Results and Discussion

20.5.1 Experimental Setup

To demonstrate the effectiveness of the presented method, flight experiment data from a ring laser gyroscope (RLG) based POS is applied. The experiment was carried out for aerial mapping with one aircraft shown in Fig. 20.2a. The installation of the major experimental equipments are shown in Fig. 20.2b. The IMU of the POS was mounted rigidly on an inertial stable platform (ISP), which is used to isolate the imaging load and IMU from disturbing motions of aircraft. The experiment lasted for nearly 6000 s and the average mapping altitude was about 500 m. The flight trajectory from the GPS solution is illustrated in Fig. 20.2c. The sensor specifications of the POS are listed in Table 20.2. The update frequency of the INS and the GPS are 200 Hz and 10 Hz, respectively. In addition, the physical constants and the initial navigation states for the INS algorithm are listed in Table 20.3.

In the following of this experiment study, data from a normal flight mission are employed as fault free results. The GPS outlier and unexpected inertial sensor bias

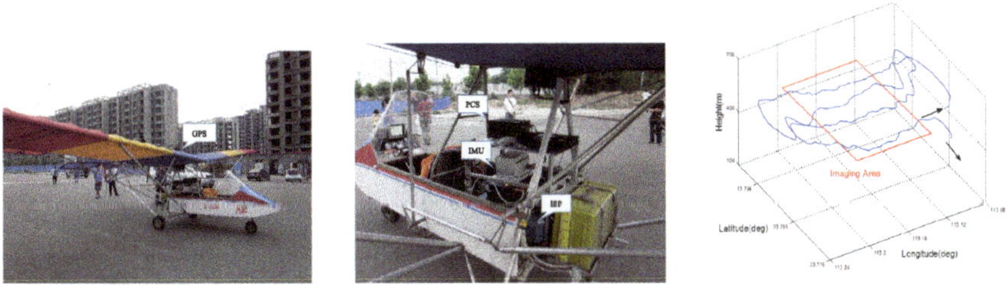

(a) (b) (c)

Fig. 20.2 **a** The A2C aircraft for flight experiment; **b** The installation of the RLG POS; **c** The flight trajectory of the experiment

Table 20.2 The specifications of INS/GPS System

Sensor	Error	Quantity
Gyroscope	Random constant	$0.01°$/h (1σ)
	Random noise	$0.01°$/h (1σ)
Accelerometer	Random constant	$100\,\mu$g (1σ)
	Random noise	$50\,\mu$g (1σ)
GPS	Velocity error	0.03 m/s (RMS)
Receiver	Position error	0.05 m (RMS)

Table 20.3 The physical constants and the initial navigation states

Term	Quantity	Term	Quantity
L_0	33.777°	ψ_0	90.168°
λ_0	113.183°	θ_0	−1.766°
h_0	110.616 m	γ_0	5.659°
ω_e	15.041 °/h	e	1/298.325
R_e	6378.135 km		

are considered as faults. Both the EKF and extended H_i/H_∞-FDF are designed to detect the occurrence of the faults respectively, i.e., the Case I and Case II.

20.5.2 Case I: EKF Based Experimental Results

Firstly, we consider the case of $\vartheta(k)$, $v(k)$ being Gaussian white noises. In this case, an EKF is used as a residual generator for detecting the GPS outlier or inertial sensor bias fault. The evaluation function defined in (20.21) follows a central χ^2 distribution with freedom 6. Set the FAR as 0.05, the threshold for FD is readily obtained from the table of χ^2 quantities as

$$J_{th,1} = \chi^2_{0.05}(6) = 12.592.$$

Set $N = 100$ and calculate the evaluation function at time step k as

$$J_{N,1}(k) = \frac{1}{N+1} \sum_{i=k-N}^{k} r^T(i) r(i). \tag{20.33}$$

In Fig. 20.3a, the time history of the GPS east velocity of this flight is shown. It is easy to see that two sets of GPS outliers occurred at about 1200 s and 2200 s, which may be a result of signal disturbances during the flight. The GPS outlier is taken as

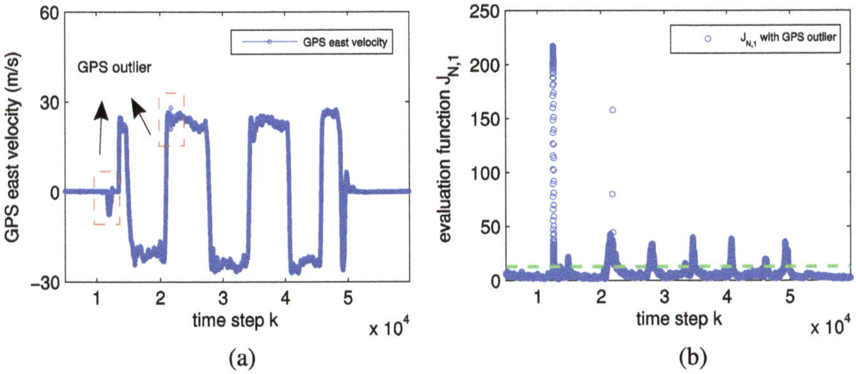

Fig. 20.3 **a** The time history of GPS east velocity; **b** The evaluation function $J_{N,1}(k)$ for the presence of GPS outliers in Case I

Fig. 20.4 The RLG drift in a temperature test

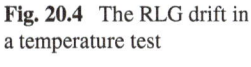

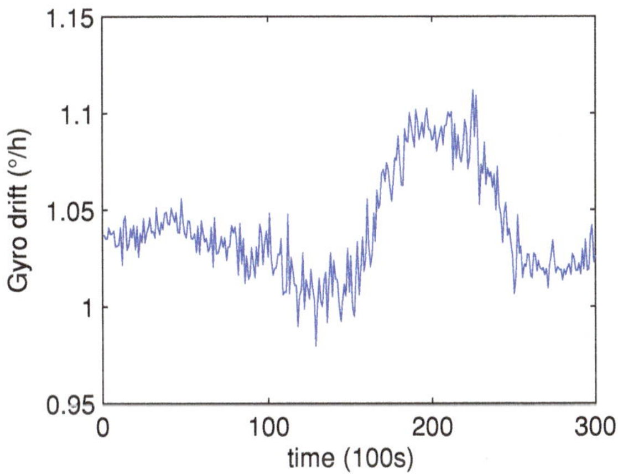

fault and the EKF based FD is applied. The evolution of evaluation function $J_{N,1}(k)$ is shown in Fig. 20.3b. It is seen that $J_{N,1}(k) > J_{th,1}$ after the appearance of GPS outliers and, therefore, fault alarms are delivered in time.

Next, we consider an unexpected inertial sensor bias caused by temperature drift. This kind of fault may be happen when ambient temperature is changed [3]. Figure 20.4 clearly exhibits a varying bias of an RLG in a typical temperature test. Thus it is of practical significance to detect the sensor fault in terms of varying bias occurred during the experiment. Without loss of generality, the inertial sensor fault is simulated as shown in Fig. 20.5a, and added to the raw gyroscope data from 3700 s to 3950 s, which is demonstrated in Fig. 20.5b. Then we can get the faulty data of inertial sensor bias. Applying the EKF based FD again, the evolution of $J_{N,1}(k)$ is shown in Fig. 20.6. Besides the peaks due to GPS outliers, one can see that the generated $J_{N,1}(k)$ arises remarkably when the simulated fault of inertial sensor bias occurred. Therefore, the inertial sensor faults can also be detected efficiently.

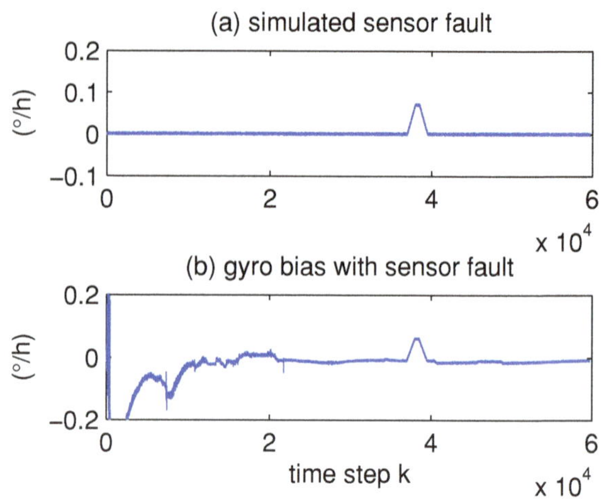

Fig. 20.5 **a** The simulated gyroscope fault; **b** The simulated fault is added to the raw gyroscope data

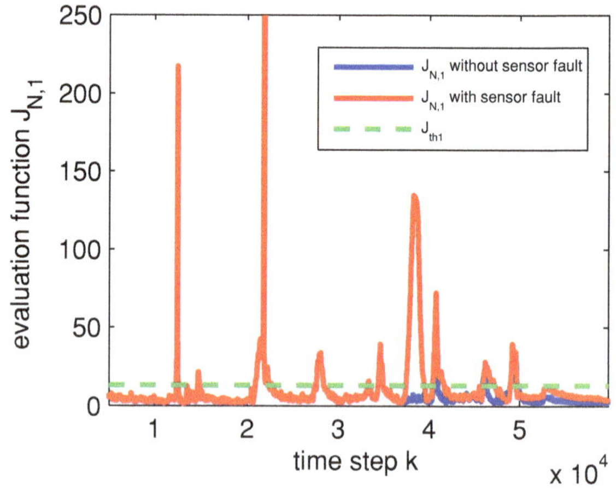

Fig. 20.6 The evaluation function $J_{N,1}(k)$ for the presence of inertial sensor bias in Case I

Remark 20.3 It is noted from Figs. 20.3 and 20.6 that, exclusive of the detection of the GPS and inertial sensor faults, this approach delivers false alarms associated with the turn-around maneuvers of the aircraft at 1400 s, 2100 s, 2700 s and so on. For real systems, the apparent maneuvers inevitably give rise to dynamic model mismatches and consequently the estimation fluctuations. The residual process reflects such fluctuations, so that the evaluation function value arises accordingly during each maneuver. In this sense, the selection of the threshold is essentially an important topic to reduce the false alarms, which directly involves the residual evaluation.

20.5.3 Case II: H_i/H_∞-FDF Based Experimental Results

In order to show the effectiveness of the extended H_i/H_∞-FDF, we now consider the case of $l_{2,[0,N]}$-norm bounded $\vartheta(k)$ and $v(k)$. For this purpose, the following vectors are simulated and add $C_b^n \Delta\vartheta(k)$, $C_b^n \Delta\vartheta(k)$ to the raw IMU and GPS data, respectively.

$$\begin{cases} \Delta\vartheta(k) = [4e^{-7}\ 4e^{-7}\ 4e^{-7}\ 5e^{-4}\ 5e^{-4}\ 5e^{-4}\ ]^T \sin 2\pi t, \\ \Delta v(k) = [0.03\ 0.03\ 0.03\ 1e^{-8}\ 8e^{-9}\ 0.05]^T \sin 2\pi t. \end{cases}$$

By applying Algorithm 20.4.16, the extended H_i/H_∞-FDF can be obtained. As a comparison, both the GPS outliers and inertial sensor fault are simulated as in Case I. The evaluation function $J_{N,2}(k)$ is calculated like identical with (20.33). The evolution of evaluation functions $J_{N,2}(k)$ in different faulty cases are shown in Figs. 20.7 and 20.8. Set a threshold as

$$J_{th,2} = \sup_{f=0} J_{N,2}(k) = 16.$$

Thus we can detect the occurrence of fault by applying (20.23). It is seen from Figs. 20.7 and 20.8 that the faults can be detected effectively for INS/GPS integrated system influenced by $l_{2,[0,N]}$-norm bounded disturbance, while the FAR is zero.

Remark 20.4 By applying (20.23), a fault alarm is delivered if $J_N > J_{th}$, which means the occurrence of a fault over time window $[0, N]$. In order to detect fault in time, it is better to choose the window width not too large. On the other hand, the extended H_-/H_∞-FDF is designed for the case of $\vartheta(k),\ v(k) \in l_{2,[0,N]}$. If a threshold is chosen so that the false alarm rate is zero, it is better the evaluation

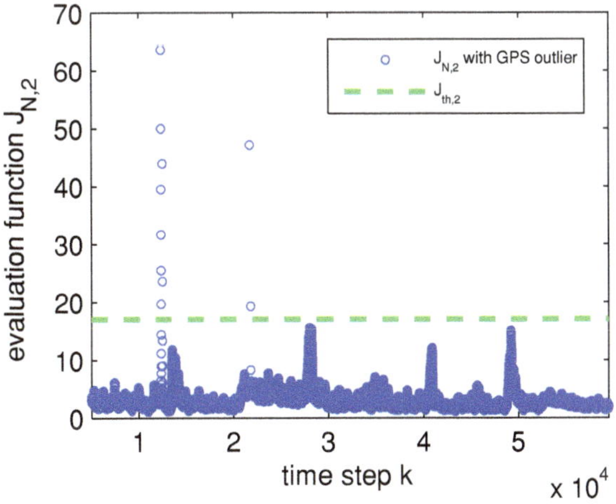

Fig. 20.7 The evolution of evaluation function $J_{N,2}(k)$ for the presence of GPS outliers in Case 2

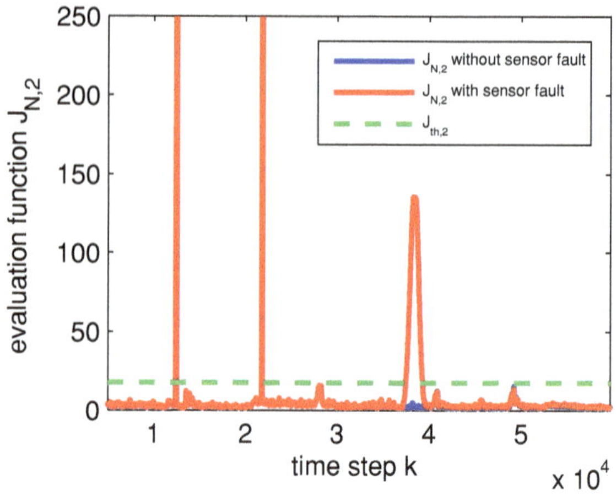

Fig. 20.8 The evaluation function $J_{N,2}(k)$ for the presence of inertial sensor bias in Case 2

function $J_N(k) = \frac{1}{N+1}\sum_{i=k-N}^{k} r^T(i)r(i)$ is sufficiently smooth in fault free case. In this case, the N should be large enough. The choice of N is not unique and, in this experiment, we set $N = 100$ for both Case I and Case II.

20.6 Conclusion

In this chapter, an extended H_i/H_∞-FDF have been applied to deal with FD issues of nonlinear INS/GPS integrated system. At the beginning, state space models of nonlinear INS/GPS integrated system have been established incorporating the potential GPS and inertial sensor faults. On this basis, an EKF based residual generator and the χ^2 test based evaluation strategy were used for FD under consideration of the disturbances being Gaussian noises. For the case of $l_{2,[0,N]}$-norm bounded disturbances, an extended H_i/H_∞-FDF has been designed, and then a norm-based evaluation function and a corresponding threshold with a zero FAR have been applied for FD. It is shown that the EKF delivers a unified form of residual generator for both the EKF based FD and the extended H_i/H_∞-optimal FD. The results on an aerial mapping experiment have demonstrated the applicability of the designed FD systems.

References

1. Bar-Itzhack, I. Y., & Berman, N. (1988). Control theoretic approach to inertial navigation systems. *Journal of Guidance, Control, and Dynamics, 11*(3), 237–245.
2. Chen, L., & Fang, J. (2014). A hybrid prediction method for bridging GPS outages in high-precision POS application. *IEEE Transactions on Instrumentation and Measurement, 63*(6), 1656–1665.
3. Cheng, J., Fang, J., Wu, W., et al. (2014). Temperature drift modeling and compensation of RLG based on PSO tuning SVM. *Measurement, 55*(3), 246–254.
4. Fang, J., & Li, J. (2009). Integrated model and compensation of thermal errors of silicon microelectromechanical gyroscope. *IEEE Transactions on Instrumentation and Measurement, 58*(9), 2923–2930.
5. Fang, J., & Sheng, Y. (2011). Study on innovation adaptive EKF for in-flight alignment of airborne POS. *IEEE Transactions on Instrumentation and Measurement, 60*(4), 1378–1388.
6. Gong, X., Zhang, R., & Fang, J. (2013). Application of unscented R-T-S smoothing on INS/GPS integration system post processing for airborne earth observation. *Measurement, 46*(3), 1074–1083.
7. Gong, X., Zhang, J., & Fang, J. (2015). A modified nonlinear two-filter smoothing for high-precision airborne integrated GPS and inertial navigation. *IEEE Transactions on Instrumentation and Measurement, 64*(12), 3315–3322.
8. Groves, P. D. (2008). *Principles of GNSS, inertial and multisensor integrated navigation systems, Boston, MA*. USA: Artech House.
9. Kim, K. H., Lee, J. G., & Park, C. G. (2009). Adaptive two-stage extended Kalman filter for a fault-tolerant INS/GPS loosely coupled system. *IEEE Transactions on Aerospace and Electronic Systems, 45*(1), 125–137.
10. Li, K., Chang, L., & Hu, B. (2015). Unscented attitude estimator based on dual attitude representations. *IEEE Transactions on Instrumentation and Measurement, 64*(12), 3564–3576.
11. Liang, Y., & Jia, Y. (2015). A nonlinear quaternion-based fault-tolerant SINS/GNSS integrated navigation method for autonomous UAVs. *Aerospace Science and Technology, 40,* 191–199.
12. Liu, J. (2012). Research on electromagnetic susceptibility of fiber optical gyroscope. In *Proceedings of the Symposium on Photonics and Optoelectronics* (pp. 1–4). May 21–23, Shanghai, China.
13. Madyastha, V. K., Ravindra, V. C., Vaitheeswaran, S. M., et al. (2012). A novel INS/GPS fusion architecture for aircraft navigation. In *Proceedings of the 15th International Conference on Information Fusion* (pp. 2132–2139). July 09–12, Singapore.
14. Reid, D. B., & Lithopoulos, E. (1998). High precision pointing system for airborne sensors. In *Proceedings of the IEEE Position Location and Navigation Symposium (Cat. No.98CH36153)* (pp. 303 308). April 20–23, Toronto, Canada.
15. Rogers, R. M. (2007). *Applied mathematics in integrated navigation systems* (3rd ed.). Reston, VA, USA: AIAA.
16. Sohne, W., Heinze, O., & Groten, E. (1994). Integrated INS/GPS system for high precision navigation applications. In *Proceedings of IEEE PLANS* (pp. 310–313). Las Vegas, USA.
17. Vitanov, I., & Aouf, N. (2014). Fault diagnosis for MEMS/INS using unscented Kalman filter enhanced by Gaussian process adaption. In *Proceedings of the Adaptive Hardware Systems* (pp. 120–126). July 14–17, Leicester, U.K.
18. Wen, X., Zhang, H., & Zhou, L. (1997). The property of the technology of inertial navigation failure diagnosis. *Journal of Chinese Inertial Technology, 5*(3), 13–16.
19. White, N. A., Maybeck, P. S., & DeVilbiss, S. L. (1998). Detection of interference/jamming and spoing in a DGPS-aided inertial system. *IEEE Transactions on Aerospace and Electronic Systems, 34*(4), 1208–1217.
20. Zarchan, P., & Musoff, H. (2009). *Fundamentals of Kalman filtering: A practical approach* (3rd ed.). Reston: AIAA.

21. Zhang, Y., Gao, Z., Wang, G., et al. (2014). Modeling of thermal-induced rate error for FOG with temperature ranging from -40°C to 60°C. *IEEE Photonics Technology Letters, 26*(1), 18–21.
22. Zhong, M., Guo, D., & Guo, J. (2015). PMI based nonlinear H_∞ estimation of unknown sensor error for INS/GPS integrated system. *IEEE Sensors Journal, 15*(5), 2785–2794.
23. Zuo, K., Zhong, M., & Liu, B. (2010). Adaptive SINS/GPS outlier detection and accommodation using innovation orthogonal. *Journal of Beijing Institute of Technology, 19*(4), 427–431.
24. Meskin, D., & Itzhack, Y. (1992). Unified approach to inertial navigation system error modeling. *Journal of Guidance Control and Dynamics, 15*(3), 648–653.

Chapter 21
Krein Space Based H_∞ Fault Estimation for Discrete-Time Nonlinear Systems

This chapter demonstrates a Krein space based finite horizon H_∞-filtering method for discrete-time nonlinear systems and, on this basis, handling H_∞ fault estimation issues for the INS/GPS integrated system by combing it with proportional and multi-integral (PMI) observer scheme. The core idea lies in converting the targeting problem into a minimum of an indefinite quadratic form such that a relationship between nonlinear filter in Hilbert space and nonlinear estimation in Krein space can be established. Then, by using first-order Taylor approximation and Krein space projection, a sufficient and necessary condition for the minimum is derived and a feasible solution of the nonlinear filter can be obtained by recursively computing Riccati recursions. Especially, for fault estimation purpose, a PMI based nonlinear fault estimator is constructed and designed in the framework of H_∞ filtering. Successful applications of the results in INS/GPS integrated system are demonstrated through experimental studies.

21.1 Introduction

Issues of H_∞ filtering based fault diagnosis for LDTV systems have been investigated in Chaps. 11–13. During the past two decades, increasing attention has been paid to applying H_∞ filtering technique to the fault diagnosis for nonlinear systems and a great progress has been made, see for example [2, 4, 7, 8, 24, 26, 29–31] and references therein. In [24], a minmax approach was proposed to H_∞ estimation and a sufficient condition was obtained by utilizing a Hamilton-Jacobi inequality. As shown in [24], however, the computation burden of H_∞ nonlinear filter was too heavy and a feasible solution of the Hamilton-Jacobi inequality was always not realized in practice. For the computational purpose, an approximate solution of Hamilton-Jacobi inequality was developed based on local linearization in [24], while [4] was in

M. Zhong et al., *Fault Diagnosis for Linear Discrete Time-Varying Systems and Its Applications*, https://doi.org/10.1007/978-981-19-5438-2_21

essence a continuous-time counterpart of [24]. In [30], the problem of robust H_∞ filtering was investigated for a class of systems described by a linear state model with known state dependent nonlinearities satisfying global Lipschitz conditions and a solution can be obtained via one Riccati equation. The approaches in [7, 8, 26, 29, 31] were available under the assumptions of nonlinear functions satisfying certain bounded conditions and, based on the Lyapunov functional, solutions of H_∞ filters were obtained in terms of LMIs. In [2], the discrete-time mixed H_2/H_∞ filtering problem for affine nonlinear systems was investigated by using a dynamic game theory approach. By finding a Nash-equilibrium solution to such a two-player nonzero-sum dynamic game, necessary and sufficient conditions for the solvability of the problem were given in terms of discrete-time Hamilton-Jacobi-Issac's equations (DHJIEs). Considering the difficulty of solving the coupled DHJIEs, a first-order Taylor series approximation approach was developed and sufficient conditions for approximation solvability of the problem were also derived. We can see that, approximate solution via linearization as well iterative algorithm have been so far the mostly used efficient way of H_∞ nonlinear filtering and endeavors on developing computationally efficient methods are still of significance.

On the other hand, the interesting connections of H_∞ estimation (including filtering, prediction and smoothing) with Kalman filtering in Krein space (see [15, 32, 33]) has led to many research efforts on Krein space based H_∞-optimization approaches to fault detection and estimation for LDTV systems, as demonstrated in the Parts II–IV. Yet, study on the Krein space based fault diagnosis for nonlinear systems remains open. This motivates us to develop a Krein space approach to H_∞ nonlinear filtering for fault diagnosis purpose.

From the viewpoint of fault detection and estimation for the INS/GPS integrated system, KF exploits a powerful synergism between the two types of navigation systems [3, 5, 20, 21, 23, 35], concerning the complementary error characteristics of INS and GPS. As a result, the improved position, velocity and attitude estimates can be obtained from the acquisition of the INS/GPS integrated system [23]. In some low-accuracy applications, the linearized model and conventional KF can satisfy the accuracy requirements [25]. For the nonlinear model of the INS/GPS integration system, minimum mean square error (MMSE) estimators such as the EKF and UKF, present a good performance of error estimation if the system nonlinear model and the statistics of additive Gaussian noise are exactly known [10]. In many practical application situations, however, the following concerns are often rose:

- the INS model contains the sensor biases that may deviate from their nominal values by unknown constants or variables and the nonlinear system dynamics are inevitable [1, 27]. These unknown biases may seriously degrade the performance of the INS/GPS integrated system;
- most of the existing INS/GPS integration need an exact model of the inertial sensor bias characteristics. However, due to the influences of operation and environmental condition changes, it is difficult to get exact a priori information of the gyroscopes and accelerometers biases. Particularly for airborne INS/GPS systems, a fault of the inertial sensor may be occasionally occurred and, if it is the case, large errors exhibited by the inertial sensors would probably cause bad performance of KF [14].

Since the characteristics of the biases or faults are usually unknown, the problem of accurate estimation of the INS/GPS integrated system remains a challenge. On the other research front, the PMI observers have been developed for linear descriptor systems to estimate the states and faults simultaneously [12, 19]. Especially in [11, 13, 22], the PMI filter has been applied to fault estimation and accommodation;
- the performance of MMSE estimators such as the EKF and UKF is actually limited to the sensitivity of the statistical changes of the process/measurement noises and model uncertainties. While the severe mechanical vibrations, electrical magnitude disturbances and especially temperature changes give rise to evident inertial sensor errors, which can be regarded as norm bounded. In this case, traditional MMSE estimators may poorly estimate sensor errors. Despite fruitful results have been reported on the INS initial alignment and INS/GPS navigation subject to multiple disturbances, such as the predictive iterated Kalman filter [9] and strong tracking filter (STF) algorithm [18, 36] robust nonlinear H_∞ estimation issues for INS/GPS system have to be further investigated concerning the computational burden and accuracy requirement.

Regarding of the above observations, the objectives of this chapter are of the following threefold.

- Firstly, H_∞ nonlinear filtering problem is handled based on Krein space estimation. By formulating the design of an H_∞ nonlinear filter into minimum of an indefinite quadratic form, a sufficient and necessary condition for the minimum is derived by using Krein space projection and innovation re-organization;
- Secondly, we consider the fault estimation issues for the INS/GPS integrated system with inertial sensor error. By regarding such error as an additive unknown fault, a PMI observer based H_∞ nonlinear estimator is designed and realized by recursively computing Riccati recursions.
- Finally, we demonstrate experimental studies to show the applicability of the presented schemes.

21.2 Problems Formulation

In this section, the problems of H_∞ filtering for nonlinear systems and PMI observer based H_∞ nonlinear fault estimation for INS/GPS interested systems are formulated.

21.2.1 Problem I

Consider the following nonlinear system

$$\begin{cases} x(k+1) = f(x(k)) + g(x(k))w(k), \ x(0) = x_0, \\ y(k) = h(x(k)) + v(k), \end{cases} \tag{21.1}$$

where $x(k) \in \mathbb{R}^n$, $y(k) \in \mathbb{R}^q$, $w(k) \in \mathbb{R}^m$, $v(k) \in \mathbb{R}^q$ are the state, measurement output, driving disturbance, and measurement disturbance, respectively, the initial state x_0 is assumed to be unknown, and $w(k),\ v(k) \in l_{2,[0,N]}$, the functions $f(\cdot)$, $g(\cdot)$ and $h(\cdot)$ are assumed to be smooth in their arguments.

Given observations $\{y(i)\}_{i=0}^{k}$, to estimate some arbitrary linear combination of the states, i.e.,

$$s(k) = L(k)x(k), \tag{21.2}$$

with given $L(k) \in \mathbb{R}^{q\times n}$, the following filter is constructed

$$\begin{cases} \hat{x}(k+1) = f(\hat{x}(k)) + H_1(\hat{x}(k))(y(k) - h(\hat{x}(k)),\ \hat{x}(0) = 0, \\ \hat{s}(k) = L(k)\hat{x}(k) + H_2(\hat{x}(k))(y(k) - h(\hat{x}(k)), \end{cases} \tag{21.3}$$

where $\hat{x}(k)$ is the state estimation, $H_1(\hat{x}(k))$ and $H_2(\hat{x}(k))$ are parameter matrices to be designed. The finite horizon H_∞ filtering problem under investigation is then stated as to find $H_1(\hat{x}(k))$ and $H_2(\hat{x}(k))$ such that

$$\sup_{(x_0,w,v)\neq 0} \frac{\sum_{k=0}^{N} \|s(k)-\hat{s}(k)\|^2}{x_0^T P_0^{-1} x_0+\sum_{k=0}^{N}(\|w(k)\|^2 + \|v(k)\|^2)} < \gamma^2, \tag{21.4}$$

where $\gamma > 0$ is a given scalar, $N > 0$ is a known integer, $P_0 > 0$ is a weighting matrix. Introduce the following quadratic form

$$J_N = x_0^T P_0^{-1} x_0+\sum_{k=0}^{N} \|w(k)\|^2 + \sum_{k=0}^{N} \|v(k)\|^2-\gamma^{-2} \sum_{k=0}^{N} \|s(k)-\hat{s}(k)\|^2. \tag{21.5}$$

Obviously, (21.4) is satisfied if and only if $J_N > 0$ for all $(x_0, w, v) \neq 0$. Therefore, the H_∞ nonlinear filtering can be formulated as the following minmax problem

$$\min_{H_i(\hat{x}(k))} \max_{(x_0,w,v)} J_N. \tag{21.6}$$

For a priori H_∞ nonlinear filter, i.e., the special case of $H_2(\hat{x}(k)) = 0$, the existence conditions of such a minimax problem have been obtained utilizing Hamilton-Jacobi inequality in [2, 24]. Since a feasible solution was almost too difficult to be available, approximate solutions based on local linearization have been so far very useful for computational purpose. it is no doubt a challenge to solve the posteriori H_∞ nonlinear filtering problem. On the other hand, studies in Chaps. 11–13 have shown that H_∞ linear filter can be obtained by applying Krein space projection. Sufficient and necessary conditions of such an estimate have been given and the solutions can be recursively calculated via Riccati recursions so as to reduce the computational efforts in finding projections and the possibility of recursive solutions.

Inspired by the achievements of H_∞ linear filter using Krein space projection and EKF for nonlinear systems, in this chapter we aims to give a Krein space approach to H_∞ nonlinear filtering for system (21.1). To facilitate subsequent discussion, we define

$$v_s(k)=\left[v^T(k)\ \tilde{s}^T(k)\right]^T,\ \tilde{s}(k)=\hat{s}(k)-s(k),\ v_{sN}=\left[v_s^T(0)\ v_s^T(1)\ \cdots\ v_s^T(N)\right]^T,$$
$$w_N=\left[w^T(0)\ w^T(1)\ \cdots\ w^T(N)\right]^T,\ Q_{v_sN}=\operatorname{diag}(Q_{v_s}, Q_{v_s}, \cdots, Q_{v_s}),$$

where

$$Q_{v_s}=\operatorname{diag}(I, -\gamma^2 I) \tag{21.7}$$

Then J_N in (21.5) can be rewritten as

$$J_N=\begin{bmatrix} x_0 \\ w_N \\ v_{sN} \end{bmatrix}^T \begin{bmatrix} P_0 & 0 & 0 \\ 0 & I & 0 \\ 0 & 0 & Q_{v_sN} \end{bmatrix}^{-1} \begin{bmatrix} x_0 \\ w_N \\ v_{sN} \end{bmatrix}. \tag{21.8}$$

Let

$$r(k)=\hat{s}(k),\ y_s(k)=\begin{bmatrix} y(k) \\ r(k) \end{bmatrix}.$$

We furthermore have

$$\begin{cases} x(k+1)-f(x(k))+g(x(k))w(k), \\ y_s(k)=\begin{bmatrix} h(x(k)) \\ L(k)x(k) \end{bmatrix}+v_s(k),\ x(0)=x_0. \end{cases} \tag{21.9}$$

Thus J_N can be considered as an indefinite quadratic form of $\{x_0, w_N, v_{sN}\}$. Similar to the linear system case, $H_1(\hat{x}(k))$ and $H_2(\hat{x}(k))$ solve the H_∞ nonlinear filtering problem if J_N in (21.8) subject to (21.3) and (21.9) has a minimum and its value at the minimum is positive.

We now reformulate the H_∞ nonlinear filtering problem as follows.

Problem 21.1 Given nonlinear system (21.1), H_∞ filter (21.3) and J_N in (21.8), find $H_i(i=1,2)$ solving

$$\min_{H_i(\hat{x}(k))} \max_{x_0, w_N, v_{sN}} J_N,\ s.t.\ (21.3) \text{ and } (21.9)$$

such that J_N at its minimum is positive.

21.2.2 Problem II

Consider an INS/GPS integrated system modeled by (20.1). Let

$$d(t) = [\,(\delta\omega_{ib}^{b})^T\ (\delta f^b)^T\,]^T$$

with δf^b and $\delta\omega_{ib}^{b}$ being the output errors of accelerometers and gyros, respectively. Define

$$x(t) = [\,(\phi)^T\ (\delta V)^T\ (\delta p)^T\,]^T,\ y(t) = [\,(\delta V)^T\ (\delta p)^T\,]^T.$$

Then, the continuous-time nonlinear dynamical equation of INS/GPS integrated system can be described by

$$\begin{cases} \dot{x}(t) = F(t)x(t) + q(x,t) + B(t)d(t), \\ y(t) = C(t)x(t) + v(t), \end{cases} \tag{21.10}$$

where $v(t)$ is the measurement noise vector, $F(t)$, $B(t)$, $q(x,t)$ and $C(t)$ are given by

$$F(t) = \begin{bmatrix} 0_{3\times 3} & F_{11} & F_{12} \\ 0_{3\times 3} & F_{21} & F_{22} \\ 0_{3\times 3} & F_{31} & F_{32} \end{bmatrix},\ B(t) = \begin{bmatrix} C_b^n & 0_{3\times 3} \\ 0_{3\times 3} & C_b^n \\ 0_{3\times 3} & 0_{3\times 3} \end{bmatrix},$$
$$q(x,t) = \begin{bmatrix} (I - C_n^{n,ins})\omega_{in}^n \\ (I - C_n^{n,ins})C_b^n \hat{f}_b \\ 0_{3\times 1} \end{bmatrix},\ C(t) = \begin{bmatrix} 0_{3\times 3} & I_{3\times 3} & 0_{3\times 3} \\ 0_{3\times 3} & 0_{3\times 3} & I_{3\times 3} \end{bmatrix}, \tag{21.11}$$

respectively, with $F_{11},\ F_{12},\ F_{21},\ F_{22},\ F_{31},\ F_{32}$ being given in (20.6).

In the existing results, a common way is to model the accelerometer and gyro errors as a random constant bias mixing Gauss white noise [28, 34], i.e.,

$$d(t) = b_c(t) + w_g(t), \tag{21.12}$$

where $b_c(t)$ denotes the unknown constant bias of sensor and $w_g(t)$ stands for the sensor noise which is zero mean Gaussian random sequences. Meanwhile, the measurement noise is also regard as zero mean Gaussian random sequences $v_g(t)$ and

$$\mathbb{E}[w_g(t)] = 0,\ \mathbb{E}[v_g(t)] = 0,\ \mathbb{E}\left[\begin{bmatrix} w_g(t) \\ v_g(t) \end{bmatrix}\begin{bmatrix} w_g(\tau) \\ v_g(\tau) \end{bmatrix}^T\right] = \begin{bmatrix} Q(t) & 0 \\ 0 & R(t) \end{bmatrix}\delta_{t\tau},$$

where $Q(t) > 0,\ R(t) > 0,\ \delta_{t\tau}$ is the Kronecker function. The initial state x_0 is assumed to be uncorrelated with the white noises. Together with (21.12), an augmented 15-dimensional state model is constructed for estimating state and bias. Thus an accurate estimation can be obtained by directly applying EKF or UKF.

However, the problem is not as straightforward as the above presentations. Due to installation at the external of carrier, the inertial sensor works with unknown disturbance caused by temperature shift, atmospheric turbulence and electromagnetic disturbance. To overcome the problem, some researches concentrate on the model of sensor errors and relevant estimation method. In [34], the sensor errors were described by

$$d(t) = b_r(t) + w_g(t). \tag{21.13}$$

Different from traditional model, the author modeled the error as one order Markov process $b_r(t)$ with white noise $w_g(t)$ for simplicity. Aiming for non-Gauss noise, STF provided another way of designing estimator for INS/GPS system, see for example [6, 17]. The study in [17] proposed a fuzzy strong tracking UKF for INS/GPS system. In [6], STF was used to estimate INS error when GPS had outrages.

Actually, in practical environment, the stochastic characteristics may not be exactly known. In this case, one can consider the sensor errors as low frequency time-varying bias $b(t)$ with noise $w(t)$, i.e.,

$$d(t) = b(t) + w(t), \tag{21.14}$$

where the time-varying bias $b(t)$ is approximated by the following polynomial function

$$b(t) = B_0 + B_1 t + B_2 t^2 + \cdots + B_{q-1} t^{q-1} \tag{21.15}$$

the noise $w(t)$ is energy bounded, and $B_i (i = 0, 1, \cdots, q - 1)$ are unknown constant vectors with $q \geq 1$.

These two parts exhibit completely various characteristics and need be handled with different methods. In POS application, the inertial sensor of INS/GPS system shows greatly sensitive to temperature variations. No matter constant bias or Markov process can not truly describe the process that sensor errors vary with temperature. Figure 21.1 shows that ring laser gyroscope outputs vary with temperature shift. It is mentioned that the normal output should be $1^\circ/h$. As is shown in the figure, besides high frequency noise, the slowly variation shows low frequency characteristic and INS/GPS system will regard it as useful output. Therefore we should estimate and compensate it. Based on above statements, only time-varying bias $b(t)$ satisfies the requirement of sensor errors. Obviously, a time-varying bias of the form (21.15) is more general than a Markov bias. It is remarkable that for the bias Markov model, if the related time parameter of Markov drift is known, the model is definitely accurate; if not, the bias model of Markov drift in [34] can be considered as a special case of the polynomial function (21.15).

In this regard, a modified nonlinear error model of INS/GPS integrated system is obtained as

$$\begin{cases} \dot{x}(t) = F(t)x(t) + q(x,t) + B(t)b(t) + B(t)w(t), \\ y(t) = C(t)x(t) + v(t). \end{cases} \tag{21.16}$$

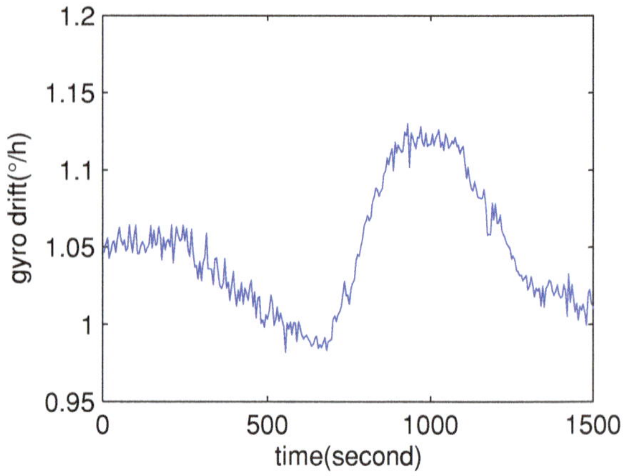

Fig. 21.1 Sensor real output variation with temperature shift

Unlike traditional model, (21.16) divides the error into two parts, low frequency random bias and high frequency unknown noise. Meanwhile (21.14) is closer to practical flight situation than (21.12)–(21.13). Now the challenging task is how to simultaneously achieve accurate estimation of time-varying bias $b(t)$ with uncertain noise $w(t)$.

Recently, the PMI with an observer structure has been applied to fault estimation and accommodation. It is then intuitive that we can extend PMI observer results to the estimation of unknown random bias for INS/GPS system. Now we get a further insight into the random bias. Define

$$\xi_i(t) = b^{(q-i)}(t) \quad (i = 1, 2, \cdots q). \tag{21.17}$$

It follows from (21.15) and (21.17) that

$$\begin{cases} \dot{\xi}_1(t) = 0, \\ \dot{\xi}_2(t) = \xi_1(t), \\ \quad\vdots \\ \dot{\xi}_q(t) = \xi_{q-1}(t). \end{cases} \tag{21.18}$$

Augmenting (21.16)–(21.18) yields

$$\begin{cases} \dot{\bar{x}}(t) = \bar{F}(t)\bar{x}(t) + \bar{q}(\bar{x}, t) + \bar{B}(t)w(t), \\ y(t) = \bar{C}(t)\bar{x} + v(t), \end{cases} \tag{21.19}$$

where

$$\bar{x}(t) = [x^T(t)\ \xi_1^T(t)\ \xi_2^T(t)\ \cdots\ \xi_q^T(t)]^T,\ \bar{q}(\bar{x},t) = [q(x,t)^T\ 0\ 0\ \cdots\ 0]^T,$$

$$\bar{F}(t) = \begin{bmatrix} F(t) & 0 \cdots 0 & B(t) \\ 0 & 0 \cdots 0 & 0 \\ 0 & 1 \cdots 0 & 0 \\ \vdots & \vdots \ddots \vdots & \vdots \\ 0 & 0 \cdots 1 & 0 \end{bmatrix},\ \bar{B}(t) = \begin{bmatrix} B(t) \\ 0 \\ 0 \\ \vdots \\ 0 \end{bmatrix},\ \bar{C}(t) = [C(t)\ 0\ 0\ \cdots\ 0].$$

Let the sampling period be T_{GPS}, and a zero-order-hold is chosen for the model (21.19). The discrete-time model can be obtained as

$$\begin{cases} \bar{x}(k+1) = \bar{F}(k)\bar{x}(k) + \bar{q}(\bar{x}(k)) + \bar{B}(k)w(k), \\ y(k) = \bar{C}(k)\bar{x}(k) + v(k), \end{cases} \tag{21.20}$$

where $w(k),\ v(k) \in l_{2,[0,N]}$, $\bar{C}(k) = [\,C(k)\ 0\ 0\ \cdots\ 0\,]$, and

$$\bar{F}(k) = \begin{bmatrix} F(k) & 0 & \cdots & 0 & B(k) \\ 0 & 1 & \cdots & 0 & 0 \\ 0 & T_{GPS} & \cdots & 0 & 0 \\ \vdots & \vdots & \ddots & \vdots & \vdots \\ 0 & 0 & \cdots & T_{GPS} & 1 \end{bmatrix},\ \bar{B}(k) = \begin{bmatrix} B(k) \\ 0 \\ 0 \\ \vdots \\ 0 \end{bmatrix}.$$

In general, we would like to estimate all navigation states and inertial bias states, say

$$z(k) = \bar{x}(k). \tag{21.21}$$

Let $\hat{z}(k|k)$ denote the estimate of $z(k)$ given observations $\{y(i)\}_{i=0}^k$. We then have the filtered error

$$\hat{z}(k|k) - x(k).$$

Aiming at (21.20), the problem under investigation can be transformed into designing of suitable nonlinear H_∞ filter

$$\begin{cases} \hat{\bar{x}}(k+1|k+1) = \hat{\bar{x}}(k+1|k) + K(k+1)(y(k) - \bar{C}(k)\hat{\bar{x}}(k+1|k)), \\ \hat{\bar{x}}(k+1|k) = \bar{F}(k)\hat{\bar{x}}(k|k) + \bar{q}(\hat{\bar{x}}(k|k)), \end{cases} \tag{21.22}$$

such that

$$\sup_{(\bar{x}_0,w,v)\neq 0} \frac{\sum_{k=0}^N \|\hat{z}(k|k) - z(k)\|^2}{\bar{x}_0^T P_0^{-1}\bar{x}_0 + \sum_{k=0}^N (\|w(k)\|^2 + \|v(k)\|^2)} < \gamma^2, \tag{21.23}$$

where disturbance attenuation level $\gamma > 0$ and known integer $N > 0$, $K(k+1)$ denotes the gain matrix to be designed.

Introduce the following quadratic form

$$J_N = \bar{x}_0^T P_0^{-1} \bar{x}_0 + \sum_{k=0}^{N} \|w(k)\|^2 + \sum_{k=0}^{N} \|v(k)\|^2 - \gamma^{-2} \sum_{k=0}^{N} \|\hat{z}(k|k) - z(k)\|^2. \tag{21.24}$$

Obviously, (21.23) is satisfied if and only if $J_N > 0$ for all $(\bar{x}_0, w(k), v(k)) \neq 0$. Hence, a PMI based nonlinear H_∞ fault estimation problem for the INS/GPS integrated system is formulated as follows.

Problem 21.2 Given the error model of INS/GPS integrated system (21.16) and J_N in form of (21.24), develop a PMI based nonlinear H_∞ filter (21.22) such that

(i) J_N has a minimum with respect to $\bar{x}_0$, $w(k)$, $v(k)$ and
(ii) a $\hat{z}(k|k)$ can be find such that the value of J_N at this minimum is positive, viz.

$$\min_{(\bar{x}_0, w(k), v(k))} J_N > 0.$$

21.3 Krein Space Based H_∞ Filtering for Nonlinear Systems

This section is dedicated to addressing Problem 21.1via Krein space projection technique. To this end, an artificial nonlinear stochastic system in Krein space is first introduced and a relationship between H_∞-filtering and the minimization of J_N is built. After that, a sufficient and necessary condition for the minimum of J_N is derived by using Krein space projection and innovation analysis. Finally, a solution to the H_∞ nonlinear filter is derived.

21.3.1 *The Conditions for a Minimum of J_N*

Similar to the well known EKF in Hilbert space, we first consider to solve the H_∞ nonlinear filtering problem in a neighborhood of the state estimated trajectory $\hat{x}(k)$.

Introduce

$$F(k) = \frac{\partial f(x)}{\partial x}|_{x=\hat{x}(k)}, \ G(k) = g(\hat{x}(k)), \ H(k) = \frac{\partial h(x)}{\partial x}|_{x=\hat{x}(k)}.$$

Then nonlinear functions $f(x(k))$, $g(x(k))$, $h(x(k))$ in (21.9) can be expanded in Talyor series about $\hat{x}(k)$ as

$$\begin{aligned} f(x(k)) &= f(\hat{x}(k)) + F(k)(x(k) - \hat{x}(k)) + \cdots, \\ g(x(k)) &= g(\hat{x}(k)) + \cdots = G(k) + \cdots, \\ h(x(k)) &= h(\hat{x}(k)) + H(k)(x(k) - \hat{x}(k)) + \cdots. \end{aligned}$$

Neglecting higher order terms yields

$$\begin{cases} x(k+1) = f(\hat{x}(k)) + F(k)(x(k) - \hat{x}(k)) + G(k)w(k), \\ y(k) = h(\hat{x}(k)) + H(k)(x(k) - \hat{x}(k)), \end{cases} \tag{21.25}$$

which enables us to respectively approximate (21.9) as

$$\begin{cases} x(k+1) = F(k)x(k) + f(\hat{x}(k)) - F(k)\hat{x}(k) + G(k)w(k), \\ y_s(k) = \begin{bmatrix} H(k) \\ L(k) \end{bmatrix} x(k) + \begin{bmatrix} h(\hat{x}(k)) - H(k)\hat{x}(k) \\ 0 \end{bmatrix} + v_s(k). \end{cases} \tag{21.26}$$

Define

$$\begin{aligned} &\Phi(j,k) = F(j-1)F(j-2)\cdots F(k), \\ &\Phi(j,j) = I,\ j = 1, 2, \cdots, N;\ k = 0, 1, \cdots, j-1, \\ &u_1(k) = f(\hat{x}(k)) - F(k)\hat{x}(k),\ u_2(k) = \begin{bmatrix} h(\hat{x}(k)) - H(k)\hat{x}(k) \\ 0 \end{bmatrix}, \\ &u_{1N} = \begin{bmatrix} u_1^T(0)\ u_1^T(1) \cdots u_1^T(N) \end{bmatrix}^T,\ u_{2N} = \begin{bmatrix} u_2^T(0)\ u_2^T(1) \cdots u_2^T(N) \end{bmatrix}^T, \\ &y_{sN} = \begin{bmatrix} y_s^T(0)\ y_s^T(1) \cdots y_s^T(N) \end{bmatrix}^T,\ v_{sN} = \begin{bmatrix} v_s^T(0)\ v_s^T(1) \cdots v_s^T(N) \end{bmatrix}^T. \end{aligned}$$

It allows us to rewrite (21.26) as

$$y_{sN} = \Omega x_0 + \Gamma_w w_N + \Gamma_1 u_{1N} + u_{2N} + v_{sN}, \tag{21.27}$$

where

$$\Gamma_1 = \begin{bmatrix} 0 & & \\ \bar{H}(1) & 0 & \\ \bar{H}(2)\Phi(2,1) & \bar{H}(2) & 0 \\ \cdots & \cdots & \cdots\cdots\cdots \end{bmatrix},\ \Omega = \begin{bmatrix} \bar{H}(0) \\ \bar{H}(1)\Phi(1,0) \\ \vdots \\ \bar{H}(N)\Phi(N,0) \end{bmatrix},$$

$$\Gamma_w = \Gamma_1 \mathrm{diag}\{G(0), G(1), \cdots, G(N)\},\ \bar{H}(k) = \begin{bmatrix} H(k) \\ L(k) \end{bmatrix}.$$

Furthermore, we make the following change of coordinates

$$\begin{bmatrix} x_0 \\ w_N \\ y_{sN} - \Gamma_1 u_{1N} - u_{2N} \end{bmatrix} = \begin{bmatrix} I & 0 & 0 \\ 0 & I & 0 \\ \Omega & \Gamma_w & I \end{bmatrix} \begin{bmatrix} x_0 \\ w_N \\ v_{sN} \end{bmatrix}$$

to obtain

$$J_N = \begin{bmatrix} x_0 \\ w_N \\ z_N \end{bmatrix}^T \left\{ \begin{bmatrix} I & 0 & 0 \\ 0 & I & 0 \\ \Omega & \Gamma_w & I \end{bmatrix} \begin{bmatrix} P_0 & 0 & 0 \\ 0 & I & 0 \\ 0 & 0 & Q_{v_sN} \end{bmatrix} \begin{bmatrix} I & 0 & 0 \\ 0 & I & 0 \\ \Omega & \Gamma_w & I \end{bmatrix}^T \right\}^{-1} \begin{bmatrix} x_0 \\ w_N \\ z_N \end{bmatrix}, \tag{21.28}$$

where

$$z_N = y_{sN} - \Gamma_1 u_{1N} - u_{2N}.$$

Then the H_∞ nonlinear filtering problem in Problem 21.1 can be further reformulated as to find $H_1(\hat{x}(k))$ and $H_2(\hat{x}(k))$ such that the J_N subject to (21.3) and (21.26) has a minimum and $J_N > 0$.

Next, we consider to find a minimum for J_N using Krein space projection and innovation re-organization. For this purpose, the following Krein space nonlinear stochastic system is considered

$$\begin{cases} \mathbf{x}(k+1) = f(\mathbf{x}(k)) + g(\mathbf{x}(k))\mathbf{w}(k), \\ \mathbf{y}_s(k) = \begin{bmatrix} h(\mathbf{x}(k)) \\ L(k)\mathbf{x}(k) \end{bmatrix} + \mathbf{v}_s(k). \end{cases} \tag{21.29}$$

where

$$\mathbf{y}_s(k) = \begin{bmatrix} \mathbf{y}(k) \\ \mathbf{r}(k) \end{bmatrix}, \left\langle \begin{bmatrix} \mathbf{x}_0 \\ \mathbf{w}(k) \\ \mathbf{v}_s(k) \end{bmatrix}, \begin{bmatrix} \mathbf{x}_0 \\ \mathbf{w}(j) \\ \mathbf{v}_s(j) \end{bmatrix} \right\rangle = \begin{bmatrix} P_0 & 0 & 0 \\ 0 & I\delta_{kj} & 0 \\ 0 & 0 & Q_{vs}\delta_{kj} \end{bmatrix}.$$

Denote by $\hat{\mathbf{x}}(k)$ the projection of $\mathbf{x}(k)$ onto space spanned by $\{\mathbf{y}_s(i)\}_{i=0}^{k-1}$. Using the first-order Taylor approximations

$$\begin{aligned} f(\mathbf{x}(k)) &\simeq f(\hat{\mathbf{x}}(k)) + F(k)(\mathbf{x}(k) - \hat{\mathbf{x}}(k)), \\ g(\mathbf{x}(k)) &\simeq G(k), \\ h(\mathbf{x}(k)) &\simeq h(\hat{\mathbf{x}}(k)) + H(k)(\mathbf{x}(k) - \hat{\mathbf{x}}(k)). \end{aligned}$$

we have

$$\begin{cases} \mathbf{x}(k+1) = f(\hat{\mathbf{x}}(k)) + F(k)(\mathbf{x}(k) - \hat{\mathbf{x}}(k)) + G(k)\mathbf{w}(k), \\ \mathbf{y}_s(k) = \begin{bmatrix} h(\hat{\mathbf{x}}(k)) + H(k)(\mathbf{x}(k) - \hat{\mathbf{x}}(k)) \\ L(k)\mathbf{x}(k) \end{bmatrix} + \mathbf{v}_s(k), \end{cases} \tag{21.30}$$

which leads to the following approximation form

$$\begin{cases} \mathbf{x}(k+1) = F(k)\mathbf{x}(k) + \mathbf{u}_1(k) + G(k)\mathbf{w}(k), \\ \mathbf{y}_s(k) = \begin{bmatrix} H(k) \\ L(k) \end{bmatrix} \mathbf{x}(k) + \mathbf{u}_2(k) + \mathbf{v}_s(k), \end{cases} \tag{21.31}$$

where

$$\mathbf{u}_1(k) = f(\hat{\mathbf{x}}(k)) - F(k)\hat{\mathbf{x}}(k), \;\; \mathbf{u}_2(k) = \begin{bmatrix} h(\hat{\mathbf{x}}(k)) - H(k)\hat{\mathbf{x}}(k) \\ 0 \end{bmatrix}.$$

Similarly, we can get

$$\mathbf{z}_N = \mathbf{y}_{sN} - \Gamma_1 \mathbf{u}_{1N} - \mathbf{u}_{2N},$$

where

$$\mathbf{u}_{1N} = \left[\mathbf{u}_1^T(0) \; \mathbf{u}_1^T(1) \cdots \mathbf{u}_1^T(N) \right]^T, \;\; \mathbf{u}_{2N} = \left[\mathbf{u}_2^T(0) \; \mathbf{u}_2^T(1) \cdots \mathbf{u}_2^T(N) \right]^T.$$

By applying Lemma 3.2.2 in [15], it is easy to know that the expression

$$\begin{bmatrix} \hat{x}_0 \\ \hat{w}_N \end{bmatrix} = \begin{bmatrix} P_0 \Omega^T \\ \Gamma_w^T \end{bmatrix} R_{\mathbf{z}_N}^{-1} z_N \tag{21.32}$$

yields a unique minimum of J_N over $\{x_0, w_N\}$, subject to (21.3) and (21.26), if and only if $R_{\mathbf{z}_N}$, $R_{\mathbf{v}_s}$ have the same inertia and the value of J_N at its minimum is

$$J_N(\hat{x}_0, \hat{w}_N, y_{sN}) = z_N^T R_{\mathbf{z}_N}^{-1} z_N.$$

With known $\hat{\mathbf{x}}(k)$, it is easy to see that

$$R_{\mathbf{z}_N} = R_{\mathbf{y}_{sN}}.$$

Thus the following proposition can be obtained.

Proposition 21.1 *The expression (21.32) yields a unique minimum of J_N over $\{x_0, w_N\}$, subject to (21.3) and (21.26), if and only if $R_{\mathbf{y}_{sN}}$ and $R_{\mathbf{v}_s}$ have the same inertia. Moreover, the value of J_N at the minimum is given by*

$$J_N(\hat{x}_0, \hat{w}_N, y_{sN}) = (y_{sN} - \Gamma_1 u_{1N} - u_{2N})^T R_{\mathbf{y}_{sN}}^{-1} (y_{sN} - \Gamma_1 u_{1N} - u_{2N}).$$

Unfortunately, it is not an easy task to directly check the inertial conditions of Proposition 21.1. In addition, it is also necessary to choose the estimation of $\hat{\mathbf{x}}(k)$ such that the value of J_N at its minimum is positive. To reduce the computational burden, recursively checking for the conditions in Proposition 21.1 and the calculation of $J_N(\hat{x}_0, \hat{w}_N, y_{sN})$ will be presented in the next subsection by applying Krein space projections and innovation reorganization.

21.3.2 Recursive Formulae of Krein Space Nonlinear Filter

Define the following innovations

$$\begin{aligned}&\mathbf{e}(k)=\mathbf{x}(k)-\hat{\mathbf{x}}(k),\ P(k)=<\mathbf{e}(k),\mathbf{e}(k)>,\ \tilde{\mathbf{y}}(k)=\mathbf{y}(k)-\hat{\mathbf{y}}(k),\\&\tilde{\mathbf{r}}(k)=\mathbf{r}(k)-\hat{\mathbf{r}}(k),\ \tilde{\mathbf{r}}(k|k)=\mathbf{r}(k)-\hat{\mathbf{r}}(k|k),\ \tilde{\mathbf{y}}_{\mathbf{s}}(k)=\mathbf{y}_s(k)-\hat{\mathbf{y}}_{\mathbf{s}}(k),\end{aligned}$$

$$\tilde{\mathbf{y}}_{\mathbf{s}1}(k)=\mathbf{y}_s(k)-\hat{\mathbf{y}}_{\mathbf{s}1}(k),\ \hat{\mathbf{y}}_{\mathbf{s}1}(k)=\begin{bmatrix}\hat{\mathbf{y}}(k)\\\hat{\mathbf{r}}(k)\end{bmatrix},\ \hat{\mathbf{y}}_{\mathbf{s}}(k)=\begin{bmatrix}\hat{\mathbf{y}}(k)\\\hat{\mathbf{r}}(k|k)\end{bmatrix}.$$

where $\hat{\mathbf{x}}(k),\ \hat{\mathbf{y}}(k),\ \hat{\mathbf{r}}(k)$ denote the projections of $\mathbf{x}(k),\ \mathbf{y}(k)$ and $\mathbf{r}(k)$ on the space spanned by $\{\mathbf{y}_s(i)\}_{i=0}^{k-1}$, respectively. $\hat{\mathbf{r}}(k|k)$ stands for the projection of $\mathbf{r}(k)$ on the space spanned by $\{\{\mathbf{y}_s(i)\}_{i=0}^{k-1},\mathbf{y}(k)\}$.

It is known that $\{\tilde{\mathbf{y}}_{\mathbf{s}1}(i)\}_{i=0}^{k-1}$ forms an orthogonal basis of $\mathcal{L}\{\{\mathbf{y}_s(i)\}_{i=0}^{k-1}\}$ due to the construction of innovation $\tilde{\mathbf{y}}_{\mathbf{s}1}(i)$. We thus have

$$\mathcal{L}\{\{\tilde{\mathbf{y}}_{\mathbf{s}1}(i)\}_{i=0}^{k-1}\}=\mathcal{L}\{\{\mathbf{y}_s(i)\}_{i=0}^{k-1}\}.$$

Furthermore,

$$\begin{aligned}\hat{\mathbf{x}}(k+1)&=\text{Proj}\{\mathbf{x}(k+1)|\{\mathbf{y}_s(i)\}_{i=0}^{k}\}\\&=\text{Proj}\{\mathbf{x}(k+1)|\{\tilde{\mathbf{y}}_{\mathbf{s}1}(i)\}_{i=0}^{k}\}\\&=\text{Proj}\{\mathbf{x}(k+1)|\{\tilde{\mathbf{y}}_{\mathbf{s}1}(i)\}_{i=0}^{k-1}\}+\left\langle\mathbf{x}(k+1),\tilde{\mathbf{y}}_{\mathbf{s}1}(k)\right\rangle R_{\tilde{\mathbf{y}}_{\mathbf{s}1}}^{-1}(k)\tilde{\mathbf{y}}_{\mathbf{s}1}(k)\\&=F(k)\text{Proj}\{\mathbf{x}(k)|\{\tilde{\mathbf{y}}_{\mathbf{s}1}(i)\}_{i=0}^{k-1}\}+\mathbf{u}_1(k)\qquad(21.33)\\&\quad+\left\langle\mathbf{x}(k+1),\tilde{\mathbf{y}}_{\mathbf{s}1}(k)\right\rangle R_{\tilde{\mathbf{y}}_{\mathbf{s}1}}^{-1}(k)\tilde{\mathbf{y}}_{\mathbf{s}1}(k)\\&=F(k)\hat{\mathbf{x}}(k)+\mathbf{u}_1(k)+K_p(k)\tilde{\mathbf{y}}_{\mathbf{s}1}(k)\\&=f(\hat{\mathbf{x}}(k))+K_p(k)\tilde{\mathbf{y}}_{\mathbf{s}1}(k),\qquad(21.34)\\\hat{\mathbf{y}}_{\mathbf{s}1}(k)&=\begin{bmatrix}H(k)\\L(k)\end{bmatrix}\hat{\mathbf{x}}(k)+\mathbf{u}_2(k)=\begin{bmatrix}h(\hat{\mathbf{x}}(k))\\L(k)\hat{\mathbf{x}}(k)\end{bmatrix}\qquad(21.35)\end{aligned}$$

with

$$K_p(k)=\left\langle\mathbf{x}(k+1),\tilde{\mathbf{y}}_{\mathbf{s}1}(k)\right\rangle R_{\tilde{\mathbf{y}}_{\mathbf{s}1}}^{-1}(k)=F(k)P(k)\left[\,H^T(k)\ L^T(k)\,\right]R_{\tilde{\mathbf{y}}_{\mathbf{s}1}}^{-1}(k),$$

where $P(k)$ satisfies the Riccati recursion

$$P(k+1) = F(k)P(k)F^T(k) - K_p(k)R_{\tilde{\mathbf{y}}_{\mathbf{s}1}}(k)K_p^T(k) + G(k)G^T(k), \; P(0) = P_0,$$

$$\begin{aligned} R_{\tilde{\mathbf{y}}_{\mathbf{s}1}}(k) &= \begin{bmatrix} H(k) \\ L(k) \end{bmatrix} P(k) \begin{bmatrix} H^T(k) & L^T(k) \end{bmatrix} + Q_{vs} \\ &= \begin{bmatrix} I + H(k)P(k)H^T(k) & H(k)P(k)L^T(k) \\ L(k)P(k)H^T(k) & L(k)P(k)L^T(k) - \gamma^2 I \end{bmatrix}. \end{aligned} \tag{21.36}$$

Thus, the innovations of $\mathbf{e}(k)$ and $\tilde{\mathbf{y}}_{\mathbf{s}1}(k)$ can be recursively computed as

$$\begin{aligned} \mathbf{e}(k+1) &= F(k)\mathbf{e}(k) - K_p(k)\tilde{\mathbf{y}}_{\mathbf{s}1}(k) + G(k)\mathbf{w}(k), \\ \tilde{\mathbf{y}}_{\mathbf{s}1}(k) &= \begin{bmatrix} H(k) \\ L(k) \end{bmatrix} \mathbf{e}(k) + \mathbf{v}_{\mathbf{s}}(k), \end{aligned}$$

Moreover,

$$\begin{aligned} \hat{\mathbf{r}}(k|k) &= \text{Proj}\{\mathbf{r}(k)|\{\{\mathbf{y}_{\mathbf{s}1}(i)\}_{i=0}^{k-1}, \mathbf{y}(k)\}\} \\ &= \text{Proj}\{\mathbf{r}(k)|\{\{\tilde{\mathbf{y}}_{\mathbf{s}1}(i)\}_{i=0}^{k-1}, \tilde{\mathbf{y}}(k)\}\} \\ &= \text{Proj}\{\mathbf{r}(k)|\{\{\tilde{\mathbf{y}}_{\mathbf{s}1}(i)\}_{i=0}^{k-1}\}\} + \langle \mathbf{r}(k), \tilde{\mathbf{y}}(k) \rangle \, R_{\tilde{\mathbf{y}}}^{-1}(k)\tilde{\mathbf{y}}(k) \\ &= \hat{\mathbf{r}}(k) + K_r(k)\tilde{\mathbf{y}}(k) \\ &= L(k)\hat{\mathbf{x}}(k) + K_r(k)\tilde{\mathbf{y}}(k), \end{aligned} \tag{21.37}$$

where

$$K_r(k) = \langle \mathbf{r}(k), \tilde{\mathbf{y}}(k) \rangle \, R_{\tilde{\mathbf{y}}}^{-1}(k) = L(k)P(k)H^T(k)(I + H(k)P(k)H^T(k))^{-1}.$$

Hence,

$$\tilde{\mathbf{r}}(k|k) = \mathbf{r}(k) - \hat{\mathbf{r}}(k|k),$$

which implies that

$$\tilde{\mathbf{r}}(k|k) \perp \{\{\tilde{\mathbf{y}}_{\mathbf{s}1}(i)\}_{i=0}^{k-1}, \tilde{\mathbf{y}}(k)\}.$$

As a result, we have

$$\begin{aligned} \langle \tilde{\mathbf{y}}(k), \tilde{\mathbf{r}}(k|k) \rangle = 0, \; & \mathcal{L}\{\{\tilde{\mathbf{y}}_{\mathbf{s}1}(i)\}_{i=0}^{k-1}, \tilde{\mathbf{y}}(k)\} = \mathcal{L}\{\{\tilde{\mathbf{y}}_{\mathbf{s}}(i)\}_{i=0}^{k-1}, \tilde{\mathbf{y}}(k)\}, \\ \tilde{\mathbf{r}}(k|k) & \perp \mathcal{L}\{\{\tilde{\mathbf{y}}_{\mathbf{s}}(i)\}_{i=0}^{k-1}, \tilde{\mathbf{y}}(k)\} \end{aligned}$$

and

$$\begin{aligned} R_{\tilde{\mathbf{r}}}(k|k) &= \langle \tilde{\mathbf{r}}(k|k), \tilde{\mathbf{r}}(k|k) \rangle = \begin{bmatrix} -K_r(k) & I \end{bmatrix} R_{\tilde{\mathbf{y}}_{\mathbf{s}1}}(k) \begin{bmatrix} -K_r^T(k) \\ I \end{bmatrix} \\ &= -\gamma^2 I + L(k)[P^{-1}(k) + H^T(k)H(k)]^{-1}L(k). \end{aligned}$$

It follows readily that

$$R_{\tilde{\mathbf{y}}_s}(k) = \begin{bmatrix} \langle \tilde{\mathbf{y}}(k), \tilde{\mathbf{y}}(k)\rangle & \langle \tilde{\mathbf{y}}(k), \tilde{\mathbf{r}}(k|k)\rangle \\ \langle \tilde{\mathbf{r}}(k|k), \tilde{\mathbf{y}}(k)\rangle & \langle \tilde{\mathbf{r}}(k|k), \tilde{\mathbf{r}}(k|k)\rangle \end{bmatrix} = \begin{bmatrix} R_{\tilde{\mathbf{y}}}(k) & 0 \\ 0 & R_{\tilde{\mathbf{r}}}(k|k) \end{bmatrix}. \quad (21.38)$$

Next is the derivation of an alternative inertial condition instead of Proposition 21.1 using $R_{\tilde{\mathbf{y}}_s}(k)$. Define

$$\begin{aligned} \mathbf{y}_{sk} &= \left[\mathbf{y}_s^T(0)\ \mathbf{y}_s^T(1)\ \cdots\ \mathbf{y}_s^T(k)\right]^T,\ R_{\mathbf{y}_{sk}} = \mathrm{diag}(R_{\mathbf{y}_s}(0), R_{\mathbf{y}_s}(1), \cdots, R_{\mathbf{y}_s}(k)), \\ \tilde{\mathbf{y}}_{sk} &= \left[\tilde{\mathbf{y}}_s^T(0)\ \tilde{\mathbf{y}}_s^T(1)\ \cdots\ \tilde{\mathbf{y}}_s^T(k)\right]^T,\ R_{\tilde{\mathbf{y}}_{sk}} = \mathrm{diag}(R_{\tilde{\mathbf{y}}_s}(0), R_{\tilde{\mathbf{y}}_s}(1), \cdots, R_{\tilde{\mathbf{y}}_s}(k)), \\ y_{sk} &= \left[y_s^T(0)\ y_s^T(1)\ \cdots\ y_s^T(k))\right]^T,\ \tilde{y}_{sk} = \left[\tilde{y}_s^T(0)\ \tilde{y}_s^T(1)\ \cdots\ \tilde{y}_s^T(k)\right]^T. \end{aligned}$$

where

$$\tilde{y}(k) = y(k) - \hat{y}(k),\ \tilde{r}(k|k) = r(k) - \hat{r}(k|k)$$

and $\hat{y}(k)$, $\hat{r}(k|k)$ are obtained from the Krein space projection formulae (21.34)–(21.35) and (21.37). It is easy to see that

$$\mathbf{z}(k) = \Xi(k)\tilde{\mathbf{y}}_{sk},\ z(k) = \Xi(k)\tilde{y}_{sk},\ R_{\mathbf{y}_{sk}} = \Xi(k)R_{\tilde{\mathbf{y}}_{sk}}\Xi(k)^T,$$

where

$$\Xi(k) = [\phi(i,j)]_{k+1,k+1},\ \phi(k)(i,i) = \begin{bmatrix} I & 0 \\ \langle \mathbf{r}(i), \tilde{\mathbf{y}}(i)\rangle R_{\tilde{\mathbf{y}}}^{-1}(i) & I \end{bmatrix},$$

for $i = 1, 2, \cdots, k+1$, and

$$\phi(k)(i,j) = \begin{bmatrix} \langle \mathbf{y}(i), \tilde{\mathbf{y}}(j)\rangle R_{\tilde{\mathbf{y}}}^{-1}(j) & 0 \\ \langle \mathbf{r}(i), \tilde{\mathbf{y}}(j)\rangle R_{\tilde{\mathbf{y}}}^{-1}(j) & 0 \end{bmatrix},$$

for $i = 2, 3, \cdots, k+1;\ j = 1, 2, \cdots, i-1$. Hence, $R_{\mathbf{y}_s}(k)$ has the same inertia with $R_{\tilde{\mathbf{y}}_s}(k)$. In view of (21.7) and (21.38), the inertias of $R_{\tilde{\mathbf{y}}_s}(k)$ and $Q_{vs}(k)$ coincide if and only if $R_{\tilde{\mathbf{y}}}(k) > 0,\ R_{\tilde{\mathbf{r}}}(k|k) < 0$, i.e.,

$$\begin{cases} I + H(k)P(k)H^T(k) > 0, \\ L(k)(P^{-1}(k) + H^T(k)H(k))L^T(k) - \gamma^2 I < 0. \end{cases} \quad (21.39)$$

Applying Proposition 21.1, J_N has a minimum if and only if $R_{\mathbf{y}_s}(k)$ has the same inertia with $Q_{vs}(k)$, which is equivalent to $R_{\tilde{\mathbf{y}}}(k) > 0,\ R_{\tilde{\mathbf{r}}}(k|k) < 0$. Moreover, the value of J_N at the minimum equals to

$$
\begin{aligned}
J_N(\hat{x}_0, \hat{w}_N, y_{sN}) &= z_N^T R_{\mathbf{y}_{sN}}^{-1} z_N = \tilde{y}_{sN}^T R_{\tilde{\mathbf{y}}_{sN}}^{-1} \tilde{y}_{sN} \\
&= \sum_{k=0}^{N} \tilde{y}^T(k) R_{\tilde{\mathbf{y}}}^{-1}(k) \tilde{y}(k) + \sum_{k=0}^{N-1} \tilde{r}^T(k|k) R_{\tilde{\mathbf{r}}}^{-1}(k|k) \tilde{r}(k|k).
\end{aligned} \tag{21.40}
$$

On this basis, the following proposition is straightforward.

Proposition 21.2 *The indefinite quadratic form J_N has a unique minimum over $\{x_0, w_N\}$, subject to (21.3) and (21.26), if and only if (21.39) being satisfied. Moreover, the value of J_N at the minimum is given by (21.40).*

21.3.3 The Calculation of $H_1(\hat{x}(k))$ and $H_2(\hat{x}(k))$

So far we have derived the existence conditions and obtain the unique minimum of J_N in Proposition 21.2. It is noted that (21.39) holds if and only if $R_{\tilde{\mathbf{y}}}(k|k) > 0$ and $R_{\tilde{\mathbf{r}}}(k|k) < 0$. To achieve a positive minimum, the $H_1(\hat{x}(k))$ and $H_2(\hat{x}(k))$ should be chosen such that $\tilde{r}(k|k) = 0$, i.e., $r(k) = \hat{r}(k|k)$.

In observing (21.38), the innovations $\{\tilde{\mathbf{y}}_s(i)\}_{i=0}^k$ form an orthogonal basis for $\mathcal{L}\{\{\tilde{\mathbf{y}}_{s_1}(i)\}_{i=0}^k\}$, we can further express the projection $\hat{\mathbf{x}}(k+1)$ as

$$
\begin{aligned}
\hat{\mathbf{x}}(k+1) &= \text{Proj}\{\mathbf{x}(k+1) | \{\tilde{\mathbf{y}}_{s1}(i)\}_{i=0}^k\} \\
&= \text{Proj}\{\mathbf{x}(k+1) | \{\tilde{\mathbf{y}}_{s}(i)\}_{i=0}^k\} \\
&= \text{Proj}\{\mathbf{x}(k+1) | \{\tilde{\mathbf{y}}_{s}(i)\}_{i=0}^{k-1}\} + \left\langle \mathbf{x}(k+1), \tilde{\mathbf{y}}_s(k) \right\rangle R_{\tilde{\mathbf{y}}_s}^{-1}(k) \tilde{\mathbf{y}}_s(k) \\
&= F(k)\hat{\mathbf{x}}(k) + \mathbf{u}_1(k) + \langle \mathbf{x}(k+1), \tilde{\mathbf{y}}(k) \rangle R_{\tilde{\mathbf{y}}}^{-1}(k) \tilde{\mathbf{y}}(k) \\
&\quad + \langle \mathbf{x}(k+1), \tilde{\mathbf{r}}(k|k) \rangle R_{\tilde{\mathbf{r}}}^{-1}(k|k) \tilde{\mathbf{r}}(k|k) \\
&= f(\hat{\mathbf{x}}(k)) + K_1 \tilde{\mathbf{y}}(k) + K_2(k) \tilde{\mathbf{r}}(k|k),
\end{aligned}
$$

where

$$
\begin{aligned}
K_1(k) &= \langle \mathbf{x}(k+1), \tilde{\mathbf{y}}(k) \rangle R_{\tilde{\mathbf{y}}}^{-1}(k) \\
&= \langle F(k)\mathbf{x}(k) + \mathbf{u}_1(k) + G(k)\mathbf{w}(k), \tilde{\mathbf{y}}(k) \rangle R_{\tilde{\mathbf{y}}}^{-1}(k) \\
&= F(k) P(k) H^T(k) R_{\tilde{\mathbf{y}}}^{-1}(k), \\
R_{\tilde{\mathbf{y}}}(k) &= I + H(k) P(k) H^T(k), \\
K_2(k) &= \langle \mathbf{x}(k+1), \tilde{\mathbf{r}}(k|k) \rangle R_{\tilde{\mathbf{r}}}^{-1}(k|k).
\end{aligned}
$$

Let $\tilde{\mathbf{r}}(k|k) = 0$. Now $\hat{x}(k+1)$ and $\hat{r}(k|k)$ can be calculated by the formulae of $\hat{\mathbf{x}}(k+1)$, $\hat{\mathbf{r}}(k|k)$, respectively, i.e.,

$$
\begin{cases}
\hat{x}(k+1) = f(\hat{x}(k)) + K_1(k)(y(k) - h(\hat{x}(k))), \\
\hat{r}(k|k) = L(k)\hat{x}(k) + K_{\mathbf{r}}(k)(y(k) - h(\hat{x}(k))).
\end{cases}
$$

Thus we have $\tilde{r}(k|k) = 0$. The value of J_N at its minimum is

$$J_N(\hat{x}_0, \hat{w}_N, y_{sN}) = \sum_{k=0}^{N} \tilde{y}^T(k) R_{\tilde{\mathbf{y}}}^{-1}(k) \tilde{y}(k) > 0.$$

Therefore, the underlying H_∞ a posteriori filter can be given by (21.3) with

$$\begin{aligned} H_1(\hat{x}(k)) &= K_1(k) = F(k)P(k)H^T(k)R_{\tilde{\mathbf{y}}}^{-1}(k), \\ H_2(\hat{x}(k)) &= K_{\mathbf{r}}(k) = L(k)P(k)H^T(k)R_{\tilde{\mathbf{y}}}^{-1}(k). \end{aligned}$$

Now the following proposition is readily to be concluded.

Proposition 21.3 *Given $\gamma > 0$, if (21.39) is satisfied, then the H_∞ nonlinear a posteriori filter can be given by (21.3) with*

$$\begin{aligned} H_1(\hat{x}(k)) &= F(k)P(k)H^T(k)(I + H(k)P(k)H^T(k))^{-1}, \\ H_2(\hat{x}(k)) &= L(k)P(k)H^T(k)(I + H(k)P(k)H^T(k))^{-1}, \\ P(k+1) &= F(k)P(k)F^T(k) - F(k)P(k)\begin{bmatrix} H(k) \\ L(k) \end{bmatrix}^T \\ &\quad \times R_{\tilde{\mathbf{y}}_{s1}}^{-1}(k)\begin{bmatrix} H(k) \\ L(k) \end{bmatrix} P(k)F^T(k) + G(k)G^T(k), \ \ P(0) = P_0, \end{aligned}$$

where $R_{\tilde{\mathbf{y}}_{sI}}(k)$ is given by (21.36).

Remark 21.1 It is noted that the H_∞ filter (21.3) is nonlinear, although first-order linearization is considered on designing $H_1(\hat{x}(k))$ and $H_2(\hat{x}(k))$.

Remark 21.2 The expended H_∞ filter in [24] can be considered as a special case of (21.3) with $H_2(\hat{x}(k)) = 0$. In addition, the existence conditions for the minimum of J_N are sufficient and necessary, while the ones in [24] are sufficient. From this point of view, the above H_∞ nonlinear filter is less conservative.

It is worth noting that, despite (21.3) is initially applied for filtering, we can augment the concerned faults as state variables and then adopt the H_∞ filter for state estimation so as to obtain the estimates of faults. The PMI based nonlinear H_∞ filter (21.22) can be applied to this end, as discussed in the forthcoming section.

21.4 PMI Observer Based Nonlinear H_∞ Fault Estimation

Based on the above Krein space based H_∞ filter, this section concerns to cope with the nonlinear fault estimation issue in Problem 21.2 for the INS/GPS integrated system, by combining the PMI observer with Krein space H_∞-filtering technique.

21.4.1 Krein Space Based Solution

Given INS/GPS integrated system (21.20), $z(k)$ in (21.21) and $\hat{z}(k|k)$ in (21.23), we begin with defining

$$v_s(k) = \begin{bmatrix} v(k) \\ \hat{z}(k|k) - z(k) \end{bmatrix}.$$

Then J_N in (21.24) can the be rewritten as

$$J_N = \begin{bmatrix} \bar{x}_0 \\ w_N \\ v_{sN} \end{bmatrix}^T \begin{bmatrix} P_0 & 0 & 0 \\ 0 & I & 0 \\ 0 & 0 & Q_{v_{sN}} \end{bmatrix}^{-1} \begin{bmatrix} \bar{x}_0 \\ w_N \\ v_{sN} \end{bmatrix}, \tag{21.41}$$

where $w_N = [\, w(0)^T \; w^T(1) \cdots w^T(N)\,]^T$, $v_{sN} = [\, v_s(0)^T \; v_s(1)^T \cdots v_s(N)^T\,]^T$, $Q_{v_{sN}} = \text{diag}(Q_{v_s}, Q_{v_s}, \cdots, Q_{v_s})$ and $Q_{v_s} = \text{diag}(I, -\gamma^2 I)$.

It is noted that the J_N in (21.41) has the same form with the one in (21.8). In fact, the H_∞ filter (21.22) is a special case of (21.3) with $L(k)$ in (21.2) being $L(k) = I$(i.e., $\hat{z}(k|k) = s(k)$), $H_1(k)$ and $H_2(k)$ in (21.3) being $H_1(k) = 0$ and $H_2(k) = K(k+1)$. In this regard, the above derived solution of Krein space nonlinear H_∞-filter in Sect. 21.3 can be directly applied to address the Problem 21.2. To this end, we introduce the following Krein space stochastic system

$$\begin{cases} \bar{\mathbf{x}}(k+1) = \bar{F}(k)\bar{\mathbf{x}}(k) + \bar{q}(\bar{\mathbf{x}}(k)) + \bar{B}(k)\mathbf{w}(k), \\ \mathbf{y}_s(k) = \begin{bmatrix} \bar{C}(k) \\ I \end{bmatrix} \bar{\mathbf{x}}_k + \mathbf{v}_s(k), \end{cases} \tag{21.42}$$

where

$$\mathbf{y}_s(k) = \begin{bmatrix} \mathbf{y}(k) \\ \hat{\mathbf{z}}(k|k) \end{bmatrix}, \mathbf{v}_s(k) = \begin{bmatrix} \mathbf{v}(k) \\ \hat{\mathbf{z}}(k|k) - \mathbf{z}(k) \end{bmatrix}$$

$$\left\langle \begin{bmatrix} \bar{\mathbf{x}}_0 \\ \mathbf{w}(k) \\ \mathbf{v}_s(k) \end{bmatrix}, \begin{bmatrix} \bar{\mathbf{x}}(0) \\ \mathbf{w}(j) \\ \mathbf{v}_s(j) \end{bmatrix} \right\rangle = \begin{bmatrix} P_0 & 0 & 0 \\ 0 & I\delta_{kj} & 0 \\ 0 & 0 & Q_{vs}\delta_{kj} \end{bmatrix}$$

Denote $\hat{\bar{\mathbf{x}}}(k|k)$ as the projection of $\bar{\mathbf{x}}(k)$ on the space spanned by $\{\{\mathbf{y}_s(i)\}_{i=0}^{k-1}, \mathbf{y}(k)\}$. Using the first order Taylor approximations

$$\bar{q}(\bar{\mathbf{x}}(k)) \simeq \bar{q}(\hat{\bar{\mathbf{x}}}(k|k)) + \Gamma(k)(\bar{\mathbf{x}}(k) - \hat{\bar{\mathbf{x}}}(k|k)), \;\; \Gamma(k) = \frac{\partial \bar{q}(\bar{\mathbf{x}})}{\partial \bar{\mathbf{x}}}|_{\bar{\mathbf{x}}=\hat{\bar{\mathbf{x}}}(k|k)},$$

we have

$$\begin{cases} \bar{\mathbf{x}}(k+1) = G(k)\bar{\mathbf{x}}(k) + \mathbf{u}(k) + \bar{B}_k\mathbf{w}(k), \\ \mathbf{y}_s(k) = \begin{bmatrix} \bar{C}(k) \\ I \end{bmatrix} \bar{\mathbf{x}}(k) + \mathbf{v}_s(k), \end{cases} \tag{21.43}$$

where

$$G(k) = \bar{F}_k + \Gamma_k, \;\; \mathbf{u}(k) = \bar{q}(\hat{\bar{\mathbf{x}}}(k|k)) - \Gamma(k)\hat{\bar{\mathbf{x}}}(k|k). \tag{21.44}$$

Let $R_{\tilde{\mathbf{y}}_{s,k}}$ be the Gramian matrix of $\tilde{\mathbf{y}}_s(k)$ with

$$\tilde{\mathbf{y}}_s(k) = \mathbf{y}_s(k) - \begin{bmatrix} \bar{C}_k\hat{\bar{\mathbf{x}}}(k|k-1) \\ \hat{\bar{\mathbf{x}}}(k|k-1) \end{bmatrix}. \tag{21.45}$$

According to Propositions 21.1–21.3, we directly have the following proposition.

Proposition 21.4 *Given $\gamma > 0$, if conditions*

$$I + \bar{C}(k)P(k)\bar{C}^T(k) > 0, \;\; -\gamma^2 I + (P^{-1}(k) + \bar{C}^T(k)\bar{C}(k))^{-1} < 0 \tag{21.46}$$

are satisfied with $P(k)$ being recursively computed by

$$\begin{aligned} P(k+1) = {} & G(k)P(k)G^T(k) - G(k)P(k)\begin{bmatrix} \bar{C}(k) \\ I \end{bmatrix}^T R^{-1}_{\tilde{\mathbf{y}}_{s,k}} \begin{bmatrix} \bar{C}(k) \\ I \end{bmatrix} P(k)G^T(k) \\ & + \bar{B}(k)\bar{B}(k)^T, \end{aligned} \tag{21.47}$$

where

$$R_{\tilde{\mathbf{y}}_{s,k}} = \begin{bmatrix} I + \bar{C}(k)P(k)\bar{C}^T(k) & \bar{C}(k)P(k) \\ P(k)\bar{C}^T(k) & -\gamma^2 I + P(k) \end{bmatrix}, \tag{21.48}$$

then the nonlinear H_∞ fault estimator (21.22) guaranteeing (21.23) can be obtained with

$$K(k+1) = P(k+1)\bar{C}^T(k+1)(I + \bar{C}(k+1)P(k+1)\bar{C}^T(k+1))^{-1}. \tag{21.49}$$

Due to the same derivation of Proposition 21.4 from Propositions 21.1–21.3, the proof is omitted here for simplicity.

21.4.2 Recursive Realization of PMI Based Fault Estimator

In order to get a solution of the H_∞ estimator (21.22) such that (21.23) is satisfied, one can calculate $\hat{\bar{x}}(k+1|k+1)$ by (21.22) and choose $\hat{z}(k|k) = \hat{x}(k|k)$, i.e.

Algorithm 21.4.17 PMI based H_∞ nonlinear estimator for INS/GPS integrated system

1: Let $k = 0$, set $\hat{\bar{x}}(0|0) = 0$. Choose appropriate γ and q.
2: If the conditions of (21.46) hold true, go to step 3; otherwise, exist.
3: Calculate $K(k+1)$, and $P(k+1)$ using (21.49) and (21.47), respectively.
4: Update $y(k+1)$ using (21.20).
5: Calculate $\hat{\bar{x}}(k+1|k+1)$ using (21.50).
6: Correct the solution by using $\hat{x}(k+1|k+1)$, $\bar{C}_b^n = \hat{\phi} \times C_b^n$, $\bar{V} = V - \delta V$ and $\bar{p} = p - \delta p$, respectively.
7: Reset $\hat{x}(k+1|k+1) = 0$, and let $\hat{\bar{x}}(k+1|k+1) = [\mathbf{0}_{1\times 9}, \hat{\xi}_{1,k+1}^T, \hat{\xi}_{2,k+1}^T, \cdots, \hat{\xi}_{q,k+1}^T]$.
8: Let $k = k+1$, go to Step 2 until $k = N$.

$$\begin{cases} \hat{\bar{x}}(k+1|k+1) = \hat{\bar{x}}(k+1|k) + K(k+1)(y(k+1) - \bar{C}(k+1)\hat{\bar{x}}(k+1|k)), \\ \hat{\bar{x}}(k+1|k) = \bar{F}(k)\hat{\bar{x}}(k|k) + \bar{q}(k)(\hat{\bar{x}}(k|k)), \hat{z}(k|k) = \hat{x}(k|k) \end{cases} \tag{21.50}$$

with $K(k+1)$ given in (21.49). Rewrite the gain $K(k+1)$ as

$$K(k+1) = [\,(K_{P,k+1})^T \;\; (K_{1,k+1})^T \;\; \cdots \;\; (K_{q,k+1})^T\,]^T,$$

where $K_{P,k+1}$ denotes the proportional gain, $K_{i,k+1}$ $(i = 1, 2, \cdots, q)$ denote integral gains. Then $\hat{\bar{x}}(k+1|k+1)$ can be re-expressed in the following form

$$\begin{cases} \hat{x}(k+1|k+1) = \hat{x}(k+1|k) + B(k)\hat{\xi}_{q,k+1} \\ \qquad\qquad +K_{P,k+1}(y(k+1) - C(k+1)\hat{x}(k+1|k)), \\ \hat{\xi}_{1,k+1} = K_{1,k+1}(y(k+1) - C(k+1)\hat{x}(k+1|k)) + \hat{\xi}_{1,k}, \\ \hat{\xi}_{2,k+1} = K_{2,k+1}(y(k+1) - C(k+1)\hat{x}(k+1|k)) + T_{GPS}\hat{\xi}_{1,k} + \hat{\xi}_{2,k}, \\ \qquad\vdots \\ \hat{\xi}_{q,k+1} = K_{q,k+1}(y(k+1) - C(k+1)\hat{x}(k+1|k)) + T_{GPS}\hat{\xi}_{q-1,k} + \hat{\xi}_{q,k}, \end{cases}$$

that is the thus called a PMI based H_∞ nonlinear estimator.

It is seen that there are multi-integral information included in $\hat{\xi}_{i,k+1} (i = 1, 2..., q)$. That is to say, $\hat{\xi}_{i,k+1}$ is an estimation of the $(q-i)$th derivation of the bias $b(k+1)$ in the form (21.15), and $\hat{\xi}_{q,k+1}$ is an estimation of $b(k+1)$. At the same time, the estimation of $x(k+1)$ includes innovation $y(k+1) - C(k+1)\hat{x}(k+1|k)$ and bias estimation $B(k)\hat{\xi}_{q,k+1}$. Therefore, considering some time-varying biases caused by temperature shift, choosing suitable $q \geq 1$, the proposed method can achieve accurate navigation state estimation $\hat{x}(k)$ with the disturbances of time-varying bias $\hat{b}(k+1)$. The algorithm of PMI based H_∞ nonlinear estimator for INS/GPS integrated system is summarized as Algorithm 21.4.17.

21.5 Experimental Study

To show the applicability of the developed Krein space schemes of H_∞ nonlinear filtering and fault estimation, two experimental examples on INS/GPS integrated system are given below.

21.5.1 Example I

In this example, Krein space based nonlinear H_∞ filtering method is applied to the INS/GPS integrated system and a comparison of it with the widely used EKF is made.

Consider the strap-down INS/GPS integrated system modeled by (20.3)–(20.8) (that can also be expressed as (21.16)). In the underlying flight experiment, the IMU was mounted onto the bottom of Y-12 type aircraft. The main specifications of the INS/GPS are listed in Table 21.1. The flight trajectory is shown in Fig. 21.2a. The local earth radius is $R_e = 6378135\,\text{m}$, and $R_M = R_e[1 - e(2 - 3\sin^2 L)]$, $R_N = R_e(1 + e\sin^2 L)$, $e = 0.0034$, $\omega_{ie} = 7.2921 \times 10^{-5}\,\text{rad/s}$. The test time interval is $(1000\,\text{s}, 8000\,\text{s})$. The flight speed is about 120 km/h. The sampling period of GPS receiver is $T_{GPS} = 1\,\text{s}$.

In normal flight condition, both the process noise and measurement noise are usually considered as white Gaussian noises. One can apply an EKF to the state estimation for the INS/GPS integrated system. However, due to the influences of the operation and environmental conditions, it is difficult to get the real exact priori information of the gyroscopes and accelerometers biases. Moreover, the GPS outlier may also be inevitable. For the case of the noises being $l_{2,[0,N]}$-norm bounded, the H_∞ nonlinear filter becomes a suitable choice.

We first derive the discreterized system with sampling period $T = 1\,\text{s}$. Next, let $s(k) = x(k)$, $\gamma = 1.2$ and design H_∞ nonlinear filter (21.3) using Proposition 21.3. By applying the H_∞ nonlinear filter and the EKF to the INS/GPS integrated system, we can get the estimation of $x(k)$. Here only the estimation of gyroscope and accelerometer biases are shown in Figs. 21.3, 21.4 and 21.5, i.e., the $H_\infty - 1$ filter

Table 21.1 The main specifications of INS/GPS system

Sensors	Specifications
Gyroscope	Random constant 0.1°/h, white noise 0.1°/h(1σ)
Accelerometer	Random constant 100 ug, white noise 50 ug(1σ)
GPS	Velocity: 0.05 m/s (RMS), Position 1.5 m (RMS)

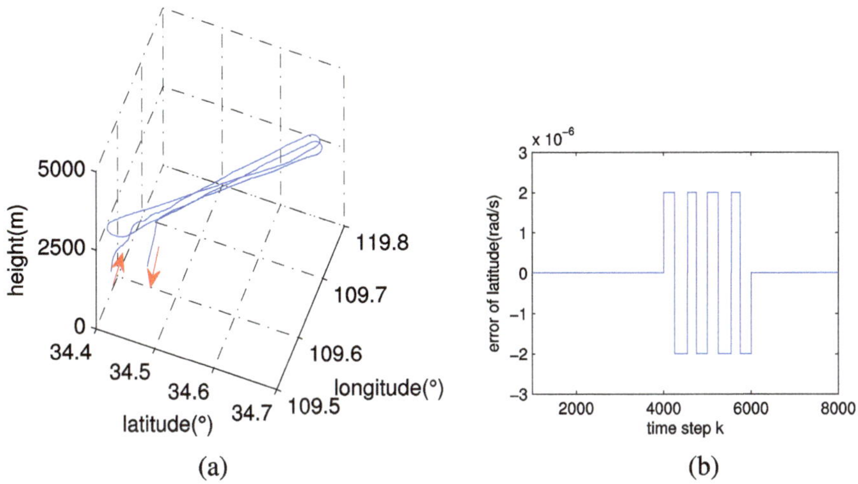

Fig. 21.2 **a** The trajectory of the flight experiment; **b** The simulated GPS outliers

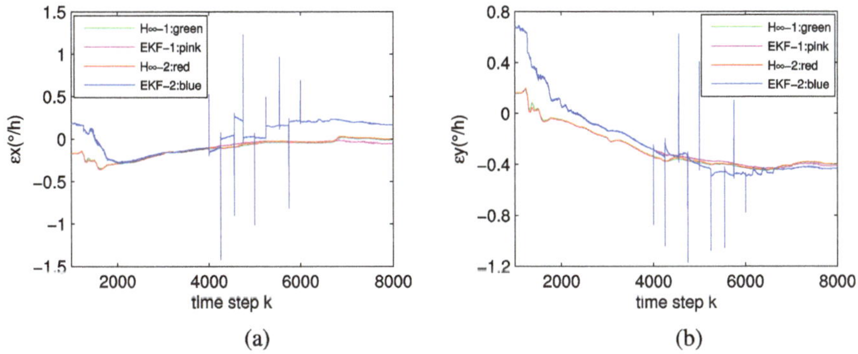

Fig. 21.3 **a** The estimation of east gyroscope bias; **b** The estimation of north gyroscope bias

and EKF-1 filter. Furthermore, we apply the developed H_∞ nonlinear filter and the EKF to the flight data with simulated GPS outlier shown in Fig. 21.2b. The estimation of gyroscope and accelerometer biases are also presented in Figs. 21.3, 21.4 and 21.5, i.e., the $H_\infty - 2$ filter and EKF-2 filter.

It can be seen from Figs. 21.3, 21.4 and 21.5 that even in the case of normal flight condition, the H_∞ nonlinear filtering results are better than the results of EKF. In the case of GPS outlier, the achieved disturbance rejection level is obvious.

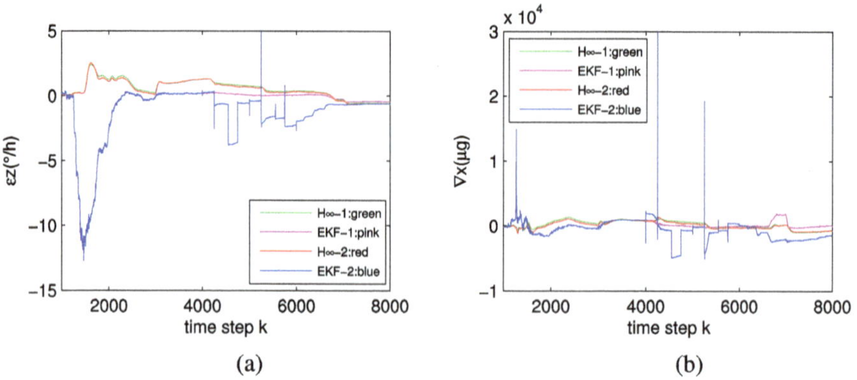

Fig. 21.4 **a** The estimation of up gyroscope bias; **b** The estimation of east accelerometer bias

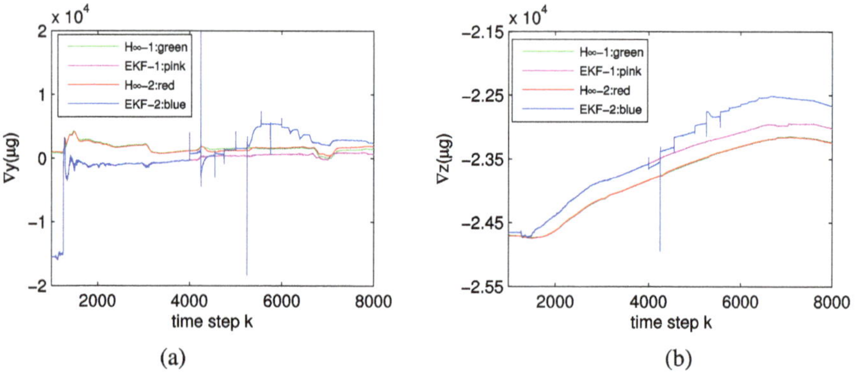

Fig. 21.5 **a** The estimation of north accelerometer bias; **b** The estimation of up accelerometer bias

21.5.2 Example II

To demonstrate the effectiveness of the PMI based H_∞ fault estimation method for the integrated INS/GPS system, a flight experiment of ring laser gyroscope POS is used, wherein the experimental setup is same with the one in Sect. 20.5.1.

Set $\gamma = 1.2$ and

$$\begin{aligned} P_0 = \mathrm{diag}\big(&(0.1^\circ)^2, (0.1^\circ)^2, (0.5^\circ)^2, (0.05\,\mathrm{m/s})^2, (0.05\,\mathrm{m/s})^2, (0.05\,\mathrm{m/s})^2, \\ &(0.1\,\mathrm{m})^2, (0.1\,\mathrm{m})^2, (0.1\,\mathrm{m})^2, \left(0.01^\circ/\mathrm{h}^3\right)^2, \left(0.01^\circ/\mathrm{h}^3\right)^2, \left(0.01^\circ/\mathrm{h}^3\right)^2, \\ &\left(10^{-13}\,\mathrm{m/s}^4\right)^2, \left(10^{-13}\,\mathrm{m/s}^4\right)^2, \left(10^{-13}\,\mathrm{m/s}^4\right)^2, \left(0.01^\circ/\mathrm{h}^2\right)^2, \left(0.01^\circ/\mathrm{h}^2\right)^2, \\ &\left(0.01^\circ/\mathrm{h}^2\right)^2, \left(10^{-8}\,\mathrm{m/s}^3\right)^2, \left(10^{-8}\,\mathrm{m/s}^3\right)^2, \left(10^{-8}\,\mathrm{m/s}^3\right)^2, (0.01^\circ/\mathrm{h})^2, \\ &(0.01^\circ/\mathrm{h})^2, (0.01^\circ/\mathrm{h})^2, (100\,\mu\mathrm{g})^2, (100\,\mu\mathrm{g})^2, (100\,\mu\mathrm{g})^2\big)\,. \end{aligned}$$

Table 21.2 The RMSE statistics of the navigation states

	PMI based H_∞	EKF	STF
Latitude (m)	0.0242	0.0243	0.0243
Longitude (m)	0.0724	0.0753	0.0740
Height (m)	0.0547	0.0652	0.0604
Heading(°)	0.0078	0.0075	0.0075
Pitch(°)	0.0044	0.0042	0.0042
Roll(°)	0.0064	0.0062	0.0063

We first design an H_∞ nonlinear estimator using the Krein space approach for INS/GPS integrated system. As a comparison, the well known EKF and STF are also designed.

In case 1, we choose a PMI based H_∞ nonlinear estimator with $q = 1$ to estimate the INS/GPS state and bias. Meanwhile, for the comparison, the 15-dimension EKF and STF are presented. We evaluate the system performance by the commonly used root mean square error (RMSE) [16], i.e.,

$$RMSE(\Delta s) = \sqrt{\frac{1}{N}\sum_{i=1}^{N} \Delta s_i^2},$$

where Δs denotes the estimation error of a navigation state in comparison with the aerial triangulation solution which is more accurate than former. N is the estimation number. The RMSE results of the system performance are listed in Table 21.2.

It could be seen that the RMSE of PMI based H_∞ nonlinear filter is equivalent to regular EKF and STF, even smaller than them in position information. Due to the constant bias, the STF can not verify the tracking ability of break wave. That is to say, the developed PMI nonlinear H_∞ estimation method can achieve the same performance with EKF when the experiment is in normal condition.

To compare the performance of estimating random bias of the inertial sensor,we present EKF, STF and proposed method with $q = 3$ in case 2. We choose a similar drift signal to simulate sensor errors like Fig. 21.1 and added it to the raw IMU data in about (2000 s,3000 s). Furthermore, we apply these estimators to the flight data with simulated GPS sinusoidal noise. Figures 21.6 and 21.7 show the navigation results of the INS/GPS integrated system random bias estimation using three different estimators.

As expected, it is shown that the east and north components of gyroscope and accelerometer errors are well estimated by the PMI based H_∞ nonlinear estimator. From the comparison between three estimators, the tracking performance of the proposed estimator is better than that of the others, and simultaneously it can guarantee uncertain noise attenuation and disturbance rejection as usual. It is mentioned that due to the vertical component of the inertial sensors bias unobservable, both the esti-

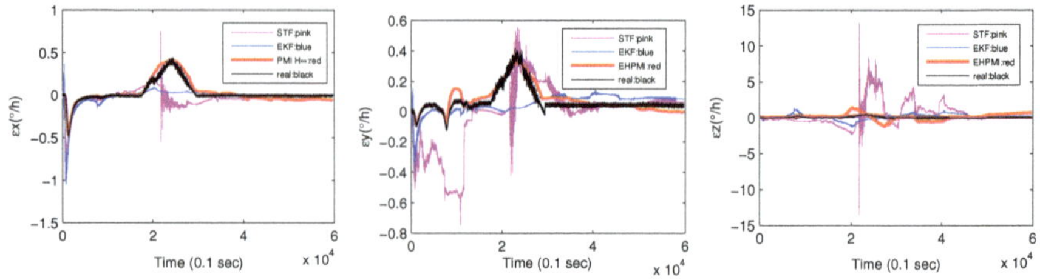

Fig. 21.6 The east, north and up gyroscope bias error

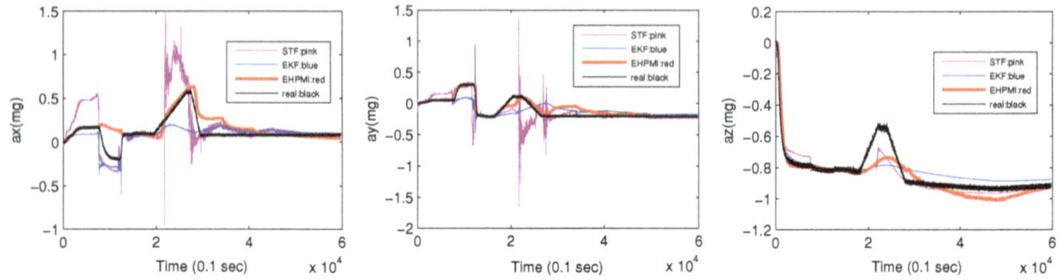

Fig. 21.7 The east, north and up accelerometer bias error

Table 21.3 The RMSE statistics of the navigation states using three estimators

	PMI based H_∞	EKF	STF
Latitude (m)	0.1992	0.2336	0.2045
Longitude (m)	0.5262	0.6742	0.5630
Height (m)	0.0978	0.1248	0.1022
Heading(°)	0.1244	0.3356	0.1592
Pitch(°)	0.0090	0.0838	0.0099
Roll(°)	0.0029	0.0493	0.0034

mation of gyroscope bias and accelerometer bias of axis z are not improved evidently [16]. The RMSE results of the system performance are listed in Table 21.3 to show the accommodation of proposed estimator.

We can see from Table 21.3 that the RMSE results of PMI based H_∞ nonlinear estimator are smaller than those of other filters, because the estimation of time-varying bias was used for the compensation of navigation state. It proves that the developed algorithm could more effectively estimate and accommodate the random bias than traditional estimators.

21.6 Conclusion

In this chapter, an extension of Krein space based approach to finite horizon H_∞ fault estimation for a class of discrete-time nonlinear systems has been studied and, on this basis, a PMI observer based nonlinear H_∞ fault estimation method to the integrated INS/GPS system has been presented. The key of these schemes lies in converting the problem of nonlinear H_∞ filtering into a minimum of an indefinite quadratic form and then finding a sufficient and necessary condition for the minimum, by establishing a relationship between nonlinear filter in Hilbert space and nonlinear estimation in Krein space. A feasible solution to the nonlinear H_∞ filter has been derived via computing Riccati recursions. Moreover, a PMI observer based H_∞ fault estimator was constructed for the integrated INS/GPS system, with respect to augmenting the bias faults as states. It is worth emphasizing that, even though the first-order Taylor approximations were used during the derivation of nonlinear H_∞ filtering, the derived filter and fault estimator are intrinsically nonlinear. The applicability of the developed approaches has been demonstrated through the experimental study on an INS/GPS integrated system.

References

1. Akeila, E., Salcic, Z., & Swain, A. (2014). Reducing low-cost INS error accumulation in distance estimation using self-resetting. *IEEE Transactions on Instrumentation and Measurement, 63*(1), 177–184.
2. Aliyu, M. D. S., & Boukas, E. K. (2011). Discrete-time mixed H_2/H_∞ nonlinear filtering. *International Journal of Robust and Nonlinear Control, 21*(11), 1257–1282.
3. Barbour, N., & Schmidt, G. (2001). Inertial sensor technology trends. *IEEE Sensors Journal, 1*(4), 332–339.
4. Berman, N., & Shaked, U. (1996). H_∞ nonlinear filtering. *International Journal of Robust and Nonlinear Control, 6*(4), 281–296.
5. Cannon, M. E., Nayak, R., Lachapelle, G., et al. (2001). Low cost INS/GPS integration: Concepts and testing. *The Journal of Navigation, 54*(1), 119–134.
6. Chen, X., Shen, C., & Zhang, W. (2013). Novel hybrid of strong tracking Kalman filter and wavelet neural network for GPS/INS during GPS outages. *Measurement, 46*(10), 3847–3854.
7. Dong, H., Lam, J., & Gao, H. (2011). Distributed H_∞ filtering for repeated scalar nonlinear systems with random packet losses in sensor networks. *International Journal of Systems Science, 42*(9), 1507–1519.
8. Dong, H., Wang, Z., & Gao, H. (2013). Distributed H_∞ filtering for a class of Markovian jump nonlinear time-delay systems over lossy sensor networks. *IEEE Transactions on Industrial Electronics, 60*(10), 4665–4672.
9. Fang, J., & Gong, X. (2010). Predictive iterated Kalman filter for INS/GPS integration and its application to SAR motion compensation. *IEEE Transactions on Instrumentation and Measurement, 59*(4), 909–915.
10. Fang, Q., & Sheng, X. (2013). UKF for integrated vision and inertial sensors based on three-view geometry. *IEEE Sensors Journal, 13*(7), 2711–2719.
11. Gao, Z., & Ding, S. X. (2007). Fault estimation and fault-tolerant control for descriptor systems via proportional, multiple-integral and derivative observer design. *IET Control Theory & Applications, 1*(5), 1208–1218.

12. Gao, Z. W., & Ho, D. W. C. (2004). Proportional multiple-integral observer design for descriptor systems with measurement output disturbances. *IEE Proceedings: Control Theory Applications, 151*(3), 279–288.
13. Gao, Z., Ding, S. X., & Ma, Y. (2007). Robust fault estimation approach and its application in vehicle lateral dynamic systems. *Optimal Control Application Methods, 28,* 143–156.
14. Groves, P. D. (2008). *Principles of GNSS, inertial, and multisensor integrated navigation systems* (pp. 363–368). Boston: Artech House.
15. Hassibi, B., Sayed, A. H., & Kailath, T. (1999). *Indefinite-quadratic estimation and control: A unified approach to H_2 and H_∞ theories*. SIAM.
16. Hong, S., Chun, H., Kwon, S. H., et al. (2008). Observability measures and their application to GPS/INS. *IEEE Transactions on Vehicular Technology, 57*(1), 97–106.
17. Jwo, D. J., & Lai, S. Y. (2009). Navigation integration using the fuzzy strong tracking unscented Kalman filter. *Journal of Navigation, 62*(2), 303–322.
18. Kim, K. H., Lee, G. J., & Park, C. G. (2009). Adaptive two-stage extended Kalman filter for a fault-tolerant INS/GPS loosely coupled system. *IEEE Transactions on Aerospace and Electronic Systems, 45*(1), 125–137.
19. Koenig, D. (2005). Unknown input proportional multiple-integral observer design for linear descriptor systems: Application to state and fault estimation. *IEEE Transactions on Automatic Control, 50*(2), 212–217.
20. Noureldin, A., Karamat, T. B., & Georgy, J. (2013). *Fundamentals of inertial navigation, satellite-based positioning and their integration* (pp. 247–271). Berlin, Germany: Springer.
21. Rogers, R. M. (2007). *Applied mathematics in integrated navigation systems* (3rd ed.). Reston, VA: AIAA.
22. Sami, M., & Patton, R. J. (2012). An FTC approach to wind turbine power maximisation via T-S fuzzy modelling and control. *IFAC Proceedings Volumes, 45*(20), 349–354.
23. Schmidt, G. T. (2010). INS/GPS technology trends. NATO, Rep. RTO-EN-SET–116.
24. Shaked, U., & Berman, N. (1995). H_∞ nonlinear filtering of discrete-time processes. *IEEE Transactions on Signal Processing, 43*(9), 2205–2209.
25. Sohne, W., Heinze, O., & Groten, E. (1994). Integrated INS/GPS system for high precision navigation applications. In *Proceedings of 1994 IEEE Position, Location and Navigation Symposium–PLANS'94* (pp. 310–313). April 11–15, Las Vegas, NV, USA.
26. Wang, Z., Shen, B., & Liu, X. (2012). H_∞ filtering with randomly occurring sensor saturations and missing measurements. *Automatica, 48*(3), 556–562.
27. Whitacre, W. W., & Campbell, M. E. (2012). Decentralized geolocation and bias estimation for uninhabited aerial vehicles with articulating cameras. *Journal of Guidance, Control, and Dynamics, 34*(2), 564–573.
28. Wu, Y., Hu, X., Wu, M., et al. (2006). Strapdown inertial navigation using dual quaternion algebra: Error analysis. *IEEE Transactions on Aerospace and Electronic Systems, 42*(1), 259–266.
29. Xia, Y., & Li, L. (2012). H_∞ filtering for nonlinear singular Markovian jumping systems with interval time-varying delays. *International Journal of Systems Science, 43*(2), 272–284.
30. Xie, L., de Souza, C. E., & Wang, Y. (1996). Robust filtering for a class of discrete-time uncertain nonlinear systems: An H_∞ approach. *International Journal of Robust and Nonlinear Control, 6*(4), 297–312.
31. Yin, Y., Shi, P., Liu, F., et al. (2013). Fuzzy model-based robust H_∞ filtering for a class of nonlinear nonhomogeneous Markov jump systems. *Signal Processing, 93*(9), 2381–2391.
32. Zhang, H., & Xie, L. (2007). *Control and estimation of systems with input/output delays*. Berlin, Germany: Springer.
33. Zhang, H., Xie, L., Soh, Y. C., et al. (2005). H_∞ fixed-lag smoothing for discrete linear time-varying systems. *Automatica, 41*(5), 839–846.
34. Zhong, Y., Gao, S., & Li, W. (2012). A quaternion-based method for SINS/SAR integrated navigation system. *IEEE Transactions on Aerospace and Electronic Systems, 48*(1), 514–523.

35. Zhou, J., Knedlik, S., & Ubolkosold, P., et al. (2008). A novel design of an adaptive hybrid low-cost GPS/INS integration system, In *Proceedings of the 2008 National Technical Meeting of The Institute of Navigation* (pp. 521–531). January 28–30, San Diego, CA.
36. Zhou, D., & Frank, P. M. (1996). Strong tracking filtering of nonlinear time-varying stochastic system with colored noise: Application to parameter estimation and empirical robustness analysis. *International Journal of Control, 65*(2), 295–307.

Chapter 22
Adaptive In-Flight Alignment of INS/GPS Systems for Aerial Mapping

The integrated INS/GPS measurement system can be used to provide attitude information and then the exterior orientation parameters can be derived for the purpose of direct georeference of the airborne imagery. In-flight alignment (IFA) plays an important role in achieving high accuracy of attitude estimation in the integrated INS/GPS measurement system. However, the statistics of INS noise is usually time-varying and it often degrades seriously the estimation accuracy in practice. In order to solve this problem, this chapter is devoted to IFA of the integrated INS/GPS system for aerial mapping applications. Firstly, an adaptive algorithm is developed to adjust the window size of data processing in IFA. Then the covariances of INS noise can be estimated and updated online, so that the resulted estimation performance can be improved. Moreover, a strong tracking filter (STF) is applied to guarantee the convergence of the IFA algorithm as well as its robustness against the parameter perturbation and trajectory maneuvers. A real aerial mapping experiment demonstrates the effectiveness of the methods.

22.1 Introduction

Nowadays, the use of INS/GPS systems in aerial mapping and surveying has been increasingly recognized and extensively accepted in the community of photogrammetry and remote sensing [1, 9, 13, 21]. With the capability of long-term high-accuracy estimation of imaging sensors' position and orientation information, this dedicated INS/GPS system is also named the POS [8, 17, 20]. One special advantage of the remote sensing missions aided by INS/GPS systems is the considerable reduction of the necessary ground control points for aerial triangulation (AT) in large-format camera-based airborne surveying [17, 20, 25]. In this way, direct georeferencing (DG) is realized by acquiring the movement information from the estimated results

M. Zhong et al., *Fault Diagnosis for Linear Discrete Time-Varying Systems and Its Applications*, https://doi.org/10.1007/978-981-19-5438-2_22

of the INS/GPS systems on-board the aircraft [13, 17]. The system solution is used to georeference the airborne imagery, i.e. to provide the exterior orientation (EO) parameters of the image data. Especially, the INS/GPS systems are of necessity for the cases of laser scanner or hyperspectral imagers where external references are not available [11, 22]. In short, obtaining the orientation information of airborne imaging sensors is a main task of the INS/GPS systems and the attitude information provided is often a basis for high resolution imaging. Due to various application scenarios, however, the systems may possibly be installed outside the aircraft cabin and inevitably are subject to environmental disturbances such as electromagnetic interference or serious thermal gradient [5, 27, 28]. There is no doubt that these disturbances may result in evident degradation of inertial senor quality and affect seriously the measurement accuracy of attitude information. Therefore, how to keep the performance of the INS/GPS systems at an acceptable level becomes a big challenge.

The in-flight alignment (IFA) has always been a reliable method to maintain the measurement accuracy of the INS/GPS systems. The development of IFA has received much attention and fruitful results have been published in decades [2, 3, 12, 26]. As stated in [3], IFA relates directly to the ability to calibrate the sensor errors and align the system itself in motion with respect to the references. Different from the alignment on ground, the primary task of IFA is the initial attitude determination of an INS in flight, which is the basis of the subsequent integral operations in the INS algorithm [26]. Discussions in [2, 7, 12] point out that the IFA comprises two alignment stages: the coarse alignment and the fine alignment. The coarse alignment is achieved by means of acceleration leveling and gyrocompassing. Then the fine alignment is implemented by Kalman filters which exploit the GPS information to estimate attitude errors so as to correct the INS result. Considering the attitude information is always required for DG, the IFA actually plays a major role in guaranteeing the attitude accuracy of the INS/GPS systems. Therefore, it is quite important to develop the filtering algorithms to promote the alignment performance.

One of the frequently encountered problems in filtering is that the statistical deviations of the inertial sensor noises from the nominal values. Take the mechanical gyroscopes for example. The flight maneuvers may frequently arouse dynamic errors that can hardly be modelled. Then they are often treated as additive random noises accounting for the covariance changes in the stochastic model of the INS/GPS system [6]. The mismatches of noise covariance will give rise to serious estimation errors in the filtering process if they are not properly addressed. Numerous adaptive IFA methods have been devoted to the online enhancement of the filtering performance when the systems are subject to unforeseen interference [7, 12, 16]. The adaptive strategy functions to adaptively estimate the noise covariances in the filter, taking them as deterministic parameters in a finite-length time window [16]. Among the adaptive algorithms, the maximum likelihood based adaptive estimation algorithm is widely accepted and used in various applications [15]. It prevails in real time implementation, unbiased estimation and asymptotically convergence to the true value [18]. Hence, it is a suitable choice for the IFA of the INS/GPS systems. However, there exist some limitations in the maximum likelihood based algorithm. A key problem has to be confronted is the reasonable selection of window size for data processing. Arguments on this issue have lasted for a long time [15, 16]. In view of the trajectory features of typical remote sensing flights, it motivates us to utilize two distinct movement modes to automatically determine the window size and eventually amend the adaptive estimation algorithm in the IFA.

On the other hand, a robust algorithm, namely the strong tracking filter (STF), should also be included for the IFA [14, 29]. It comes from the concern on the influences of unexpected dynamic errors. For the conventional EKF, the state estimation error covariance will converge and at last propagate into a steady status if the filter functions normally. As a result, it makes the filter gain converge to a minimum as well. However, when abrupt state changes occur, the EKF can hardly reflect the changes in a timely manner due to the fact that the filter gain remains minimum. On the contrary, the STF adaptively regulates the predictive estimation error covariance to overcome the problem and tunes the Kalman gain in accordance with the state changes. Therefore, as a closed-loop estimator, the STF is a proper choice to deal with the filtering fluctuations in the process of IFA.

Based on the above observations, in this chapter we strive to develop an adaptive IFA method for the aerial mapping usage by incorporating the adaptive estimation algorithm and the STF, with main attention being paid to the the improvement of the attitude estimation accuracy. In what follows, the descriptions of the system and IFA will be first given. And then a modified adaptive algorithm is developed for the real time use and, together with the STF, an improved IFA method is designed. A flight test is finally presented to illustrate the effectiveness of the developed system.

22.2 Preliminaries and Problem Statement

In this section we give a brief description of the INS/GPS system in the remote sensing applications. Then the conventional IFA is introduced on the basis of adaptive covariance estimation. At last we will show the existing problems in the IFA when it is applied to aerial mapping missions.

22.2.1 System Description

In general, a typical INS/GPS system for aerial mapping is installed with the imaging sensor inside the aircraft cabin shown in Fig. 22.1. The INS/GPS system consists of an IMU, a computer system (CS), a GPS receiver and affiliated cables. The GPS receiver usually is assembled in the CS. The IMU is tightly mounted on the imaging sensor, so that the attitude measurement from IMU can be easily transformed into the orientation information of the imaging sensor. Further, to isolate the external vibrations, a gyro-stabilized sensor mount (GSM) is used to which the imaging sensor is fixed rigidly. The GPS antenna is mounted on top of the cabin. During

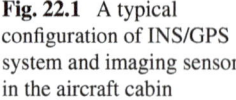

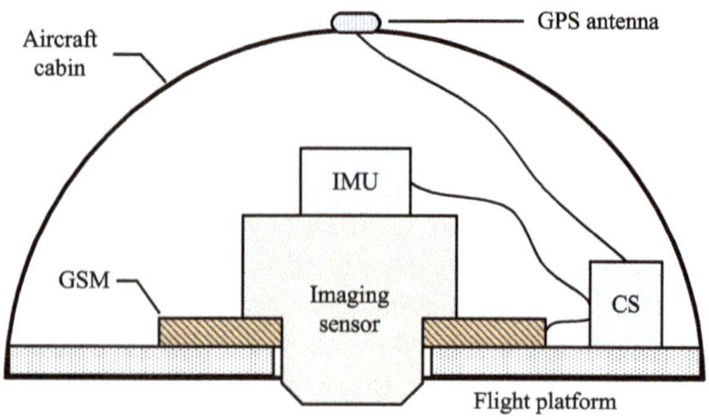

Fig. 22.1 A typical configuration of INS/GPS system and imaging sensor in the aircraft cabin

the flight, IMU and GPS data are processed and fused in the CS and an optimal estimation of the motion states is then provided via on-board computer for the DG of the imaging sensor. Besides it is also delivered as an attitude feedback to the GSM. By this means, the EO parameters of each image can be directly derived from the results of the INS/GPS system in real time. In addition, the photoelectric encoders along the GSM gimbal axes transmit the signals of the angular changes of the mounting surface to the onboard computer in real time. By applying the dynamic lever arm compensation method, the lever arm errors among the GPS antenna, the IMU and imaging center can be effectively removed [4].

Unlike ordinary navigation system, the INS/GPS system dedicated for the mapping applications places much attention on the long-term high-accuracy measurement of the attitude. A constant concern comes from the doubt that the system could hardly sustain the measurement accuracy due to a variety of unexpected disturbances. This may be possible since the engine vibrations and thermal gradients are frequently occurred phenomena. As a result, the gyroscopes and accelerometers in the IMU will inevitably output data contaminated by the perturbances. To mitigate this, an IFA operation therefore becomes quite necessary to align and amend the system itself through the entire estimation process.

22.2.2 A Brief Review of the Existing IFA

Normally the IFA is realized by implementing the EKF to use the GPS data and correct the INS errors. In the whole alignment operation, one can easily notice that the alignment accuracy is impacted directly by the filtering performance [12]. Further the filter performance hinges greatly on the correct knowledge of the noise statistics [16]. In practice, the gyroscopes and accelerometers of the INS are subject to dynamic and thermal disturbances resulting in statistically variable noise, which is represented as the covariance changes of the process noise in the filter. Thus a promising approach to solve the problem is the real-time adaptive estimation of noise covariance based on the maximum likelihood theory. In the following we will state the system model, the EKF and the adaptive algorithm for IFA in sequence.

A. Nonlinear System Model and EKF

In this chapter, we consider the system dynamic model given in Sect. 20.2.1, i.e., choosing the ENU coordinate frame as the navigation frame n, the nonlinear INS equations is given in (20.14) and the state space description of the system is described by (20.15) with denoting ϑ by w. For ease of reading, we rewrite the model as follows

$$\begin{cases} \dot{x}(t) = f(t, x(t)) + G(t)w(t) \\ \qquad = f_N(t, x(t)) + f_L(t)x(t) + G(t)w(t), \\ y(t) = C(t)x(t) + v(t), \end{cases} \tag{22.1}$$

with $C(t)$ given in (20.9). The process noise $w(t)$ is composed of the random errors of inertial sensors and follows the Gaussian distribution with

$$\mathbb{E}[w(t)] = 0, \ \mathbb{E}[w(t)w^T(\tau)] = Q(t)\delta_{t\tau}. \tag{22.2}$$

and the measurement noise $v(t)$ is Gaussian distributed with

$$\mathbb{E}[v(t)] = 0, \ \mathbb{E}[v(t)v^T(\tau)] = R(t)\delta_{t\tau}. \tag{22.3}$$

$\delta_{t\tau} = 1$ if $t = \tau$ and $\delta_{t\tau} = 0$ otherwise.

In the model (22.1), $f(t, x(t))$ is a nonlinear function of $x(t)$. For the purpose of linearization, a first-order approximation is used for the systems dynamics matrix

$$F(t) = \left. \frac{\partial f(t, x(t))}{\partial x(t)} \right|_{x(t)=\hat{x}(t_k)}. \tag{22.4}$$

With given sample period T_s, the state transition matrix for the discrete system can thus be approximated by the Taylor-series expansion as

$$\Phi(k+1|k) \approx I + F(t_k)T_s \tag{22.5}$$

and

$$C(k) = C(kT_s). \tag{22.6}$$

To compute the Kalman gain $K(k)$, the Riccati equations are identical to those of the standard Kalman filter. For the sake of brevity, we will not present these equations here. The old estimate can be directly propagated by integrating the actual nonlinear differential equations forward as

$$\hat{x}(k|k-1) = \hat{x}(k-1) + f(\hat{x}(k-1|k-1)T_s. \tag{22.7}$$

Define the innovation as

$$e(k) = y(k) - H(k)\hat{x}(k|k-1). \tag{22.8}$$

Then a new estimate of the filter is readily obtained by incorporating (22.7) and (22.8) as

$$\hat{x}(k) = \hat{x}(k|k-1) + K(k)e(k) \tag{22.9}$$

and it is right the EKF.

B. Adaptive INS Noise Covariance Estimation

The reason to choose the maximum likelihood based adaptive estimation algorithm stems from the fact that for the case of i.i.d measurements, an asymptotically unbiased estimate with finite covariance can always be found through the maximum likelihood condition such that no other unbiased estimate with a lower covariance exists [16]. In our case, the INS/GPS measurements are assumed to be independent with Gaussian distribution. The maximum likelihood based algorithm utilizes the innovation sequence of the filter to estimate noise covariances online, such that a proper performance level can be sustained in face of statistical changes of the noise process. Thus it is commonly used in the IFA.

The likelihood function is the conditional density of the measurements and conditioned upon the value of the uncertain parameters. Denote by α the parameter vector which consists of the unknown diagonal elements in the process noise covariance matrix $Q(k)$, and assume it constant in an arbitrary series of N measurements $Y_N = \{y(1), y(2), \cdots, y(N)\}$. We resort to construct a likelihood function based on the the conditional probability density, i.e.

$$\mathcal{L}(Y_N, \alpha) = p(Y_N|\alpha),$$

so the maximum likelihood estimate of α will be the one which maximizes $\mathcal{L}(\cdot)$. $p(Y_N|\alpha)$ can be obtained as

$$p(Y_N|\alpha) = \prod_{i=1}^{N} p(y_i|\alpha).$$

Recall that the INS/GPS measurements follow the Gaussian distribution. Thus the likelihood function at time step k can be expressed as

$$\mathcal{L}(Y_N, \alpha)_k = \frac{1}{\sqrt{(2\pi)^N|\hat{R}_e(k)|}} \exp\left\{-\frac{1}{2}e^T(k)\hat{R}_e^{-1}(k)e(k)\right\}. \tag{22.10}$$

where $\hat{R}_e(k) = \frac{1}{N}\sum_{i=k-N+1}^{k} e(i)e^T(i)$. Through a series of proper manipulations on (22.10), we can finally get the formula to adaptively estimate $Q(k)$ as

$$\hat{Q}(k) = G^{-}(k)\Bigg[P(k) + \sum_{j=k-N+1}^{N} \Delta x(j)\Delta x^{T}(j) - \Phi(k|k-1)P(k-1)\Phi^{T}(k|k-1)\Bigg]\left(G^{-}(k)\right)^{T}, \tag{22.11}$$

where $\Delta x(k) = \hat{x}(k) - \hat{x}(k|k-1)$ and $G^{-}(k)$ is the pseudo-inverse matrix of $G(k)$ with $G^{-}(k) = [G^{T}(k)G(k)]^{-1}G^{T}(k)$. Note that the estimation process inside the window of N is required to be basically in steady state. The interested reader is referred to [16] for more details.

22.2.3 Problem Statement

It should be pointed out that the maximum likelihood estimate converges, in a probabilistic sense, to the true value of the variable as the number of sample data grows without bound. However, for the case with small sample sizes, such an estimate will in general be biased. The sample size puts additional restriction on the choice of the estimation window size. In conventional maximum likelihood based adaptive algorithms, the estimation window is of fixed length N in which the covariance matrix $Q(k)$ is assumed as constant. Otherwise the inconstancy will influence the accuracy of the maximum likelihood estimation [15].

When it applies to remote sensing missions, however, this prerequisite can be hardly guaranteed. The frequent turn-around maneuvers are an inevitable portion of the flight trajectory during the mapping. Unfortunately unknown dynamic errors of the inertial sensors are also aroused due to mechanical imperfections and sensitivity to motion-related interference. As the errors are usually of high orders and difficult to model, a practical way is to take them as additive process noises and estimate the varying noise covariance accurately so that the influences caused by these errors could be alleviated as possible [10]. Nevertheless, the coefficient matrix of the filter in a fixed-length window is no longer constant and then the accuracy of maximum likelihood estimate is certainly decreased.

On the other hand, as shown in Fig. 22.2, a conclusion can be drawn from conventional aerial mapping trajectories that the movements during mapping are basically composed of the turn-around maneuvers and the straight flights in consecutive order. This enlightens us to implement the IFA by taking advantage of the cyclical shift of movements, namely, to reduce the window size during turn-around maneuvers or to enlarge it during the straight flights. In this way, it is possible to approximately ensure the prerequisite of constant parameters in a window and maintain the algorithm performance.

The selection of N is essentially a trade-off between the unbiasedness of the estimate and the tracking ability of the parameter changes resulted by flight dynamics [15]. On the one hand, larger window size will definitely bring less estimation bias.

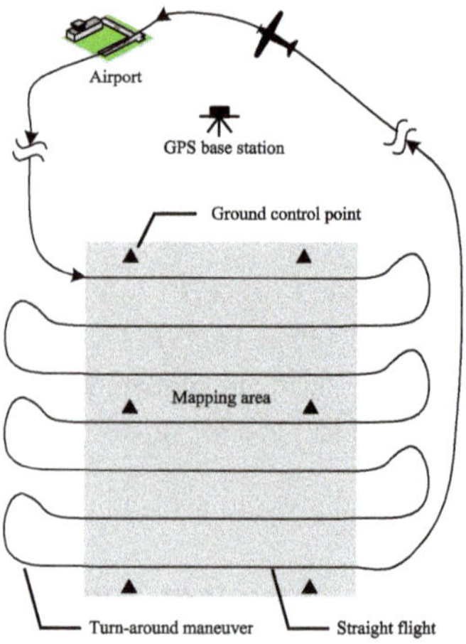

Fig. 22.2 Schematic diagram of typical flight trajectory in aerial mapping missions

Table 22.1 The comparison of distinct window size change trends

Window	Advantage	Disadvantage
Enlarge N	▹ unbiased estimate	▹ weak tracking ability
	▹ eliminate potential impacts of single outliers	▹ more need of on-board storage resource
Reduce N	▹ good tracking ability	▹ biased estimate
	▹ less on-board resource requirement	

On the other hand, it is more tractable to track the dynamic parameter changes by reducing the window size. Table 22.1 has compared the different effects of the opposite trends. Hence, how to properly select N on the basis of actual conditions at the time becomes a challenging problem [16].

To solve the problem, an intuitively more rational strategy balancing the two goals is to

$$\begin{cases} \text{increase the value of } N, & \text{if straight flight;} \\ \text{decrease the value of } N, & \text{if turn-around maneuver.} \end{cases}$$

Now the challenge is altered to that how to find an appropriate bound of the window size for the maximum likelihood algorithms reconciling the contradictions between estimation accuracy and dynamic response. As there is yet no relevant literature on selecting N, we will in the next section propose a selection criterion in the framework of sample complexity analysis for probability estimation. On the other hand, the maneuvers of the aircraft arouse more serious dynamic errors of the INS. In order to attenuate the influences of the errors and the divergence trend during the alignment, an STF is included for the IFA.

22.3 An Adaptive IFA Method for Aerial Mapping

In this section, we first give a sample bound to determine the minimum window size. Then an adaptive strategy of window selection is given. Next, we apply the STF to enhance the robustness of the estimates. Finally, the IFA method is summarized for the implementation in aerial mapping.

22.3.1 The Determination of Minimum Window Size

It is well known that the whiteness of the innovation sequence is a dominant indicator of the optimum of the Kalman filter. It also reflects the complete extraction of the measurement information. In most researches, therefore, the evaluation functions are constructed with the innovation sequence. In this subsection, we will follow this standpoint and construct the evaluation function to specify the effective sample number in a window of the maximum likelihood estimate from a probabilistic point of view.

Define the following innovation based evaluation function

$$J\big(e(k)\big) = \sum_{j=k-N_e+1}^{k} e^T(j)R_e^{-1}(j)e(j). \tag{22.12}$$

where $R_e(k)$ is the covariance matrix of innovation. Ideally, the innovations are independent normally distributed and $J\big(e(k)\big)$ follows central χ^2 distribution with $6N_e$ degrees of freedom, indicating the complete utilization of the information from GPS measurement and consequently the best estimation performance of the filter. Set a level of significance η and the threshold value χ^2_η can be obtained from the χ^2 distribution table. Then given the critical region $\{\chi^2 > \chi^2_\eta\}$, the null and alternative hypotheses are

$$\begin{aligned} H_0 &: J\big(e(k)\big) \le \chi^2_\eta, \ \forall k; \\ H_1 &: J\big(e(k)\big) > \chi^2_\eta, \ \exists k. \end{aligned}$$

A conventional way to evaluate estimation performance of the filter is to implement the χ^2 test. However, considering the unexpected dynamic errors, a probabilistic approach is more compatible to alter the meaning of optimal estimation from its usual deterministic sense to a probabilistic one in face of the uncertainties in the system. In this sense, the innovation $e(k)$ can be viewed as random variable bounded in a specific set $\mathcal{B}_\mathcal{D}$, while the threshold χ^2_η is equivalent to a performance level γ. Then the performance evaluation is transformed to guarantee that $J\big(e(k)\big) \le \gamma$ for all $e(k) \in \mathcal{B}_\mathcal{D}$. It can be analyzed by introducing two sets, i.e., the good set $\mathcal{B}_\mathcal{G}$ and the bad set $\mathcal{B}_\mathcal{B}$, defined as

$$\mathcal{B}_{\mathcal{G}} = \Big\{ e(k) \in \mathcal{B}_{\mathcal{D}} : \ J\big(e(k)\big) \leq \gamma \Big\}, \tag{22.13}$$

$$\mathcal{B}_{\mathcal{B}} = \Big\{ e(k) \in \mathcal{B}_{\mathcal{D}} : \ J\big(e(k)\big) > \gamma \Big\}. \tag{22.14}$$

In a probabilistic context, a measure of optimality can instead be linked to a requirement that the volume of $\mathcal{B}_{\mathcal{G}}$ be "adequately large". From this perspective, the optimality of the estimation is shifted, in a probability sense, to test if "most" innovations satisfy (22.13). In other words, it can be supposed that the risk that $e(k)$ belongs to $\mathcal{B}_{\mathcal{B}}$ happens with a small probability. If it is the case, one can conclude that the obtained estimation is "practically optimal" [24].

Note that the innovation $e(k)$ is a random variable, it is natural in subsequent to carry out an interval estimation of $e(k)$ and clarify the neighborhood with a probability measure. If we have the following function

$$\mathrm{II}_{\mathcal{B}_{\mathcal{G}}}\big(e(k)\big) = \begin{cases} 1, \text{ if } e(k) \in \mathcal{B}_{\mathcal{G}}; \\ 0, \text{ otherwise.} \end{cases} \tag{22.15}$$

one can see that $\mathrm{II}_{\mathcal{B}_{\mathcal{G}}}(e(k))$ actually follows the binomial distribution.

Inspired by the idea of randomized algorithms in Chap. 18, we can introduce the probability inequalities for the sample complexity to derive the minimal value of window size N for estimation [24]. Specifically, by using (22.15), we can compute the reliability of the estimate in an arbitrary window of length N as

$$\hat{p}_N(\gamma) = \frac{1}{N} \sum_{i=1}^{N} \mathrm{II}_{\mathcal{B}_{\mathcal{G}}}\big(e(i)\big).$$

Meanwhile, the true probability is

$$p(\gamma) = \mathrm{Pr}_e\big\{ J(e) \leqslant \gamma \big\}.$$

Thus the reliability is represented by the so-called "closeness" of $\hat{p}_N(\gamma)$ to $p(\gamma)$; that is, with given accuracy $\epsilon \in (0, 1)$, we can assume that

$$\big| \hat{p}_N(\gamma) - p(\gamma) \big| < \epsilon$$

holds with high probability. With a known confidence $\delta \in (0, 1)$, it is required that

$$\mathrm{Pr}_{e(1),\cdots,e(N)} \Big\{ \big| \hat{p}_N(\gamma) - p(\gamma) \big| < \epsilon \Big\} > 1 - \delta. \tag{22.16}$$

holds for a finite value of N. Then the problem is changed to find the minimum sample number with which (22.16) is satisfied for given ϵ and δ.

According to the Chernoff bound originally used for the sample complexity analysis, we have for any $\epsilon \in (0, 1)$ and $\delta \in (0, 1)$, if

$$N \geq \frac{1}{2\epsilon^2} \log \frac{2}{\delta}, \tag{22.17}$$

then $|\hat{p}_N(\gamma) - p(\gamma)| < \epsilon$ holds with a probability greater than $1 - \delta$. Without loss of generality, choose $\epsilon = 0.05$ and $\delta = 0.01$, and a minimum of sample number is readily obtained from (22.17) as $N_0 = 1060$. Naturally, it could be used as the baseline of the window size for the maximum likelihood estimation in IFA.

Remark 22.1 We have cast the problem of determining the minimal window size of the maximum likelihood algorithm into the sample complexity analysis of the probability estimation of an evaluation function. The use of the Chernoff bound gives us a very instrumental tool to find the minimal window size for the adaptive algorithm in a probabilistic sense. The determination of the window size associates intimately with the reliability analysis of the probabilistic estimations, so the minimum window size can be computed a priori and explicitly with given ϵ and δ. A detailed treatment can be found in [24].

22.3.2 Adaptive Selection of the Window Size

According to the trajectory features of typical aerial mapping flights shown in Fig. 22.2, we will design a strategy to automatically select N dependent on the movement segment which is depicted in Fig. 22.3. When it is in the stage of turn-around maneuver, N is reduced gradually till N_0. On the contrary, when the aircraft flies level and straight, N is enlarged accordingly with an increment added to N_0.

In the aerial mapping missions, the aircraft is often constrained to fly at a certain altitude so as to preserve the image resolution. Therefore, we can efficiently detect the changes in the flight trajectory by simply monitoring the changes of the heading angle $\Delta\psi(k)$.

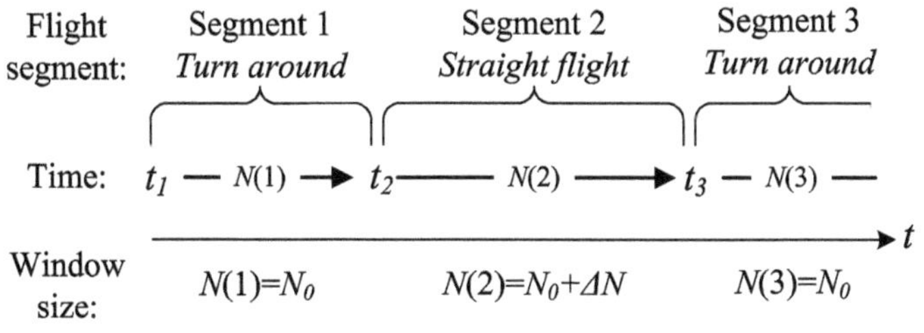

Fig. 22.3 Schematic diagram of the adaptive selecting of window size

Suppose the maximum heading angle change between two successive time steps in all straight flight segments is $\Delta\Psi$, and then the magnitude of the heading changes can be calculated with

$$\zeta(k) = \frac{\|\Delta\psi(k)\|}{\Delta\Psi}, \tag{22.18}$$

where

$$\Delta\psi(k) = \psi(k) - \psi(k-1). \tag{22.19}$$

When the aircraft turns around, apparently $\Delta\psi(k)$ is greater than $\Delta\Psi$ rendering $\zeta(k) > 1$ and hence N should be reduced according to the devised strategy. Otherwise, N should be enlarged. Besides, noticing that the aligning process is a time series, the window size is also a function of the filtering frequency ϖ of the integrated system. Therefore, the following equation is proposed to select the integer values of N at time step k for adaptive IFA as:

$$N(k) = \frac{\varpi}{\varpi_A}(N_0 + N) = \frac{\varpi}{\varpi_A}\left(N_0 + \lceil N_0\kappa(k)\rceil\right), \tag{22.20}$$

where ϖ_A is determined empirically depending on aerial mapping plan and actual system, $\lceil\cdot\rceil$ denotes the ceiling function and

$$\kappa(k) = \begin{cases} a(1-\zeta(k)), & \text{if } \zeta(k) \le 1, \\ 0, & \text{otherwise.} \end{cases}$$

with a the line slope. One can easily see that N continuously declines till N_0 in the maneuver stage ($\zeta(k) > 1$), or linearly augments from N_0 in the straight flight stage ($\zeta(k) \le 1$). Consequently, (22.11) is rewritten as

$$\hat{Q}(k) = G^{-}(k)\Bigg[P(k) + \sum_{j=k-N(k)+1}^{k} \Delta x(j)\Delta x^T(j) - \Phi(k|k-1)P(k-1)\Phi^T(k|k-1)\Bigg]\left(G^{-}(k)\right)^T. \tag{22.21}$$

So far, we have obtained the adaptive selecting algorithm of N to implement the IFA of INS/GPS system.

Remark 22.2 From the viewpoint of probabilistic estimation, the minimum window size is determined based on the reliability analysis of the estimate, which simply involves the sample size. But when it pertains to the real INS/GPS integration operations, one has to consider the speed to collect enough innovations for the window size update. Therefore, taking the influence of sampling time into account, the coefficient ϖ_A is added to (22.20) to adjust the window size N, which is often assigned to a value equal to or less than the filtering frequency.

Based on the determination of N_0, we are able to automatically select the window size, which is directly in accordance to the practical movement characteristics of the aircraft. In this way, a trade-off is attained balancing the estimation unbiasedness and the tracking ability. In addition, one can see that (22.20) is prone to implement by on-board computer for real time use. But it has to be pointed out that the filtering stability might fall victim to the varying disturbances, so we will discuss how to improve the robustness in the following subsection.

22.3.3 The Strong Tracking Filter

Apart from the benefits the adaptive estimation brought for the IFA, it is necessary to introduce a robust algorithm considering a latent trouble that the motion-related errors may give rise to estimation divergence. The introduction of the STF is an effective means to cope with it. The mechanism of STF is realized through automatically calculating the fading factors in face of estimation errors [29]. It weighs more on recent measurements to attenuate the negative influences aroused historically by model mismatch. The STF has identical filtering structures to the EKF except that the predictive estimation error covariance is regulated online. Thus the EKF used for the IFA of the INS/GPS system can be replaced by the STF with the predicted covariance matrix of estimation error computed as

$$\begin{aligned} P(k|k-1) =& \lambda(k)\Phi(k|k-1)P(k-1)\Phi^T(k|k-1) \\ &+ G(k-1)Q(k-1)G^T(k-1), \end{aligned} \tag{22.22}$$

where $P(k-1)$ is the covariance matrix of estimation error at time step $k-1$ and the fading factor $\lambda(k)$ is obtained as

$$\lambda(k) = \max\left\{1, \frac{\text{Tr}(M(k))}{\text{tr}(N(k))}\right\} \tag{22.23}$$

with

$$\begin{aligned} M(k) &= C_0(k) - C(k)Q(k-1)C^T(k) - R(k), \\ N(k) &= C(k)\Phi(k|k-1)P(k-1)C^T(k)\Phi^T(k|k-1). \end{aligned}$$

Note that it is desired to put more weight on the recent measurement information to compute $P(k|k-1)$, so $C_0(k)$ is actually calculated as

$$C_0(k) = \begin{cases} e(k)e^T(k), & \text{if } k = 1; \\ \dfrac{\rho C_0(k-1) + e(k)e^T(k)}{1+\rho}, & \text{if } k > 1. \end{cases}$$

where $\rho \in (0, 1]$ is the forgetting factor.

Remark 22.3 The discrepancies between the measurements and model predictions is used to tune $\lambda(k)$ to enlarge the a priori error covariance $P(k|k-1)$, and eventually to enlarge the filter gain $K(k)$. In comparison, when the EKF reaches a steady estimation status, the filter gain approaches a minimum and loses the ability to track abrupt state changes. From this viewpoint, it is the reason why the STF owns excellent tracking ability in IFA and is capable of tackling the divergence threats.

22.3.4 Design of the IFA Method for Aerial Mapping

Based on the above algorithms, we can now demonstrate the schematic diagram of the system, which is shown in Fig. 22.4. It is clear that the heading state obtained in the INS algorithm is used to calculate an indicator for window size update. Meanwhile the position and velocity states are used to yield the measurement vector for the STF. By incorporating the updated noise covariance matrix, the IFA is executed by the STF which provides the navigation results to the imaging sensors and simultaneously delivers the corrections back to the INS algorithm.

Also presented is the work flow of the proposed algorithm shown in Fig. 22.5. The core idea is to update the window length according to the flight dynamics before the adaptive process noise covariance estimation. Further the inclusion of the STF ensures the robustness of the IFA against latent estimation divergence. The realization of IFA method is summarized in Algorithm 22.3.18.

As a short summary, in the above developed IFA method, an adaptive estimation algorithm with window size update is given, which is dedicated for the aerial mapping applications. The discussion of the Chernoff bound to determine the minimum of the window size may provide a novel perspective for related maximum likelihood based algorithm studies. Additionally the STF is chosen as well in lieu of the EKF to adaptively regulate the predictive estimation error covariance to avoid the divergency problem. Therefore, the above mentioned alignment method can be called an adaptive IFA method which is adjustable and feasible to use for aerial mapping.

Algorithm 22.3.18 IFA method for aerial mapping

1: Initialize the window length $N(1)$, coefficients a and ρ. Let k=0, $\hat{x}(0)$=0, and $P(0) = P_0$;
2: If $N(k+1) > N_0$, calculate $\Delta\psi(k+1)$ using (22.19) and update $N(k+1)$ with (22.20);
3: Calculate $Q(k+1)$ using (22.20);
4: Update $F(k+1)$, $\Phi(k+1)$ and $H(k+1)$ using (22.4), (22.5) and (22.6) respectively;
5: Calculate the fading factor $\lambda(k+1)$ with (22.23) and update $P(k+1, k)$ using (22.22);
6: Calculate the other filter matrices;
7: Calculate $e(k)$ with (22.8) and update $\hat{x}(k+1)$ using (22.9);
8: Correct the navigation states as $\bar{C}_b^n = (I - \hat{\phi}) \times C_b^n$, $\bar{v} = v - \delta\hat{v}$ and $\bar{p} = p - \delta\hat{p}$;
9: Reset $\hat{x}(k+1)$, i.e. $\hat{x}(k+1) = [0 \quad \hat{\varepsilon}^T \quad \hat{\nabla}^T]^T$;
10: Let $k = k+1$, and go to Step 2 until the end.

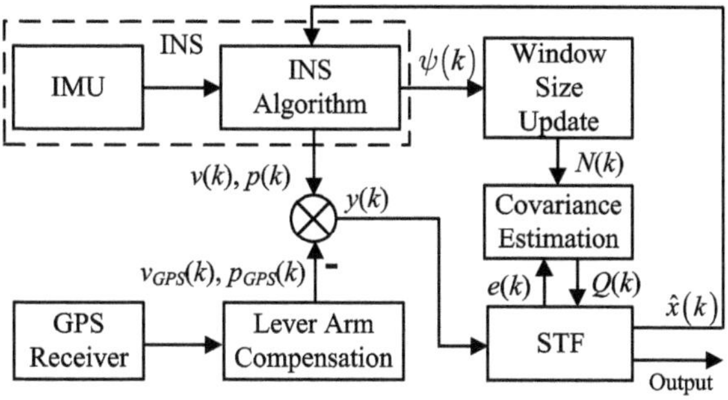

Fig. 22.4 Schematic diagram of the INS/GPS system for aerial mapping

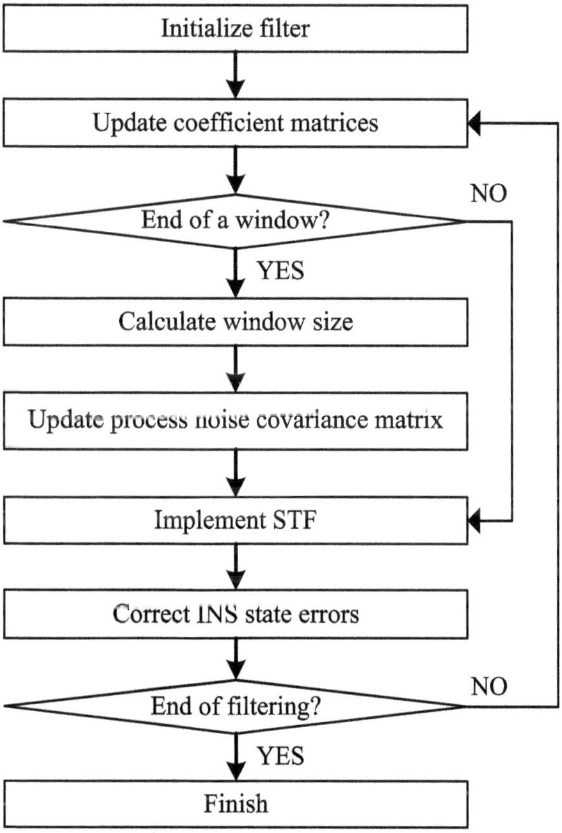

Fig. 22.5 Work flow of the IFA of the INS/GPS system

22.4 Experimental Study

To validate the presented method, we choose to use a flight experiment data for aerial mapping obtained from a dynamically tuned gyroscope (DTG) based POS. Then four filtering algorithms are applied to implement the IFA and results are compared with the AT solution.

22.4.1 Experimental Setup

The experiment was carried out for an aerial mapping mission and a light aircraft was selected for the flight shown in Fig. 22.6a. All the experimental equipments were installed in the aircraft cabin, which is shown in Fig. 22.6b. The DTG POS is mainly composed of a DTG IMU and a GPS receiver which is used for direct georeferencing. The IMU was rigidly fixed to a Hasselblad digital camera and together they were mounted to a small-scale GSM. The GPS antenna was mounted on the middle of the aircraft's wings and the GPS base station was set up near the runway. Through the experiment, the update frequencies of the IMU and GPS are 100 Hz and 10 Hz, respectively. The coefficients in the Algorithm 22.3.18 are set as $\varpi_A = 10, a = -2$ and $\rho = 0.95$. Other coefficients can be derived from the sensor specifications of the POS listed in Table 22.2.

In this experiment, we consider the following four algorithms for the IFA of the system:

- *Algorithm 1*: the standard EKF;
- *Algorithm 2*: the EKF with conventional adaptive estimation algorithm (fixed window size $N = 1060$);

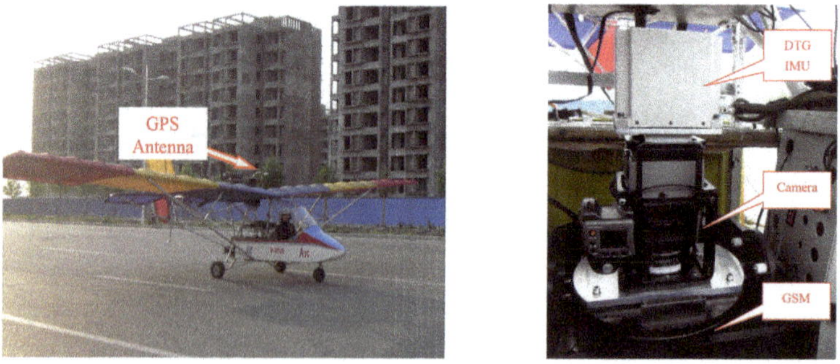

(a) (b)

Fig. 22.6 **a** The light aircraft of the experimental equipments; **b** The installation of the experimental equipments

Table 22.2 The sensor specifications of the POS

Sensors	Terms
Gyroscope	Random constant: $0.1°/\mathrm{h}(1\sigma)$
	Bias stability: $0.1°/\mathrm{h}(1\sigma)$
Accelerometer	Random constant: $100\,\mu\mathrm{g}(1\sigma)$
	Bias stability: $100\,\mu\mathrm{g}(1\sigma)$
GPS receiver	Velocity: 0.03 m/s (RMS)
	Position: 0.05–0.3 m (RMS)

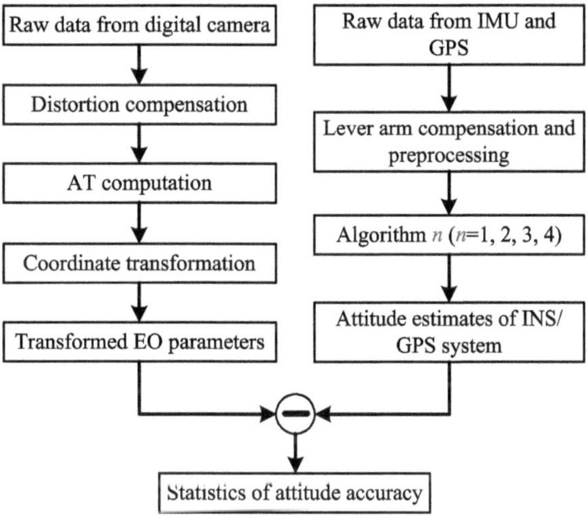

Fig. 22.7 The testing scheme for the accuracy statistics of the INS/GPS system

- *Algorithm 3*: the EKF with the proposed adaptive estimation algorithm (adaptive selection of window size);
- *Algorithm 4*: the STF with the proposed adaptive estimation algorithm (adaptive selection of window size).

It is easy to see that the Algorithm 4 is right the algorithm in this chapter.

Meanwhile, the ground control points were also deployed in the mapping area and precisely measured in advance for the purpose of AT. The results of AT, as a reference data, provided the EO parameters of the camera at each exposure moment that can be analytically transformed to the attitude information of the POS. The testing scheme is illustrated in Fig. 22.7. Thus it gives us an obvious advantage to compare the estimation results of the POS with the AT solution so as to validate the proposed IFA method.

Remark 22.4 It is widely known that the DTG is a mechanically spinning gyroscope, and the maneuvers of the aircraft would probably bring in dynamic errors, which can be described as random noises. This directly leads to motion-related variation of the noise covariance in the filter. If it is not properly handled, the system performance will be affected seriously. The IFA attempts to extract adequate useful information from the GPS measurements, so that the noise covariance can be adaptively estimated in real time. In this way, the attitude accuracy of the POS can be maintained at an acceptable performance level in face of dynamic disturbances.

22.4.2 Experimental Results

The experiment lasted for more than 4700 s and the average flight altitude was about 560 m. The flight trajectory obtained from the GPS solution is concisely shown in Fig. 22.8. It is easy to see that there were 4 flight strips for mapping. And 156 images associated with the EO parameters were obtained from the AT calculations.

Compared with the AT results, we can calculate the root-mean-square (RMS) errors of the four algorithms. The RMS errors are listed in Table 22.3 and the attitude estimation errors are shown in Fig. 22.9. The window size used for covariance estimation is automatically altered dependent upon the corresponding movement status, which is depicted in Fig. 22.10. The window size is reduced when the aircraft is maneuvering, whereas it is enlarged in the straight flight strips.

From Table 22.3, one can find that the Algorithm 2 basically exhibits an equivalent performance level to the Algorithm 1. It implies the conventional adaptive estimation method does not work effectively as the estimated covariances may have already

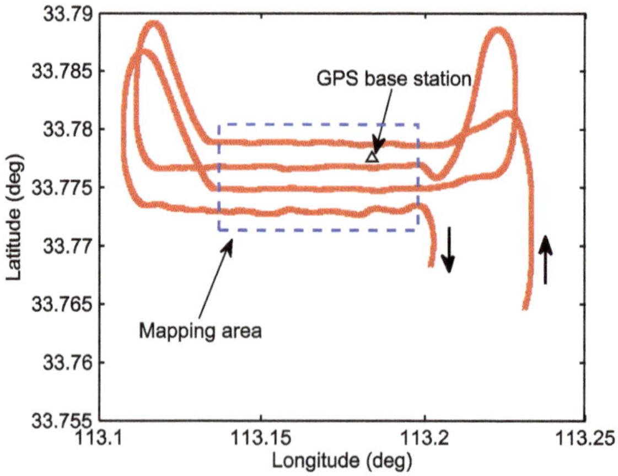

Fig. 22.8 The flight trajectory

Table 22.3 The statistics of the attitude RMS errors

RMS	Algorithm 1	Algorithm 2	Algorithm 3	Algorithm 4
Heading (°)	0.098	0.099	0.073	0.060
Pitch (°)	0.006	0.006	0.007	0.006
Roll (°)	0.008	0.008	0.007	0.007

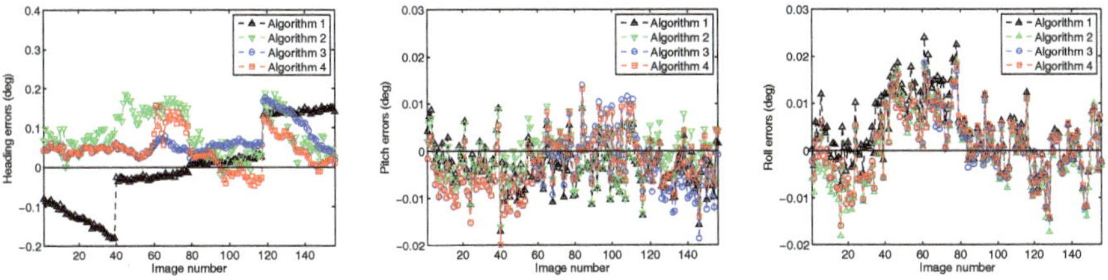

Fig. 22.9 The heading, pitch and roll errors of the four algorithms compared with the AT solution

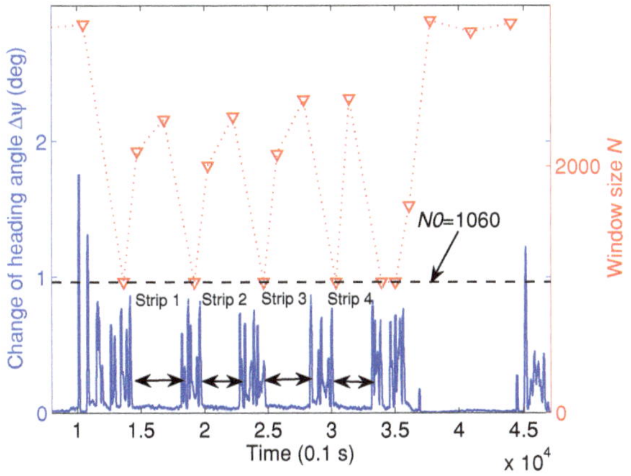

Fig. 22.10 The correspondence is straightforward between the heading angle change and the updated window size. The window size (red downward-pointing triangle) is reduced to N_0 (black dashed line) when the heading angle change (blue solid line) is apparent indicating the turning maneuver of the aircraft. On the contrary, the window size is enlarged accordingly in the four flight strips for mapping

been biased in the fixed-length window. In comparison, the Algorithm 3 behaves better especially in the heading accuracy, which demonstrates that the change of window size has promoted the overall performance. In Fig. 22.11, there are three short intervals where the fading factor $\lambda(k)$ is greater than 1. It means that the STF functions in time and regulates the filtering gain to track the filter parameter variation. As a result, the Algorithm 4 has realized remarkable performance improvement especially in heading accuracy and prevented possible divergence at the same time.

From the first subgraph in Fig. 22.9, it is noted that for some images (approximately from No. 60 to No. 78) the Algorithm 4 has relatively larger heading errors. This is for the reason that the STF tuned the predictive estimation error covariance matrix and gave rise to sudden changes of the estimates. Nevertheless, one must admit that the adaption of the STF has eventually improved the global performance of the Algorithms 4, as depicted in Table 22.3. Besides it is apparent that the heading error of each image after the deviation and the RMS error of heading are both

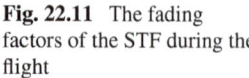

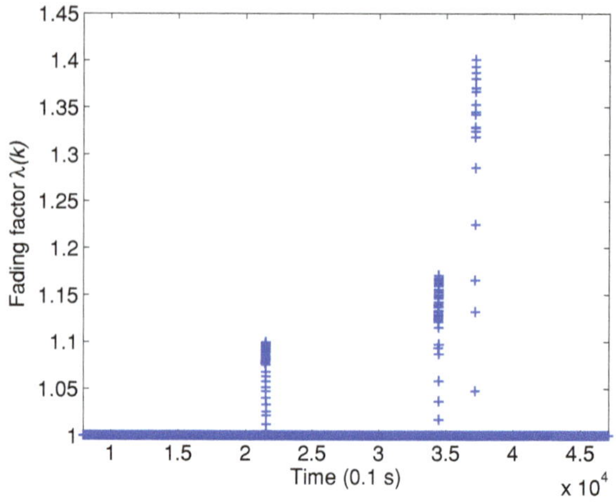

Fig. 22.11 The fading factors of the STF during the flight

smaller than those of the other algorithms. On the other hand, it reminds us to pay more attention on the use of STF and further modification is still needed.

Remark 22.5 The main influencing factor of the pitch and roll errors is the accelerometer biases, while the heading errors are dominated by the gyroscope biases [19]. The marginal improvement of the pitch and roll accuracies implies that the accelerometers exhibit a relatively stable and reliable performance throughout the entire experimentation. Moreover, the achieved pitch and roll accuracies have already fulfilled the mapping demands that the least required accuracy is no more than 0.03° [23]. By contrast, the DTG is more vulnerable to external impacts giving rise to random errors that deteriorate the heading accuracy. This kind of errors can be equivalent to additive gyroscope noises that deviate the covariances from the nominal values previously determined [10]. Therefore, the IFA method in this chapter effectively estimates the changed noise covariances and noticeably improves the heading accuracy.

22.5 Conclusion

In this chapter, an adaptive IFA method has been demonstrated aiming at improving the measurement accuracy of the INS/GPS system for aerial mapping applications. Our attention has been mainly paid to address the varying INS noises and ensure the robustness of state estimation. To deal with the varying INS noise, an adaptive covariance estimation algorithm based on the automatic selection of the window size has been developed so as to promote the attitude measurement performance. The STF has been applied for data processing in the IFA method such that the robustness of state estimation can be guaranteed. A real flight test has demonstrated the effectiveness of the developed IFA method.

References

1. Ahn, H.-S., & Won, C.-H. (2006). Fast alignment using rotation vector and adaptive Kalman filter. *IEEE Transactions on Aerospace and Electronic Systems, 42*(1), 70-C83.
2. Bar-Itzhack, I. Y., & Porat, B. (1980). Azimuth observability enhancement during inertial navigation system in-flight alignment. *Journal of Guidance Control Dynamics, 3*(4), 337–344.
3. Baziw, J., & Leondes, C. T. (1972). In-flight alignment and calibration of inertial measurement units-Part I: General formulation. *IEEE Transactions on Aerospace and Electronic Systems, AES–8*(4), 439–449.
4. Cao, Q., Zhong, M., & Zhao, Y. (2015). Dynamic lever arm compensation of SINS/GPS integrated system for aerial mapping. *Measurement, 60,* 39–49.
5. Cheng, J., Fang, J., Wu, W., et al. (2014). Temperature drift modeling and compensation of RLG based on POS tuning SVM. *Measurement, 55*(3), 246–254.
6. Choukroun, D., Bar-Itzhack, I. Y., & Oshman, Y. (2006). Novel quaternion Kalman filter. *IEEE Transactions on Aerospace Electronic Systems, 42*(1), 174–190.
7. Fang, J., & Yang, S. (2011). Study on innovation adaptive EKF for in-flight alignment of airborne POS. *IEEE Transactions on Instrumentation and Measurement, 60*(4), 1378–1388.
8. Fang, J., Chen, L., & Yao, J. (2014). An accurate gravity compensation method for high-precision airborne POS. *IEEE Transactions on Geoscience and Remote Sensing, 52*(8), 4564–4573.
9. Grejner-Brzezinska, D. A., Toth, C. K., Sun, H., et al. (2011). A robust solution to high-accuracy geolocation: Quadruple integration of GPS, IMU, pseudolite, and terrestrial laser scanning. *IEEE Transactions on Instrumentation and Measurement, 60*(11), 3694-C3708.
10. Groves, P. D. (2008). *Principles of GNSS, inertial, and multisensor integrated navigation systems*. Boston: Artech House.
11. Hernandez-Lopez, D., Gomez-Laho, J., Fernandez-Hernandez, J., et al. (2016). Performance quality control of pushbroom sensors in photogrammetric flights. *IEEE Sensors Journal, 16*(2), 365-C374.
12. Hong, W., Han, K.-J., & Lee, C., et al. (2010). Three stage in-flight alignment with covariance shaping adaptive filter for the strap-down inertial navigation system. *AIAA Guidance Navigation Control Conference*, Toronto, Ontario: AIAA.
13. Kocaman, S. (2003). *GPS and INS integration with Kalman filtering for direct georeferencing of airborne imagery*. ETH, Zurich, Switzerland, Geodetic Seminar Report.
14. Lin, C., Chang, Y., Hung, C., et al. (2015). Position estimation and smooth tracking with a fuzzy-logic-based adaptive strong tracking Kalman filter for capacitive touch panels. *IEEE Transactions on Industrial Electronics, 62*(8), 5097–5108.
15. Maybeck, P. S. (1982). *Stochastic models, estimation, and control* (Vol. 2). New York: Academic Press.
16. Mohamed, A. H., & Schwarz, K. P. (1999). Adaptive Kalman filtering for INS/GPS. *Journal of Geodesy, 73,* 193–203.
17. Mostafa, M., Hutton, J., & Lithopoulos, E. (2001). Airborne direct georeferencing of frame imagery: An error budget. In *proceedings of the 3rd International Symposium on Mobile Mapping Technology*, January 3–5, Cairo, Egypt, ID: 55116179.
18. Nikolic, J., Furgale, P., Melzer, A., et al. (2016). Maximum likelihood identification of inertial sensor noise model parameters. *IEEE Sensors Journal, 16*(1), 163–176.
19. Park, H. W., Lee, J. G., & Park, C. G. (1995). Covariance analysis of strapdown INS considering gyrocompass characteristics. *IEEE Transactions on Aerospace and Electronic Systems, 31*(1), 320–328.
20. Shen, X., Zhang, Y., Lu, X., et al. (2015). An improved method for transforming GPS/INS attitude to national map projection frame. *IEEE Geoscience and Remote Sensing Letters, 12*(6), 1302–1306.
21. Sjanic, Z., & Gustafsson, F. (2015). Navigation and SAR focusing with map aiding. *IEEE Transactions on Aerospace and Electronic Systems, 51*(3), 1652–1663.

22. Soloviev, A., & de Haag, M. U. (2010). Monitoring of moving features in laser scanner-based navigation. *IEEE Transactions on Aerospace and Electronic Systems, 46*(4), 1699–1715.
23. *Specifications for IMU/GPS supported aerial photography, Standard GB/T27 919* (2011).
24. Tempo, R., Calafiore, G., & Dabbene, F. (2013). *Randomized algorithms for analysis and control of uncertain systems with applications* (2nd ed.). London, U.K.: Springer.
25. Toth, C. K. (2002). Sensor integration in airborne mapping. *IEEE Transactions on Instrumentation and Measurement, 51*(6), 1367-C1373.
26. Wu, Y., & Pan, X. (2013). Velocity/position integration formula Part I: Application to in-flight coarse alignment. *IEEE Transactions on Aerospace and Electronic Systems, 49*(2), 1006–1023.
27. Zhong, M., Guo, D., & Guo, J. (2015). PMI-based nonlinear H_∞ estimation of unknown sensor error for INS/GPS integrated system. *IEEE Sensors Journal, 15*(5), 2785–2794.
28. Zhong, M., Guo, J., & Yang, Z. (2016). On real time performance evaluation of the inertial sensors for INS/GPS integrated systems. *IEEE Sensors Journal, 16*(17), 6652–6661.
29. Zhou, D., & Frank, P. M. (1996). Strong tracking filtering of nonlinear time-varying stochastic systems with coloured noise: Application to parameter estimation and empirical robustness analysis. *International Journal of Control, 65*(2), 295–307.

Chapter 23
On Real-Time Performance Evaluation of the Inertial Sensors in INS/GPS Systems

Inertial sensors in the INS are the key components for motion measurement. However, they are vulnerable to external disturbances and performance degradation is likely to happen. A performance evaluation method is thus essential for real time monitoreal-timering of motion measurement. On the other hand, the assumption that the sensor noises are Gaussian white in conventional evaluation methods may be violated in practice. The statistics of sensor noise may be unknown yet norm-bounded. Hence, a more general evaluation method to cover this case is needed. In response to these requirements, in this chapter we attempt to apply a unified design scheme of evaluation function for the INS/GPS integrated system with either stochastic or norm-bounded disturbances. The core idea lies in converting the evaluation issue into finding a minimum of an evaluation function of a quadratic form. For ease of online implementation, a recursive algorithm is developed by means of projection and innovation analysis. The application of developed scheme to the monitoring of inertial sensors in the INS/GPS system is demonstrated.

23.1 Introduction

Increasing applications of the INS/GPS integrated system in a wide range such as navigation, location and motion measurement is self-evident [5, 8, 20]. The inertial sensors (namely gyros and accelerometers) are the key components of the INS/GPS system. To fuse the data from the INS and GPS for continuous estimation of the navigation states, the qualities of inertial sensor are thus intimately of great importance. Hence, inertial sensor errors must be properly handled before its usage into the INS/GPS systems [11, 19]. Apart from the deterministic errors calibrated and compensated by manufacturers, the performance of the inertial sensors hinges on the stochastic characteristics which are often described by the sensor noise statistics. Normally the sensor noises are assumed as Gaussian white, based on which the

M. Zhong et al., *Fault Diagnosis for Linear Discrete Time-Varying Systems and Its Applications*, https://doi.org/10.1007/978-981-19-5438-2_23

Kalman filter is able to obtain optimal estimates of the states. For many engineering applications, however, the assumptions may hardly accord with the practical situations. If inertial sensors behave in a disturbed manner, the statistical changes of the sensor noises are likely to happen. This is unavoidable especially when it is aroused by the complex perturbation of the vehicles' environments. Without prompt warnings, all these unexpected defects will inevitably result in performance degradation of the Kalman filter or even mission failure. It is, therefore, necessary to monitor and evaluate the inertial sensors' performance through the whole operation of the INS/GPS system.

As the performance of inertial sensors is directly linked to the underlying sensor characteristics, numerous methods have been developed to model and identify the inertial sensor errors [2, 13]. An IEEE standard has described the inertial sensor behaviors with two types of models, i.e., the dynamic and stochastic models [22]. The former involves the input and output relationships while the latter is hypothesized to have the same output characteristics excited by white noise. Formally, the deterministic errors in dynamic models can be calibrated in laboratory or estimated by the filter [9, 12, 16]. The stochastic models describe the random errors in a probabilistic sense, which also provide key parameters for the Kalman filter [4]. Assumed as Gaussian white, the noise statistics of the inertial sensors are routinely determined in laboratory using power spectral density (PSD), root mean-square (RMS) or Allan variance [4, 17, 23]. By this means, complete a priori knowledge of the sensor noise statistics ensures the optimal estimates of the INS/GPS system by Kalman filtering [8]. It is easy to note that all the methods are applicable only when the inertial sensors are subject to Gaussian noises. However, as pointed out in [7], new generation of navigation and positioning systems emphasizes highly on the reliability as well as the accuracy. Diverse application contexts give rise to new problems such as system complexity and environmental variation when applying INS/GPS systems. On the one hand, the rise of system complexity will reduce the entire reliability and bring extra maintenance burden particularly for systems in the field. Then an evaluation method becomes necessary providing real-time warnings to any latent system degradations. On the other hand, the above mentioned environmental disturbances usually account for the non-Gaussian sensor noises. Thus, it is meaningful to consider the non-Gaussian case when designing the evaluation algorithms. In this sense, the real-time performance evaluation is of practical significance yet remains a challenge.

The evaluation is a systematic determination of the sensor's quality using criteria governed by a set of standards, which is primarily used to give a reliable prediction and identification of the future sensor changes [21]. From this viewpoint, PSD, RMS and Allan variance, which are commonly accepted in the community, can be viewed as evaluation methods for quality grading. Under fine static test conditions, they are utilized to identify the stochastic model of a single sensor, which is supposed driven by white noise and the output characteristics remains [22]. Hence, they can be regarded as the sensor level methods. Although sensor level evaluation brings an insight to the inherent sensor behaviors, it places strict limitations on the test conditions, such as well controlled environments and offline processing, making it difficult to be carried out in the field. Another evaluation strategy resorts to assess certain statistical quantities, by using the system level information, to monitor the system operation situation [3, 14, 18]. A typical method is the χ^2 test. This associates the state estimator with constructing statistical formulae for assessment. Apparently system level evaluation has already incorporated the environmental impacts into the statistical quantities, and thereby it is more reasonable to monitor the sensor performance under external disturbances in real time. From this perspective, the system level evaluation is of practical significance to the INS/GPS system for online use. Therefore, it is our motivation to develop an effective evaluation method as a systematic determination of the inertial sensor's qualities.

In this chapter, we concern the airborne INS/GPS system used for remote sensing and strive to apply a unified evaluation function to monitor the inertial sensors corrupted by unknown yet norm-bounded noises. It can be regarded as a typical application of the unified evaluation method given in Chap. 18 in the INS/GPS system.

23.2 Preliminaries and Problem Statement

23.2.1 Kalman Filter Based State Estimation for the INS/GPS System

Choose the local ENU coordinate frame as the navigation frame n. The state space model of the INS/GPS system can be described by (20.11). By defining

$$x(t) = [\varphi^T \ \delta V^T \ \delta p^T \ \varepsilon^T \ \nabla^T]^T, \ y(t) = [(\delta V)^T (\delta p)^T]^T,$$

a state space model of the INS/GPS system is further represented by a linear continuous time-varying process in form of (20.13), i.e.,

$$\begin{cases} \dot{x}(t) = F(t)x(t) + G(t)w(t), \\ y(t) = H(t)x(t) + v(t), \end{cases} \tag{23.1}$$

where $F(t)$, $G(t)$ and $H(t)$ are referred to Sect. 20.2.2, $w(t)$ and $v(t)$ are white noises with the zero means and covariance $Q(t)$ and $R(t)$, respectively. As we focus mainly on the inertial sensor qualities and the GPS signals for aircrafts are usually good, $R(t)$ is regarded as constant matrix.

With given sample period T_s, the LDTV system model corresponding to (23.1) can be obtained as

$$\begin{cases} x(k+1) = A(k)\,x(k) + B_w(k)\,w(k), \ x(0) = x_0, \\ y(k) = C(k)\,x(k) + v(k), \end{cases} \tag{23.2}$$

where

$$A(k) = \Phi((k+1)T_s, kT_s), \tag{23.3}$$

$$B_w(k) = \int_{kT_s}^{(k+1)T_s} \Phi((k+1)T_s, \tau)G(\tau)d\tau, \tag{23.4}$$

$$C(k) = H(kT_s) \tag{23.5}$$

with $\Phi((k+1)T_s, kT_s)$ being the state transition matrix. Then, by means of Kalman filter, the system states can be estimated by

$$\begin{cases} \hat{x}(k|k-1) = A(k-1)\hat{x}(k-1), x(0) = x_0, \\ \hat{x}(k) = \hat{x}(k|k-1) + K(k)(y(k) - C(k)\hat{x}(k|k-1)), \end{cases} \tag{23.6}$$

with

$$K(k) = P(k|k-1)C^T(k)R_E^{-1}(k), \tag{23.7}$$

$$R_e(k) = C(k)P(k|k-1)C^T(k) + R(k), \tag{23.8}$$

$$\begin{aligned} &P(k|k-1) = A(k-1)P(k-1)A^T(k-1) + B_w(k-1)Q(k-1)B_w(k-1)^T, \\ &P(k) = (I - K(k)C(k))P(k|k-1)(I - K(k)C(k))^T + K(k)R(k)K^T(k). \end{aligned} \tag{23.9}$$

It is ideal that the sensor noise process is taken as Gaussian white and the statistical knowledge is available. In this case, there is no need to evaluate sensor performance, since all the noise terms in its stochastic model just need to be identified once prior to the applications in the field. However, it cannot always be the case in the engineering context. Based on the given system model and filter structure, we will further discuss the evaluation methods in the following subsection.

23.2.2 Existing Evaluation Methods

Inertial sensors of different accuracy grades are primarily classified by the stochastic error characteristics [8]. Theoretically, the sensor noise process is assumed following the Gaussian distribution, based on which most evaluation methods are developed. In sensor level, for example, the Allan variance is a broadly used approach for the identification and evaluation of noise processes. From the viewpoint of system operation, on the other hand, χ^2 test is widely accepted and utilizes the filter's results to calculate a statistical quantity to detect the system defects. In the following, we will show the two methods in sequence and our research is related to the χ^2 test.

A. Allan Variance

The Allan variance quantifies the noise terms in the stochastic models such as the bias instability, angle random walk, quantization noise and so on. Therefore, it is an important tool for characterizing the inertial sensor noises. The Allan variance is a time-domain method and literature has provided a standard procedure to compute it using sampled data with small variations [22].

Denote by z the random sensor noise process. The Allan variance $\sigma_z^2(\tau_A)$ of the random process z is a function of the average time τ_A, which describes variation range of the integrated process as a function of τ_A. The mean of the variances of successive blocks is then the Allan variance over τ_A. It can be briefly written as follows

$$\bar{z}_k(\tau_A) = \frac{1}{\tau_A}\int_{k\tau_A}^{(k+1)\tau_A} z(t)\mathrm{d}t,$$

$$\sigma_z^2(\tau_A) = \frac{1}{2}\mathbb{E}[(\bar{z}_{k+1}(\tau_A) - \bar{z}_k(\tau_A))^2].$$

The Allan variance method can characterize various types of error terms in the data with the above formula.

It is remarkable that the Allan variance test obviously appeals to benign test conditions like static mechanical and thermal environments, which could hardly be satisfied in the field. Although it exactly quantifies many noise terms in the stochastic model for evaluation, it is not realistic to perform the Allan variance test of the INS/GPS system for each flight mission. It is definitely an offline method realized under the lab-scale conditions, rather than an online evaluation means.

B. χ^2 Test

The χ^2 test relates to the case that the sensor noise processes, in spite of Gaussian, have been changed in terms of covariances, leading to latent estimation accuracy losses. This in reverse can be used to detect the noise statistics changes, and becomes an indicator of the sensor performance.

If the sensor noise covariances are no longer consistent with the values previously identified, it conflicts with the assumptions set in Sect. 23.2.1, i.e.,

$$\mathbb{E}[w(i)w(j)^T] = Q(i)\delta_{ij},\ \mathbb{E}[w(i)] = 0.$$

where $Q(i)$ turns out to be time variant. In this case, the χ^2 test is an effective evaluation method for the detection of the noise statistics changes [15].

To detect the occurrence of the noise covariance variation, the primary task is to construct an appropriate statistical formula. Recall the system model (23.2) and Kalman filter (23.6), and define the innovation sequence of the system as

$$r(k) = y(k) - C(k)\hat{x}(k), \tag{23.10}$$

and a statistical function can be designed as

$$\eta(k) = r(k)R_e^{-1}(k)r(k)^T. \tag{23.11}$$

with

$$R_e(k) = \mathbb{E}[r(i), r(j)] = (C(i)P(i)C(i) + R(i))\delta_{ij},$$

one can find that $\eta(k)$ follows the χ^2 distribution with $q = \dim(r) = 6$ degrees of freedom.

Next, with the level of significance $\alpha \in (0, 1)$, the critical region is given as

$$\Pr\{\chi^2 > \chi^2_\alpha(q)\}.$$

The threshold value $\chi^2_\alpha(q)$ can be obtained from the χ^2 distribution table. Then we can use the following decision logic to test the statistical changes of the sensor noise process:

$$\begin{array}{ll} \text{if } \forall k,\ \eta(k) \le \chi^2_\alpha(q), & \text{sensor works normally,} \\ \text{if } \exists k,\ \eta(k) > \chi^2_\alpha(q), & \text{noise statistics have changed.} \end{array}$$

It should be pointed out that as a system level method, the χ^2 test is feasible to implement in real time during the flight. It utilizes the system information from the Kalman filter, and constructs an evaluation function. By continuously comparing with the threshold, it achieves real-time assessment of the system. Nevertheless, the χ^2 test is restricted in use merely for the Gaussian distributed cases. When the sensor noise statistics deviate from the assumption, the estimation accuracy declines and (23.11) no longer follows χ^2 distribution, which limits the use of χ^2 test in practice. Therefore, to be more general, the case of norm-bounded noise processes should be taken into consideration.

23.2.3 *Problem Statement*

In a more general application context, the inertial sensor noises are possibly non-Gaussian and the statistical information is not always available. It is no longer rational to apply the above mentioned methods for the INS/GPS system. However, in the subsection, we will first stick to the Gaussian noise assumption and give an alternative for the χ^2 testing, which is the basis of subsequent discussions. Then we will extend to the non-Gaussian case and show a potentially workable way to tackle the evaluation problem in such case.

Recall that an input-output model of the LDTV system (23.2) can be obtained as

$$y_N = \Omega x_0 + \Gamma_w w_N + v_N = \Omega x_0 + \Gamma_d d_N. \tag{23.12}$$

where $\Gamma_d = [\Gamma_w \quad I]$ and

$$\begin{aligned} &y_N = [y^T(0)\ y^T(1)\ \cdots\ y^T(N)]^T, w_N = [w^T(0)\ w^T(1)\ \cdots\ w^T(N)]^T, \\ &v_N = [v^T(0)\ v^T(1)\ \cdots\ v^T(N)]^T, d_N = [w_N^T \quad v_N^T]^T, \\ &\Phi(i, j) = A(i-1)A(i-2)\cdots A(j), \Psi(i, i) = I, \end{aligned}$$

$$i = 1, 2, \cdots N; j = 0, 1, \cdots, i-1,$$

$$\Omega = \begin{bmatrix} C(0) \\ C(1)\Phi(1,0) \\ \vdots \\ C(N)\Phi(N,0) \end{bmatrix}, \quad \begin{array}{l} \Gamma_w = \left[\Gamma_w(j,k)\right], \\ \Gamma_w(j,k) = 0 \quad \text{for} \quad \mathrm{j} \le \mathrm{k}, \\ \Gamma_w(j,k) = C(j)\Psi(j,k)B_w(k) \quad \text{for} \quad \mathrm{j} > \mathrm{k} \ge 1. \end{array}$$

Denote by Π_0 the covariance matrix of x_0, and the covariance matrix of d_N is

$$R_d = \mathbb{E}[d_N d_N^T] = \begin{bmatrix} \mathcal{Q} & 0 \\ 0 & \mathcal{R} \end{bmatrix},$$

where $\mathcal{Q} = \mathrm{diag}(Q(0), \ldots, Q(N))$, $\mathcal{R} = \mathrm{diag}(R(0), \ldots R(N))$.

Let $\bar{d}_N = [x_0 \quad d_N^T]^T$, so (23.12) is changed to

$$y_N = \Gamma_{\bar{d}} \bar{d}_N,$$

where $\Gamma_{\bar{d}} = [\Omega \quad \Gamma_d]$. And its covariance matrix is readily obtained as

$$R_y = \mathbb{E}[y_N y_N^T] = \Gamma_{\bar{d}} R_{\bar{d}} \Gamma_{\bar{d}}^T, \tag{23.13}$$

where

$$R_{\bar{d}} = \begin{bmatrix} \Pi_0 & 0 \\ 0 & R_d \end{bmatrix}.$$

Consider the Gaussian assumption and it is easy to find that $\{x_0, d_N\}$ is with jointly Gaussian distribution. If $(\Gamma_{\bar{d}} \Gamma_{\bar{d}}^T)$ is positive definite, then $\bar{d}_N$ can be estimated using the least square estimation as

$$\hat{\bar{d}}_N = \Gamma_{\bar{d}}^T \left(\Gamma_{\bar{d}} \Gamma_{\bar{d}}^T\right)^{-1} y_N.$$

Since $\hat{\bar{d}}_N$ is an approximation of d_N comprising all the noise sequences in $[0, N]$, a natural object to study is the following evaluation function

$$J_y(N) = \hat{\bar{d}}_N^T R_{\bar{d}}^{-1} \hat{\bar{d}}_N = y_N^T \left(\Gamma_{\bar{d}} R_{\bar{d}} \Gamma_{\bar{d}}^T\right)^{-1} y_N. \tag{23.14}$$

As y_N follows jointly Gaussian distribution with

$$\mathbb{E}\langle y_N, y_N \rangle = \Gamma_{\bar{d}} R_{\bar{d}} \Gamma_{\bar{d}}^T, \quad \mathbb{E}[y_N] = 0,$$

it is easy to note that $J_y(N)$ follows the χ^2 distribution with $6(N+1)$ degrees of freedom. Under this assumption, we can evaluate the inertial sensors' performance with χ^2 testing if $\{x_0, d_N\}$ is jointly Gaussian distributed. Here it can be viewed as an alternative of (23.11) for χ^2 test.

Now, we consider the case that the noise is non-Gaussian distribution and, without loss of generality, assume the random noise process of inertial sensor is l_2-norm bounded within a given time length, i.e., $w(k) \in l_{2,[0,N]}$.

Recall that $\bar{d}_N = [x_0 \;\; d_N^T]^T$, and instead we introduce the following scalar quadratic form

$$J(x_0, d_N) = \bar{d}_N^T R_{\bar{d}}^{-1} \bar{d}_N = \begin{bmatrix} x_0 \\ w_N \\ v_N \end{bmatrix}^T \begin{bmatrix} \Pi_0 & 0 & 0 \\ 0 & \mathcal{Q} & 0 \\ 0 & 0 & \mathcal{R} \end{bmatrix}^{-1} \begin{bmatrix} x_0 \\ w_N \\ v_N \end{bmatrix}. \tag{23.15}$$

Note that the above formula directly makes use of the sensor noise processes despite the unknown noise statistics, avoiding the underlying distribution issues. Within a given window length N and the noise being l_2-norm bounded, it clearly suggests that the value of (23.15) has a upper bound. The quadratic form of the random processes can be actually viewed as a representation of the noise "energy". If a supremum of (23.15) is determined by analyzing the normal sensor data, the evaluation can then be realized by comparing the function values with the supremum.

One can see that, although (23.15) covers a wider range of non-Gaussian cases, it is not feasible to calculate by computers because it contains the unknown random noise processes. Additionally, the matrix computation in (23.15) would be a heavy burden for on-board computers. At this point, the remaining tasks are twofold:

- find an executable form of the function equivalent to (23.15) and,
- simplify the evaluation function as possible for online calculation.

Based on these, we are then able to develop a dedicated evaluation scheme for the INS/GPS system.

It is noted that the key of these two tasks lies in developing an online realization algorithm to estimate the evaluation function $J(x_0, d_N)$ of the quadratic form (23.15), which has been address in Chap. 18 by virtue of Krein space projection. Inspired by this, below focus on addressing the above two tasks on this basis.

23.3 Real-Time Performance Evaluation of the INS/GPS System

In this section, we first formulate the problem of performance evaluation as finding a minimum, and then simplify the evaluation function for real-time use. At last, we design an evaluation scheme specifically for the INS/GPS system.

23.3.1 Design of the Evaluation Function

To facilitate on-board computations, we make the change of coordinates from (23.12) that

$$\begin{bmatrix} x_0 \\ w_N \\ y_N \end{bmatrix} = \begin{bmatrix} I & 0 & 0 \\ 0 & I & 0 \\ \Omega & \Gamma_w & I \end{bmatrix} \begin{bmatrix} x_0 \\ w_N \\ v_N \end{bmatrix}.$$

Substituting it into (23.15), it is easy to get

$$J(x_0, w_N, y_N) = \begin{bmatrix} x_0 \\ w_N \\ y_N \end{bmatrix}^T \left\{ \begin{bmatrix} I & 0 & 0 \\ 0 & I & 0 \\ \Omega & \Gamma_w & I \end{bmatrix} \begin{bmatrix} \Pi_0 & 0 & 0 \\ 0 & \mathcal{Q} & 0 \\ 0 & 0 & \mathcal{R} \end{bmatrix} \begin{bmatrix} I & 0 & 0 \\ 0 & I & 0 \\ \Omega & \Gamma_w & I \end{bmatrix}^T \right\}^{-1} \begin{bmatrix} x_0 \\ w_N \\ y_N \end{bmatrix}. \quad (23.16)$$

One can find that (23.16) is a frequently encountered quadratic form, as seen in Sect. 18.4. From Lemma 3.2.2. in [10], it has a unique minimum over $\hat{x}_0 = \Pi_0 \Omega^T R_y^{-1} y_N$ and $\hat{w}_N = \mathcal{Q}\Gamma_w R_y^{-1} y_N$, if and only if

$$\Pi_0 - \Pi_0 \Omega^T (\Gamma_{\bar{d}} R_{\bar{d}} \Gamma_{\bar{d}}^T)^{-1} \Omega \Pi_0 > 0$$

and the minimum value of (23.16) subject to (23.2) equals to

$$J_y(\hat{x}_0, \hat{w}_N, y_N) = y_N^T R_y^{-1} y_N.$$

Recall (23.13) and (23.14), and it is interesting to find that

$$J(\hat{x}_0, \hat{w}_N, y_N) = J_y(N).$$

Define the minimums of (23.16) as the evaluation function for sensor performance evaluation, and $J_y(N)$ can be viewed as an estimation of $J(x_0, w_N, y_N)$. If the sensor noise "energy" augments, the minimum $J(\hat{x}_0, \hat{w}_N, y_N)$ would be probably surpassed, which indicates statistical changes of the sensor noises and consequently the performance degradation. To this end, the problem of performance evaluation is actually changed to find a minimum of $J(x_0, w_N, y_N)$ over x_0 and w_N. And we can use (23.14) as an executable alternative of (23.16) for the computer processing.

When the inertial sensors are with good qualities in a normally operating INS/GPS system, we can determine the supremum of (23.14), namely, the threshold for performance evaluation, viz.,

$$J_{th,y} = \sup J_y(N).$$

Then the presence of sensor performance degradation can be tested by

$$\begin{cases} J_y(N) < J_{th,y}, \text{ no degradation,} \\ J_y(N) \geq J_{th,y}, \text{ degradation occurs.} \end{cases}$$

Obviously $J_{th,y}$ represents the maximum influence of the noise "energy" in the normal system. When the value of $J_y(N)$ exceeds this threshold, we can assert that the degradation occurs.

Remark 23.1 As long as the sensor noise process is random and l_2-norm bounded, there is always a noise "energy" limitation, so it is possible to measure the "energy" in terms of function values of certain quadratic formula. Motivated by this, the above evaluation function is constructed of a quadratic form of the noise process and we compares with the upper bound of the statistical values. Then the "energy" changes directly result in exceeding the evaluation thresholds. By this means we realize the performance evaluation of the inertial sensors.

23.3.2 Simplification of the Evaluation Function

It should be noted that the computation burden of (23.14) is still heavy, particularly when N is large. For the purpose of real-time calculations, we need to reformulate this function into a recursive form so as to alleviate the computation complexity.

For the INS/GPS systems, the Kalman filter (23.6) is used as the state estimator, from which the innovation sequence is obtained. Based on Algorithm 18.4.11, we know that $J_y(N)$ can be reformulated by using the innovation re-organization and projection. The innovations give a useful geometric insight into the recursive projection problem and enables the recursive estimation. In this way, the formula can be simplified considerably, prone to real-time recursive calculation. Detailed derivation is given as follows.

First, define the innovation in Kalman filter as

$$e(k) = y(k) - \hat{y}(k|k-1), \quad 0 < k \leq N, \tag{23.17}$$

where $\hat{y}(k|k-1)$ is the projection of $y(k)$ onto the linear subspace spanned by $\{y(i)\}_{i=0}^{k-1}$. As $\hat{y}(k|k-1) = C(k)\hat{x}(k|k-1)$ and $\tilde{x}(k) = x(k) - \hat{x}(k|k-1)$, we have

$$\begin{aligned} R_e(k) = \langle e(k), e(k)\rangle &= \langle C(k)\tilde{x}(k), C(k)\tilde{x}(k)\rangle + \langle v(k), v(k)\rangle \\ &= C(k)P(k|k-1)C^T(k) + R(k), \end{aligned}$$

which is same as (23.8). Note that (23.17) is a Gram-Schmidt orthogonalization procedure and the innovations form an orthogonal basis of linear subspace $\mathcal{L}\{\{y(i)\}_{i=0}^{k-1}\}$. Therefore, according to Sect. 4.4, one can change (23.13) to

$$R_y = \mathbb{E}[Le_N(Le_N)^T] = LR_eL^T,$$

where $R_e = \text{diag}\{R_e(0), \ldots, R_e(N)\}$. And (23.14) could be expressed in terms of the innovations as

$$J_y(N) = y_N^T R_y^{-1} y_N = e_N^T R_e^{-1} e_N = \sum_{i=0}^{N} e^T(i) R_e^{-1}(i) e(i). \tag{23.18}$$

It is apparent to find that $J_y(N)$ can now be computed via recursions, and the calculation of (23.18) is remarkably simplified compared to (23.14). Hence it becomes feasible for the real-time implementation of the on-board computers. Specifically, in course of the online filtering process of the INS/GPS integration, the time window slides by each time step, and $J_y(N)$ ought to be updated accordingly. For the sake of successive computations, we rewrite (23.18) at the time step k as

$$J_y(N, k) = \sum_{i=k-N+1}^{k} e^T(i) R_e^{-1}(i) e(i). \tag{23.19}$$

A recursive solution for the on-board computation of the INS/GPS system is thus achieved.

Remark 23.2 From above discussions, it is clear that (23.19) is essentially an evaluation function to represent the inertial sensor noise "energy". When external disturbances impose influences on the gyros or accelerometers, the dramatic changes of (23.19) rapidly reflect the performance degradation in a straightforward manner. From this point of view, it is an effective method for the performance evaluation in a more general application context. It is also applicable for the cases that the sensor noises are Gaussian. Thus the presented method can be used to efficiently evaluate the sensor performance, and further the system performance.

Remark 23.3 As to the airborne INS/GPS systems for remote sensing, performance degradation may possibly result in serious impact or mission failure, which engineers try to prevent. Hence, a timely warning from the performance evaluation method is quite useful and necessary. In this sense, this method actually constitutes a significant proportion in the framework of the system health management. Moreover, although the system model (23.2) is linear, the evaluation method in this chapter can also be expanded to nonlinear INS/GPS systems.

23.3.3 Evaluation Algorithm for the INS/GPS System

In the remote sensing applications, the INS/GPS systems are used for motion compensation [1, 6, 9]. In particular, a typical flight trajectory for the aerial mapping usually includes two distinct movement modes. They are the straight flight during the imaging operation, and the turn-round maneuver to alter the mapping strip.

The innovations unavoidably enlarge during the maneuvers. It comes for the reason that the nonlinear change of the coefficient matrix will lead to the estimation error oscillation lasting for a short period before convergence. Thus we have to introduce two thresholds denoted by $J_{th,1}$ and $J_{th,2}$ to represent the maximum values in straight flight stage and maneuver stage, respectively. Similarly, the presence of the performance degradation can be tested by

$$\begin{cases} \text{if } J_y(N,k) \le J_{th,1}, & \text{then straight flight,} \\ \text{if } J_{th,1} < J_y(N,k) \le J_{th,2}, & \text{then turn-round maneuver,} \\ \text{if } J_y(N,k) > J_{th,2}, & \text{then degradation alarm.} \end{cases} \tag{23.20}$$

Notice that the actual values of $J_{th,1}$ and $J_{th,2}$ are system-specific which should be empirically determined. At each time step, the evaluation function (23.19) is updated to monitor the current sensor quality, so we are able to realize the performance evaluation online by testing (23.20).

The data flow diagram of the evaluation process is shown in Fig. 23.1, and now we summarize the sensor performance evaluation procedure in Algorithm 23.3.19.

Remark 23.4 The developed scheme is designed on the basis that the flights in remote sensing applications commonly contain two typical sorts of movements. Then two thresholds are set to avoid false alarms according to the flight features, and the degradation alarm is delivered only when $J_{th,2}$ is exceeded.

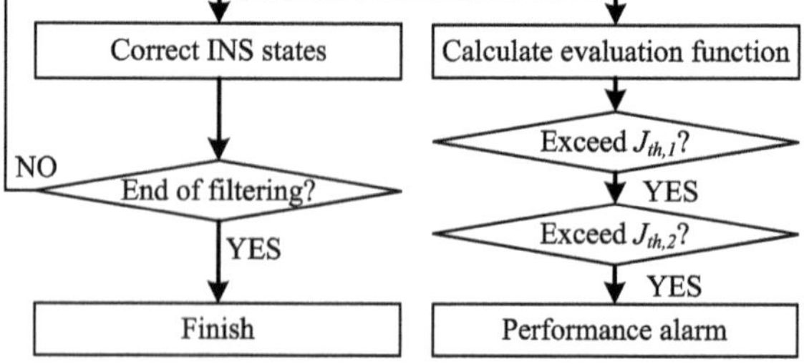

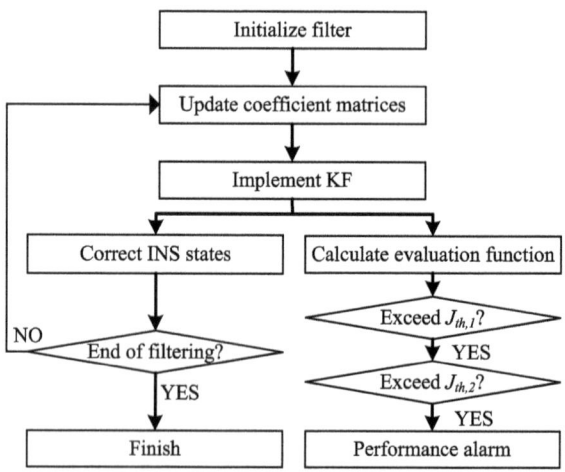

Fig. 23.1 Data flow of the evaluation process

Algorithm 23.3.19 Sensor performance evaluation procedure

1: Set the window length N, thresholds $J_{th,1}$ and $J_{th,2}$ and let k=0, $\hat{x}_0$=0, and $P_0 = \Pi_0$;
2: Update $A(k+1)$, $B(k+1)$ and $C(k+1)$ using (23.3), (23.4) and (23.5) respectively;
3: Calculate $R_e(k+1)$, $K(k+1)$ and $P(k+1)$ using (23.8), (23.7) and (23.9) respectively;
4: Calculate $\hat{x}(k+1)$ using (23.6);
5: Calculate $e(k)$ using (23.17);
6: If $k > N$, calculate $J_y(N,k)$ using (23.19);
7: Evaluate the sensor performance by testing (23.20) and alarm if $J_{th,2}$ is exceeded;
8: Correct the navigation states as $\bar{C}_b^n = (I - \hat{\phi}) \times C_b^n$, $\bar{v} = v - \delta\hat{v}$ and $\bar{p} = p - \delta\hat{p}$;
9: Reset $\hat{x}(k+1)$, i.e., $\hat{x}(k+1) = [0 \quad \hat{\varepsilon}^T \quad \hat{\nabla}^T]^T$;
10: Let $k = k+1$, and go to Step 2 until the end.

23.4 Experimental Study

23.4.1 Experimental Setup

In this section, two flight tests of an INS/GPS system for aerial mapping are considered. We select typical trajectory segments in the mapping missions of a synthetic aperture radar containing one straight flight strip and two turn-round maneuvers. The INS/GPS system we used functions well in the normal operation circumstances, and the system accuracies fulfill the mapping requirements of the radar. The experimental data in two flight tests are derived from one identical INS/GPS system. We design a normal case and a disturbed case in each flight test to show the efficacy of the developed method. Take the original experimental data of the INS/GPS system as the normal case data. Then the disturbed case data is obtained by adding a set of norm-bounded disturbance data to the normal case data.

The system mainly includes a dynamically tuned gyro based IMU and a NovAtel GPS receiver. The IMU is rigidly fixed to the radar antenna beneath the aircraft while the GPS antenna is mounted on the top of the craft cabin. Figure 23.2a shows the installation of the radar antenna. The update frequencies of the IMU and GPS are 100 Hz and 1 Hz, respectively. The navigation constants are set as $\omega_{ie} = 15.04°/\mathrm{h}$, $R_e = 6378135\,\mathrm{m}$ and $e = 1/298.257$. The sensor specifications are listed in Table 20.2 with the white noise in accelerometer being 100 μg (1σ), the velocity and position in GPS receiver being 0.03 m/s (RMS) and 0.05–0.3 m (RMS), respectively. Set the window length as $N = 100$, and $J_y(N,k)$ is calculated using (23.19).

23.4.2 Example I

A flight segment of the time interval (4100 s, 5400 s) is selected. The flight trajectory of this interval is shown in Fig. 23.2b with flight height around 3000 m. A simulated sensor noise process is designed driven by the trigonometric functions at the interval of (4600 s, 4800 s), representing unexpected sensor disturbances. It is given as

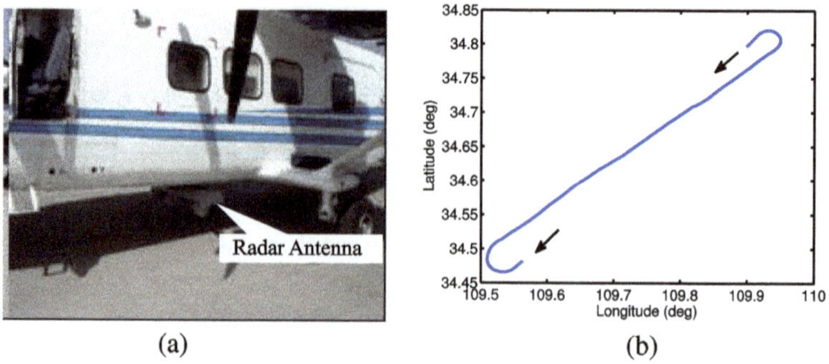

Fig. 23.2 **a** The IMU is installed inside the radar antenna; **b** A segment of the flight trajectory

$$\begin{cases} \text{noise: } 40\sin(k/150)\text{ mg}, & \text{when } 4600 < k \leq 4750, \\ \text{noise: } 23\cos(k/50)\text{ mg}, & \text{when } 4750 < k \leq 4800. \end{cases}$$

Add the simulated data to the real accelerometer data of axis x, and thus we get the disturbed case data, which is concisely shown in Fig. 23.3.

Now we consider the developed evaluation method. Denote by $J_{y1}(k)$ and $J_{y2}(k)$ the evaluation functions in the two cases, respectively. It can be seen from Fig. 23.4 that both the evaluation function values increase when the aircraft turns round approximately at the intervals of (4165 s, 4285 s) and (5270 s, 5330 s). Thereby, we choose the thresholds as $J_{th,1} = 10$ and $J_{th,2} = 16$ to indicate the straight flight mode and maneuver mode, respectively. In the normal case, the evaluation function propagates smoothly below the dash-dot line of $J_{th,2}$. As the unexpected disturbances occur in the IMU , the value of $J_{y2}(k)$ continuously increases. At $k = 4623$, $J_{y2}(4623) = 16.3 > J_{th,2}$ and so a degradation alarm is delivered. When the simulated error ends, $J_{y2}(4812) = 15.6 < J_{th,2}$ and the alarm is turned off at $k = 4812$.

Fig. 23.3 The raw sensor data of the normal and disturbed cases

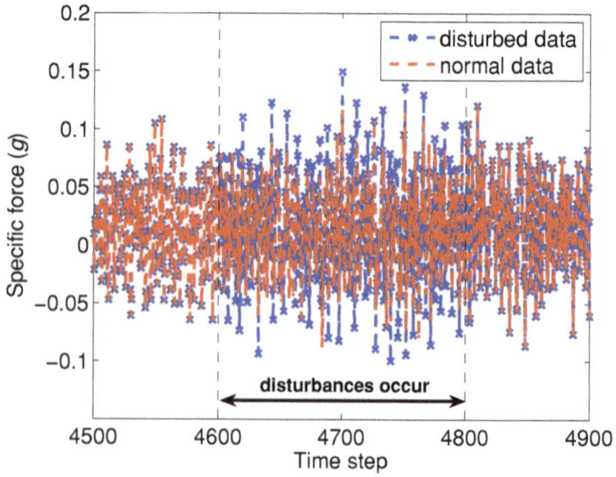

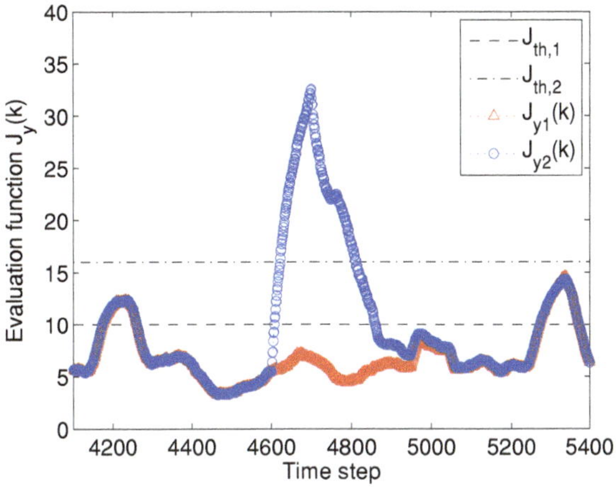

Fig. 23.4 The function values in the two cases using the developed method in example I

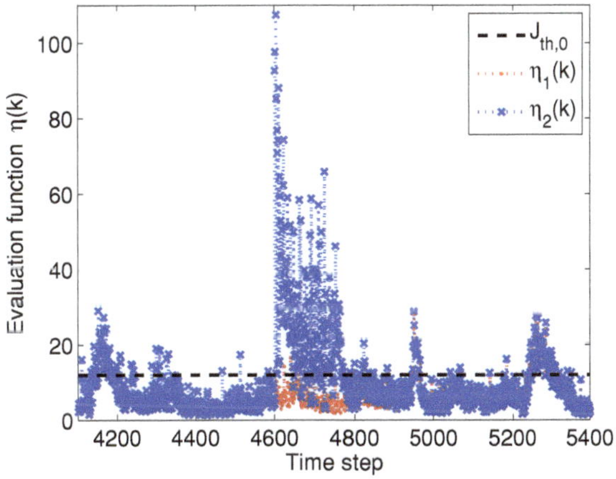

Fig. 23.5 The function values in the two cases using the χ^2 test in example I

It means the developed method reveals the degradation happening inside the IMU correctly in the disturbed case.

We consider the χ^2 test for comparison, and (23.11) is used to calculate the function values. The threshold is determined directly from the χ^2 distribution table as $J_{th,0} = 12.6$. Denote by $\eta_1(k)$ and $\eta_2(k)$ the statistical functions of the χ^2 test in the normal case and the disturbed case, respectively. The results in the two cases are shown in Fig. 23.5. It is easy to note that the function values frequently exceed the threshold in the normal case, which implies the inapplicability of the χ^2 test for the INS/GPS system.

23.4.3 Example II

A typical flight segment of the interval (1800 s, 3400 s) is selected in this case. The simulated disturbances occur in the interval of (2500 s, 2700 s) and are governed by

$$\begin{cases} \text{noise: } 40\sin(k/150)\text{ mg}, & \text{when } 2500 < k \leq 2650, \\ \text{noise: } 23\cos(k/50)\text{ mg}, & \text{when } 2650 < k \leq 2700. \end{cases}$$

The results of developed method are shown in Fig. 23.6. $J_{th,1}$ and $J_{th,2}$ are same as those in Example I. The function still propagates below the line of $J_{th,2}$ in the normal case, indicating the system performance sustains at an acceptable level. In the disturbed case, $J_{y2}(2520) = 16.2 > J_{th,2}$ and an alarm begins at $k = 2520$; $J_{y2}(2727) = 15.8 < J_{th,2}$ and the alarm is turned off at $k = 2727$. The evaluation function reacts in time to deliver an alarm when the simulated disturbances occur.

Similarly, the results of the χ^2 test are shown in Fig. 23.7. Evident fluctuations around $J_{th,0}$ through the entire estimation process illustrate that the χ^2 test cannot reflect the system performance properly. It thus is not an applicable choice for the performance evaluation of the INS/GPS system.

To sum up, the improved method is capable of delivering an alarm when the system performance is degraded by disturbances, whereas the χ^2 test performs badly in examples I and II since the function values exceed the threshold extensively even in the normal case. Therefore we can conclude that the method in this chapter outperforms the conventional χ^2 test, and is especially suitable for the INS/GPS system used for aerial surveying and mapping.

Results have shown that (23.19) could be used to indicate the unexpected sensor degradations effectively during the flight. Hence, it proves that the inertial sensor performance could be evaluated in real-time with the developed evaluation method. It also should be pointed that there is a time lag between the occurrence of the simulated

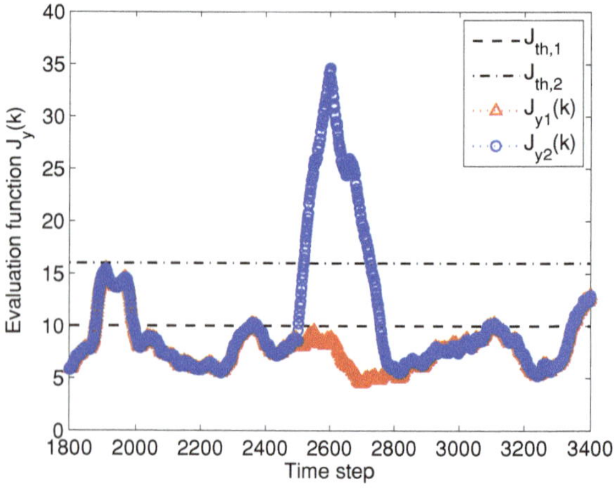

Fig. 23.6 The function values in the two cases using the developed method in example II

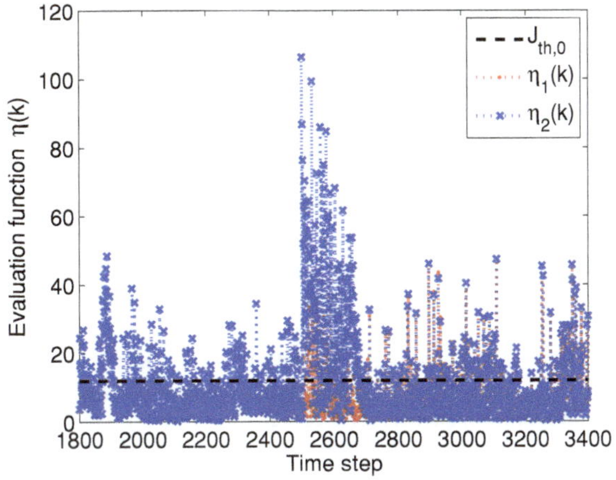

Fig. 23.7 The function values in the two cases using the χ^2 test in example II

noise and the alarm of sensor degradation, for the reason that it needs to take some time before the function could reflect the errors aroused by the sensor degradation. In addition, it has to be noted that the threshold values are always system-specific and should be determined prior to practical uses.

Moreover, the developed performance evaluation method is also useful in evaluating the signal quality of GPS measurements. GPS outliers or noise statistics variations have a straightforward reflection in the innovation sequence, and then the evaluation function values present evident changes in a timely manner.

23.5 Conclusion

In this chapter, we have studied performance evaluation issues of inertial sensors in the INS/GPS integrated systems. It can be regarded as an application of the unified evaluation method in Chap. 18 to the INS/GPS integrated systems, which was developed by virtue of projection and innovation analysis techniques in Krein spaces. A simplified recursive algorithm has been also developed for real-time realization. Two remote sensing tests have validated the effectiveness of the developed algorithm.

References

1. Ahn, H.-S., & Won, C.-H. (2009). DGPS/IMU integration-based geolocation system: Airborne experimental test results. *Aerospace Science and Technology, 13*(6), 316–324.
2. Barbour, N., & Schmidt, G. (2001). Inertial sensor technology trends. *IEEE Sensors Journal, 1*(4), 332–339.
3. Da, R. (1994). Failure detection of dynamical systems with the state chi-square test. *Journal of Guidance, Control, and Dynamics, 17*(2), 271–277.

4. El-Sheimy, N., Hou, H., & Niu, X. (2008). Analysis and modeling of inertial sensors using allan variance. *IEEE Transactions on Instrumentation and Measurement, 57*(1), 140–149.
5. Fang, J., & Yang, S. (2011). Study on innovation adaptive EKF for in-flight alignment of airborne POS. *IEEE Transactions on Instrumentation and Measurement, 60*(4), 1378–1388.
6. Fang, J., Chen, L., & Yao, J. (2014). An accurate gravity compensation method for high-precision airborne POS. *IEEE Transactions on Geoscience and Remote Sensing, 52*(8), 4564–4573.
7. Groves, P. D., Wang, L., & Walter, D., et al. (2014). The four key challenges of advanced multi-sensor navigation and positioning. In *IEEE/ION Position, Location and Navigation Symposium* (pp. 773–792). May 05–08, Monterey, CA, USA.
8. Groves, P. D. (2008). *Principles of GNSS, inertial, and multisensor integrated navigation systems*. Boston: Artech House.
9. Guo, J., & Zhong, M. (2013). Calibration and compensation of the scale factor errors in DTG POS. *IEEE Transactions on Instrumentation and Measurement, 62*(10), 2784–2794.
10. Hassibi, B., Sayed, A. H., & Kailath, T. (1999). *Indefinite-quadratic estimation and control: A unified approach to H_2 and H_∞ theories*. SIAM.
11. Lv, P., Lai, J., & Liu, J., et al. (2014). The compensation effects of gyrosy stochastic errors in a rotational inertial navigation system. *The Journal of Navigation, 67*(6), 1069–1088.
12. Noureldin, A., Karamat, T. B., Georgy, J., et al. (2013). *Fundamentals of inertial navigation, satellite-based positioning and their integration*. Berlin, Germany: Springer.
13. Savage, P. G. (2013). Blazing gyros: The evolution of strapdown inertial navigation technology for aircraft. *Journal of Guidance, Control and Dynamics, 36*(3), 637–655.
14. Soken, H. E., & Hajiyev, C. (2010). Pico satellite attitude estimation via robust unscented Kalman filter in the presence of measurement faults. *ISA Transactions, 49*(3), 249–256.
15. Soken, H. E., & Hajiyev, C. (2012). Fault tolerant attitude estimation for pico satellites using robust adaptive UKF. *IFAC Proceedings Volumes, 45*(20), 726–731.
16. Syed, Z. F., Aggarwal, P., Goodall, C., et al. (2007). A new multi-position calibration method for MEMS inertial navigation systems. *Measurement Science and Technology, 18*(7), 1897–1907.
17. Vaccaro, R. J., & Zaki, A. S. (2012). Statistical modeling of rate gyros. *IEEE Transactions on Instrumentation and Measurement, 61*(3), 673–684.
18. Yun, S. H., Kang, C. W., & Park, C. G. (2014). Reducing the computation time in the state chi-square test for IMU fault detection. In *Proceedings of the 14th International Conference on Control, Automation and Systems (ICCAS)* (pp. 879–883). Goyang, South Korea.
19. Zhong, M., Guo, D., & Guo, J. (2015). PMI-based nonlinear H_∞ estimation of unknown sensor error for INS/GPS integrated system. *IEEE Sensors Journal, 15*(5), 2785–2794.
20. Zhong, M., Liu, H., & Song, N. (2015). On designing an extended H_-/H_∞-FDF for a class of nonlinear systems. *IFAC-PapersOnLine, 48*(21), 707–712.
21. *Evaluation* (2015). https://en.wikipedia.org/wiki/evaluation.
22. *IEEE Standard Specification Format Guide and Test Procedure for Single-Axis Interferometric Fiber Optic Gyros, IEEE Standard 952–1997* (1998).
23. *Test Methods for Fibre Optic Gyroscopes, Chinese National Military Standard GJB 2426A-2004* (2004).

Milton Keynes UK
Ingram Content Group UK Ltd.
UKHW020631061123
432050UK00002B/8

9 789811 954405